Lecture Notes in Computer Science 10226

More information about this series at http://www.springer.com/series/7407

Peter Höfner · Damien Pous
Georg Struth (Eds.)

Relational and Algebraic Methods in Computer Science

16th International Conference, RAMiCS 2017
Lyon, France, May 15–18, 2017
Proceedings

Springer

Editors
Peter Höfner
Data61, CSIRO
Sydney, NSW
Australia

Georg Struth
University of Sheffield
Sheffield
UK

Damien Pous
CNRS
Lyon
France

ISSN 0302-9743 ISSN 1611-3349 (electronic)
Lecture Notes in Computer Science
ISBN 978-3-319-57417-2 ISBN 978-3-319-57418-9 (eBook)
DOI 10.1007/978-3-319-57418-9

Library of Congress Control Number: 2017938136

LNCS Sublibrary: SL1 – Theoretical Computer Science and General Issues

Printed on acid-free paper

This Springer imprint is published by Springer Nature
The registered company is Springer International Publishing AG
The registered company address is: Gewerbestrasse 11, 6330 Cham, Switzerland

Preface

This volume contains the proceedings of the 16th International Conference on Relational and Algebraic Methods in Computer Science (RAMiCS 2017), which was held at ENS Lyon, France, during May 15–18, 2017.

The RAMiCS conferences aim to bring a community of researchers together to advance the development and dissemination of relation algebras, Kleene algebras, and similar algebraic formalisms. Topics covered range from mathematical foundations to applications as conceptual and methodological tools in computer science and beyond. More than 25 years after its foundation in 1991 in Warsaw, Poland—initially as "Relational Methods in Computer Science"—RAMiCS remains a main venue in this field. The series merged with the workshops on Applications of Kleene Algebra in 2003 and adopted its current name in 2009. Previous events were organized in Dagstuhl, Germany (1994), Paraty, Brazil (1995), Hammamet, Tunisia (1997), Warsaw, Poland (1998), Québec, Canada (2000), Oisterwijk, The Netherlands (2001), Malente, Germany (2003), St. Catharines, Canada (2005), Manchester, UK (2006), Frauenwörth, Germany (2008), Doha, Qatar (2009), Rotterdam, The Netherlands (2011), Cambridge, UK (2012), Marienstatt, Germany (2014), and Braga, Portugal (2015).

RAMiCS 2017 attracted 32 submissions, of which 17 were selected for presentation by the Program Committee. Each submission was evaluated according to high academic standards by at least three independent reviewers, and scrutinized further during two weeks of intense electronic discussion. The organizers are very grateful to all Program Committee members for this hard work, including the lively and constructive debates, to the external reviewers for their generous help and expert judgments, and especially to Wolfram Kahl, Martin E. Müller, and Michael Winter for shepherding three submissions towards acceptance. Without this dedication we could not have assembled such a high-quality program; we hope that all authors have benefitted from these efforts.

Apart from the submitted articles, this volume features the contributions of three invited speakers. The article on an *"Algebra for Quantitative Information Flow"* by Annabelle McIver and her co-authors presents a new model for reasoning about confidentiality in security applications. Jean-Éric Pin's paper on the *"Dual Space of a Lattice as the Completion of a Pervin Space"* introduces Pervin spaces as useful tools for computing dual spaces of lattices, with applications in language theory. Alexandra Silva has contributed an abstract of her talk on *"A (Co)Algebraic Theory of Succinct Acceptors."* We are delighted that all three invited speakers accepted our invitation to present their work at the conference.

Last, but not least, we would like to thank the members of the RAMiCS Steering Committee for their support and advice. We gratefully acknowledge financial support by the Laboratoire de l'Informatique du Parallélisme (LIP), the Ecole Normale Supérieure de Lyon (ENS de Lyon), and the Laboratoire d'excellence en mathématique et informatique fondamentale (Labex MILYON) of the University of Lyon; and

Catherine Desplanches, Enric Cosme-Llópez, Anupam Das, Christian Doczkal, and Valeria Vignudelli for their help with organizing this conference. We also appreciate the excellent facilities offered by the EasyChair conference administration system, and Alfred Hofmann and Anna Kramer's help in publishing this volume with Springer. Finally, we are indebted to all authors and participants for supporting this conference.

May 2017

Peter Höfner
Damien Pous
Georg Struth

Organization

Organizing Committee

Damien Pous (Conference Chair)	CNRS, France
Peter Höfner (PC Chair)	Data61, CSIRO, Australia
Georg Struth (PC Chair)	University of Sheffield, UK

Program Committee

Luca Aceto	Reykjavik University, Iceland
Rudolf Berghammer	Kiel University, Germany
Filippo Bonchi	CNRS, France
Jules Desharnais	Laval University, Canada
Hitoshi Furusawa	Kagoshima University, Japan
Tim Griffin	University of Cambridge, UK
Walter Guttmann	University of Canterbury, New Zealand
Robin Hirsch	University College London, UK
Peter Höfner	Data61, CSIRO, Australia
Marcel Jackson	La Trobe University, Australia
Jean-Baptiste Jeannin	Samsung Research America, USA
Peter Jipsen	Chapman University, USA
Christian Johansen	University of Oslo, Norway
Wolfram Kahl	McMaster University, Canada
Dexter Kozen	Cornell University, USA
Szabolcs Mikulas	Birkbeck University of London, UK
Bernhard Möller	University of Augsburg, Germany
José N. Oliveira	University of Minho, Portugal
Damien Pous	CNRS, France
Georg Struth	University of Sheffield, UK
Pascal Weil	CNRS, France
Michael Winter	Brock University, Canada

Steering Committee

Rudolf Berghammer	Kiel University, Germany
Jules Desharnais	Laval University, Canada
Ali Jaoua	Qatar University, Qatar
Peter Jipsen	Chapman University, USA
Bernhard Möller	University of Augsburg, Germany
José N. Oliveira	University of Minho, Portugal
Ewa Stella Orlowska	National Institute of Telecommunications, Poland
Gunther Schmidt	Bundeswehr University of Munich, Germany
Michael Winter	Brock University, Canada

Additional Reviewers

Alasdair Armstrong
Roland Glück
Ian Hayes
Tom Hirschowitz
Simon Huber
Barbara König
Dietrich Kuske
Jérôme Lang
Kamal Lodaya
Martin E. Müller
Koki Nishizawa
Jean-Éric Pin
Jurriaan Rot
Gunther Schmidt
Insa Stucke
Harrie De Swart
Norihiro Tsumagari

Sponsors

Laboratoire de l'Informatique du Parallélisme (LIP)
École Normale Supérieure de Lyon (ENS de Lyon)
Labex MILYON/ANR-10-LABX-0070

Local Organizers

Catherine Desplanches
Enric Cosme-Llópez
Christian Doczkal
Damien Pous
Anupam Das
Valeria Vignudelli

Abstracts of Invited Talks

A (Co)Algebraic Theory of Succinct Acceptors

Alexandra Silva

Department of Computer Science, University College London, London, UK

Abstract. The classical subset construction connects deterministic and non-deterministic automata that accept the same language. This construction can be interpreted in a more general setting in which non-determinism is replaced by other side-effects captured by a monad. The more general construction has interesting algorithmic aspects, having led, for example, Bonchi and Pous to devise a very efficient algorithm to check for language equivalence of non-deterministic automata. At the core of their algorithm is the fact that both the state space of the determinized automaton and its semantics — languages over an alphabet — have an algebraic structure: they are Eilenberg-Moore algebras for the monad. Not unexpectedly, in the case of the powerset monad, Eilenberg-Moore algebras are join-semilattices. In this talk, we will start by reviewing the general powerset construction and several applications thereof. We will then show that the dual question to determinization also has interesting algorithmic aspects. In particular, we will look at succinct acceptors of (generalized) languages based on different algebraic structures of their state space. For instance, for classical regular languages the construction will yield a non-deterministic automaton where the states represent the join-irreducibles of the language accepted by a (potentially) larger deterministic automaton. Other examples include weighted and nominal languages. Some of the material in the talk is based on the article 'Fuzzy machines in a category' (Arbib and Manes, 1975).

Algebra for Quantitative Information Flow

A.K. McIver[1], C.C. Morgan[2], and T. Rabehaja[1]

[1] Department of Computing, Macquarie University, Sydney, Australia
annabelle.mciver@mq.edu.au
[2] School of Computer Science and Engineering,
UNSW and Data61, Sydney, Australia

Abstract. A core property of program semantics is that local reasoning about program fragments remains sound even when the fragments are executed within a larger system. Mathematically this property corresponds to *monotonicity of refinement*: if A refines B then $\mathcal{C}(A)$ refines $\mathcal{C}(B)$ for any (valid) context defined by $\mathcal{C}(\cdot)$.

In other work we have studied a *refines* order for information flow in programs where the comparison defined by the order preserves both functional *and* confidentiality properties of secrets. However the semantic domain used in that work is only sufficient for scenarios where either the secrets are static (i.e. once initialised they never change), or where contexts $\mathcal{C}(\cdot)$ never introduce *fresh* secrets.

In this paper we show how to extend those ideas to obtain a model of information flow which supports local reasoning about confidentiality. We use our model to explore some algebraic properties of programs which contain secrets that can be updated, and which are valid in arbitrary contexts made up of possibly freshly declared secrets.

Contents

Invited Papers

Contributed Papers

Invited Papers

Algebra for Quantitative Information Flow

A.K. McIver[1(✉)], C.C. Morgan[2], and T. Rabehaja[1]

[1] Department of Computing, Macquarie University, Sydney, Australia
annabelle.mciver@mq.edu.au

[2] School of Computer Science and Engineering, UNSW and Data61, Sydney, Australia

Abstract. A core property of program semantics is that local reasoning about program fragments remains sound even when the fragments are executed within a larger system. Mathematically this property corresponds to *monotonicity of refinement*: if A refines B then $\mathcal{C}(A)$ refines $\mathcal{C}(B)$ for any (valid) context defined by $\mathcal{C}(\cdot)$.

In other work we have studied a *refines* order for information flow in programs where the comparison defined by the order preserves both functional *and* confidentiality properties of secrets. However the semantic domain used in that work is only sufficient for scenarios where either the secrets are static (i.e. once initialised they never change), or where contexts $\mathcal{C}(\cdot)$ never introduce *fresh* secrets.

In this paper we show how to extend those ideas to obtain a model of information flow which supports local reasoning about confidentiality. We use our model to explore some algebraic properties of programs which contain secrets that can be updated, and which are valid in arbitrary contexts made up of possibly freshly declared secrets.

Keywords: Refinement · Information flow · Security · Monotonicity · Probabilistic semantics · Compositional reasoning · Dalenius desideratum

1 Introduction

Algebras are powerful tools for describing and reasoning about complex behaviours of programs and algorithms. The effectiveness of algebraic reasoning is founded on the principle that equalities between expressions mean that those expressions are interchangeable: if P and Q are algebraic expressions representing programs that are considered to have "the same" behaviours, then $\mathcal{C}(P)$ and $\mathcal{C}(Q)$ must also exhibit "the same" behaviours for any program context $\mathcal{C}(\cdot)$ represented in the algebra. In theories of non-interference security this principle poses a surprising challenge in models describing properties of programs containing secrets which can both be updated during program execution, *and*

T. Rabehaja—We acknowledge the support of the Australian Research Council Grant DP140101119. This work was carried out while visiting the Security Institute at ETH Zürich.

P. Höfner et al. (Eds.): RAMiCS 2017, LNCS 10226, pp. 3–23, 2017.
DOI: 10.1007/978-3-319-57418-9_1

which can be partially observed by a passive but curious adversary. Although there are many semantic models for reasoning about information flow, they typically support only a subset of these behaviours. For example [1,4,24] assume that the secrets once set never change. Our more recent work [13,16] does allow updates to secrets, however it also assumes a "closed system model" for program execution, where there is a single global secret type which must be declared at the outset.

In this paper we show how to extend the applicability of algebraic reasoning for all contexts and behaviours, in particular we remove the assumption of a closed system model of operation. On a technical level this requires generalising our earlier model [13,16] based on Hidden Markov Models (*HMM*'s) to include not only information flow about some declared secret, but also information flow that can potentially have an impact on third-party secrets – undeclared in a given program fragment – but introduced later as part of a context $\mathcal{C}(\cdot)$. In terms of practical verification this theoretical extension is crucial: it means that local reasoning about program fragments remains sound even when those program fragments are executed in contexts which could contain arbitrary secrets.

The surprise here is that our extension of standard *HMM*'s is related to an old problem in privacy in "read-only" statistical databases, first articulated by Dalenius [6] and later developed by Dwork [7]. It says that third-party information flow is possible if a database's contents are known to be correlated with data not in the databases; in this case, information revealed by a query could also lead to information leaked through the correlation.

Our approach rests fundamentally on Goguen and Meseguer's original model for *qualitative* non-interference [9] and on more recent work in *quantitative* information flow of communication channels [24, and its citations]; we combine them into a denotational program-semantics in the style of [21, for qualitative] and [13, for quantitative].

In [9] the program state is divided into high- and low-security portions, and a program is said to be "non-interfering" if the low variables' final values cannot be affected by the high variables' initial values. Note that this is a qualitative judgement: a program either suffers from interference or it doesn't. Here instead we follow others in *quantifying* the interference in a program [24], since it has been recognised for some time that absolute noninterference is in practice too strong: even a failed password guess leaks what the password is not. Then a more personal choice is that we address leaks wrt. the high variables' *final* values, not their initial ones; we explain below why we believe this view is important for refinement.

We embed both of the above features in a denotational semantics based on *HMM*'s supporting a refinement relation in the style of [3,12,20]: our programs here are probabilistic, without demonic choice, and include a special statement `leak` that passes information directly to an adversary. A result is that the apparently "exotic" problems we highlighted above become surprisingly mundane. For example, the program state is entirely secret (all of it is "high"), and leaks are not through "low variables" but rather are explicit via the special `leak` statements.

Furthermore, conventional refinement – the tradition on which we draw – compares programs' final states, not their initial ones: and thus so do we here. Finally, the "Dalenius" problem, of our potential effect on third parties unknown to us, is merely the issue of preserving refinement when extra variables are declared that were not mentioned in our original program.

We make the following contributions.

(a) We note that the Dalenius issue is simply that security of data can be affected by programs that do not refer to it, and we illustrate it by example;
(b) We review Hidden Markov Models and explain how they can be used as the basis for an abstract model of programs that model information flows to secrets that can be updated;
(c) We show, by considering *HMM*'s as transformers of correlations, that they can also model information flows of possibly correlated secrets;
(d) We define a partial order on *abstract HMM*'s based on the information order defined elsewhere [1,2,18] and show that it is general in the sense that equality is maintained in arbitrary contexts.

We begin in Sect. 2 with an example addressing (a) just above. In Sect. 3 we show how to model program fragments as *HMM*'s and, in doing so, we show that to address the issue at (a) it is sufficient to model only the correlation between initial and final states of secrets in local reasoning in order to predict general information flows about arbitrary secrets in arbitrary contexts. In Sect. 4 we show how to define a partial order on *HMM*'s as correlation transformers resulting in a general law of equivalence (Theorem 1). Finally in Sect. 5 we prove some general algebraic laws valid for abstract *HMM*'s as correlation transformers.

2 Getting Real: Updating Secrets and Third-Party "Collateral" Damage in Everyday Programs

Figure 1 illustrates the difference between running a program in isolation versus in a system where there are multiple possibly correlated secrets. We adopt the working assumption of a passive, but curious adversary, by which we mean an adversary who is trying to guess secrets. She does so by observing the program as it executes and matching observations to (possible) values of the secret. The adversary is able to do this because we assume that she has a copy of the program code. The adversary is not actively malicious, however: she cannot change data nor affect normal program operation.

In both scenarios in Fig. 1 there is a loop which is subject to a timing attack and, for simplicity, we assume that it can be performed by counting how many times the guard is executed. We use an explicit statement `leak` as a signal that our passive adversary can observe when the loop body is executed, even though she cannot observe the exact value of the secret `X`. When `leak (X>1)` is executed, the adversary learns whether the current value of `X` is strictly greater than 1 or not. She cannot deduce anything else about `X`, but by accumulating

```
// Secret X is initially uniformly distributed over 𝒳 = {0..N − 1}.
```

Program run in isolation		Program run in context	
`while(X > 1) {`		`Z:= X`	*
`   leak (X > 1)`	†	`while(X > 1) {`	
`   X:= X-2`		`   leak (X > 1)`	†
`}`		`   X:= X-2`	
		`}`	

* *The program is run in a context where new secret* Z *is freshly declared, and then set to the initial value of* X

† `leak (...)` *This models the timing leak, by allowing the adversary to count the number of times the loop checks the loop guard as it executes.*

Fig. 1. Timing attacks in isolation and in context

all her observations, and her knowledge of the program code, over time she can deduce many facts about the initial and final states of X.

First of all, since the adversary knows the program code, without even executing it, she deduces that the final value of X will be either 0 or 1. To learn something about the *initial* state, she must observe the program as it runs: if she observes that $(\mathtt{X} > 1)$ was true three times in total then the initial value of X *must have been* either 6 or 7; if it was true only twice, then the initial value of X *must have been* either 4 or 5.

What can the adversary do with this analysis?

Suppose that the adversary also knows that there are no other secrets, i.e. even if the loop is part of a larger piece of code, the only secret variable referred to in that code is X. This means that her knowledge about the initial state of X is not useful because X is no longer equal to its initial state (unless it was 0 or 1). Thus the information leak about the initial value of X is not useful to the adversary who tries to guess the current value of X.

On the other hand, suppose that in a different scenario there is a second secret Z and it is initially correlated with X, as in the program at the right in Fig. 1. Now the fact that X was initially 6 or 7 is highly significant because it tells the adversary that the *current* value of Z is either 6 or 7. And the adversary might actually be more interested in Z than in X. This is the Dalenius problem referred to above: the impact of some information leaks become manifest only when programs are executed in some context with fresh secrets.

In sequential program semantics, it is usual to focus on the final state because the aim is to establish some goal by updating the variables, and so when we integrate security we still need to consider final values. Indeed it is the concentration on final values that allows small state-modifying programs, whether secure or not, to be sequentially composed to make larger ones [13, 15, 16]. But here we have demonstrated by example that this is not enough if we want a semantics which is compositional when contexts introduce new secrets, because if the semantics only captures the uncertainty about the final states, we have potentially lost the

Dalenius effect i.e. how much uncertainty remains about the correlated secrets. For example, if we only consider scenarios where there is a single secret `X`, then we would only need to consider the remaining uncertainty of final states. We could then confidently argue that the loop on the left at Fig. 1 is equivalent to the program $\mathtt{X} := \mathtt{X} \mod 2$. However, consider context $\mathcal{C}(P)$ defined by $\mathtt{Z} := \mathtt{X}; P$, where `Z` is a secret (and P is some program fragment). We must now consider confidentiality properties of *both* `Z` and `X`, and for monotonicity of refinement since $\mathcal{C}(\mathtt{X} := \mathtt{X} \mod 2)$ leaks less than $\mathcal{C}(\mathtt{leak}(\mathtt{X} > \mathtt{1}); \mathtt{X} := \mathtt{X} \mod 2)$, then our semantics must distinguish between $(\mathtt{X} := \mathtt{X} \mod 2)$ and $\mathtt{leak}\ (\mathtt{X} > \mathtt{1}); \mathtt{X} := \mathtt{X} \mod 2$, even though all their properties concerning the final state of `X` are the same.

In the remainder of the paper we describe a semantics which combines information flow and state updates in which refinement between program fragments can be determined by local reasoning alone (i.e. only about `X` when the program fragments only refer to `X`). Crucially the refinement relation satisfies monotonicity when contexts can include fresh secrets (as in $\mathcal{C}(\cdot)$ above).

3 A Denotational Model for Quantitative Information Flow

3.1 Review of the Probability Monad and Hyper-Distributions

We model secrets as probability distributions reflecting the uncertainty about their values (if indeed they are secret). Our semantics for programs computing with secrets is based on the "probability monad", which we now review; and in Sect. 3.2 we explain how it can be used to define a model for information flow.

Given a state space $\mathcal{X}$ we write $\mathbb{D}\mathcal{X}$ for the set of probability distributions over $\mathcal{X}$, which we assume here to be finite so that we consider only discrete distributions that assign a probability to every individual element in $\mathcal{X}$. For some distribution δ in $\mathbb{D}\mathcal{X}$ we write δ_x, between 0 and 1, for the probability that δ assigns to x in particular.[1]

The *probability monad* [8] is based on $\mathbb{D}$ as a "type constructor" that obeys a small collection of laws shared by other, similar constructors like say the powerset operator $\mathbb{P}$. Each monad has two polymorphic functions η for "unit", and μ for "multiply", that interact with each other in elegant ways. For example in $\mathbb{P}$, unit η has type $\mathcal{X}\to\mathbb{P}\mathcal{X}$, generically in $\mathcal{X}$, and $\eta(x)$ is the singleton set $\{x\}$; in $\mathbb{D}$ the unit has type $\mathcal{X}\to\mathbb{D}\mathcal{X}$ and $\eta(x)$ is $[x]$, the point-distribution on x.[2]

For multiply, in $\mathbb{P}$ it is distributed union, taking a set of sets to the one set that is the union of them all, having thus the type $\mathbb{P}^2\mathcal{X}\to\mathbb{P}\mathcal{X}$. In $\mathbb{D}$ we can construct $\mathbb{D}(\mathbb{D}\mathcal{X})$ or $\mathbb{D}^2\mathcal{X}$ which is the set of "distributions of distributions"

[1] Mostly we use the conventional $f(x)$ for application of function f to argument x. Exceptions include δ_x for δ applied to x and $\mathbb{D}f$ for functor $\mathbb{D}$ applied to f and $f.x.y$ for function $f(x)$, or $f.x$, applied to argument y, and $[\![H]\!].\pi$, when H is an *HMM* inside semantic brackets $[\![\cdot]\!]$.

[2] The point distribution on x assigns probability 1 to x alone, and probability 0 to everything else; we write it $[x]$.

(just as $\mathbb{P}^2$ was sets of sets), equivalently distributions over $\mathbb{D}\mathcal{X}$. We call these "hyper-distributions", and use them below to model information flow; we normally use capital Greeks like Δ for hyper-distributions. In $\mathbb{D}$, the multiply has type $\mathbb{D}^2\mathcal{X}\rightarrow\mathbb{D}\mathcal{X}$ — it "squashes" distributions of distributions back to a single one, and is defined $\mu(\Delta)_x := \sum_{\delta:\,\mathbb{D}\mathcal{X}} \Delta_\delta \times \delta_x$, giving the probability assigned to x as the sum of the inner probabilites (of x), each scaled by their corresponding outers. We also write avg for the μ of $\mathbb{D}\mathcal{X}$.[3]

Monadic type-constructors like $\mathbb{P}$ and $\mathbb{D}$ are *functors*, meaning they can be applied to functions as well as to objects: thus for f in $\mathcal{X}\rightarrow\mathcal{Y}$ the function $\mathbb{P}f$ is of type $\mathbb{P}\mathcal{X}\rightarrow\mathbb{P}\mathcal{Y}$ so that for X in $\mathbb{P}\mathcal{X}$ we have $\mathbb{P}f(X) = \{f(x) \mid x{\in}X\}$ in $\mathbb{P}\mathcal{Y}$. In $\mathbb{D}$ instead we get the *push forward* of f, so that for π in $\mathbb{D}\mathcal{X}$ we have $(\mathbb{D}f)(\pi)_y = \sum_{f(x)=y} \pi_x$.

We shall write $\mathcal{X}$ and $\mathcal{Z}$ for secret types and $\mathcal{Y}$ for the type of observations used to enable us to model information flow. Given a joint distribution J in $\mathbb{D}(\mathcal{X}{\times}\mathcal{Y})$ we define hyper-distributions of secrets by abstracting from the values of observations as follows: we retain only the probabilities that an observation occurs, together with the residual uncertainty in $\mathcal{X}$ related to those observations. The probability of observation y is computed from $\overrightarrow{J}$ the marginal on $\mathcal{Y}$ (relative to the joint distribution J in $\mathbb{D}(\mathcal{X}{\times}\mathcal{Y})$), so that $\overrightarrow{J}_y := \sum_{x:\,\mathcal{X}} J_{xy}$. Next, for each observation y, we can compute $J^y{:}\,\mathbb{D}\mathcal{X}$, the conditional probability distribution over x given this y. It is $J^y_x := J_{xy}/\overrightarrow{J}_y$. This conditional probability distribution represents the *residual uncertainty* of x by taking the observation y into account. We now write $[J]{:}\,\mathbb{D}^2\mathcal{X}$ so that δ is in the support of $[J]$ provided that there is some $y{:}\,\mathcal{Y}$ such that $\delta = J^y$. Finally $[J]_\delta$ is equal to the sum $\sum_{\delta=J^y} \overrightarrow{J}_y$.

3.2 Review of "Traditional" vs. More Recent Quantitative Information Flow Semantics for (Non-)interference

Goguen and Meguer's treatment of non-interference separated program variables into high- and low-security, and defined "non-interference" of high inputs with low outputs: that a change in a high input-value should never cause a consequential change in a low output-value [9]. A hostile observer of final low values in that case could never learn anything about initial high values. Subsequent elaborations of this allowed more nuanced measurements, determining "how much" information was revealed by low variables about the initial values of high variables. The measurements can be of many different kinds: Shannon Entropy was until recently the default choice, but that has now been significantly generalised [2].

[3] We are aware that in $\mathbb{D}(\mathbb{D}\mathcal{X})$ the outer $\mathbb{D}$ is not acting over a finite type: indeed $\mathbb{D}\mathcal{X}$ is non-denumerable even when $\mathcal{X}$ is finite, so a fully general treatment would use proper measures as we have done elsewhere [14,16]. Here however we use the fact that, for programs, the only members of $\mathbb{D}^2\mathcal{X}$ we encounter have finite support (i.e. finitely many $\mathbb{D}\mathcal{X}$'s within them), and constructions like $\sum_{\delta:\,\mathbb{D}\mathcal{X}} \Delta_\delta \delta_x$ remain meaningful.

In the traditional style, the side channel attack of Fig. 1 would be modelled by an explicit assignment to some low-security variable `L` say, actually in the program text; and the program's security would be assessed in terms of how much final observations of `L` could tell you about the original secret value `X`. In particular, the program's action on `X` and `L` together would be described as a joint distribution, and standard Bayesian reasoning would be used to ask (and answer) questions like "Given this particular final value of `L`, how do we change our *prior* belief of the distribution on `X` to an *a-posteriori* distribution on `X`?"

Our more recent style here is instead to make the whole of the program's state-space hidden, and to model information flow to the hostile observer via an explicit `leak` statement. Execution of a statement `leak` E, for some expression E in the program variables (here just `X`), models the emission of E's value at that moment directly to an observer: from it, she makes deductions about `X`'s possible values at that point. Usually E will not be injective (since otherwise she would learn `X` exactly); but, unless E is constant, she will still learn something. But how much? Assume in the following program that `X` is initially one of $0, 1, 2$ with equal probability:

$$\texttt{X:=}\ \texttt{X} + 1;\quad \texttt{leak}\ (\texttt{Xmod2});\quad \texttt{X:=}\,2 * \texttt{X}.$$

Informal reasoning would say that after the first statement `X` is uniformly distributed over 1–3; after the second statement it would (via Bayesian reasoning) either be uniform over 1,3 (if a 1 was leaked) or it would certainly be 2 (if a zero was leaked) — and the first case would occur with probability $2/3$, the second with probability $1/3$. After the third statement, then, our observer would $2/3$ of the time believe `X` to be uniform over $2, 6$ and $1/3$ of the time know that `X` was 4. A key feature of this point of view is that her final "belief state" can be summarised in a hyper-distribution introduced above. In this case we would have the distribution "uniformly either 2 or 6" itself with probability $2/3$, and the distribution "certainly 4" itself with probability $1/3$.

Hyper-distributions (objects of type $\mathbb{D}^2\mathcal{X}$), or *hypers* for short, explicitly structure the relationship between *a-posteriori* distributions ("2 or 6" and "certainly 4" above) and the probabilities with which those "posteriors" occur ($2/3$ and $1/3$ resp.) — we call the a-posteriori, or posterior distributions "inners" of the hyper; and we call the distribution over them the "outer". In our model for security programs we use this two-layered feature to provide a clean structure for information flow. It is based on our general conviction that the value of an observation itself is not important; what matters is how much change that observation induces in the probability distribution of a secret value [18]. Therefore the observations' values do not need to participate in the semantics of information flow, and its formalisation becomes much simpler.

That semantic simplification also enables a calculus of information flow, explored in other work [13], and allows the use of monads, a very general semantic tool for rigorous reasoning about computations [19] and even the implementation of analysis tools [23].

Definition 1 [16]. *Given a state space $\mathcal{X}$ of hidden (i.e. high-security) variables, a denotational model of quantitative non-interference secure-sensitive programs consists of functions from prior (input) distributions on the state space to hyper (output) distributions on the same space — the domain is $\mathbb{D}\mathcal{X}\to\mathbb{D}^2\mathcal{X}$.*

Given two abstract programs $h^1, h^2{:}\,\mathbb{D}\mathcal{X}\to\mathbb{D}^2\mathcal{X}$ we define their composition as $h^1; h^2{:=}\,\mathsf{avg}\circ\mathbb{D}h^2\circ h^1$, which is also of type $\mathbb{D}\mathcal{X}\to\mathbb{D}^2\mathcal{X}$.[4]

In other work we have shown that Definition 1 is an abstraction of *HMM*'s and works well in *closed* systems where there is exactly one secret $\mathcal{X}$ and that the composition defined using the Giry constructors correspond exactly to composition of *HMM*'s. However, as illustrated by Fig. 1, modelling only the residual uncertainty of the final state does not enable us to draw conclusions about behaviour of the program fragment running in the larger context in which fresh secrets participate in some larger computation. It turns out however that we are able to predict the behaviour of a program fragment in such larger contexts by preserving the uncertainty with respect to the correlation between initial *and* final states. We do this by viewing *HMM*'s as *correlation transformers* and we show that this is sufficient to obtain a compositional model suitable for open systems where contexts of execution can contain arbitrary fresh secrets.

3.3 *HMM*'s as Correlation Transformers

The basic step of an *HMM* consists of a secret type $\mathcal{X}$, and two stochastic matrices[5], one to describe the updates to the secret (called a Markov matrix) and the other to describe the information flowing about the secret (called a Channel matrix). A Markov matrix M (over $\mathcal{X}\times\mathcal{X}$) defines $M_{xx'}$ to be the transition probability for an initial value of the secret x updated to x'. A Channel matrix C (over $\mathcal{X}\times\mathcal{Y}$) defines C_{xy} to be the probability that y is observed given that the secret is currently x, where recall that we use $\mathcal{Y}$ for the type of observations. A Hidden Markov Model step is also a stochastic matrix, and is determined by first a Channel step followed by a Markov step, denoted here by $(C{:}M)$. The execution of the Markov step is independent of the observation, and so $(C{:}M)_{xyx'}{:=}\,C_{xy}\times M_{xx'}$ (where "×" here means multiplication). In general we write $\mathcal{A}\twoheadrightarrow\mathcal{B}$ for the type of stochastic matrix over $\mathcal{A}\times\mathcal{B}$, so that rows are labelled by type $\mathcal{A}$. Thus $M{:}\,\mathcal{X}\twoheadrightarrow\mathcal{X}$, $C{:}\,\mathcal{X}\twoheadrightarrow\mathcal{Y}$ and $(C{:}M){:}\,\mathcal{X}\twoheadrightarrow\mathcal{Y}\times\mathcal{X}$.

We can compose *HMM* steps to obtain the result of executing several leak-update steps one after another. Let $H^1{:}\,\mathcal{X}\twoheadrightarrow\mathcal{Y}^1\times\mathcal{X}$ and $H^2{:}\,\mathcal{X}\twoheadrightarrow\mathcal{Y}^2\times\mathcal{X}$. We define their composition $H^1; H^2{:}\,\mathcal{X}\twoheadrightarrow(\mathcal{Y}^1\times\mathcal{Y}^2)\times\mathcal{X}$, which now has observation type $\mathcal{Y}^1\times\mathcal{Y}^2$ so that information leaks accumulate.

$$(H^1; H^2)_{x(y_1,y_2)x'} \quad := \quad \sum_{x''{:}\,\mathcal{X}} {H^1}_{xy_1x''}\times{H^2}_{x''y_2x'}\,. \tag{1}$$

[4] This is the standard method of composing functions defined by a monad.

[5] A matrix is stochastic if its rows sum to 1.

We use the term *HMM* to mean both an *HMM* step, and more generally some composition of steps. In the latter case, the observation type $\mathcal{Y}$ will actually consist of a product of observation types arising from the observations of the component steps. Given an *HMM* $H{:}\,\mathcal{X} \twoheadrightarrow \mathcal{Y}{\times}\mathcal{X}$ and an initial distribution $\pi{:}\,\mathbb{D}\mathcal{X}$ we write $(\pi\rangle H)$ for the joint distribution $\mathbb{D}(\mathcal{X}{\times}\mathcal{Y}{\times}\mathcal{X})$ defined by $(\pi\rangle H)_{xyx'}{:=}\,\pi_x{\times}H_{xyx'}$. It is the probability that the initial state was x, that the final state is now x' and that the adversary observed y. Similarly, as special cases, we write $(\pi\rangle C)$ and $(\pi\rangle M)$ for the result of applying a prior π to respectively a pure Channel and a pure Markov.

In the next sections (Sects. 3.2 and 4) we describe a modification of Definition 1 based on *HMM*'s; it focusses on tracking correlations between initial and final states. We begin by illustrating how the loop body of Fig. 1 can be represented as an *HMM*-style matrix. Recall its definition as a program fragment:

$$\texttt{leak}\ (\texttt{X} > \texttt{1});\quad \texttt{X} := \texttt{X} - \texttt{2}. \tag{2}$$

The first statement of (2) –`leak (X > 1)`– corresponds to a channel matrix in $\mathcal{X}{\twoheadrightarrow}\mathcal{Y}$, where the observation type $\mathcal{Y}$ consists of two values, one for when the secret is strictly greater than 1 and one where the secret is no more than 1.

$$C:\quad \begin{array}{c} \\ 0\\1\\2\\3 \end{array} \begin{array}{c} \begin{array}{cc} \circ\mathtt{G} & \circ\mathtt{L} \end{array} \\ \begin{pmatrix} 0 & 1\\ 0 & 1\\ 1 & 0\\ 1 & 0 \end{pmatrix} \end{array}$$

The labels $\circ$G, $\circ$L denote the observations that X is strictly greater than 1, or no more than 1 respectively. C_{xy} is the chance that y will be observed given that the secret is x. Observe that this is a deterministic channel.

If $\pi{:}\,\mathbb{D}\mathcal{X}$ is a prior distribution over $\mathcal{X}$ we create a joint distribution in $\mathbb{D}(\mathcal{X}{\times}\mathcal{Y})$ defined by $(\pi\rangle C)$. In our example, we take π to the uniform distribution over $\{0,1,2,3\}$; for each observation, we learn something about this initial value of X — if $\circ$G is observed, then we can use Bayesian reasoning to compute the residual uncertainty of the secret. It is the conditional distribution over $\mathcal{X}$ of $(\pi\rangle C)$ given the observation $\circ$G; we call this posterior $(\pi\rangle C)^{\circ\mathtt{G}}$. If $\circ$L is observed instead we can similarly define the posterior $(\pi\rangle C)^{\circ\mathtt{L}}$ as the conditional distribution over $\mathcal{X}$ of $(\pi\rangle C)$ given the observation $\circ$L. Notice that both posteriors occur with probability 1/2.

The second statement of (2) is only executed in a context when $\texttt{X} > \texttt{1}$ and therefore is equivalent to `if (X>1) then X:= X-2`. It corresponds to a Markov matrix $\mathcal{X}{\twoheadrightarrow}\mathcal{X}$:

$$M:\quad \begin{array}{c} \\ 0\\1\\2\\3 \end{array} \begin{array}{c} \begin{array}{cccc} 0 & 1 & 2 & 3 \end{array} \\ \begin{pmatrix} 1 & 0 & 0 & 0\\ 0 & 1 & 0 & 0\\ 1 & 0 & 0 & 0\\ 0 & 1 & 0 & 0 \end{pmatrix} \end{array}$$

Each entry of the Markov matrix $M_{xx'}$ provides the probability that the final state will be x' given that it is x initially. Note that it is impossible for x' to be either 2 or 3.

The combination of the two statements yields an *HMM* to form the composition $(C{:}M){:}\,\mathcal{X} \twoheadrightarrow \mathcal{Y}{\times}\mathcal{X}$, where recall that $(C{:}M)_{xyx'}{:=}\,C_{xy}{\times}M_{xx'}$ (where "$\times$"

here means multiplication). For our example y is one of the observations $\circ\texttt{G}$ or $\circ\texttt{L}$. The combination as an *HMM* matrix becomes:

$$
(C{:}M):\quad
\begin{array}{c}
\\ \\ \texttt{0}\\ \texttt{1}\\ \texttt{2}\\ \texttt{3}
\end{array}
\begin{array}{c}
\overbrace{\texttt{0}\quad \texttt{1}\quad \texttt{2}\quad \texttt{3}}^{\circ\texttt{G}}\quad \overbrace{\texttt{0}\quad \texttt{1}\quad \texttt{2}\quad \texttt{3}}^{\circ\texttt{L}}\\
\begin{pmatrix} 0&0&0&0&1&0&0&0\\ 0&0&0&0&0&1&0&0\\ 1&0&0&0&0&0&0&0\\ 0&1&0&0&0&0&0&0\end{pmatrix}
\end{array}
$$

The labels $\circ\texttt{G}$ and $\circ\texttt{L}$ denote the observations corresponding to those from C, and the other column labels come from the column labels in M. Thus each column is labelled by a pair in $\{\circ\texttt{G}, \circ\texttt{L}\}{\times}\mathcal{X}$.

Notice that the rows are *not* identical, because M updates the state in a way dependent on its incoming value.

Consider now the initial (uniform) prior $\pi{:}\,\mathbb{D}\mathcal{X}$ combined with the matrix $(C{:}M)$ above. The combination is a joint distribution $(\pi\rangle(C{:}M))$ of type $\mathbb{D}(\mathcal{X}{\times}\mathcal{Y}{\times}\mathcal{X})$. We can now take the conditional probability with respect to an observation $y{:}\,\mathcal{Y}$ to obtain the corresponding residual uncertainty over the correlation in $\mathbb{D}(\mathcal{X}^2)$ between the initial and final state of $\texttt{X}$. For example, given that $\circ\texttt{G}$ was observed, the probability that the initial value of the secret was 2 and the final value is 0 is $1/4 \div 1/2 = 1/2$. The probability that the initial value of the secret was 3 and its final value is 0 is 0.

We summarise all the posterior distributions by forming the hyper-distribution $[\pi\rangle(C{:}M)]{:}\,\mathbb{D}^2\mathcal{X}^2$. The outers in our example are both $1/2$, because each observation occurs with the same probability; the corresponding inners are distributions of correlations modelling the adversary's residual uncertainty. These correlations retain just enough detail about the relationship between initial and final states to explain the behaviour of (2) in arbitrary contexts. In particular $\delta{:}\,\mathbb{D}\mathcal{X}^2$ is in the support of $[\pi\rangle(C{:}M)]$ means that for some observation, the adversary can deduce the likelihood between the initial and final states of the secret, and from that the likelihood of its initial value, and the likelihood of its current value.

The hyper-distribution over correlations for our example at (2) is as follows.

$$
[\pi\rangle(C{:}M)] : \quad
\begin{array}{ccc}
 & \boxed{1/2} & \boxed{1/2}\\[1ex]
\begin{array}{c}\texttt{0}\\ \texttt{1}\\ \texttt{2}\\ \texttt{3}\end{array} &
\boxed{\begin{array}{cc}0&0\\ 0&0\\ 1/2&0\\ 0&1/2\end{array}} &
\boxed{\begin{array}{cc}1/2&0\\ 0&1/2\\ 0&0\\ 0&0\end{array}}\\[1ex]
 & \texttt{0}\ \ \texttt{1} & \texttt{0}\ \ \texttt{1}
\end{array}
$$

The labels along the outside of the boxes represent the possible initial states ($\texttt{0,1,2,3}$ on the left column), and the possible final states ($\texttt{0,1}$ twice along the bottom). The large boxes represent the two distinct posterior distributions (one for each observation in $\{\circ\texttt{G}, \circ\texttt{L}\}$), and the small boxes ($1/2$ each) are the marginal probabilities for each observation.

Notice that we no longer need labels of type $\mathcal{Y}$, in fact they have been replaced by outers (the small boxes containing $1/2$ each). By comparing with $(C{:}M)$ above, we also see now that only the relevant effect of the observations has been preserved in $[\pi\rangle(C{:}M)]$ – for example there are no columns only containing zeros because they represent events that cannot occur. Only the relevant posteriors are retained, together with the chance that they are observed. For example with probability 1/2 the observer can now deduce when the initial state was in the set $\{\mathtt{0},\mathtt{0}\}$ or in the set $\{\mathtt{2},\mathtt{3}\}$.

Next we show that $[\pi\rangle(C{:}M)]$ is all that is required for computing the behaviour when we introduce a correlated fresh secret Z, as in this program fragment:

$$\mathtt{Z} := \mathtt{X};\ \ \mathtt{leak}\ (\mathtt{X} > 1);\ \ \mathtt{X} := \mathtt{X} - 2. \tag{3}$$

Take the *HMM* $(C{:}M)$ above representing the program fragment at (2), but now consider executing it with extra secret Z as at (3). The initialisation of Z creates an interesting correlation between X and Z given by $\mathit{\Pi}^*{:}\,\mathbb{D}(\mathcal{Z}{\times}\mathcal{X})$ which is defined to be $\mathit{\Pi}^*_{zx}{:=}\,\pi_x$ if and only if $x = z$, where recall π is the uniform initialisation of X. We now compute the final joint distribution $(\mathit{\Pi}^*\rangle(C{:}M))$ of type $\mathbb{D}(\mathcal{Z}{\times}\mathcal{Y}{\times}\mathcal{X})$; in this case because of the definition of $\mathit{\Pi}^*$, it is

$$(\mathit{\Pi}^*\rangle(C{:}M))_{zyx'} \quad := \quad \mathit{\Pi}_{zz} \times (C{:}M)_{zyx'},$$

with corresponding hyper-distribution in $\mathbb{D}^2(\mathcal{Z}{\times}\mathcal{X})$ given as follows:

$[\mathit{\Pi}^*\rangle(C{:}M)]$:	1/2	1/2
(0,0)	0	1/2
(1,1)	0	1/2
(2,0)	1/2	0
(3,1)	1/2	0

The labels along the outside of the boxes now represent the final posteriors in $\mathbb{D}(\mathcal{Z}{\times}\mathcal{X})$. Notice that the posteriors can be computed directly from $(\pi\rangle(C{:}M))$ as explained above. In fact, as we describe below, it can also be computed directly from the abstraction $[\pi\rangle(C{:}M)]$.

More generally we can study the behaviour of an *HMM* $H{:}\,\mathcal{X} \rightarrow\!\!\triangleright\, \mathcal{Y}{\times}\mathcal{X}$ when it is executed in the context of some arbitrary correlation $\mathit{\Pi}{:}\,\mathcal{Z}{\times}\mathcal{X}$. The joint distribution $\mathit{\Pi}\rangle H$ is computed explicitly as

$$(\mathit{\Pi}\rangle H)_{zyx'} \quad := \quad \sum_{x{:}\,\mathcal{X}} \mathit{\Pi}_{zx}{\times}H_{xyx'}. \tag{4}$$

Just as in the special case above, we can calculate the associated hyper-distribution from H's posterior correlations on the initial and final value of the secret type $\mathcal{X}$. We first define a matrix $Z{:}\,\mathcal{Z} \rightarrow\!\!\triangleright\, \mathcal{X}$ as $Z_{zx}{:=}\,\mathit{\Pi}_{zx}/\overrightarrow{\mathit{\Pi}}_x$, and $\overrightarrow{\mathit{\Pi}}^*{:}\,\mathbb{D}\mathcal{X}^2$ as:

$$\mathit{\Pi}^*{}_{xx'}{:=}\,\overrightarrow{\mathit{\Pi}}_x \qquad \textit{if and only if} \quad x = x'.$$

With these in place we have that $\mathit{\Pi} = Z \cdot \mathit{\Pi}^*$, where we are using $(Z\cdot)$ as matrix product, as in $(Z \cdot \Upsilon)_{zx'}{:=} \sum_{x{:}\,\mathcal{X}} Z_{zx}{\times}\Upsilon_{xx'}$. We can now express (4) equivalently as an equation between hyper-distributions:

$$[\mathit{\Pi}\rangle H] \quad = \quad \mathbb{D}(Z\cdot)[\overrightarrow{\mathit{\Pi}}\rangle H]. \tag{5}$$

Notice that we have applied $\mathbb{D}$ to the function $(Z\cdot)$, since it must act on the inners of the hyper-distribution $[\overrightarrow{\Pi}\rangle H]$ to re-install correlations of type $\mathbb{D}(\mathcal{Z}{\times}\mathcal{X})$. But $[\overrightarrow{\Pi}\rangle H]$ is the special case of H applied to a prior in $\mathbb{D}\mathcal{X}$, which models the hyper-distribution of the correlation between the initial and final value of the secret (type $\mathcal{X}$). This allows us to define a family of liftings of *HMM*'s which, by construction, can be computed from hyper-distributions of the form $[\overrightarrow{\Pi}\rangle H]$.

Definition 2. *Given an HMM* $H{:}\mathcal{X} \rightarrow\!\!\!\!\triangleright \mathcal{Y}{\times}\mathcal{X}$. *We say that* $\mathcal{X}$ *is its mutable type; for fresh secret type* $\mathcal{Z}$ *called the correlation type define* $[\![H]\!]^{\mathcal{Z}}{:}\mathbb{D}(\mathcal{Z}{\times}\mathcal{X}) \rightarrow \mathbb{D}^2(\mathcal{Z}{\times}\mathcal{X})$ *as*

$$[\![H]\!]^{\mathcal{Z}}.\Pi \quad := \quad \mathbb{D}(Z\cdot)[\overrightarrow{\Pi}\rangle H],$$

where Z *is defined relative to* Π *as at* (5).

Definition 2 divides the secrets up into a mutable part $\mathcal{X}$ and correlated part $\mathcal{Z}$, where the former can have information leaked about it, and then subsequently be updated by H, whilst the latter cannot be changed (by H), but can have information about its value leaked through a correlation with $\mathcal{X}$. Our next definition, modifies Definition 1 to describe an abstract semantics taking the correlated part into account, consistent with the way a concrete *HMM* leaks information about correlated secrets. Our abstraction enforces the behaviour of the correlated secret to be determined by the behaviour of the mutable state as in Definition 2, via the commuting diagram summarised in Fig. 2.

Definition 3. *Given a state space* $\mathcal{X}$ *of hidden (i.e. high-security) variables, a context-aware denotational model of quantitative non-interference is a family of functions from prior (input) distribution correlations on* $\mathbb{D}(\mathcal{Z}{\times}\mathcal{X})$ *to (output) hyper-distributions on the same space, where* $\mathcal{Z}$ *is any correlated type, and* $\mathcal{X}$ *is the mutable type. For a given correlated type* $\mathcal{Z}$, *function* $h^{\mathcal{Z}}$ *has type* $\mathbb{D}(\mathcal{Z}{\times}\mathcal{X}){\rightarrow}\mathbb{D}^2(\mathcal{Z}{\times}\mathcal{X})$. *Moreover* $h^{\mathcal{X}}$ *and* $h^{\mathcal{Z}}$ *must satisfy the commuting diagram in Fig. 2.*

We define the composition of $h_1^{\mathcal{Z}}, h_2^{\mathcal{Z}}$ *to be* $h_1^{\mathcal{Z}}; h_2^{\mathcal{Z}} = \mathsf{avg} \circ \mathbb{D}h_2^{\mathcal{Z}} \circ h_1^{\mathcal{Z}}$.

Figure 2 describes the commuting diagram which captures exactly the effect on collateral secrets as described by the concrete situation of *HMM*'s at (5). The function $h^{\mathcal{X}}$ can preserve the correlation between the initial and final values of the mutable variables, after which the correlated variable can be reinstalled by applying $\mathbb{D}(Z\cdot)$ to the hyper-distribution $\mathbb{D}^2\mathcal{X}^2$.

The next lemma shows that the standard Giry composition also satisfies the healthiness condition of Fig. 2.

Lemma 1. *Let* $h_{1,2}^{\mathcal{Z}}{:}\mathbb{D}(\mathcal{Z}{\times}\mathcal{X}) \rightarrow \mathbb{D}^2(\mathcal{Z}{\times}\mathcal{X})$ *satisfy the commuting diagram in Fig. 2. Then* $h_1^{\mathcal{Z}}; h_2^{\mathcal{Z}}$ *also satisfies it.*

At left: Given a correlation $\Pi{:}\,\mathbb{D}(\mathcal{Z}{\times}\mathcal{X})$ define a matrix $Z{:}\,\mathcal{Z}{\times}\mathcal{X}$ and $\mathcal{X}$-prior $\pi{:}\,\mathbb{D}\mathcal{X}$ to be such that $\Pi_{z,x} = Z_{z,x}\pi_x$. This is arranged so that $\Pi = Z{\cdot}\pi^*$ where $\pi^*{:}\,\mathbb{D}\mathcal{X}^2$ with $\pi^*_{x'x} = \pi_x$ iff $x = x'$.

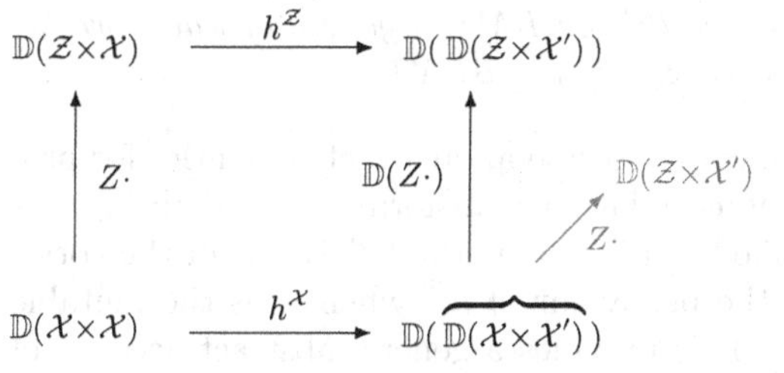

At right (in grey): For an initial/final distribution $\delta{:}\,\mathbb{D}(\mathcal{X}{\times}\mathcal{X}')$, the left-multiplication $Z{\cdot}\delta$ produces a distribution in $\mathbb{D}(\mathcal{Z}{\times}\mathcal{X}')$, just as matrix multiplication would (with δ as a matrix of probabilities). *At right (in black):* The $\mathbb{D}$-lifting (push forward) of the multiplication $(Z{\cdot})$ thus takes an initial-final hyper in $\mathbb{D}^2(\mathcal{X}{\times}\mathcal{X}')$ to a hyper in $\mathbb{D}^2(\mathcal{Z}{\times}\mathcal{X}')$.

Summary: A collateral Z is linked to our state X by joint distribution $\Pi{:}\,\mathbb{D}(\mathcal{Z}{\times}\mathcal{X})$. This Π can be decomposed into its right marginal $\pi{:}\,\mathcal{X}$ on our state space, and a "collateral stochastic channel matrix" $Z{:}\,\mathcal{Z}{\twoheadleftarrow}\mathcal{X}$ between it and $\mathcal{Z}$, i.e a right conditional of Π. For each x the matrix Z gives the conditional distribution over $\mathcal{Z}$, as in "the probability that Z is some value, given that X is x". The original joint distribution Π is restored from π and Z by matrix multiplication. Since π is not presented as a matrix, but Z is, we use the notation $Z\langle\pi$ to reconstruct Π from the components. (Note that right-conditionals are not necessarily unique; but the variation on x's where $\pi.x{=}0$ does not affect $\mathbb{D}(Z{\cdot})$ at right.)

Fig. 2. Healthiness condition for h: general collateral correlation $\Pi{:}\,\mathbb{D}(\mathcal{Z}{\times}\mathcal{X})$ can be computed from the effect of h on initial and final states.

Proof. Let $\Pi{:}\,\mathbb{D}(\mathcal{Z}{\times}\mathcal{X})$ and Z, π^* be defined as in Fig. 2 so that $\Pi = Z\cdot\pi^*$. We now reason:

$$
\begin{array}{lll}
 & (h_1^{\mathcal{Z}}; h_2^{\mathcal{Z}}).\Pi & \\
= & \mathsf{avg}\circ\mathbb{D}h_2^{\mathcal{Z}}\circ h_1^{\mathcal{Z}}.\Pi & \text{“Definition 3”} \\
= & \mathsf{avg}\circ\mathbb{D}h_2^{\mathcal{Z}}\circ\mathbb{D}(Z\cdot)\circ h_1^{\mathcal{X}}.\pi^* & \text{“Figure 2 for } h_1\text{: } h_1^{\mathcal{Z}}.\Pi = \mathbb{D}(Z\cdot)\circ h_1^{\mathcal{X}}.\pi^*\text{”} \\
= & \mathsf{avg}\circ\mathbb{D}(h_2^{\mathcal{Z}}\circ(Z\cdot))\circ h_1^{\mathcal{X}}.\pi^* & \text{“Function composition: } \mathbb{D}(f\circ g) = \mathbb{D}f\circ\mathbb{D}g\text{”} \\
= & \mathsf{avg}\circ\mathbb{D}(\mathbb{D}(Z\cdot)\circ h_2^{\mathcal{X}})\circ h_1^{\mathcal{X}}.\pi^* & \text{“Figure 2 for } h_2\text{: } h_2^{\mathcal{Z}}\circ(Z\cdot) = \mathbb{D}(Z\cdot)\circ h_2^{\mathcal{X}}\text{”} \\
= & \mathsf{avg}\circ\mathbb{D}^2(Z\cdot)\circ\mathbb{D}h_2^{\mathcal{X}}\circ h_1^{\mathcal{X}}.\pi^* & \text{“Function composition”} \\
= & \mathbb{D}(Z\cdot)\circ\mathsf{avg}\circ\mathbb{D}h_2^{\mathcal{X}}\circ h_1^{\mathcal{X}}.\pi^* & \text{“Monad law: } \mathsf{avg}\circ\mathbb{D}^2 f = \mathbb{D}f\circ\mathsf{avg}\text{”} \\
= & \mathbb{D}(Z\cdot)\circ(h_2^{\mathcal{X}}; h_1^{\mathcal{X}}).\pi^*. & \text{“Definition 3”}
\end{array}
$$

By construction, the action of *HMM*'s defined above at Definition 2 satisfy the commuting diagram of Fig. 2, because for π^* defined in Fig. 2 we have

$$[\![H]\!]^{\mathcal{X}}.\pi^* \quad = \quad [\overrightarrow{\pi^*}\rangle H]. \tag{6}$$

Finally we note that *HMM* composition given above at (1) is consistent with the abstract semantics.

Lemma 2. *Let $H^{1,2}{:}\,\mathcal{X} \twoheadrightarrow \mathcal{Y}{\times}\mathcal{X}$ be HMM's, with mutable type $\mathcal{X}$; further let $\mathcal{Z}$ be any collateral type. Then $[\![H^1; H^2]\!]^{\mathcal{Z}} = [\![H^1]\!]^{\mathcal{Z}}; [\![H^2]\!]^{\mathcal{Z}}$, where composition of*

HMM's (inside $[\![\cdot]\!]$) is defined at (1), and composition in the semantics (outside $[\![\cdot]\!]$) is defined at Definition 3.

Proof. *This follows from [16][Theorem 12] for HMM's generally where we set the secret type explicitly to be $\mathcal{Z}\times\mathcal{X}$ (rather than just $\mathcal{X}$).*

In this section we have shown how to describe an abstract semantics for programs based on viewing *HMM*'s as correlation transformers, generalising our previous work [16]. We have identified in Fig. 2 how the behaviour of the correlation transformer $h^{\mathcal{X}}$ determines the behaviour of $h^{\mathcal{Z}}$ when $\mathcal{X}$ is the mutable type and $\mathcal{Z}$ is the correlation type. This provides a general abstract account of how programs modelled as *HMM*'s update and leak information about secrets.

So far we have not defined a refinement order on abstract *HMM*'s which takes information flow into account. We do that next.

4 Generalising Entropy: Secure Refinement

We quantify our ignorance of hidden variables' unknown exact value using *uncertainty measures* over their (known) distribution, a generalisation of (e.g.) Shannon entropy and others [1,2,13,18,24]. These measures are continuous, concave functions in $\mathbb{D}\mathcal{X}\to\mathbb{R}$ [16]. With them, programs' security behaviours can be compared wrt. the average uncertainty of their final (probabilistic) state when run from the same prior distribution; and for programs that don't update their state (e.g. the information channels of Shannon, intensively studied in current security research), the *amount of information* flowing due to a single program's execution can be measured by looking for a *change* in uncertainty, i.e. by comparing the program's prior uncertainty with its average posterior uncertainty. Such comparisons between uncertainties are used to define secure refinement.

It is assumed that the adversary knows the program text (and for us this usually means some *HMM*), and that he observes the values emitted by (for example) `leak` statements as described above. Given a hyper-distribution produced by some program, each inner is a posterior distribution having some uncertainty; and the (outer) probability of that inner represents the probability with which that uncertainty occurs.

For general $\mathcal{S}$, a distribution $\sigma{:}\,\mathbb{D}\mathcal{S}$ and a real-valued (measurable) function $f{:}\,\mathcal{S}\to\mathbb{R}$, we write $\mathcal{E}_\sigma(f)$ for the expected value of f over σ. Typical cases are when $\mathcal{S}{=}\mathcal{X}$ and $f{:}\,\mathcal{X}\to\mathbb{R}$ is over the initial state, and when $\mathcal{S}{=}\mathbb{D}\mathcal{X}$ and we are taking expected values of some $f{:}\,\mathbb{D}\mathcal{X}\to\mathbb{R}$ over an output hyper in $\mathbb{D}^2\mathcal{X}$: in that case $\mathcal{E}_\Delta(\mathtt{se})$ would e.g. be the *conditional* Shannon Entropy of a hyper Δ, where $\mathtt{se}.\pi = -\sum_{x{:}\,\mathcal{X}} \pi_x \log(\pi_x)$.

An important class of uncertainty measures, more appropriate for security applications than Shannon entropy alone, are the "loss functions" [16].

Definition 4. *A* loss function *ℓ is of type $I\to\mathcal{X}\to\mathbb{R}$ for some index set I, with the intuitive meaning that $\ell.i.x$ is the cost to the adversary of using "attack strategy" i when the hidden value turns out actually to be x. Her expected cost* *for an attack planned but not yet carried out is then* $\mathcal{E}_\delta(\ell.i)$ *if* δ *is the distribution in* $\mathbb{D}\mathcal{X}$ *she knows to be governing* x *currently.*[6]

From such an ℓ *we define an uncertainty measure* $U_\ell(\rho) := \inf_{i:I}\ \mathcal{E}_\rho(\ell.i)$. *When* I *is finite, the* inf *can be replaced by* min.

The inf represents a rational strategy where the adversary minimises her cost or risk under the ignorance expressed by her knowing only the distribution ρ and not an exact value: she will choose the strategy i whose expected cost $\mathcal{E}_\rho(\ell.i)$ is the least. If ρ is the prior δ, then $U_\ell(\delta)$ is her expected cost if attacking without running the program, i.e. she attacks the input. If she does run the program, producing output hyper $\Delta=[\![P]\!].\delta$, then her expected cost is $\mathcal{E}_\Delta(U_\ell)$; here she attacks the program's output, taking advantage of the observations she made as the program ran.

For example, consider the following loss function for which the index set I is the same as $\mathcal{X}$, and the adversary is trying to guess the secret. If she chooses some i which turns out to be the same as the secret, then her losses are 0, otherwise if her guess is wrong, then she loses \$1. Formalised, this becomes:

$$\ell.i.x := (0\ \underline{\text{if}}\ i{=}x\ \underline{\text{else}}\ 1). \tag{7}$$

In Fig. 1 if the prior $\pi{:}\,\mathbb{D}\mathcal{X}$ is uniform over `X` then $U_\ell(\pi) = 3/4$, since whichever i is picked there is only 1/4 chance that it is equal to the value of the variable. After executing the program however, the hyper-distribution Δ on `X` alone has a single posterior which assigns equal probability to 0 or 1, thus $\mathcal{E}_\Delta(U_\ell) = 1/2$, showing that the adversary is better able to guess the secret after executing the program than she was before.

Loss functions have been studied extensively elsewhere [2], where they have been shown to describe more accurately than Shannon entropy the adversary's intentions and losses versus benefits involved in attacks [24]. The crucial property of the derived uncertainty measures is that they are *concave functions* of distributions — this feature embodies the idea that when information is leaked then the losses to the attacker will be reduced. Thus if C is Channel matrix, and I is the identity Markov matrix then we always obtain the inequality:

$$\mathcal{E}_{[\![(C:I)]\!]^{z}.\Pi} U_\ell \quad \leq \quad U_\ell(\Pi) \tag{8}$$

where the expression on the left gives the adversary's losses relative to the release of information through C and loss function ℓ, and on the right are the losses without any release of information.

In other work [13,16,18] we have shown that loss functions (equivalently their dual "gain functions") are sufficient to determine hyper-distributions, that is (remarkably) that if we know $\mathcal{E}_\Delta(U_\ell)$ for all ℓ then we know Δ itself.[7] They therefore define a "secure refinement" relation between programs (Definition 5 below), based on their output hyper-distributions. Any program that, for all loss functions, can only cost more for the adversary, never less, is regarded as being more secure:

[6] Here $\ell.i$ is the function $\ell(i)$ of type $\mathcal{X}{\rightarrow}\mathbb{R}$ — we are using Currying.

[7] This was called the *Coriaceous Conjecture* in [2].

Definition 5. [13] *Let* $H^{1,2}$ *be programs represented by HMM's in* $\mathcal{X} \twoheadrightarrow \mathcal{Y}{\times}\mathcal{X}$ *so that* $\mathcal{X}$ *is the mutable type. We say that* $H^1{\sqsubseteq}H^2$ *just when, for any correlated type* $\mathcal{Z}$ *we have* $\mathcal{E}_{[\![H^1]\!]^{\mathcal{Z}}.\Pi}(U_\ell) \leq \mathcal{E}_{[\![H^2]\!]^{\mathcal{Z}}.\Pi}(U_\ell)$ *for all priors* $\Pi{:}\,\mathbb{D}(\mathcal{Z}{\times}\mathcal{X})$ *and loss functions* ℓ *on* $\mathcal{Z}{\times}\mathcal{X}$.

Notice that Definition 5 captures both functional and information flow properties: when the loss function is derived from a single choice, it behaves like a standard "probabilistic predicate" and the refinement relation for this subset of loss functions reduces to the well known functional refinement of probabilistic programs [12].

In order to determine when $H_1 \sqsubseteq H_2$ it might seem as though all contexts need to be considered. Fortunately the healthiness condition summarised in Lemma 2 means that general refinement in all correlation contexts can follow from local reasoning relative to $[\![H]\!]^{\mathcal{X}}$. *Context-aware refinement* is defined with respect only to correlations between initial and final states.

Definition 6. *Let* $H^{1,2}$ *be two HMM's both with mutable type* $\mathcal{X}$. *We say that* $H^1\tilde{\sqsubseteq}H^2$, *whenever* $\mathcal{E}_{[\![H^1]\!]^{\mathcal{X}}.\delta}(U_\ell) \leq \mathcal{E}_{[\![H^2]\!]^{\mathcal{X}}.\delta}(U_\ell)$ *for all* $\delta{:}\,\mathbb{D}\mathcal{X}^2$ *and* $\ell{:}\,I{\to}\mathcal{X}^2{\to}\mathbb{R}$.

Our principal monotonicity result concerns state extension: it shows that context-aware refinement is preserved within any context — even if fresh variables have been declared.

Theorem 1. *Let* $H^{1,2}$ *be HMM's with mutable type* $\mathcal{X}$. *Then*

$$H^1\tilde{\sqsubseteq}H^2 \quad \textit{iff} \quad H^1 \sqsubseteq H^2.$$

Proof. If $H^1 \sqsubseteq H^2$ *holds then it is clear that* $H^1\tilde{\sqsubseteq}H^2$.

Alternatively, we observe that from Fig. 2 we deduce that $\mathbb{D}(Z\cdot)\circ[\![H]\!]^{\mathcal{X}}.\pi^* = [\![H]\!]^{\mathcal{Z}}.\Pi$, *where* Z, π^* *are determined by* Π. *This means that for any* $\delta{:}\,\mathbb{D}(\mathcal{Z}{\times}\mathcal{X})$ *in the support of* $[\![H]\!]^{\mathcal{Z}}.\Pi$, *there is a corresponding* $\delta^*{:}\,\mathbb{D}\mathcal{X}^2$ *in the support of* $\mathbb{D}(Z\cdot)\circ[\![H]\!]^{\mathcal{X}}.\pi^*$ *such that* $\delta = Z\cdot\delta^*$.

Now given $\ell{:}\,I \to \mathcal{Z}{\times}\mathcal{X} \to \mathbb{R}$ *we calculate:*

$$\begin{array}{lll}
 & \sum_{x',z} \ell.i.z.x' \times \delta_{zx'} & \\
= & \sum_{x',z} \ell.i.z.x' \times \sum_x Z_{zx} \times \delta^*_{xx'} & \text{“}\delta = Z\cdot\delta^*\text{”} \\
= & \sum_{x',x,z} \ell.i.z.x' \times Z_{zx} \times \delta^*_{xx'} & \text{“Rearrange”} \\
= & \sum_{x',x} \sum_z (\ell.i.z.x' \times Z_{zx}) \times \delta^*_{xx'} & \text{“Rearrange”} \\
= & \sum_{x',x} \ell^*.i.x.x' \times \delta^*_{xx'}. & \text{“Define } \ell^*.i.x.x' := \sum_z \ell.i.z.x' {\times} Z_{zx}\text{”}
\end{array}$$

We see now that $\mathcal{E}_{[\![H]\!]^{\mathcal{X}}.\pi^*}(U_{\ell^*}) = \mathcal{E}_{[\![H]\!]^{\mathcal{Z}}.\Pi}(U_\ell)$, *where on the left we have an expression that only involves the mutable type. Thus if* $H^1 \not\sqsubseteq H^2$ *we can find some* ℓ *and some* $\mathcal{Z}$ *such that* $\mathcal{E}_{[\![H^1]\!]^{\mathcal{Z}}.\Pi}(U_\ell) > \mathcal{E}_{[\![H^2]\!]^{\mathcal{Z}}.\Pi}(U_\ell)$ *which, by the above, means that* $\mathcal{E}_{[\![H^1]\!]^{\mathcal{X}}.\pi^*}(U_{\ell^*}) > \mathcal{E}_{[\![H^2]\!]^{\mathcal{X}}.\pi^*}(U_{\ell^*})$ *implying that* $H^1 \tilde{\not\sqsubseteq} H^2$.

Theorem 1 now restores the crucial monotonicity result for reasoning about information flow for sequential programs modelled as *HMM*'s. In particular it removes the quantification over all priors in $\mathbb{D}(\mathcal{Z}{\times}\mathcal{X})$, replacing it with a quantification over all priors in $\mathbb{D}\mathcal{X}^2$. We comment on how this applies to practical program analysis in Sect. 6.

5 Some Algebraic Inequalities

In this section we illustrate some useful algebraic laws for security-aware programs modelled as abstract *HMM*'s denoted by Definition 2, and equality determined by the partial order at Definition 5.

Given $\delta^1, \delta^2 : \mathbb{D}\mathcal{S}$ we define convex summation for distributions by $\delta^1 \ {}_p{+}\ \delta^2$, also in $\mathbb{D}\mathcal{S}$, as $(\delta^1 \ {}_p{+}\ \delta^2)_s := \delta^1{}_s{\times}p + \delta^2{}_s{\times}(1{-}p)$, where $0 \le p \le 1$. Similarly, given $\Delta^1, \Delta^2 : \mathbb{D}^2\mathcal{S}$ we define convex summation between hyper-distributions as $\Delta^1 \ {}_p{\oplus}\ \Delta^2$, also in $\mathbb{D}^2\mathcal{S}$, as $(\Delta^1 \ {}_p{\oplus}\ \Delta^2)_\delta := \Delta^1{}_\delta{\times}p + \Delta^2{}_\delta{\times}(1{-}p)$.

5.1 Basic Laws for Information Flow

We write $h : \mathbb{D}(\mathcal{Z}{\times}\mathcal{X}) \to \mathbb{D}^2(\mathcal{Z}{\times}\mathcal{X})$, where $\mathcal{X}$ is the mutable type and $\mathcal{Z}$ is some correlated type. The laws in Theorem 2 describe some basic monotonicity relationships between *HMM*'s. (1) says that if there is more information available in the prior, then there will be more information flow. Similarly (2) says that if all the observations are suppressed, then less information flows: recall that avg applied to a hyper-distribution averages the inners (in our case the posteriors) and therefore summarises the state updates only. (3–4) say that refinement is preserved by sequential composition. Finally (5) says that if both h^1, h^2 simply release information but don't update the state, then the order in which that information is released is irrelevant.

Theorem 2. *Let h, h^1, h^2 be instances of HMM's respectively $[\![H]\!]^{\mathcal{Z}}, [\![H_1]\!]^{\mathcal{Z}}, [\![H_2]\!]^{\mathcal{Z}}$ with mutable type $\mathcal{X}$ and correlated type $\mathcal{Z}$. Further, let $\Pi : \mathbb{D}(\mathcal{Z}{\times}\mathcal{X})$, and $0 \le p \le 1$. The following refinements hold.*

1. $h.\Pi \ {}_p{\oplus}\ h.\Pi' \quad \sqsubseteq \quad h.(\Pi \ {}_p{+}\ \Pi')$
2. $h \sqsubseteq \eta \circ \mathsf{avg} \circ h$
3. $h^1 \sqsubseteq h^2$ *implies* $h; h^1 \sqsubseteq h; h^2$
4. $h^1 \sqsubseteq h^2$ *implies* $h^1; h \sqsubseteq h^2; h$
5. *If* h^1, h^2 *correspond to channels, then* $h^1; h^2 = h^2; h^1$.

Proof. *(1–4) have appeared elsewhere for an HMM model without collateral variables (see [13, 16] for example), and the proof here is similar for each possible correlated state, and relies on the concavity of loss functions. (5) also follows directly from the definition of channels.*

5.2 Information Flows Concerning the Collateral State

In some circumstances we can summarise simply the behaviour of a complex *HMM* matrix formed by sequentially composing some number of leak-update steps. We look at two cases here, and both result in summarising the overall effect as a single step of an *HMM*, i.e. as a leak of information concerning the mutable type, followed by a Markov update.

Let H be an *HMM* matrix with mutable type $\mathcal{X}$, and recall $\mathsf{dup}\colon \mathcal{X} \to \mathcal{X}^2$ is defined by $\mathsf{dup}.x = (x, x)$. Now define $\mathsf{chn}.[\![H]\!]^{\mathcal{X}}\colon \mathbb{D}(\mathcal{X}^2) \to \mathbb{D}^2\mathcal{X}^2$

$$\mathsf{chn}.[\![H]\!]^{\mathcal{X}} \quad := \quad \mathbb{D}^2(\mathsf{dup}) \circ \mathbb{D}(\overleftarrow{\cdot}) \circ [\![H]\!]^{\mathcal{X}}, \tag{9}$$

which ignores the update of the final state, and records the information flow concerning the initial state only [17].

Similarly we can define the overall Markov state change $\mathsf{mkv}.[\![H]\!]^{\mathcal{X}}\colon \mathbb{D}(\mathcal{X}^2) \to \mathbb{D}^2\mathcal{X}^2$, which simply ignores the information flow. Its output is therefore a point distribution in $\mathbb{D}^2\mathcal{X}$:

$$\mathsf{mkv}.[\![H]\!]^{\mathcal{X}} \quad := \quad \eta \circ \mathsf{avg} \circ [\![H]\!]^{\mathcal{X}}. \tag{10}$$

If the effect of an *HMM* can be summarised as its associated channel followed by its associated Markov update then it can be written in the form $\mathsf{chn}.[\![H]\!]^{\mathcal{X}}; \mathsf{mkv}.[\![H]\!]^{\mathcal{X}}$. Next we illustrate two circumstances when this can (almost) happen.

We say that $\mathsf{chn}.[\![H]\!]^{\mathcal{X}}$ is *standard* if it does not leak information probabilistically. For example the leak statement in Fig. 1 is standard — informally this means any information it does leak is not "noisy" and corresponds to the adversary deducing exactly some predicate. We can express standard leaks equationally by saying that if we leak the information (about the initial state) first, and then run the program, we learn nothing more — this is not true if the information released is noisy because each time a noisy channel is executed, a little more information is released. Thus non-probabilistic leaks have associated channel satisfying the following:

$$\mathsf{chn}.[\![H]\!]^{\mathcal{X}}; [\![H]\!]^{\mathcal{X}} \quad = \quad [\![H]\!]^{\mathcal{X}}. \tag{11}$$

Similarly we say that $\mathsf{mkv}.[\![H]\!]^{\mathcal{X}}$ is *standard* if the relation between the initial and final values for all inners in $[\![H]\!]^{\mathcal{X}}$ is functional, which can be expressed equationally as:

$$\mathsf{mkv}.[\![H]\!]^{\mathcal{X}} \circ \mathbb{D}(\mathsf{dup}) \circ (\overleftarrow{\cdot}).\delta \quad = \quad \eta(\delta), \tag{12}$$

for any δ in the support of $[\![H]\!]^{\mathcal{X}} \circ \mathbb{D}\mathsf{dup}.\pi$.

Theorem 3 says that if either $\mathsf{chn}.[\![H]\!]^{\mathcal{X}}$ or $\mathsf{mkv}.[\![H]\!]^{\mathcal{X}}$ is standard then H is refined by a leak step followed by a Markov update. In the latter case where $\mathsf{mkv}.[\![H]\!]^{\mathcal{X}}$ is standard the refinement goes both ways.

Theorem 3. *Let H be an HMM with mutable type $\mathcal{X}$.*

1. *If $\mathsf{mkv}.[\![H]\!]^{\mathcal{X}}$ is standard, then*
   $[\![H]\!]^{\mathcal{X}} \circ \mathbb{D}\mathsf{dup} = (\mathsf{chn}.[\![H]\!]^{\mathcal{X}}; \mathsf{mkv}.[\![H]\!]^{\mathcal{X}}) \circ \mathbb{D}\mathsf{dup}$

2. *If $\mathsf{chn}.[\![H]\!]^{\mathcal{X}}$ is standard, then*
   $[\![H]\!]^{\mathcal{X}} \circ \mathbb{D}\mathsf{dup} \mathrel{\widetilde{\sqsubseteq}} (\mathsf{chn}.[\![H]\!]^{\mathcal{X}}; \mathsf{mkv}.[\![H]\!]^{\mathcal{X}}) \circ \mathbb{D}\mathsf{dup}$[8]

[8] We overload $\widetilde{\sqsubseteq}$ defined on *HMM*'s directly to be defined similarly for the abstract semantics: $h^1 \widetilde{\sqsubseteq} h^2$ of type $\mathbb{D}\mathcal{X}^2 \to \mathbb{D}^2\mathcal{X}^2$ if $\mathcal{E}_{h^1(\delta)}(U_\ell) \le \mathcal{E}_{h^2(\delta)}(U_\ell)$ for all ℓ.

Proof. Suppose that $\mathsf{mkv}.[\![H]\!]^{\mathcal{X}}$ *is standard. We reason as follows:*

$$
\begin{array}{ll}
 & (\mathsf{chn}.[\![H]\!]^{\mathcal{X}}; \mathsf{mkv}.[\![H]\!]^{\mathcal{X}}) \circ \mathbb{D}(\mathsf{dup}) \\
= & \mathsf{avg} \circ \mathbb{D}(\mathsf{mkv}.[\![H]\!]^{\mathcal{X}}) \circ \mathbb{D}^2(\mathsf{dup}) \circ \mathbb{D}(\overleftarrow{\cdot}) \circ [\![H]\!]^{\mathcal{X}} \circ \mathbb{D}(\mathsf{dup}) \quad \text{“Definition 3 and (9)”} \\
= & \qquad \text{“Function composition: } \mathbb{D}(f \circ g) = \mathbb{D}f \circ \mathbb{D}g\text{”} \\
 & \mathsf{avg} \circ \mathbb{D}(\mathsf{mkv}.[\![H]\!]^{\mathcal{X}} \circ \mathbb{D}(\mathsf{dup}) \circ (\overleftarrow{\cdot})) \circ [\![H]\!]^{\mathcal{X}} \circ \mathbb{D}(\mathsf{dup}) \\
= & \mathsf{avg} \circ \mathbb{D}(\eta) \circ [\![H]\!]^{\mathcal{X}} \circ \mathbb{D}(\mathsf{dup}) \quad \text{“(12)”} \\
= & [\![H]\!]^{\mathcal{X}} \circ \mathbb{D}(\mathsf{dup}). \quad \text{“Monad law: } \mathsf{avg} \circ \mathbb{D}(\eta) \text{ is the identity”}
\end{array}
$$

Now suppose that $\mathsf{chn}.[\![H]\!]^{\mathcal{X}}$ *is standard. We reason as follows*

$$
\begin{array}{ll}
 & [\![H]\!]^{\mathcal{X}} \\
= & \mathsf{chn}.[\![H]\!]^{\mathcal{X}}; [\![H]\!]^{\mathcal{X}} \quad \text{“(11)”} \\
\tilde{\sqsubseteq} & \mathsf{chn}.[\![H]\!]^{\mathcal{X}}; \mathsf{mkv}.[\![H]\!]^{\mathcal{X}}. \quad \text{“Theorem 2(2), (10) and (3)”}
\end{array}
$$

Recall our program in Fig. 2 — since the change to X is functional, it means that overall the *HMM* model for the loop is standard in its Markov component. Thus by Theorem 3 (2) we can summarise its behaviour as a single *HMM*-style step, which we can also write as

$$
\mathtt{leak}(\mathtt{X} \div 2); \mathtt{X} := \mathtt{X} - (\mathtt{X} \bmod 2). \tag{13}
$$

The inclusion of the `leak` statement now ensures that the possible impact on third-parties is now accurately recorded.

6 Related Work and Discussion

In this paper we have studied an abstract semantic model suitable for reasoning about information flow in a general sequential programming framework. A particular innovation is to use hyper-distributions over correlations of initial and final states. Hyper-distributions summarise the basic idea in quantitative information flow that the *value* of the observation is not important, but only the effect it induces on change of uncertainty wrt. the secret. An important aspect is that our context-aware refinement order means that local reasoning is now sufficient to deduce that the behaviours of `leak (X); X:=0` are not the same as those of `X:=0`: even though all confidentiality properties concerning *only* the final value of X are the same in both program fragments. This is because they leak differing kinds of information about he initial state, and this could become significant when the program fragments are executed within contexts containing fresh secrets correlated with X.

We have illustrated the model by proving some algebraic properties; further work is required to develop the equational theory, and to apply it to a semantics for a general programming language.

Classical analyses of quantitative information flow assume that the secret does not change, and early approaches to measuring insecurities in programs are based on determining a “change in uncertainty” of some “prior” value of the secret — although how to measure the uncertainty differs in each approach. For example Clark et al. [4] use Shannon entropy to estimate the number of bits being leaked; and Clarkson et al. [5] model a change in belief. Smith [24] demonstrated the importance of using measures that have some operational significance, and the idea was developed further [2] by introducing the notion of g-leakage to express such significance in a very general way. The partial order used here on programs is the same as the g-leakage order introduced by Alvim et al. [2], but it appeared also in even earlier work [13]. Its properties have been studied extensively [1].

Others have investigated information flow for dynamic secrets, for example Marzdiel et al. [11] use probabilistic automata. Our recent work similarly explored dynamic secrets, but allows only a single secret type $\mathcal{X}$ [13,16].

The abstract treatment of probabilistic systems with the introduction of a "refinement order" was originally due to the probabilistic powerdomain of Jones and Plotkin [10]; and those ideas were extended to include demonic nondeterminism (as well as probability) by us [22]. In both cases the order (on programs) corresponds to an order determined by averaging over "probabilistic predicates" which are random variables over the state space. The compositional refinement order for information flow appeared in [13] for security programs expressed in a simple programming language and in [1] for a channel model.

Our work here is essentially the Dalenius scenario presented in a programming-language context where `X` is the statistical database and the correlation with `Z` is "auxiliary information" [7] except that, unlike in the traditional presentation, ours allows the "database" (the password) to be updated. This model can be thought of as a basis for developing a full semantics for context-aware refinement for a programming language with the aim of reasoning about and developing information flow analysis which is valid generally for all operating scenarios.

References

1. Alvim, M.S., Chatzikokolakis, K., McIver, A., Morgan, C., Palamidessi, C., Smith, G.: Additive and multiplicative notions of leakage, and their capacities. In: IEEE 27th Computer Security Foundations Symposium, CSF 2014, Vienna, Austria, 19–22 July 2014, pp. 308–322. IEEE (2014)
2. Alvim, M.S., Chatzikokolakis, K., Palamidessi, C., Smith, G.: Measuring information leakage using generalized gain functions. In: Proceedings of the 25th IEEE Computer Security Foundations Symposium (CSF 2012), pp. 265–279, June 2012
3. Back, R.-J.R., von Wright, J.: Refinement Calculus: A Systematic Introduction. Springer, Heidelberg (1998)
4. Clark, D., Hunt, S., Malacaria, P.: Quantitative analysis of the leakage of confidential data. Electr. Notes Theor. Comput. Sci. **59**(3), 238–251 (2001)
5. Clarkson, M.R., Myers, A.C., Schneider, F.B.: Belief in information flow. In: 18th IEEE Computer Security Foundations Workshop, (CSFW-18 2005), 20–22 June 2005, Aix-en-Provence, France, pp. 31–45 (2005)
6. Dalenius, T.: Towards a methodology for statistical disclosure control. Statistik Tidskrift **15**, 429–444 (1977)

7. Dwork, C.: Differential privacy. In: Bugliesi, M., Preneel, B., Sassone, V., Wegener, I. (eds.) ICALP 2006. LNCS, vol. 4052, pp. 1–12. Springer, Heidelberg (2006). doi:10.1007/11787006_1
8. Giry, M.: A categorical approach to probability theory. In: Banaschewski, B. (ed.) Categorical Aspects of Topology and Analysis. LNM, vol. 915, pp. 68–85. Springer, Heidelberg (1981). doi:10.1007/BFb0092872
9. Goguen, J.A., Meseguer, J.: Unwinding and inference control. In: Proceedings of IEEE Symposium on Security and Privacy, pp. 75–86. IEEE Computer Society (1984)
10. Jones, C., Plotkin, G.: A probabilistic powerdomain of evaluations. In: Proceedings of the IEEE 4th Annual Symposium on Logic in Computer Science, Los Alamitos, California, pp. 186–195. Computer Society Press (1989)
11. Mardziel, P., Alvim, M.S., Hicks, M.W., Clarkson, M.R.: Quantifying information flow for dynamic secrets. In: 2014 IEEE Symposium on Security and Privacy, SP 2014, Berkeley, CA, USA, 18–21 May 2014, pp. 540–555 (2014)
12. McIver, A.K., Morgan, C.C.: Abstraction, Refinement and Proof for Probabilistic Systems. Monographs in Computer Science. Springer, New York (2005)
13. McIver, A., Meinicke, L., Morgan, C.: Compositional closure for bayes risk in probabilistic noninterference. In: Abramsky, S., Gavoille, C., Kirchner, C., Meyer auf der Heide, F., Spirakis, P.G. (eds.) ICALP 2010. LNCS, vol. 6199, pp. 223–235. Springer, Heidelberg (2010). doi:10.1007/978-3-642-14162-1_19
14. McIver, A., Meinicke, L., Morgan, C.: A Kantorovich-monadic powerdomain for information hiding, with probability and nondeterminism. In: Proceedings of LiCS 2012 (2012)
15. McIver, A., Meinicke, L., Morgan, C.: Hidden-Markov program algebra with iteration. Mathematical Structures in Computer Science (2014)
16. McIver, A., Morgan, C., Rabehaja, T.: Abstract hidden Markov models: a monadic account of quantitative information flow. In: Proceedings of LiCS 2015 (2015)
17. McIver, A., Morgan, C., Rabehaja, T., Bordenabe, N.: Reasoning about distributed secrets. Submitted to FORTE 2017
18. McIver, A., Morgan, C., Smith, G., Espinoza, B., Meinicke, L.: Abstract channels and their robust information-leakage ordering. In: Abadi, M., Kremer, S. (eds.) POST 2014. LNCS, vol. 8414, pp. 83–102. Springer, Heidelberg (2014). doi:10.1007/978-3-642-54792-8_5
19. Moggi, E.: Computational lambda-calculus and monads. In: Proceedings of 4th Symposium on LiCS, pp. 14–23 (1989)
20. Morgan, C.C.: Programming from Specifications, 2nd edn. Prentice-Hall, Upper Saddle River (1994). web.comlab.ox.ac.uk/oucl/publications/books/PfS/
21. Morgan, C.C.: *The Shadow Knows*: refinement of ignorance in sequential programs. In: Uustalu, T. (ed.) MPC 2006. LNCS, vol. 4014, pp. 359–378. Springer, Heidelberg (2006). doi:10.1007/11783596_21
22. Morgan, C.C., McIver, A.K., Seidel, K.: Probabilistic predicate transformers. ACM Trans. Program. Lang. Syst. **18**(3), 325–353 (1996). doi.acm.org/10.1145/229542.229547
23. Schrijvers, T., Morgan, C.: `Hypers.hs` Haskell code implementing quantitative non-interference monadic security semantics (2015). http://www.cse.unsw.edu.au/~carrollm/Hypers.pdf
24. Smith, G.: On the foundations of quantitative information flow. In: Alfaro, L. (ed.) FoSSaCS 2009. LNCS, vol. 5504, pp. 288–302. Springer, Heidelberg (2009). doi:10.1007/978-3-642-00596-1_21

Dual Space of a Lattice as the Completion of a Pervin Space
Extended Abstract

Jean-Éric Pin(✉)

IRIF, University Paris-Diderot and CNRS, Paris, France
Jean-Eric.Pin@irif.fr

We assume the reader is familiar with basic topology on the one hand and finite automata theory on the other hand. No proofs are given in this extended abstract.

1 Introduction

The original motivation of this paper, as presented in [15], was to compute the dual space of a lattice of subsets of some free monoid A^*. According to Stone-Priestley duality, the dual space of a lattice can be identified with the set of its prime filters, but it is not always the simplest way to describe it. Consider for instance the Boolean algebra generated by the sets of the form uA^*, where u is a word. Its dual space is equal to the completion of A^* for the prefix metric and it can be easily identified with the set of finite or infinite words on A, a more intuitive description than prime filters.

Elaborating on this idea, one may wonder whether the dual space of a given lattice of subsets of a space can always be viewed as a completion of some sort. The answer to this question is positive and known for a long time: for Boolean algebras, the solution is detailed as an exercise in Bourbaki [7, Exercise 12, p. 211]. In the lattice case, the appropriate setting for this question is a very special type of spaces, the so-called Pervin spaces, which form the topic of this paper.

A Pervin space is a set X equipped with a set of subsets, called the *blocks* of the Pervin space. Blocks are closed under finite intersections and finite unions and hence form a lattice of subsets of X. Pervin spaces are thus easier to define than topological spaces or (quasi)-uniform spaces. As a consequence, most of the standard topological notions, like convergence and cluster points, specialisation order, filters and Cauchy filters, complete spaces and completion are much easier to define for Pervin spaces.

The second motivation of this paper, also stemming from language theory, is the characterisation of classes of languages by inequations, which is briefly reviewed in Sect. 2. For regular languages on A^*, these inequations are of the form $u \leqslant v$ where u and v are elements of the free profinite monoid $\widehat{A^*}$. The

J. Pin—Funded by the European Research Council (ERC) under the European Unions Horizon 2020 research and innovation programme (grant agreement No. 670624) and by the DeLTA project (ANR-16-CE40-0007).

P. Höfner et al. (Eds.): RAMiCS 2017, LNCS 10226, pp. 24–40, 2017.
DOI: 10.1007/978-3-319-57418-9_2

main result of [14] states that any lattice of regular languages can be defined by a set (in general infinite) of such inequalities. The more general result of [15] states any lattice of languages (not necessarily regular) can be defined by a set of inequations of the form $u \leqslant v$, where u and v are now elements of βA^*, the Stone-Čech compactification of A^*.

It turns out that it is possible to give a simple proof of these two results using Pervin spaces. Let $\mathcal{L}$ be the set of blocks of a Pervin space X. Then the completion of X can be defined as the set of valuations on $\mathcal{L}$. A valuation on $\mathcal{L}$ is simply a lattice morphism from $\mathcal{L}$ to the two-element Boolean algebra $\{0, 1\}$. In particular, if $\mathcal{L}$ is the lattice of regular languages on A^*, then the completion of A^* is $\widehat{A^*}$. If $\mathcal{L}$ is the lattice of all languages on A^*, then the completion of A^* is βA^*.

Of course, valuations and prime ideals are just the same thing, but we prefer to use valuations, because they come with a very natural order relation: $v \leqslant w$ if and only if $v(L) \leqslant w(L)$ for all $L \in \mathcal{L}$. It is also natural to say that a set of blocks $\mathcal{K}$ satisfies the inequation $v \leqslant w$ if, for every $K \in \mathcal{K}$, $v(K) \leqslant w(K)$. Now, the characterisation of lattices by inequations takes the following form:

A set of blocks is a sublattice of $\mathcal{L}$ if and only if it can be defined by a set of inequations.

Taking for $\mathcal{L}$ the lattice of regular languages on A^*, one recovers the result of [14] and taking for $\mathcal{L}$ the lattice of all languages on A^*, one finds again the main result of [15]. Another result is worth mentioning. Let $\mathcal{L}$ be a lattice of subsets of X and let $\mathcal{K}$ be a sublattice of $\mathcal{L}$. Then the following property holds:

The Pervin space $(X, \mathcal{L})$ is a subspace of the Pervin space $(X, \mathcal{K})$ and the completion of $(X, \mathcal{K})$ is a quotient of the completion of $(X, \mathcal{L})$.

Although this result looks like a contravariant property of duality theory, one has to be careful when defining a quotient space. This is fully discussed in Sect. 7.

2 Formal Languages

In this section, we briefly review the results on languages that motivated this paper. A *lattice of languages* is a set $\mathcal{L}$ of languages of A^* containing $\emptyset$ and A^* and closed under finite unions and finite intersections. It is *closed under quotients*[1] if, for each $L \in \mathcal{L}$ and $u \in A^*$, the languages $u^{-1}L$ and Lu^{-1} are also in $\mathcal{L}$. A lattice of languages is a *Boolean algebra* if it is closed under complement.

An important object in this theory is the free profinite monoid $\widehat{A^*}$. It admits several equivalent descriptions, but we will only describe two of them. The reader is referred to [4,5,29] for more details.

[1] Recall that $u^{-1}L = \{x \in A^* \mid ux \in L\}$ and $Lu^{-1} = \{x \in A^* \mid xu \in L\}$.

The Free Profinite Monoid as the Completion of a Metric Space.
A monoid M *separates* two words u and v of A^* if there exists a monoid morphism $\varphi : A^* \to M$ such that $\varphi(u) \neq \varphi(v)$. One can show that two distinct words can always be separated by a finite monoid.

Given two words $u, v \in A^*$, we set

$$r(u,v) = \min\{|M| \mid M \text{ is a monoid that separates } u \text{ and } v\}$$
$$d(u,v) = 2^{-r(u,v)}$$

with the usual conventions $\min\emptyset = +\infty$ and $2^{-\infty} = 0$. Then d is an ultrametric, that is, satisfies the following properties, for all $u, v, w \in A^*$,

(1) $d(u,v) = d(v,u)$,
(2) $d(uw,vw) \leqslant d(u,v)$ and $d(wu,wv) \leqslant d(u,v)$,
(3) $d(u,w) \leqslant \max\{d(u,v), d(v,w)\}$.

Thus (A^*, d) is a metric space. Its completion, denoted by $\widehat{A^*}$, is called the *free profinite monoid* on A and its elements are called *profinite words.* The term "monoid" needs to be justified. In fact, the multiplication on A^* (the concatenation product) is uniformly continuous and hence can be extended in a unique way to a uniformly continuous operation on $\widehat{A^*}$. This operation makes $\widehat{A^*}$ a compact topological monoid. Recall that a *topological monoid* is a monoid M equipped with a topology on M such that the multiplication $(x,y) \to xy$ is a continuous map from $M \times M \to M$.

It is not so easy to give examples of profinite words which are not words, but here is one. In a compact monoid, the smallest closed subsemigroup containing a given element x has a unique idempotent, denoted x^ω. This is true in particular in a finite monoid and in the free profinite monoid. Thus if x is a (profinite) word, so is x^ω. Alternatively, one can define x^ω as the limit of the converging sequence $x^{n!}$. More details can be found in [3,23].

The Free Profinite Monoid as a Projective Limit.
Given a monoid morphism $f : A^* \to M$, we denote by $\sim_f$ the *kernel congruence* of f, defined on A^* by $u \sim_f v$ if and only if $f(u) = f(v)$. For each pair of surjective morphisms $f : A^* \to M$ and $g : A^* \to N$ such that $\sim_f \subseteq \sim_g$, there is a unique surjective morphism $\pi_{f,g} : M \to N$ such that $g = \pi_{f,g} \circ f$. Moreover $\pi_{f,h} = \pi_{g,h} \circ \pi_{f,g}$ and $\pi_{f,f} = Id_M$.

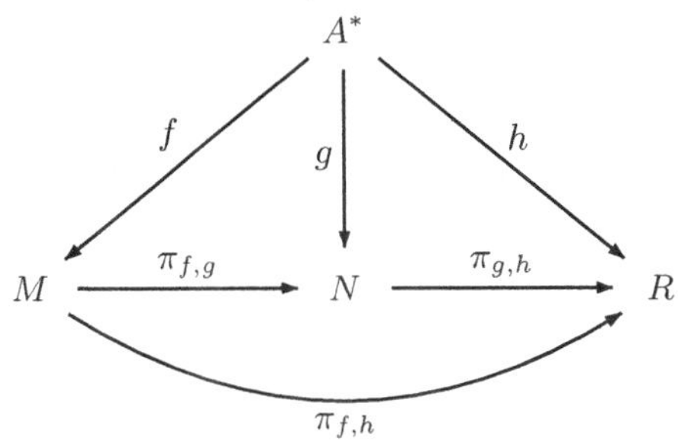

The monoid $\widehat{A^*}$ can be defined as the projective limit of the directed system formed by the surjective morphisms between finite A-generated monoids. A possible construction is to consider the compact monoid

$$P = \prod_{f:A^* \to M_f} M_f$$

where the product runs over all monoid morphisms f from A^* to some finite monoid M_f, equipped with the discrete topology. An element $(s_f)_{f:A^* \to M}$ of P is *compatible* if $\pi_{f,g}(s_f) = s_g$. The set of compatible elements is a closed submonoid of P, which is equal to $\widehat{A^*}$.

We now come to profinite inequations. Let $u, v \in \widehat{A^*}$. A regular language L of A^* satisfies the inequation $u \leqslant v$ if the condition $u \in \overline{L}$ implies $v \in \overline{L}$, where $\overline{L}$ denotes the closure of L in $\widehat{A^*}$. Here is the main result of [14]:

Proposition 2.1. *Any lattice of regular languages can be defined by a set (in general infinite) of profinite inequalities.*

This result is useful to analyse the expressive power of various fragments of monadic second order logic interpreted on finite words. It is of particular interest for lattices of regular languages closed under quotients. In this case the inequations can be directly interpreted in the ordered syntactic monoid. This notion was first introduced by Schützenberger in 1956 [25], but thereafter, he apparently only used the syntactic monoid.

Let L be a language of A^*. The *syntactic preorder* of L is the relation $\leqslant_L$ defined on A^* by $u \leqslant_L v$ if and only if, for every $x, y \in A^*$,

$$xuy \in L \implies xvy \in L.$$

The *syntactic congruence* of L is the associated equivalence relation $\sim_L$, defined by $u \sim_L v$ if and only if $u \leqslant_L v$ and $v \leqslant_L u$.

The *syntactic monoid* of L is the quotient $M(L)$ of A^* by $\sim_L$ and the natural morphism $\eta_L : A^* \to A^*/{\sim_L}$ is called the *syntactic morphism* of L. The syntactic preorder $\leqslant_L$ induces an order on the quotient monoid $M(L)$. The resulting ordered monoid is called the *syntactic ordered monoid* of L. The syntactic morphism admits a unique continuous extension $\widehat{\eta} : \widehat{A^*} \to M$. For instance, if $\eta(u) = x$, then $\widehat{\eta}(u^\omega) = x^\omega$, where x^ω is the unique idempotent power of x in M.

For instance, if L is the language $\{a, aba\}$, its syntactic monoid is the monoid $M = \{1, a, b, ab, ba, aba, 0\}$ presented by the relations $a^2 = b^2 = bab = 0$. Its syntactic order is $0 < ab < 1$, $0 < ba < 1$, $0 < aba < a$, $0 < b$.

Let $\mathcal{L}$ be a lattice of regular languages closed under quotients. One can show that $\mathcal{L}$ satisfies the profinite inequation $u \leqslant v$ if and only if, for each $L \in \mathcal{L}$, $\eta_L(u) \leqslant \eta_L(v)$. This allows one to characterise the languages of $\mathcal{L}$ by a property of their ordered syntactic monoid. Here are three examples of such results, but many more can be found in the literature [23, 24].

(1) A regular language is finite if and only if its ordered syntactic monoid satisfies the inequations $yx^\omega = x^\omega = x^\omega y$ and $x^\omega \leqslant y$ for all profinite words $x \in \widehat{A^*} - \{1\}$ and $y \in \widehat{A^*}$.

(2) A famous result of Schützenberger [26] states that a regular language is star-free if and only if its syntactic monoid satisfies the equations $xx^\omega = x^\omega$ for all profinite words $x \in \widehat{A^*}$.
(3) Our third example is related to Boolean circuits. Recall that AC^0 is the set of unbounded fan-in, polynomial size, constant-depth Boolean circuits. One can show [6,27,28] that a regular language is recognised by a circuit in AC^0 if and only if its syntactic monoid satisfies the equations $(x^{\omega-1}y)^\omega = (x^{\omega-1}y)^{\omega+1}$ for all words x and y of the same length.

It is also possible to give an inequational characterisation of lattices of languages that are not regular [15]. The price to pay is to replace the profinite monoid by an even larger space, the *Stone-Čech compactification* of A^*, usually denoted βA^*. One can define βA^* as the set of *ultrafilters* on the discrete space A^*. A second way to define it is to take the closure of the image of A^* in the product space $\prod K$ where the product runs over all maps from A^* into a compact Hausdorff space K whose underlying set is $\mathcal{P}(\mathcal{P}(A^*))$. Both spaces $\widehat{A^*}$ and βA^* are compact, but only $\widehat{A^*}$ is a compact monoid.

Let $u, v \in \beta A^*$. We say that L *satisfies the ultrafilter inequality* $u \to v$ if $u \in \overline{L}$ implies $v \in \overline{L}$, where $\overline{L}$ now denotes the closure of L in βA^*. The main result of [15] can be stated as follows:

Proposition 2.2. *Any lattice of languages can be defined by a set (in general infinite) of ultrafilter inequalities.*

In Sect. 8, we will recover Propositions 2.1 and 2.2 as a special case of Theorem 8.3. See also [13] for a duality point of view of these results.

3 Pervin Spaces

It is time to introduce the main topic of this article. Let X be a set. A *lattice of subsets* of X is a subset of $\mathcal{P}(X)$ containing $\emptyset$ and X and closed under finite intersections and finite unions. A *Boolean algebra of subsets* of X is a lattice of subsets of X closed under complement.

Given a lattice $\mathcal{L}$ of subsets of X, we denote by $\mathcal{L}^s$ the Boolean algebra generated by $\mathcal{L}$. There is a simple description of $\mathcal{L}^s$ using the set

$$D(\mathcal{L}) = \{L_1 - L_0 \mid L_0, L_1 \in \mathcal{L}\}.$$

of differences of members of $\mathcal{L}$. Indeed, Hausdorff [16] has shown that the Boolean algebra $\mathcal{L}^s$ consists of the finite unions of elements of $D(\mathcal{L})$.

3.1 The Category of Pervin Spaces

A *Pervin structure* on a set X is a lattice $\mathcal{L}$ of subsets of X. The elements of $\mathcal{L}$ are called the *blocks* of the Pervin structure. A Pervin space is a set endowed with a Pervin structure. More formally, a *Pervin space* is a pair $(X, \mathcal{L})$ where $\mathcal{L}$

is a lattice of subsets of X. A *Boolean Pervin space* is a Pervin space in which $\mathcal{L}$ is a Boolean algebra.

Let $(X, \mathcal{K})$ and $(Y, \mathcal{L})$ be two Pervin spaces. A map $\varphi : X \to Y$ is said to be *morphism* if, for each $L \in \mathcal{L}$, $\varphi^{-1}(L) \in \mathcal{K}$. In other words, a map is a morphism if the preimage of a block is a block. It is readily seen that the composition of two morphisms is again a morphism and that the identity function on X is a morphism. Pervin spaces together with their morphisms form the category **Pervin** of Pervin spaces.

Two Pervin spaces, both defined on the two-element set $\{0, 1\}$, play an important role in this theory. The first one, the *Boolean space* $\mathbb{B}$, is defined by the lattice of all subsets of $\{0, 1\}$. The second one, the *Sierpiński space* $\mathbb{S}$, is defined by the lattice $\{\emptyset, \{1\}, \{0, 1\}\}$. Note that the identity on $\{0, 1\}$ is a morphism from $\mathbb{B}$ to $\mathbb{S}$ but it *is not a morphism* from $\mathbb{S}$ to $\mathbb{B}$. More examples of Pervin spaces are given in Sect. 5.

3.2 Pervin Spaces as Preordered Sets

A Pervin space $(X, \mathcal{L})$ is naturally equipped with a preorder $\leqslant_{\mathcal{L}}$ on X defined by $x \leqslant_{\mathcal{L}} y$ if, for each $L \in \mathcal{L}$,

$$x \in L \implies y \in L.$$

The associated equivalence relation $\sim_{\mathcal{L}}$ is defined on X by $x \sim_{\mathcal{L}} y$ if, for each $L \in \mathcal{L}$,

$$x \in L \iff y \in L.$$

When the lattice $\mathcal{L}$ is understood, we will drop the index $\mathcal{L}$ and simply denote by $\leqslant$ and $\sim$ the preorder on $(X, \mathcal{L})$ and its associated equivalence relation. For instance, in the Boolean space, the preorder is the equality relation and in the Sierpiński space, the preorder is $0 \leqslant 1$.

It is easy to see that any morphism of Pervin spaces is order-preserving.

3.3 Pervin Spaces as Topological Spaces

There are two topologies of interest on a Pervin space $(X, \mathcal{L})$. The first one, simply called the *topology of* $(X, \mathcal{L})$, is the topology based on the blocks of $\mathcal{L}$. The second one, called the *symmetrical topology* of $(X, \mathcal{L})$, is the topology based on the blocks of $\mathcal{L}^s$. In view of Hausdorff's result, these definitions can be summarized as follows:

Definition 3.1. *The blocks of a Pervin space form a base of its topology. The differences of two blocks form a base of clopen sets of its symmetrical topology.*

It follows immediately from the definition of the topology that the blocks containing a point x form a basis of the filter $\mathcal{N}(x)$ of neighbourhoods of x and that

$$\mathcal{N}(x) \cap \mathcal{L} = \{L \in \mathcal{L} \mid x \in L\} \tag{1}$$

The *specialisation preorder*, defined on any topological space X, is the relation $\leqslant$ defined on X by $x \leqslant y$ if and only if $\overline{\{x\}} \subseteq \overline{\{y\}}$ or, equivalently, if and only if $x \in \overline{\{y\}}$. It turns out that in a Pervin space $(X, \mathcal{L})$, the specialisation preorder coincides with the preorder $\leqslant_{\mathcal{L}}$.

Recall that a topological space X is a *Kolmogorov space* (or T_0-*space*) if for any two distinct points of X, there is an open set which contains one of these points and not the other. Kolmogorov Pervin spaces are easy to describe:

Proposition 3.2. *Let $(X, \mathcal{L})$ be a Pervin space. The following conditions are equivalent:*

(1) *The preorder $\leqslant$ is a partial order,*
(2) *The relation $\sim$ is the equality relation,*
(3) *The space $(X, \mathcal{L})$ is Kolmogorov,*
(4) *The space $(X, \mathcal{L}^s)$ is Hausdorff.*

Being Kolmogorov is a very desirable property for a Pervin space. Fortunately, it is easy to make a Pervin space Kolmogorov by taking its quotient by the relation $\sim$. That is, one considers the quotient space $X/{\sim}$ and one defines the blocks of $X/{\sim}$ to be the sets of the form $L/{\sim}$, for $L \in \mathcal{L}$. Slightly abusing notation, we set

$$\mathcal{L}/{\sim} = \{L/{\sim} \mid L \in \mathcal{L}\}$$

The Pervin space $(X/{\sim}, \mathcal{L}/{\sim})$ is called the *Kolmogorov quotient* of X. Note that the natural map from X to $X/{\sim}$ is a morphism of Pervin spaces which induces a lattice isomorphism from $\mathcal{L}$ to $\mathcal{L}/{\sim}$.

Compact Pervin spaces have some further interesting properties. First of all, blocks and compact open subsets are closely related. Note however that since our spaces are not necessarily Hausdorff, compact subsets are not necessarily closed.

Theorem 3.3. *A compact open subset of a compact Pervin space is a block. If a Pervin space is compact for the symmetrical topology, then every block is compact open (for the usual topology). In particular, a subset of a compact Boolean Pervin space is a block if and only if it is compact open.*

When X is Kolmogorov and compact for the symmetrical topology, a few more characterisations of its blocks are available.

Theorem 3.4. *Let $(X, \mathcal{L})$ be a Kolmogorov Pervin space that is compact for the symmetrical topology and let L be a subset of X. Then the following conditions are equivalent:*

(1) *L is a block of $\mathcal{L}$,*
(2) *L is compact open in $(X, \mathcal{L})$,*
(3) *L is an upset in $(X, \mathcal{L})$ and is clopen in $(X, \mathcal{L}^s)$,*
(4) *L is an upset in $(X, \mathcal{L})$ and a block in $(X, \mathcal{L}^s)$.*

4 Complete Pervin Spaces

We have already seen that several topological definitions become much simpler in the case of a Pervin space. This is again the case for the notions studied in this section: Cauchy filters, complete spaces and completion.

Definition 4.1. *A filter $\mathcal{F}$ on a Pervin space X is Cauchy if and only if, for every block L, either $L \in \mathcal{F}$ or $L^c \in \mathcal{F}$.*

One can show that it makes no difference to consider the symmetrical Pervin structure. More precisely, if $(X, \mathcal{L})$ is a Pervin space, then a filter on X is Cauchy on $(X, \mathcal{L})$ if and only if it is Cauchy on $(X, \mathcal{L}^s)$. One can show, as in the case of a metric space, that a cluster point of a Cauchy filter is a limit point.

Definition 4.2. *A Pervin space is* complete *if every Cauchy filter converges in the symmetrical topology.*

Complete Pervin spaces admit the following characterisations.

Theorem 4.3. *Let $(X, \mathcal{L})$ be a Pervin space. The following conditions are equivalent:*

(1) *$(X, \mathcal{L})$ is complete,*
(2) *$(X, \mathcal{L}^s)$ is complete,*
(3) *$(X, \mathcal{L}^s)$ is compact.*

If these conditions are satisfied, then $(X, \mathcal{L})$ is compact.

Note however that a compact Pervin space need not be complete as shown in Example 5.2. Just like in the case of a metric space, it is easy to describe the complete subspaces of a complete Pervin space.

Proposition 4.4. *Every subspace of a complete Pervin space that is closed in the symmetrical topology is complete. A complete subspace of a Kolmogorov Pervin space is closed in the symmetrical topology.*

We now come to the formal definition of the completion of a Pervin space.

Definition 4.5. *A* completion *of a Pervin space X is a* complete Kolmogorov Pervin space *$\widehat{X}$ together with a morphism $\imath\colon X \to \widehat{X}$ satisfying the following universal property: for each morphism $\varphi : X \to Y$, where Y is a complete Kolmogorov Pervin space, there exists a* unique *morphism $\widehat{\varphi} : \widehat{X} \to Y$ such that $\widehat{\varphi} \circ \imath = \varphi$.*

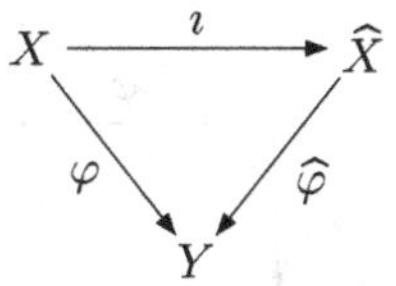

By standard categorical arguments, these conditions imply the *unicity* of the completion (up to isomorphism). The actual construction of the completion relies on the notion of valuation on a lattice.

Definition 4.6. *A* valuation *on a lattice $\mathcal{L}$ is a lattice morphism from $\mathcal{L}$ into the Boolean lattice* $\{0, 1\}$.

In other words, a valuation is a function v from $\mathcal{L}$ into $\{0, 1\}$ satisfying the following properties, for all $L, L' \in \mathcal{L}$:

(1) $v(\emptyset) = 0$ and $v(X) = 1$,
(2) $v(L \cup L') = v(L) + v(L')$,
(3) $v(L \cap L') = v(L)v(L')$,

where the addition and the product denote the Boolean operations. Valuations are naturally ordered by setting $v \leqslant v'$ if and only if $v(L) \leqslant v'(L)$ for all $L \in \mathcal{L}$.

The completion of a Pervin space can now be constructed as follows. For each block L, let

$$\widehat{L} = \{v \mid v \text{ is a valuation on } \mathcal{L} \text{ such that } v(L) = 1\}.$$

In particular, $\widehat{X}$ is the set of all valuations on $\mathcal{L}$ and one can show that the map $L \to \widehat{L}$ defines a lattice morphism from $\mathcal{L}$ to the lattice of subsets of $\widehat{X}$. Consequently, the set

$$\widehat{\mathcal{L}} = \{\widehat{L} \mid L \in \mathcal{L}\}$$

is a lattice and $(\widehat{X}, \widehat{\mathcal{L}})$ is a Pervin space. We now have a candidate for the completion, but we still need a candidate for the map $\imath : X \to \widehat{X}$. For each $x \in X$, we define $\imath(x)$ as the valuation on $\mathcal{L}$ such that $\imath(x)(L) = 1$ if and only if $x \in L$.

Theorem 4.7. *The Pervin space $(\widehat{X}, \widehat{\mathcal{L}})$ is Kolmogorov and complete and the pair $(\imath, (\widehat{X}, \widehat{\mathcal{L}}))$ is the completion of $(X, \mathcal{L})$.*

A Pervin space and its Kolmogorov quotient have isomorphic completions. Furthermore, completion and symmetrization are two commuting operations. More precisely, the *symmetrical completion* $(\widehat{X}, (\widehat{\mathcal{L}})^s)$ of $(X, \mathcal{L})$ can also be obtained as the completion of $(X, \mathcal{L}^s)$.

A nice feature of completions is that they extend to morphisms.

Theorem 4.8. *Let $(X, \mathcal{L}_X)$ and $(Y, \mathcal{L}_Y)$ be two Pervin spaces and let $\varphi : X \to Y$ be a morphism.*

(1) *There exists a unique morphism $\widehat{\varphi}$ from $(\widehat{X}, \widehat{\mathcal{L}}_X)$ to $(\widehat{Y}, \widehat{\mathcal{L}}_Y)$ such that $\imath_Y \circ \varphi = \widehat{\varphi} \circ \imath_X$.*

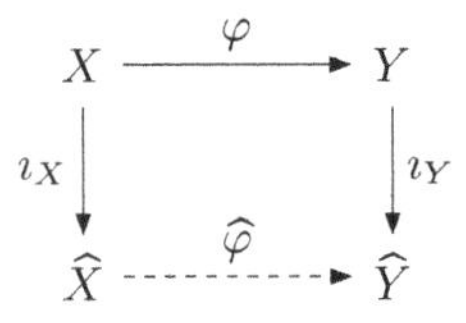

(2) *The following formulas hold for all* $v \in \widehat{X}$ *and all* $L \in \mathcal{L}_Y$*:*

$$\widehat{\varphi}(v)(L) = v(\varphi^{-1}(L))$$
$$\widehat{\varphi}^{-1}(\widehat{L}) = \widehat{\varphi^{-1}(L)}$$

Note also that in the category of Pervin spaces, completions preserve surjectivity, a property that does not hold for metric spaces. We now give a useful consequence of Theorem 4.8.

Corollary 4.9. *Let* φ_1 *and* φ_2 *be two morphisms from* X *to* Y *and let* $\widehat{\varphi}_1$ *and* $\widehat{\varphi}_2$ *be their extensions from* $\widehat{X}$ *to* $\widehat{Y}$*. If* $\varphi_1 \leqslant \varphi_2$*, then* $\widehat{\varphi}_1 \leqslant \widehat{\varphi}_2$*.*

The previous corollary is often used under a slightly different form, analogous to the *Principle of extensions of identities* of Bourbaki [7, Chapter I, Sect. 8.1, Corollary 1].

Corollary 4.10. *Let* φ_1 *and* φ_2 *be two morphisms from* $(\widehat{X}, \widehat{\mathcal{L}}_X)$ *to* $(\widehat{Y}, \widehat{\mathcal{L}}_Y)$*. If, for all* $x \in X$*,* $\varphi_1(x) \leqslant \varphi_2(x)$*, then* $\varphi_1 \leqslant \varphi_2$*. In particular, if* φ_1 *and* φ_2 *coincide on* X*, then they are necessarily equal.*

5 Examples of Pervin Spaces

In this series of examples, $(X, \mathcal{L})$ denotes a Pervin space.

Example 5.1 (Finite sets). Let $X = \mathbb{N}$ and $\mathcal{L}$ be the lattice formed by X and the finite subsets of X. This space is Hausdorff but is neither compact nor complete. Indeed, the valuation v given by $v(X) = 1$ and $v(L) = 0$ for each finite set L defines a new element, denoted $-\infty$. The completion of $(X, \mathcal{L})$ is $(\widehat{X}, \widehat{\mathcal{L}})$, where $\widehat{X} = X \cup \{-\infty\}$ and $\widehat{\mathcal{L}}$ is the lattice formed by $\widehat{X}$ and the finite subsets of X. This lattice is isomorphic to $\mathcal{L}$. The order on X is the equality relation, but in $\widehat{X}$, the order is given by $-\infty \leqslant x$ for each $x \in \widehat{X}$.

Example 5.2 (Cofinite sets). Let $X = \mathbb{N}$ and let $\mathcal{L}$ be the lattice formed by the empty set and the cofinite subsets of X. This space is Kolmogorov and compact, but it is neither Hausdorff nor complete. Indeed, the valuation v given by $v(L) = 1$ for each cofinite set L defines a new element, denoted ∞. The completion of $(X, \mathcal{L})$ is $(\widehat{X}, \widehat{\mathcal{L}})$, where $\widehat{X} = X \cup \{\infty\}$ and $\widehat{\mathcal{L}}$ is the lattice formed by the empty set and the cofinite subsets of $\widehat{X}$ containing ∞. This lattice is isomorphic to $\mathcal{L}$. The order on X is the equality relation, but in $\widehat{X}$, the order is given by $x \leqslant \infty$ for each $x \in \widehat{X}$.

Example 5.3 (Finite or cofinite sets). Let $X = \mathbb{N}$ and let $\mathcal{L}$ be the Boolean algebra of all finite or cofinite subsets of X. This space is Hausdorff but it is neither compact nor complete. Indeed, the valuation v given by $v(L) = 1$ if L is cofinite and $v(L) = 0$ if L is finite defines a new element, denoted ∞. The completion of $(X, \mathcal{L})$ is $(\widehat{X}, \widehat{\mathcal{L}})$, where $\widehat{X} = X \cup \{\infty\}$ and $\widehat{\mathcal{L}}$ is the Boolean algebra formed by the finite subsets of X and by the cofinite subsets of $\widehat{X}$ containing ∞. This Boolean algebra is isomorphic to $\mathcal{L}$.

Example 5.4 (Finite sections). Let

$$X = \left\{ \frac{1}{n} \mid n \text{ is a positive integer} \right\}$$

and $\mathcal{L}$ be the lattice formed by X and the subsets $L_n = \{\frac{1}{k} \mid 0 < k \leqslant n\}$, for $n \geqslant 0$. This space is Kolmogorov and compact, but it is neither Hausdorff nor complete since the Cauchy filter $\mathcal{L} - \{\emptyset\}$ does not converge in $(X, \mathcal{L}^s)$. The valuation v given by $v(X) = 1$ and $v(L_n) = 0$ for each n defines a new element, denoted 0. The completion of $(X, \mathcal{L})$ is $(\widehat{X}, \widehat{\mathcal{L}})$, where $\widehat{X} = X \cup \{0\}$ and $\widehat{\mathcal{L}}$ is the lattice formed by the empty set, $\widehat{X}$ and the finite subsets of $\widehat{X}$ containing 0. This lattice is isomorphic to $\mathcal{L}$. The order on $\widehat{X}$ is the chain $0 \leqslant \cdots \leqslant \frac{1}{n} \leqslant \cdots \leqslant \frac{1}{2} \leqslant 1$. Every filter has 1 as a converging point. Indeed, if $L \in \mathcal{L}$ and $1 \in L$, then $L = X$ and X is a member of all filters.

Example 5.5. Let $X = \{0, 1, 2\}$ and

$$\mathcal{L} = \{\emptyset, \{1\}, \{2\}, \{1,2\}, \{0,1,2\}\}.$$

The preorder on $(X, \mathcal{L})$ is given by $0 \leqslant 1$ and $0 \leqslant 2$. Let $\mathcal{F} = \{\{1,2\}, \{0,1,2\}\}$ be the filter generated by $\{1, 2\}$. Then $\mathcal{F}$ is converging to 0 but it is not Cauchy since neither $\{1\}$ nor its complement are in $\mathcal{F}$.

Example 5.6. Let $X = \mathbb{N}$ and

$$\mathcal{L} = \{\text{finite subsets of } \mathbb{N}\} \cup \{\{0\}^c\}.$$

The preorder on $(X, \mathcal{L})$ is given by $0 < n$ for each positive integer n. Let $\mathcal{F}$ be the Cauchy filter of all cofinite subsets of $\mathbb{N}$. Then 0 is the unique limit point of $\mathcal{F}$ in $(X, \mathcal{L})$ but $\mathcal{F}$ has no limit point in $(X, \mathcal{L}^s)$.

Example 5.7. Let $(X_1, \mathcal{L}_1)$ be the Pervin space considered in Example 5.1 and let $(X_2, \mathcal{L}_2)$ be the Pervin space considered in Example 5.4. Let $\varphi : X_1 \to X_2$ be the map defined by $\varphi(n) = \frac{1}{n+1}$. Then φ is a morphism and its completion $\widehat{\varphi} : \widehat{X}_1 \to \widehat{X}_2$ is given by $\widehat{\varphi}(-\infty) = 0$.

6 Duality Results

This section presents the links between Pervin spaces and duality theory. It relies on more advanced topological notions.

Duality theory provides three different representations of bounded distributive lattices via Priestley spaces, spectral spaces and pairwise Stone spaces [8,10,12,17]. Is it possible to recover these results using Pervin spaces? Well, not quite. Indeed, while duality is concerned with *abstract* distributive lattices, we only consider *concrete* ones, already given as a lattice of subsets. However, Pervin spaces allow one to recover these three representations for concrete distributive lattices. Let us first recall the definitions.

A topological space is *zero-dimensional* if it has a basis consisting of clopen subsets. It is *totally disconnected* if its connected components are singletons. It is well known that a compact space is zero-dimensional if and only if it is totally disconnected.

A *Stone space* is a compact totally disconnected Hausdorff space. A *pairwise Stone space* is a bitopological space $(X, \mathcal{T}_1, \mathcal{T}_2)$ which is pairwise compact, pairwise Hausdorff, and pairwise zero-dimensional.

A *Priestley space* is an ordered compact topological space $(X, \leqslant)$ satisfying the following *separation property*: if $x \nleqslant y$, then there exists a clopen upset U of X such that $x \in U$ and $y \notin U$.

A subset S of a topological space X is *irreducible* if and only if, for each finite family $(F_i)_{i \in I}$ of closed sets, the condition $S \subseteq \bigcup_{i \in I} F_i$ implies that there exists $i \in I$ such that $S \subseteq F_i$. A topological space X is *sober* if every irreducible closed subset of X is the closure of exactly one point of X.

A topological space is *spectral* if it is Kolmogorov and sober and the set of its compact open subsets is closed under finite intersection and form a basis for its topology.

The relevance of Pervin spaces to duality theory is summarized in the following result:

Theorem 6.1. *The completion of a Pervin space $(X, \mathcal{L})$ is the Stone dual of $\mathcal{L}$.*

To complete this result, it just remains to requalify complete Pervin spaces as Priestley spaces, spectral spaces and pairwise Stone spaces.

Priestley Spaces. If $(X, \mathcal{L})$ is a complete Pervin space, then $(X, \mathcal{L}, \leqslant_{\mathcal{L}})$ is a Priestley space. The proof relies on the following variation on the prime filter theorem from order theory.

Proposition 6.2. *Let $\mathcal{K}$ be a sublattice of a lattice $\mathcal{L}$ and let L be an element of $\mathcal{L} - \mathcal{K}$. Then there exist two valuations v_0 and v_1 on $\mathcal{L}$ such that $v_0(L) = 0$, $v_1(L) = 1$ and $v_1 \leqslant v_0$ on $\mathcal{K}$.*

By the way, this result not only gives the separation property of Priestley spaces, but it is also of frequent use in the theory of Pervin spaces.

Spectral Spaces. A compact Kolmogorov Pervin space need not be sober. However, if $(X, \mathcal{L})$ is a Kolmogorov Pervin space and if $(X, \mathcal{L}^s)$ is compact, then $(X, \mathcal{L})$ is spectral. In particular, a complete Pervin space is spectral.

Pairwise Stone Spaces. Let $\mathcal{L}^c = \{L^c \mid L \in \mathcal{L}\}$ be the set of complements of blocks of X. Then $(X, \mathcal{L}^c)$ is also a Pervin space and if $(X, \mathcal{L})$ is complete, then $(X, \mathcal{L}, \mathcal{L}^c)$ is a pairwise Stone space.

Let $(X, \mathcal{L})$ be a Pervin space and let $(\widehat{X}, \widehat{\mathcal{L}})$ be its completion. Then $\widehat{\mathcal{L}}$ is the set of all compact open subsets of $\widehat{X}$. It is also the set of upsets of $\widehat{X}$ that are clopen in the symmetrical topology. Moreover, the lattices $\mathcal{L}$ and $\widehat{\mathcal{L}}$ are isomorphic lattices.

These isomorphisms can be given explicitly. We just describe here the Kolmogorov case, which is simpler. Indeed, if X is Kolmogorov, then the preorder on X is an order and $\imath$ defines an embedding from X into $\widehat{X}$. We tacitly make use of this embedding to identify X with a subset of $\widehat{X}$.

Theorem 6.3. *Let $(X, \mathcal{L})$ be a Kolmogorov Pervin space. Then the maps $L \mapsto \widehat{L}$ and $K \mapsto K \cap X$ are mutually inverse lattice isomorphisms between $\mathcal{L}$ and $\widehat{\mathcal{L}}$.*

If $\mathcal{L}$ is a Boolean algebra, then all previous results simplify greatly. First, the preorders on X and on $\widehat{X}$ are equivalence relations. Next, we have:

Proposition 6.4. *Let $(X, \mathcal{L})$ be a Boolean Pervin space. Then $\mathcal{L}$ forms a basis of clopen sets. Furthermore, X is Kolmogorov if and only if it is Hausdorff. Moreover, $(\widehat{X}, \widehat{\mathcal{L}})$ is a Hausdorff compact space and $\widehat{\mathcal{L}}$ is the Boolean algebra of clopen sets of $\widehat{X}$.*

Furthermore, Theorem 6.3 can be restated as follows.

Theorem 6.5. *Let $(X, \mathcal{L})$ be a Hausdorff Boolean Pervin space. Then the formulas $\widehat{L} = \overline{L}$ and $\overline{L} \cap X = L$ hold for all $L \in \mathcal{L}$. The maps $L \mapsto \overline{L}$ and $K \mapsto K \cap X$ are mutually inverse isomorphisms of Boolean algebra between $\mathcal{L}$ and $\widehat{\mathcal{L}}$.*

In particular, the following formulas hold for all $L, L_1, L_2 \in \mathcal{L}$:

$$\overline{L_1 \cup L_2} = \overline{L}_1 \cup \overline{L}_2, \quad \overline{L_1 \cap L_2} = \overline{L}_1 \cap \overline{L}_2 \quad \text{and} \quad \overline{L^c} = \overline{L}^c.$$

7 Quotient Spaces

Let $(X, \mathcal{L})$ be a Pervin space and let $\mathcal{K}$ be a sublattice of $\mathcal{L}$. Denote by $\widehat{X}^{\mathcal{L}}$ the completion of $(X, \mathcal{L})$ and by $\widehat{X}^{\mathcal{K}}$ the completion of $(X, \mathcal{K})$. Then the following result holds.

Theorem 7.1. *Let $\mathcal{L}$ be a lattice of subsets of X and let $\mathcal{K}$ be a sublattice of $\mathcal{L}$. Then the identity function on X is a morphism from $(X, \mathcal{L})$ to $(X, \mathcal{K})$ and its completion is a quotient map from $\widehat{X}^{\mathcal{L}}$ onto $\widehat{X}^{\mathcal{K}}$.*

This theorem looks like an almost immediate consequence of Theorem 4.8 but there is a missing bit: we did not yet define the notion of a quotient map in the category of Pervin spaces.

A natural attempt would be to mimic the definition used for topological spaces (and for quasi-uniform spaces): a quotient map $\varphi : X \to Y$ should be a surjective morphism[2] such that Y is equipped with the final Pervin structure induced by φ. The second condition states that a subset S of Y is a block if and only if $\varphi^{-1}(S)$ is a block. However, Theorem 7.1 does not work under this definition. Indeed, we have already seen that the identity function I on $\{0, 1\}$

[2] Formally, an epimorphism, but it is easy to see that in the category **Pervin** epimorphisms coincide with surjective morphisms.

induces a morphism from the Boolean space $\mathbb{B}$ to the Sierpiński space $\mathbb{S}$. It is easy to see that these two spaces are isomorphic to their completion and that $\widehat{I} = I$. Consequently, I should be a quotient map from $\mathbb{B}$ to $\mathbb{S}$. However, $\{0\}$ is a subset of $\mathbb{S}$ such that $I^{-1}(0)$ is a block of $\mathbb{B}$, but it is not a block of $\mathbb{S}$. Thus our definition of a quotient map has to be improved as follows:

Definition 7.2. *Let X and Y be Pervin spaces. A surjective morphism φ from X to Y is a quotient map if and only if each upset U of Y such that $\varphi^{-1}(U)$ is a block of X is a block of Y.*

Following [1], let **Prost** denote the category of preordered sets, with order-preserving maps as morphisms. In the language of category theory, we viewed the category **Pervin** as a *concrete category over* the category **Prost**. That is, the forgetful functor now maps a Pervin space $(X, \mathcal{L})$ not to the set X, but to the preordered set $(X, \leqslant_{\mathcal{L}})$. This definition can be rephrased in purely categorical terms, in which the role of the category **Prost** is even more apparent.

Proposition 7.3. *Let $\varphi : X \to Y$ be a surjective morphism of Pervin spaces. The following conditions are equivalent:*

(1) *φ is a quotient map,*
(2) *for each Pervin space Z, any preorder-preserving map $\psi : Y \to Z$ such that $\psi \circ \varphi$ is a morphism is a morphism,*
(3) *every preorder-preserving map ψ from Y to the Sierpiński space such that $\psi \circ \varphi$ is a morphism is also a morphism.*

8 Inequations

We now give a general formulation of the results [14,15]. We first need an abstract definition of the notion of inequations.

Definition 8.1. *Let $(X, \mathcal{L})$ be a Pervin space, let L be a block of X and let (v, w) be a pair of valuations on $\mathcal{L}$. Then L* satisfies the inequation *$v \leqslant w$ if $v(L) \leqslant w(L)$. More generally, a set of blocks $\mathcal{K}$* satisfies the inequation *$v \leqslant w$ if, for every $K \in \mathcal{K}$, $v(K) \leqslant w(K)$.*

Definition 8.2 can be easily extended to a set of inequalities as follows:

Definition 8.2. *Given a set S of inequations, a block L* satisfies *S if it satisfies all the inequations of S. Similarly, a set of blocks $\mathcal{K}$* satisfies *S if it satisfies all the inequations of S. Finally the set of all blocks of X satisfying S is called the* set of blocks defined by *S.*

Formally, an *inequation* is thus a pair (v, w) of valuations on $\mathcal{L}$. We are now ready to state the main result of this section.

Theorem 8.3. *Let $(X, \mathcal{L})$ be a Pervin space. A set of blocks of X is a sublattice of $\mathcal{L}$ if and only if it can be defined by a set of inequations.*

Back to Languages. Let $X = A^*$ and let $\mathrm{Reg}(A^*)$ be the Boolean algebra of all regular languages on A. Almeida [2] has proved that the dual space of this Boolean algebra is the profinite monoid $\widehat{A^*}$. In other words, $\widehat{A^*}$ is the completion of the Pervin space $(A^*, \mathrm{Reg}(A^*))$. This result can be briefly explained as follows. In one direction, each profinite word v defines a valuation on $\mathrm{Reg}(A^*)$ defined by $v(L) = 1$ if and only if $v \in \overline{L}$, where $\overline{L}$ denotes the closure of L in $\widehat{A^*}$.

In the opposite direction, let M be a finite monoid, $\varphi : A^* \to M$ be a monoid morphism and v a valuation on $\mathrm{Reg}(A^*)$. Since $v(A^*) = 1$ and $A^* = \bigcup_{m \in M} \varphi^{-1}(m)$, one gets

$$1 = v(A^*) = \sum_{m \in M} v(\varphi^{-1}(m)).$$

Consequently, there exists an $m \in M$ such that $v(\varphi^{-1}(m)) = 1$. This m is unique since if $v(\varphi^{-1}(m')) = 1$ for some $m' \neq m$, then, as $\varphi^{-1}(m) \cap \varphi^{-1}(m) = \emptyset$, one gets

$$v(\emptyset) = v(\varphi^{-1}(m) \cap \varphi^{-1}(m)) = v(\varphi^{-1}(m))v(\varphi^{-1}(m')) = 1,$$

a contradiction. Therefore, if v is a valuation on $\mathrm{Reg}(A^*)$, there exists a unique profinite word u such that, for each monoid morphism $\varphi : A^* \to M$, $\widehat{\varphi}(u)$ is the unique element $m \in M$ such that $v(\varphi^{-1}(m)) = 1$.

Since $\widehat{A^*}$ is the completion of the Pervin space $(A^*, \mathrm{Reg}(A^*))$, a direct application of Theorem 8.3 gives back Proposition 2.1.

To recover Proposition 2.2, let us first recall that βA^* is the set of ultrafilters of the Boolean algebra $\mathcal{P}(A^*)$. Since, in the Boolean case, valuations and ultrafilters are essentially the same thing[3], the completion of the Pervin space $(A^*, \mathcal{P}(A^*))$ is isomorphic to βA^* and Theorem 8.3 gives back Proposition 2.2.

9 Bibliographic Notes

Pervin spaces were originally introduced by Pervin [22] to prove that every topological space can be derived from a quasi-uniform space. Since then, they have been regularly used to provide examples or counterexamples on quasi-uniform spaces but surprisingly, only two short articles seem to have been specifically devoted to their study, one by Levine in 1969 [21] and another one by Császár in 1993 [9]. In fact, Pervin spaces are so specific that their properties mostly appear in the literature as corollaries of more general results on quasi-uniform spaces. For instance, a quasi-uniform space is isomorphic to a Pervin space if and only if it is transitive and totally bounded. The reader interested in quasi-uniform spaces is refereed to the remarkable surveys written by Künzi [18–20]. Most notions introduced in this paper are actually adapted from the corresponding notions on quasi-uniform spaces, but they often become much simpler in the context of Pervin spaces.

Finally, the inspiring article of Erné [11], which sheds additional light on Pervin spaces, is highly recommended.

[3] Let us define the *characteristic function* of an ultrafilter $\mathcal{U}$ as the map from $\mathcal{P}(A^*)$ to $\{0, 1\}$ taking value 1 on $\mathcal{U}$ and 0 elsewhere. It is easy to see that it is a valuation on $\mathcal{P}(A^*)$. Conversely, if v is a valuation on $\mathcal{P}(A^*)$, then $v^{-1}(1)$ is an ultrafilter.

10 Conclusion

As we explained in the introduction, our original motivation was to find a simple way to describe the dual space of a lattice of subsets by a suitable completion. Metric spaces did not cover our needs, even in the case of Boolean algebras, except in the case of countable Boolean algebras. Uniform spaces did not suffice when dealing with lattices. Quasi-uniform spaces, on the other hand, while fulfilling our requirements, seemed to be too general a tool for our purpose. However, we soon realised that we only needed a very special class of quasi-uniform spaces, the Pervin spaces. To our surprise, turning to Pervin spaces did not only simplify a number of results and proofs, but also lead us to an other point of view on Stone's duality. Moreover, it led us to a notion of quotient space which seems to be more appropriate in the ordered case.

Acknowledgements. I would like to thank Mai Gehrke and Serge Grigorieff for many fruitful discussions on Pervin spaces. I would also like to thank Daniela Petrişan for her critical help on categorical notions used in this paper. Encouragements from Hans-Peter A. Künzi and Marcel Erné were greatly appreciated.

References

1. Adámek, J., Herrlich, H., Strecker, G.E.: Abstract and concrete categories: the joy of cats. Repr. Theory Appl. Categ. **17**, 1–507 (2006). Reprint of the 1990 original [Wiley, New York; MR1051419]
2. Almeida, J.: Residually finite congruences and quasi-regular subsets in uniform algebras. Portugaliæ Math. **46**, 313–328 (1989)
3. Almeida, J.: Finite Semigroups and Universal Algebra. World Scientific Publishing Co. Inc., River Edge (1994). Translated from the 1992 Portuguese original and revised by the author
4. Almeida, J.: Profinite semigroups and applications. In: Kudryavtsev, V.B., Rosenberg, I.G., Goldstein, M. (eds.) Structural Theory of Automata, Semigroups and Universal Algebra, vol. 207, pp. 1–45. Springer, Dordrecht (2005)
5. Almeida, J., Weil, P.: Relatively free profinite monoids: an introduction and examples. In: Fountain, J. (ed.) NATO Advanced Study Institute Semigroups, Formal Languages and Groups, vol. 466, pp. 73–117. Kluwer Academic Publishers, Dordrecht (1995)
6. Barrington, D.A.M., Compton, K., Straubing, H., Thérien, D.: Regular languages in NC^1. J. Comput. System Sci. **44**(3), 478–499 (1992)
7. Bourbaki, N.: General Topology. Chapters 1–4. Elements of Mathematics, vol. 18. Springer, Berlin (1998)
8. Clark, D.M., Davey, B.A.: Natural Dualities for the Working Algebraist. Cambridge Studies in Advanced Mathematics, vol. 57. Cambridge University Press, Cambridge (1998)
9. Császár, A.: D-completions of Pervin-type quasi-uniformities. Acta Sci. Math. **57**(1–4), 329–335 (1993)
10. Davey, B.A., Priestley, H.A.: Introduction to Lattices and Order, 2nd edn. Cambridge University Press, Cambridge (2002)

11. Erné, M.: Ideal completions and compactifications. Appl. Categ. Struct. **9**(3), 217–243 (2001)
12. Gehrke, M.: Canonical extensions, esakia spaces, and universal models. In: Bezhanishvili, G. (ed.) Leo Esakia on Duality in Modal and Intuitionistic Logics. OCL, vol. 4, pp. 9–41. Springer, Dordrecht (2014). doi:10.1007/978-94-017-8860-1_2
13. Gehrke, M.: Stone duality, topological algebra, and recognition. J. Pure Appl. Algebra **220**(7), 2711–2747 (2016)
14. Gehrke, M., Grigorieff, S., Pin, J.É.: Duality and equational theory of regular languages. In: Aceto, L., Damgård, I., Goldberg, L.A., Halldórsson, M.M., Ingólfsdóttir, A., Walukiewicz, I. (eds.) ICALP 2008. LNCS, vol. 5126, pp. 246–257. Springer, Heidelberg (2008). doi:10.1007/978-3-540-70583-3_21
15. Gehrke, M., Grigorieff, S., Pin, J.É.: A topological approach to recognition. In: Abramsky, S., Gavoille, C., Kirchner, C., Meyer auf der Heide, F., Spirakis, P.G. (eds.) ICALP 2010. LNCS, vol. 6199, pp. 151–162. Springer, Heidelberg (2010). doi:10.1007/978-3-642-14162-1_13
16. Hausdorff, F.: Set Theory. Chelsea Publishing Company, New York (1957). Translated by Aumann, J.R., et al.
17. Johnstone, P.T.: Stone Spaces. Cambridge Studies in Advanced Mathematics, vol. 3. Cambridge University Press, Cambridge (1986). Reprint of the 1982 edition
18. Künzi, H.-P.A.: Quasi-uniform spaces in the year 2001. In: Recent Progress in General Topology, II, pp. 313–344. North-Holland, Amsterdam (2002)
19. Künzi, H.-P.A.: Uniform structures in the beginning of the third millenium. Topol. Appl. **154**(14), 2745–2756 (2007)
20. Künzi, H.-P.A.: An introduction to quasi-uniform spaces. In: Beyond Topology. Contemporary Mathematics, vol. 486, pp. 239–304. American Mathematical Society, Providence (2009)
21. Levine, N.: On Pervin's quasi uniformity. Math. J. Okayama Univ. **14**, 97–102 (1969/70)
22. Pervin, W.J.: Quasi-uniformization of topological spaces. Math. Ann. **147**, 316–317 (1962)
23. Pin, J.-É.: Profinite methods in automata theory. In: Albers, S., Marion, J.-Y. (eds.) 26th International Symposium on Theoretical Aspects of Computer Science (STACS 2009), pp. 31–50, Internationales Begegnungs- und Forschungszentrum für Informatik (IBFI), Schloss Dagstuhl, Germany (2009)
24. Pin, J.É.: Equational descriptions of languages. Int. J. Found. Comput. Sci. **23**, 1227–1240 (2012)
25. Schützenberger, M.-P.: Une théorie algébrique du codage. In: Séminaire Dubreil-Pisot, année 1955–56, Exposé No. 15, 27 février 1956, 24 p. Inst. H. Poincaré, Paris (1956). http://igm.univ-mlv.fr/berstel/Mps/Travaux/A/1956CodageSemDubreil.pdf
26. Schützenberger, M.-P.: On finite monoids having only trivial subgroups. Inf. Control **8**, 190–194 (1965)
27. Straubing, H.: Finite Automata, Formal Logic, and Circuit Complexity, Progress in Theoretical Computer Science. Birkhäuser Boston Inc., Boston (1994)
28. Straubing, H.: On Logical Descriptions of Regular Languages. In: Rajsbaum, S. (ed.) LATIN 2002. LNCS, vol. 2286, pp. 528–538. Springer, Heidelberg (2002). doi:10.1007/3-540-45995-2_46
29. Weil, P.: Profinite methods in semigroup theory. Int. J. Alg. Comput. **12**, 137–178 (2002)

Contributed Papers

Relations as Images

Mathieu Alain and Jules Desharnais(✉)

Département d'informatique et de génie logiciel,
Université Laval, Québec, QC, Canada
mathieu.alain.2@ulaval.ca, jules.desharnais@ift.ulaval.ca

Abstract. Boolean matrices constitute an immediate representation of black and white images, with 1 and 0 representing the black and white pixels, respectively. We give relational expressions for calculating two morphological operations on images, namely dilation and erosion. These operations have been implemented under RelView and we compare the performance of RelView with that of Matlab and Mathematica, which have a package for computing various morphological operations. Heijmans et al. have defined dilation and erosion for undirected graphs with vertices weighted by grey-level values. Graphs generalise images by allowing irregular "grids". We propose a definition of dilation and erosion for nonweighted directed graphs (i.e., relations) along the same lines. These operations have been implemented under RelView too.

1 Introduction

The theory of mathematical morphology has many applications, in particular for the analysis and transformation of digital images [8,12]. Two of its basic operations are dilation and erosion. Considering that Boolean matrices constitute an immediate representation of black and white images, with 1 and 0 representing the black and white pixels, respectively, we give relational expressions for calculating dilation and erosion on relations as images. The relational expressions have easy generalisations to geometries other than the standard finite 2-D grid.

We have implemented dilation and erosion under RelView [18], because it allows a direct implementation of relational formulas, so that one can quickly visualise the results. Although we were looking more for flexibility than for performance, we have compared RelView with Matlab [16] and Mathematica [17]. We were, of course, not expecting RelView to be faster than commercial software with dedicated packages, but RelView did really well.

The initial concepts of mathematical morphology have been generalised in the more abstract framework of lattice theory, where dilations and erosions are simply adjoints of a Galois connection [5]. Extensions to graph theory have also been developed [6,7,13,14]. In [6], Heijmans et al. define dilation and erosion for undirected graphs with vertices weighted by grey-level values. We propose a definition of dilation and erosion for nonweighted directed graphs (i.e., relations) along the same lines. These operations were implemented under RelView too.

P. Höfner et al. (Eds.): RAMiCS 2017, LNCS 10226, pp. 43–59, 2017.
DOI: 10.1007/978-3-319-57418-9_3

Section 2 presents the relational background. Section 3 gives some notation for describing images in a relational way. The relational expressions for dilation and erosion are given in Sect. 4, as well as their RelView programs and the performance comparisons. Graph morphology is treated in Sect. 5, before a conclusion in Sect. 6.

The RelView programs and the data used in this article are available [15].

2 Mathematical Background

As for instance do Gries and Schneider [4], we write quantifications in the format ($\star$*variables* | *range* : *quantified expression*), where $\star$ is the quantifier, *variables* is the list of variables bound by the quantification, *range* is the constraint imposed on the bound variables and *quantified expression* speaks for itself. The range is omitted if there is no constraint.

For relations, we use the notation of [1]. The identity, empty and universal relations are denoted respectively by I, O and L, respectively; these symbols are overloaded, in the sense that they may denote constant relations of different types. The operations on relations are union ($\cup$), intersection ($\cap$), composition ($;$), transposition/conversion ($^\mathsf{T}$), complementation ($\overline{}$) and reflexive transitive closure (*). We also use left ($/$) and right ($\backslash$) residual operations.

The following laws and their generalisation to arbitrary unions and intersections are used in the sequel. See [10,11] for details on them and most of the definitions of this section.

$$R\,;\mathsf{I} = \mathsf{I}\,;R = R \tag{1}$$

$$P\,;(Q \cup R) = P\,;Q \cup P\,;R \qquad (Q \cup R)\,;P = Q\,;P \cup R\,;P \tag{2}$$

$$\mathsf{L}\,;\mathsf{L} = \mathsf{L} \qquad Q\,;\mathsf{L}\,;R = Q\,;\mathsf{L} \cap \mathsf{L}\,;R \tag{3}$$

$$(P\,;\mathsf{L} \cap Q)\,;R = P\,;\mathsf{L} \cap Q\,;R \qquad P\,;(\mathsf{L}\,;Q \cap R) = \mathsf{L}\,;Q \cap P\,;R \tag{4}$$

$$P\,;(Q\,;\mathsf{L} \cap R) = (\mathsf{L}\,;Q^\mathsf{T} \cap P)\,;R \qquad (\mathsf{L}\,;P \cap Q)\,;R = Q\,;(P^\mathsf{T}\,;\mathsf{L} \cap R) \tag{5}$$

$$R^{\mathsf{T}\mathsf{T}} = R \qquad (Q\,;R)^\mathsf{T} = R^\mathsf{T}\,;Q^\mathsf{T} \tag{6}$$

$$R \neq \mathsf{O} \Leftrightarrow \mathsf{L}\,;R\,;\mathsf{L} = \mathsf{L} \qquad \text{(Tarski rule)} \tag{7}$$

$$P\,;Q \subseteq R \Leftrightarrow P \subseteq R/Q \Leftrightarrow Q \subseteq P\backslash R \tag{8}$$

$$Q/R = \overline{\overline{Q}\,;R^\mathsf{T}} \qquad Q\backslash R = \overline{Q^\mathsf{T}\,;\overline{R}} \tag{9}$$

$$R^* = (\bigcup n{:}\mathbb{N} \mid: R^n) \tag{10}$$

A relation R is *total* iff $\mathsf{I} \subseteq R\,;R^\mathsf{T}$ (equivalently, $R\,;\mathsf{L} = \mathsf{L}$), *surjective* iff $\mathsf{I} \subseteq R^\mathsf{T};R$ (equivalently, $\mathsf{L};R = \mathsf{L}$), *univalent* iff $R^\mathsf{T};R \subseteq \mathsf{I}$, *injective* iff $R;R^\mathsf{T} \subseteq \mathsf{I}$. A *mapping* is a total and univalent relation.

$$\text{If } P \text{ is a mapping, then } P^\mathsf{T}\,;Q \subseteq R \Leftrightarrow Q \subseteq P\,;R, \qquad Q\,;P \subseteq R \Leftrightarrow Q \subseteq R\,;P^\mathsf{T}. \tag{11}$$

A relation v is a *(column) vector* iff $v = v \,;\, \mathsf{L}$. A vector v is a *point* iff it is nonempty and injective.

$$\text{If } R \text{ is a mapping and } v \text{ a point, then } R^{\mathsf{T}} \,;\, v \text{ is a point.} \tag{12}$$

$$\text{If } u \text{ is a vector and } v \text{ a point, then } u \cap v \neq \mathsf{O} \Leftrightarrow v \subseteq u. \tag{13}$$

The type of a relation R between sets S and T is stated as $R\colon S \leftrightarrow T$; if R is a vector, then its type is stated as $R\colon S$ and sometimes as $R\colon S \leftrightarrow \{\bullet\}$ for one-column vectors, since only the domain side matters.

Relational *direct products* are axiomatised as a pair (π_1, π_2) of projections satisfying the following equations:

$$\text{(a) } \pi_1^{\mathsf{T}} \,;\, \pi_1 = \mathsf{I}, \quad \text{(b) } \pi_2^{\mathsf{T}} \,;\, \pi_2 = \mathsf{I}, \quad \text{(c) } \pi_1^{\mathsf{T}} \,;\, \pi_2 = \mathsf{L}, \quad \text{(d) } \pi_1 \,;\, \pi_1^{\mathsf{T}} \cap \pi_2 \,;\, \pi_2^{\mathsf{T}} = \mathsf{I}.$$

Definition 2.1. *Let (π_1, π_2) be a direct product. We use it to define three operations on relations R_1 and R_2.*

1. Tupling: $\langle R_1, R_2] = R_1 \,;\, \pi_1^{\mathsf{T}} \cap R_2 \,;\, \pi_2^{\mathsf{T}}$.
2. Cotupling: $[R_1, R_2\rangle = \pi_1 \,;\, R_1 \cap \pi_2 \,;\, R_2$.
3. Parallel product: $[R_1, R_2] = \pi_1 \,;\, R_1 \,;\, \pi_1^{\mathsf{T}} \cap \pi_2 \,;\, R_2 \,;\, \pi_2^{\mathsf{T}}$.

We assume *sharpness*, which holds for concrete relations, i.e.,

$$\langle Q_1, Q_2] \,;\, [R_1, R_2\rangle = Q_1 \,;\, R_1 \cap Q_2 \,;\, R_2. \tag{14}$$

Properties of Tupling, Cotupling and Parallel Product

$$\langle Q_1, Q_2] \,;\, [R_1, R_2] = \langle Q_1 \,;\, R_1, Q_2 \,;\, R_2] \tag{15}$$

$$[Q_1, Q_2] \,;\, [R_1, R_2] = [Q_1 \,;\, R_1, Q_2 \,;\, R_2] \tag{16}$$

$$[Q_1, Q_2] \,;\, [R_1, R_2\rangle = [Q_1 \,;\, R_1, Q_2 \,;\, R_2\rangle \tag{17}$$

$$\pi_1 = [\mathsf{I}, \mathsf{L}\rangle \quad \pi_2 = [\mathsf{L}, \mathsf{I}\rangle \quad \pi_1^{\mathsf{T}} = \langle \mathsf{I}, \mathsf{L}] \quad \pi_2^{\mathsf{T}} = \langle \mathsf{L}, \mathsf{I}] \tag{18}$$

$$\langle Q_1, Q_2]^{\mathsf{T}} = [Q_1^{\mathsf{T}}, Q_2^{\mathsf{T}}\rangle \quad [Q_1, Q_2]^{\mathsf{T}} = [Q_1^{\mathsf{T}}, Q_2^{\mathsf{T}}] \quad [Q_1, Q_2\rangle^{\mathsf{T}} = \langle Q_1^{\mathsf{T}}, Q_2^{\mathsf{T}}] \tag{19}$$

$$[\mathsf{I}, \mathsf{I}] = \mathsf{I} \quad \langle \mathsf{L}, \mathsf{L}] = \mathsf{L} \quad [\mathsf{L}, \mathsf{L}] = \mathsf{L} \quad [\mathsf{L}, \mathsf{L}\rangle = \mathsf{L} \tag{20}$$

$$[P \cup Q, R\rangle = [P, R\rangle \cup [Q, R\rangle \qquad [P, Q \cup R\rangle = [P, Q\rangle \cup [P, R\rangle \tag{21}$$

$$\mathsf{L} \,;\, P \cap [Q_1, Q_2\rangle = [\mathsf{L} \,;\, P \cap Q_1, Q_2\rangle = [Q_1, \mathsf{L} \,;\, P \cap Q_2\rangle \tag{22}$$

$$[P \,;\, \mathsf{L}, Q\rangle \,;\, R = [P \,;\, \mathsf{L}, Q \,;\, R\rangle \quad [P, Q \,;\, \mathsf{L}\rangle \,;\, R = [P \,;\, R, Q \,;\, \mathsf{L}\rangle \tag{23}$$

The above direct products can be generalised to n-ary direct products, and similarly for tupling, cotupling and parallel product.

Direct products can be used to transform a relation R into a vector. This is called *vectorisation*. The vectorisation of a relation R is obtained by

$$\mathsf{vec}(R) = [R, \mathsf{I}\rangle \,;\, \mathsf{L}. \tag{24}$$

3 Representing an Image in the Plane by a Relation

To apply morphological operations to images (Sect. 4), we need an operation of addition in order, for instance, to displace images along coordinate axes.

We want to define a relation A of addition that takes two arguments and produces their sum. Assume two successor relations $S_1 : T_1 \leftrightarrow T_1$ and $S_2 : T_2 \leftrightarrow T_2$, two origin vectors $o_1 : T_1$ and $o_2 : T_2$, and the identity relation $\mathsf{I}_1 : T_1 \leftrightarrow T_1$. Let (π_1, π_2) be a direct product typed as $\pi_1 : T_1 \times T_2 \leftrightarrow T_1$ and $\pi_2 : T_1 \times T_2 \leftrightarrow T_2$.

The relation A has type $T_1 \times T_2 \leftrightarrow T_1$ and is defined as $A = ([S_1, S_2^{\mathsf{T}}]^* \sqcup [S_1^{\mathsf{T}}, S_2]^*) ; [\mathsf{I}_1, o_2 ; \mathsf{L}\rangle$. In words, a sum is obtained either by decreasing the second component until it is at an origin, while increasing the first component synchronously, or by increasing the second component until it is at an origin, while decreasing the first component synchronously. If $T_1 = T_2 = \mathbb{Z}$, $o_2 = \{0\}$ and S_1 and S_2 are the successor on the integers, then A adds two integers and outputs the result of the addition. But addition may be more general by allowing more than one origin and arbitrary successors. On a finite interval, the addition is "truncated". For instance, if $T_1 = \{0, 1, 2, 3\}$, $T_2 = \{0, 1, 2\}$, $o_2 = \{0\}$ and $S_1 = \{(0, 1), (1, 2), (2, 3)\}$, $S_2 = \{(0, 1), (1, 2)\}$, then

$$\begin{aligned} A = \{&((0,0),0),((0,1),1),((0,2),2),((1,0),1),((1,1),2),\\ &((1,2),3),((2,0),2),((2,1),3),((3,0),3)\}. \end{aligned}$$

In order to deal with shorter formulas, we assume that an origin has no predecessor (like o_2 in the above example) and choose

$$A = [S_1, S_2^{\mathsf{T}}]^* ; [\mathsf{I}_1, o_2 ; \mathsf{L}\rangle = \{((x_1, x_2), x_1') \mid (\exists n \mid : x_1 S_1^n x_1' \wedge o_2 S_2^n x_2)\} \qquad (25)$$

as the definition of addition[1]. Treating the more general case where an origin may have predecessors poses no difficulty. This restriction means we will be considering images in the first quadrant. But rather than use the standard convention about coordinates, we use the standard convention about matrices, that row numbering increases when going down.

The binary Boolean matrix representation of a (possibly heterogeneous) relation R *is* the representation of an image in the plane. We add to this two notions.

1. Homogeneous *successor relations* S_{d} and S_{c} for the domain (rows) and codomain (columns) of R, respectively. The products $S_{\mathsf{d}} ; R$ and $R ; S_{\mathsf{c}}$ are well typed. The successor relations are typically functions mapping a row (column) to a successor row, but we leave the door open to more general successor relations.
2. Vector *origin relations* o_{d} and o_{c} that designate origins for the domain (rows) and codomain (columns) of R, respectively. In the examples and the RelView programs below, origins are points (there is a single origin). The

[1] In these expressions, the first o_2 is a vector and the second one is an origin; the vector o_2 is composed with an L of the appropriate type to ensure the compatibility of the codomain of $o_2 ; \mathsf{L}$ with that of I_1. We write xRy for $(x, y) \in R$.

products $o_{\mathsf{d}}^{\mathsf{T}} ; R$ and $R ; o_{\mathsf{c}}$ are well typed. We use the abbreviation $O := o_{\mathsf{d}} ; o_{\mathsf{c}}^{\mathsf{T}}$. Since o_{d} and o_{c} are vectors,

$$O = o_{\mathsf{d}} ; o_{\mathsf{c}}^{\mathsf{T}} = o_{\mathsf{d}} ; \mathsf{L} ; o_{\mathsf{c}}^{\mathsf{T}} = o_{\mathsf{d}} ; \mathsf{L} \cap \mathsf{L} ; o_{\mathsf{c}}^{\mathsf{T}}, \tag{26}$$

where the last step follows by (3).

As an example, consider the following relations.

$$\underset{R}{\begin{bmatrix} 0\,0\,1\,0 \\ 0\,1\,1\,0 \\ 0\,1\,1\,0 \end{bmatrix}} \quad \underset{S_{\mathsf{d}}}{\begin{bmatrix} 0\,1\,0 \\ 0\,0\,1 \\ 0\,0\,0 \end{bmatrix}} \quad \underset{S_{\mathsf{c}}}{\begin{bmatrix} 0\,1\,0\,0 \\ 0\,0\,1\,0 \\ 0\,0\,0\,1 \\ 0\,0\,0\,0 \end{bmatrix}} \quad \underset{o_{\mathsf{d}}}{\begin{bmatrix} 1 \\ 0 \\ 0 \end{bmatrix}} \quad \underset{o_{\mathsf{c}}}{\begin{bmatrix} 1 \\ 0 \\ 0 \\ 0 \end{bmatrix}} \quad \underset{O}{\begin{bmatrix} 1\,0\,0\,0 \\ 0\,0\,0\,0 \\ 0\,0\,0\,0 \end{bmatrix}}$$

In this example, the relation R can be written as

$$R = O ; S_{\mathsf{c}}^{2} \cup S_{\mathsf{d}}^{\mathsf{T}} ; O ; S_{\mathsf{c}} \cup S_{\mathsf{d}}^{\mathsf{T}} ; O ; S_{\mathsf{c}}^{2} \cup S_{\mathsf{d}}^{2\mathsf{T}} ; O ; S_{\mathsf{c}} \cup S_{\mathsf{d}}^{2\mathsf{T}} ; O ; S_{\mathsf{c}}^{2}.$$

Thus R is the join of expressions of the form $S_{\mathsf{d}}^{i\mathsf{T}} ; O ; S_{\mathsf{c}}^{j}$.

Now suppose

$$o_{\mathsf{d}} : T_{\mathsf{d}} \leftrightarrow \{\bullet\}, \quad S_{\mathsf{d}} : T_{\mathsf{d}} \leftrightarrow T_{\mathsf{d}}, \quad o_{\mathsf{c}} : T_{\mathsf{c}} \leftrightarrow \{\bullet\}, \quad S_{\mathsf{c}} : T_{\mathsf{c}} \leftrightarrow T_{\mathsf{c}}, \quad R : T_{\mathsf{d}} \leftrightarrow T_{\mathsf{c}},$$

where T_{d} is $\mathbb{N}$ or an initial interval $[0, n_{\mathsf{d}}]$ of $\mathbb{N}$ and T_{c} is $\mathbb{N}$ or an initial interval $[0, n_{\mathsf{c}}]$ of $\mathbb{N}$. Suppose in addition that

$$\begin{aligned} o_{\mathsf{d}} &= \{(0, \bullet)\}, S_{\mathsf{d}} = \{(x, x+1) \mid x \in T_{\mathsf{d}}\}, \\ o_{\mathsf{c}} &= \{(0, \bullet)\}, S_{\mathsf{c}} = \{(x, x+1) \mid x \in T_{\mathsf{c}}\}, \end{aligned} \tag{27}$$

where $n_{\mathsf{d}} + 1$ may be undefined or equal to 0 (by addition modulo $n_{\mathsf{d}} + 1$) and similarly for $n_{\mathsf{c}} + 1$. We first show that, for the case $T_{\mathsf{d}} = T_{\mathsf{c}} = \mathbb{N}$,

$$S_{\mathsf{d}}^{i\mathsf{T}} ; O ; S_{\mathsf{c}}^{j} = \{(i, j)\}, \tag{28}$$

$$o_{\mathsf{d}}^{\mathsf{T}} ; S_{\mathsf{d}}^{i} ; R ; S_{\mathsf{c}}^{j\mathsf{T}} ; o_{\mathsf{c}} \neq \mathsf{O} \Leftrightarrow iRj. \tag{29}$$

1. Proof of (28):

$$\begin{aligned} & S_{\mathsf{d}}^{i\mathsf{T}} ; O ; S_{\mathsf{c}}^{j} \\ = \;& \{(x, y) \mid x(S_{\mathsf{d}}^{i\mathsf{T}} ; O ; S_{\mathsf{c}}^{j})y\} \\ = \;& \{(x, y) \mid (\exists u, v \mid: x S_{\mathsf{d}}^{i\mathsf{T}} u \wedge u O v \wedge v S_{\mathsf{c}}^{j} y)\} \\ = \;& \qquad \langle O = o_{\mathsf{d}} ; o_{\mathsf{c}}^{\mathsf{T}} = \{(0,0)\} \rangle \\ & \{(x, y) \mid x S_{\mathsf{d}}^{i\mathsf{T}} 0 \wedge 0 S_{\mathsf{c}}^{j} y\} \\ = \;& \{(x, y) \mid x = i \wedge y = j\} \\ = \;& \{(i, j)\}. \end{aligned}$$

2. Proof of (29):

$$o_{\mathsf{d}}^{\mathsf{T}} ; S_{\mathsf{d}}^{i} ; R ; S_{\mathsf{c}}^{j\mathsf{T}} ; o_{\mathsf{c}} \neq \mathsf{O} \;\Leftrightarrow\; \bullet(o_{\mathsf{d}}^{\mathsf{T}} ; S_{\mathsf{d}}^{i} ; R ; S_{\mathsf{c}}^{j\mathsf{T}} ; o_{\mathsf{c}})\bullet \;\Leftrightarrow\; 0(S_{\mathsf{d}}^{i} ; R ; S_{\mathsf{c}}^{j\mathsf{T}})0$$
$$\Leftrightarrow (\exists u, v \mid: 0S_{\mathsf{d}}^{i}u \wedge uRv \wedge vS_{\mathsf{c}}^{j\mathsf{T}}0) \;\Leftrightarrow\; (\exists u, v \mid: u = i \wedge uRv \wedge v = j)$$
$$\Leftrightarrow (\exists u, v \mid: u = i \wedge iRj \wedge v = j) \;\Leftrightarrow\; iRj.$$

We now look at some of the other cases.

1. If $T_{\mathsf{d}} = \mathbb{N}$ and $T_{\mathsf{c}} = [0, n_{\mathsf{c}}]$ with addition modulo $n_{\mathsf{c}} + 1$, then (i, j) in (28) and (29) should be replaced by $(i, j \;\mathsf{mod}\; (n_{\mathsf{c}} + 1))$.
2. If $T_{\mathsf{d}} = \mathbb{N}$ and $T_{\mathsf{c}} = [0, n_{\mathsf{c}}]$ with $n_{\mathsf{c}} + 1$ undefined, then
   - $\{(i, j)\}$ in (28) should be replaced by $\{(i, j)\}$ if $j \leq n_{\mathsf{c}}$ and $\{\}$ otherwise;
   - iRj in (29) may stand as is, since $o_{\mathsf{d}} ; S_{\mathsf{d}}^{i} ; R ; S_{\mathsf{c}}^{\mathsf{T}j} ; o_{\mathsf{c}} = \mathsf{O}$ and $\neg(iRj)$ if j exceeds the bounds of the interval.

The remaining combinations are treated similarly.

Because of (28) and the subsequent discussion about the other cases,

$$R = (\bigcup i, j \mid iRj : S_{\mathsf{d}}^{i\mathsf{T}} ; O ; S_{\mathsf{c}}^{j}), \tag{30}$$

provided (27) holds.

Let us find the vector expression $\mathsf{vec}(R)$ corresponding to the expression for R given in (30). We start by proving a more general form:

$$\mathsf{vec}(P ; R ; Q^{\mathsf{T}}) \;=\; (\bigcup i, j \mid iRj : [P ; S_{\mathsf{d}}^{i\mathsf{T}},\, Q ; S_{\mathsf{c}}^{j\mathsf{T}}]) \;;\; [o_{\mathsf{d}}, o_{\mathsf{c}}\rangle. \tag{31}$$

$$\begin{aligned}
& \mathsf{vec}(P ; R ; Q^{\mathsf{T}}) \\
= & \quad \langle (24) \;\;\&\;\; (30) \rangle \\
& [P ; (\bigcup i, j \mid iRj : S_{\mathsf{d}}^{i\mathsf{T}} ; O ; S_{\mathsf{c}}^{j}) ; Q^{\mathsf{T}},\, \mathsf{I}\rangle ; \mathsf{L} \\
= & \quad \langle \text{Distributivity of } ; \text{ over } \cup \text{ (2)} \;\;\&\;\; (21) \rangle \\
& (\bigcup i, j \mid iRj : [P ; S_{\mathsf{d}}^{i\mathsf{T}} ; O ; S_{\mathsf{c}}^{j} ; Q^{\mathsf{T}},\, \mathsf{I}\rangle ; \mathsf{L}) \\
= & \quad \langle (26) \rangle \\
& (\bigcup i, j \mid iRj : [P ; S_{\mathsf{d}}^{i\mathsf{T}} ; o_{\mathsf{d}} ; \mathsf{L} ; o_{\mathsf{c}}^{\mathsf{T}} ; S_{\mathsf{c}}^{j} ; Q^{\mathsf{T}},\, \mathsf{I}\rangle ; \mathsf{L}) \\
= & \quad \langle (3) \rangle \\
& (\bigcup i, j \mid iRj : [P ; S_{\mathsf{d}}^{i\mathsf{T}} ; o_{\mathsf{d}} ; \mathsf{L} \cap \mathsf{L} ; o_{\mathsf{c}}^{\mathsf{T}} ; S_{\mathsf{c}}^{j} ; Q^{\mathsf{T}},\, \mathsf{I}\rangle ; \mathsf{L}) \\
= & \quad \langle (22) \rangle \\
& (\bigcup i, j \mid iRj : ([P ; S_{\mathsf{d}}^{i\mathsf{T}} ; o_{\mathsf{d}} ; \mathsf{L},\, \mathsf{I}\rangle \cap \mathsf{L} ; o_{\mathsf{c}}^{\mathsf{T}} ; S_{\mathsf{c}}^{j} ; Q^{\mathsf{T}}) ; \mathsf{L}) \\
= & \quad \langle (5) \;\;\&\;\; (6) \;\;\&\;\; \text{Boolean algebra} \rangle \\
& (\bigcup i, j \mid iRj : [P ; S_{\mathsf{d}}^{i\mathsf{T}} ; o_{\mathsf{d}} ; \mathsf{L},\, \mathsf{I}\rangle ; Q ; S_{\mathsf{c}}^{j\mathsf{T}} ; o_{\mathsf{c}} ; \mathsf{L}) \\
= & \quad \langle (23) \;\;\&\;\; (1) \rangle \\
& (\bigcup i, j \mid iRj : [P ; S_{\mathsf{d}}^{i\mathsf{T}} ; o_{\mathsf{d}} ; \mathsf{L},\, Q ; S_{\mathsf{c}}^{j\mathsf{T}} ; o_{\mathsf{c}} ; \mathsf{L}\rangle)
\end{aligned}$$

$$= \quad \langle(17)\rangle$$
$$(\bigcup i,j \mid iRj : [P ; S_{\mathsf{d}}^{i\mathsf{T}},\ Q ; S_{\mathsf{c}}^{j\mathsf{T}}] ; [o_{\mathsf{d}} ; \mathsf{L},\ o_{\mathsf{c}} ; \mathsf{L}\rangle)$$
$$= \quad \langle o_{\mathsf{d}} \text{ and } o_{\mathsf{c}} \text{ are vectors } \& \text{ Distributivity of ; over } \cup \text{ (2)}\rangle$$
$$(\bigcup i,j \mid iRj : [P ; S_{\mathsf{d}}^{i\mathsf{T}},\ Q ; S_{\mathsf{c}}^{j\mathsf{T}}]) ; [o_{\mathsf{d}},\ o_{\mathsf{c}}\rangle$$

Then, using $P = \mathsf{I}_{\mathsf{d}}$ and $Q = \mathsf{I}_{\mathsf{c}}$ in (31), we find

$$\mathsf{vec}(R) = (\bigcup i,j \mid iRj : [S_{\mathsf{d}}^{i\mathsf{T}},\ S_{\mathsf{c}}^{j\mathsf{T}}])\ ;\ [o_{\mathsf{d}}, o_{\mathsf{c}}\rangle. \tag{32}$$

This is a nice form that is easily extended to n-ary relations.

4 Dilation and Erosion

The dilation $R \oplus P$ of an image $R \subseteq \mathbb{Z} \times \mathbb{Z}$ by a pattern $P \subseteq \mathbb{Z} \times \mathbb{Z}$ is often defined as the pointwise addition (also called the Minkowski addition) of R and P:

$$R \oplus P = \{(x_R + x_P, y_R + y_P) \mid (x_R, y_R) \in R \wedge (x_P, y_P) \in P\}.$$

Using the addition functions A_X and A_Y, this can be expressed as

$$R \oplus P = A_X^{\mathsf{T}} ; [R, P] ; A_Y, \tag{33}$$

where the typing of the relations is

$$P : X_P \leftrightarrow Y_P, \quad R : X_R \leftrightarrow Y_R, \quad A_X : X_R \times X_P \leftrightarrow X_R, \quad A_Y : Y_R \times Y_P \leftrightarrow Y_R.$$

If $X_P = Y_P = X_R = Y_R = \mathbb{Z}$, then dilation is commutative and this is a good reason for choosing a symmetric symbol like $\oplus$ for it. However, if the image R is finite and of a fixed size and if, as is normally the case, the dilation of R by P has the same size as R, then commutativity does not hold. This is why we use the symbol $\rhd$ for it.

In the sequel, the general rule for forming the names and the typing of the relations is that, for any index i,

$$\begin{array}{llll} R_i : T_{\mathsf{d}i} \leftrightarrow T_{\mathsf{c}i}, & \mathsf{I}_{\mathsf{d}i}, S_{\mathsf{d}i} : T_{\mathsf{d}i} \leftrightarrow T_{\mathsf{d}i}, & o_{\mathsf{d}i} : T_{\mathsf{d}i}, & A_{\mathsf{d}i} : T_{\mathsf{d}i} \times T_{\mathsf{d}j} \leftrightarrow T_{\mathsf{d}i}, \\ & \mathsf{I}_{\mathsf{c}i}, S_{\mathsf{c}i} : T_{\mathsf{c}i} \leftrightarrow T_{\mathsf{c}i}, & o_{\mathsf{c}i} : T_{\mathsf{c}i}, & A_{\mathsf{c}i} : T_{\mathsf{c}i} \times T_{\mathsf{c}j} \leftrightarrow T_{\mathsf{c}i}, \end{array}$$

where $T_{\mathsf{d}j}$ and $T_{\mathsf{c}j}$ are determined by the context.

The dilation of R_1 by R_2 is then

$$R_1 \rhd R_2 = A_{\mathsf{d}1}^{\mathsf{T}} ; [R_1, R_2] ; A_{\mathsf{c}1}. \tag{34}$$

This expression for dilation can be implemented quite simply on RelView. Computing A can take a long time, but this can be done once and then A can be used as a global variable. However, this takes much space, due to the parallel product $[R_1, R_2]$. We thus transform the expression.

The following derivation proves

$$R_1 \rhd R_2 = (\bigcup i,j : \mathbb{N} \mid o_{\mathsf{d}2}^{\mathsf{T}} ; S_{\mathsf{d}2}^{i} ; R_2 ; S_{\mathsf{c}2}^{j\mathsf{T}} ; o_{\mathsf{c}2} \neq \mathsf{O} : S_{\mathsf{d}1}^{i\mathsf{T}} ; R_1 ; S_{\mathsf{c}1}^{j}). \tag{35}$$

$$
\begin{aligned}
& R_1 \rhd R_2 \\
= \quad & \langle (34) \ \& \ \text{Definition of addition (25)} \ \& \ (6) \rangle \\
& \langle \mathsf{I}_{d1}, \mathsf{L} \mathbin{;} o_{d2}^{\mathsf{T}}] \mathbin{;} [S_{d1}^{\mathsf{T}}, S_{d2}]^* \mathbin{;} [R_1, R_2] \mathbin{;} [S_{c1}, S_{c2}^{\mathsf{T}}]^* \mathbin{;} [\mathsf{I}_{c1}, o_{c2} \mathbin{;} \mathsf{L} \rangle. \\
= \quad & \langle (10) \rangle \\
& \langle \mathsf{I}_{d1}, \mathsf{L} \mathbin{;} o_{d2}^{\mathsf{T}}] \mathbin{;} (\bigcup i \!:\! \mathbb{N} \mid: [S_{d1}^{\mathsf{T}}, S_{d2}]^i) \mathbin{;} [R_1, R_2] \mathbin{;} (\bigcup j \!:\! \mathbb{N} \mid: [S_{c1}, S_{c2}^{\mathsf{T}}]^j) \mathbin{;} [\mathsf{I}_{c1}, o_{c2} \mathbin{;} \mathsf{L} \rangle \\
= \quad & \langle \text{Distributivity of ; over } \cup \ (2) \rangle \\
& (\bigcup i, j \!:\! \mathbb{N} \mid: \langle \mathsf{I}_{d1}, \mathsf{L} \mathbin{;} o_{d2}^{\mathsf{T}}] \mathbin{;} [S_{d1}^{\mathsf{T}}, S_{d2}]^i \mathbin{;} [R_1, R_2] \mathbin{;} [S_{c1}, S_{c2}^{\mathsf{T}}]^j \mathbin{;} [\mathsf{I}_{c1}, o_{c2} \mathbin{;} \mathsf{L} \rangle) \\
= \quad & \langle (16) \text{ generalised by induction to the powers } i \text{ and } j \rangle \\
& (\bigcup i, j \!:\! \mathbb{N} \mid: \langle \mathsf{I}_{d1}, \mathsf{L} \mathbin{;} o_{d2}^{\mathsf{T}}] \mathbin{;} [S_{d1}^{i\mathsf{T}} \mathbin{;} R_1 \mathbin{;} S_{c1}^j, S_{d2}^i \mathbin{;} R_2 \mathbin{;} S_{c2}^{j\mathsf{T}}] \mathbin{;} [\mathsf{I}_{c1}, o_{c2} \mathbin{;} \mathsf{L} \rangle) \\
= \quad & \langle (15) \ \& \ (14) \ \& \ (1) \rangle \\
& (\bigcup i, j \!:\! \mathbb{N} \mid: S_{d1}^{i\mathsf{T}} \mathbin{;} R_1 \mathbin{;} S_{c1}^j \ \cap \ \mathsf{L} \mathbin{;} o_{d2}^{\mathsf{T}} \mathbin{;} S_{d2}^i \mathbin{;} R_2 \mathbin{;} S_{c2}^{j\mathsf{T}} \mathbin{;} o_{c2} \mathbin{;} \mathsf{L}) \\
= \quad & \langle \text{By (7), } \mathsf{L} \mathbin{;} o_{d2}^{\mathsf{T}} \mathbin{;} S_{d2}^i \mathbin{;} R_2 \mathbin{;} S_{c2}^{j\mathsf{T}} \mathbin{;} o_{c2} \mathbin{;} \mathsf{L} \text{ is either } \mathsf{L} \text{ or } \mathsf{O} \rangle \\
& (\bigcup i, j \!:\! \mathbb{N} \mid \mathsf{L} \mathbin{;} o_{d2}^{\mathsf{T}} \mathbin{;} S_{d2}^i \mathbin{;} R_2 \mathbin{;} S_{c2}^{j\mathsf{T}} \mathbin{;} o_{c2} \mathbin{;} \mathsf{L} = \mathsf{L} : S_{d1}^{i\mathsf{T}} \mathbin{;} R_1 \mathbin{;} S_{c1}^j) \\
= \quad & \langle \text{Tarski rule (7)} \rangle \\
& (\bigcup i, j \!:\! \mathbb{N} \mid o_{d2}^{\mathsf{T}} \mathbin{;} S_{d2}^i \mathbin{;} R_2 \mathbin{;} S_{c2}^{j\mathsf{T}} \mathbin{;} o_{c2} \neq \mathsf{O} : S_{d1}^{i\mathsf{T}} \mathbin{;} R_1 \mathbin{;} S_{c1}^j)
\end{aligned}
$$

Of all the formulas of this derivation, the last one is particularly suited for an implementation under RelView, since it involves no products and the range expression is a relation of size 1×1.

By (29), if (27) holds, (35) can be written more succinctly as

$$R_1 \rhd R_2 = (\bigcup i, j \!:\! \mathbb{N} \mid i R_2 j : S_{d1}^{i\mathsf{T}} \mathbin{;} R_1 \mathbin{;} S_{c1}^j). \tag{36}$$

The next step is to calculate left erosion as a left residual of dilation. We want to find an operator $\not\rhd$ such that

$$R_1 \rhd R_2 \subseteq R_3 \ \Leftrightarrow \ R_1 \subseteq R_3 \not\rhd R_2. \tag{37}$$

The left residual $R_3 \not\rhd R_2$ is the erosion of R_3 by R_2. From (34), the type of $R_1 \rhd R_2$ is the same as that of R_1, and so by (37) the type of R_3 and $R_3 \not\rhd R_2$ is that of R_1 too. Thus, $T_{d3} = T_{d1}$ and $T_{c3} = T_{c1}$. We assume that there is only one successor relation for a given type, so that

$$S_{d3} = S_{d1} \quad \text{and} \quad S_{c3} = S_{c1}. \tag{38}$$

$$
\begin{aligned}
& R_1 \rhd R_2 \subseteq R_3 \\
\Leftrightarrow \quad & \langle (35) \rangle \\
& (\bigcup i, j \!:\! \mathbb{N} \mid o_{d2}^{\mathsf{T}} \mathbin{;} S_{d2}^i \mathbin{;} R_2 \mathbin{;} S_{c2}^{j\mathsf{T}} \mathbin{;} o_{c2} \neq \mathsf{O} : S_{d1}^{i\mathsf{T}} \mathbin{;} R_1 \mathbin{;} S_{c1}^j) \subseteq R_3 \\
\Leftrightarrow \quad & \langle \text{Definition of } \cup \rangle \\
& (\forall i, j \!:\! \mathbb{N} \mid o_{d2}^{\mathsf{T}} \mathbin{;} S_{d2}^i \mathbin{;} R_2 \mathbin{;} S_{c2}^{j\mathsf{T}} \mathbin{;} o_{c2} \neq \mathsf{O} : S_{d1}^{i\mathsf{T}} \mathbin{;} R_1 \mathbin{;} S_{c1}^j \subseteq R_3)
\end{aligned}
$$

$$
\begin{aligned}
\Leftrightarrow \quad & \langle (8) \rangle \\
& (\forall i,j\!:\!\mathbb{N} \mid o_{\mathsf{d}2}^{\mathsf{T}} ; S_{\mathsf{d}2}^{i} ; R_2 ; S_{\mathsf{c}2}^{j\mathsf{T}} ; o_{\mathsf{c}2} \neq \mathsf{O} : R_1 \subseteq S_{\mathsf{d}1}^{i\mathsf{T}} \backslash R_3 / S_{\mathsf{c}1}^{j}) \\
\Leftrightarrow \quad & \langle \text{Definition of } \cap \rangle \\
& R_1 \subseteq (\textstyle\bigcap i,j\!:\!\mathbb{N} \mid o_{\mathsf{d}2}^{\mathsf{T}} ; S_{\mathsf{d}2}^{i} ; R_2 ; S_{\mathsf{c}2}^{j\mathsf{T}} ; o_{\mathsf{c}2} \neq \mathsf{O} : S_{\mathsf{d}1}^{i\mathsf{T}} \backslash R_3 / S_{\mathsf{c}1}^{j}) \\
\Leftrightarrow \quad & \langle (38) \rangle \\
& R_1 \subseteq (\textstyle\bigcap i,j\!:\!\mathbb{N} \mid o_{\mathsf{d}2}^{\mathsf{T}} ; S_{\mathsf{d}2}^{i} ; R_2 ; S_{\mathsf{c}2}^{j\mathsf{T}} ; o_{\mathsf{c}2} \neq \mathsf{O} : S_{\mathsf{d}3}^{i\mathsf{T}} \backslash R_3 / S_{\mathsf{c}3}^{j})
\end{aligned}
$$

Hence, $R_3 \not\triangleright R_2 = (\bigcap i,j\!:\!\mathbb{N} \mid o_{\mathsf{d}2}^{\mathsf{T}} ; S_{\mathsf{d}2}^{i} ; R_2 ; S_{\mathsf{c}2}^{j\mathsf{T}} ; o_{\mathsf{c}2} \neq \mathsf{O} : S_{\mathsf{d}3}^{i\mathsf{T}} \backslash R_3 / S_{\mathsf{c}3}^{j})$. Renaming R_3 yields

$$R_1 \not\triangleright R_2 = (\bigcap i,j\!:\!\mathbb{N} \mid o_{\mathsf{d}2}^{\mathsf{T}} ; S_{\mathsf{d}2}^{i} ; R_2 ; S_{\mathsf{c}2}^{j\mathsf{T}} ; o_{\mathsf{c}2} \neq \mathsf{O} : S_{\mathsf{d}1}^{i\mathsf{T}} \backslash R_1 / S_{\mathsf{c}1}^{j}). \tag{39}$$

Now we deal with right erosion. We want to find an operator $\mathbin{\backslash\!\triangleright}$ such that

$$R_1 \triangleright R_2 \subseteq R_3 \Leftrightarrow R_2 \subseteq R_1 \mathbin{\backslash\!\triangleright} R_3. \tag{40}$$

The right residual $R_1 \mathbin{\backslash\!\triangleright} R_3$ is the right erosion of R_3 by R_1. Since by (34) the type of $R_1 \triangleright R_2$ is the same as that of R_1, the type of R_3 in (40) is the same as that of R_1. Hence, there is no information in $R_1 \mathbin{\backslash\!\triangleright} R_3$ about the type of R_2, so it seems there is no way to get the right erosion. However, the result of a calculation similar to the one for left erosion is

$$R_1 \mathbin{\backslash\!\triangleright} R_3 = (\bigcap i,j\!:\!\mathbb{N} \mid \neg(S_{\mathsf{d}1}^{i\mathsf{T}} ; R_1 ; S_{\mathsf{c}1}^{j} \subseteq R_3) : (o_{\mathsf{d}2}^{\mathsf{T}} ; S_{\mathsf{d}2}^{i}) \backslash \mathsf{O} / (S_{\mathsf{c}2}^{j\mathsf{T}} ; o_{\mathsf{c}2})) \tag{41}$$

and we see that in this equation, $R_1 \mathbin{\backslash\!\triangleright} R_3$ depends on o_2, $S_{\mathsf{d}2}$ and $S_{\mathsf{c}2}$, but not on R_2. This means that one can fix the geometry by providing the type of R_2, the origin o_2 and the successors $S_{\mathsf{d}2}$ and $S_{\mathsf{c}2}$, and then it is possible to calculate the pattern R_2.

In the integer grid $\mathbb{Z} \times \mathbb{Z}$, this problem does not occur, because dilation is commutative, so that there is only one residual. In the literature, as far as we know, the only erosion considered for images is our left erosion. This is to be expected, since one wants to calculate the erosion of an image by a given pattern, not a pattern. Although computing a pattern R_2 from an image R_1 and a dilated image R_3 possibly has no interesting application, it is something that may deserve further investigation.

It is a simple matter to write RelView programs implementing the calculation of dilation (35) and erosion (39). The programs are given in Fig. 1. Figure 2 gives an example as captures of RelView windows. The pattern R_2 consists of four black pixels (four 1s in the matrix). The dilated image $R_1 \triangleright R_2$ consists of the original image (because the origin of R_2 is black) and three other images, one displaced by one pixel to the right, one by one pixel to the bottom and one by one pixel diagonally to the bottom right. Some pixels are moved out of the grid. The erosion $R_1 \not\triangleright R_2$ has black pixels where the pattern R_2 can be fully included in the image R_1. The black bars at the bottom and at the right are due to border effects that appear because the effect of the calculation of the erosion amounts to considering that the area outside of the visible grid is filled up with black pixels. These border effects also explain why the results of dilation and erosion are not symmetrical although R_1 and R_2 are; as mentioned at the beginning of Sect. 3, addition is "truncated"; to fully display dilation and erosion, the grid would have to be enlarged.

```
{
Input: relations R1, R2.
Output: dilation of R1 by R2.
}
Dilation(R1, R2)
DECL Sd1, Sc1, Sd2, Sc2, od2, oc2,
     i1, j1, i2, j2, res, cond
BEG

Sd1 = succ(R1);
Sc1 = succ(R1^);
Sd2 = succ(R2);
Sc2 = succ(R2^);
od2 = init(On1(R2));
oc2 = init(On1(R2^));
i1 = I(Sd1);
j1 = I(Sc1);
i2 = I(Sd2);
j2 = I(Sc2);
res = O(R1);

WHILE -empty(i2) DO
  WHILE -empty(j2) DO
    cond = od2^*i2*R2*j2^*oc2
    IF -empty(cond) THEN
      res = res|i1^*R1*j1 FI
    j1 = j1*Sc1;
    j2 = j2*Sc2 OD
  j1 = I(Sc1);
  j2 = I(Sc2);
  i1 = i1*Sd1;
  i2 = i2*Sd2 OD
RETURN res
END.
```

```
{
Input: relations R1, R2.
Output: erosion of R1 by R2.
}
Erosion(R1, R2)
DECL Sd1, Sc1, Sd2, Sc2, od2, oc2,
     i1, j1, i2, j2, res, cond
BEG

Sd1 = succ(R1);
Sc1 = succ(R1^);
Sd2 = succ(R2);
Sc2 = succ(R2^);
od2 = init(On1(R2));
oc2 = init(On1(R2^));
i1 = I(Sd1);
j1 = I(Sc1);
i2 = I(Sd2);
j2 = I(Sc2);
res = L(R1);

WHILE -empty(i2) DO
  WHILE -empty(j2) DO
    cond = od2^*i2*R2*j2^*oc2
    IF -empty(cond) THEN
      res = res & (i1^\R1/j1) FI
    j1 = j1*Sc1;
    j2 = j2*Sc2 OD
  j1 = I(Sc1);
  j2 = I(Sc2);
  i1 = i1*Sd1;
  i2 = i2*Sd2 OD
RETURN res
END.
```

Fig. 1. RelView programs for dilation and erosion

Table 1 gives the execution times for a sample of images and patterns. Times are rounded up to the second, except to the second decimal for Matlab. The programs were run on a Mac Pro with 64 Gb of RAM and a 3.5 GHz 6-core Intel processor; RelView was run on an Ubuntu Linux 14.04 virtual machine

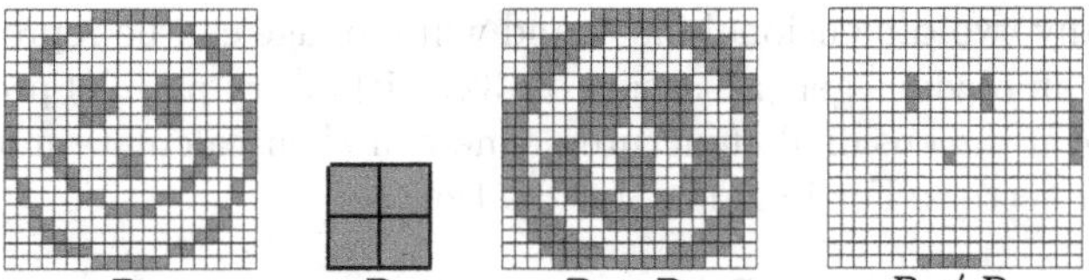

R_1 R_2 $R_1 \rhd R_2$ $R_1 \not\rhd R_2$

Fig. 2. Dilation and erosion of R_1 by R_2

(installed inside Parallels) with 8 Gb of RAM. The times given are RelView's CPU time, Matlab's timeit function time and Mathematica's Timing function time. The relations were randomly generated by RelView with a filling of 23% (the default value). The same relations were used with RelView, Matlab and Mathematica. In applications, pattern sizes of 2×2 and 3×3 are typical. We nevertheless made tests with larger patterns (100×100 in the table).

Table 1. CPU times. Size of R_1: $n_1 \times n_1$. RelV: RelView. Matl: Matlab. Math: Mathematica.

	Size of R_2: 3×3							Size of R_2: 100×100					
	Dilation			Erosion				Dilation			Erosion		
n_1	RelV	Matl	Math	RelV	Matl	Math	n_1	RelV	Matl	Math	RelV	Matl	Math
1000	2	.01	21	5	.01	20	100	3	.10	6	2	.10	6
2000	11	.02	81	25	.02	81	150	5	.10	8	4	.10	8
3000	26	.05	183	59	.05	182	200	10	.10	10	6	.10	10
4000	50	.09	327	119	.09	332	250	19	.14	13	11	.13	13
5000	85	.12	509	201	.12	509	300	35	.18	16	19	.17	16
6000	134	.20	731	315	.20	733	350	58	.23	19	36	.23	19
7000	197	.25	1007	418	.26	990	400	84	.22	23	60	.22	23
8000	246	.34	1310	545	.34	1314	450	107	.27	26	112	.27	26
9000	322	.42	1739	558	.44	1765	500	150	.34	30	199	.34	30
10000	469	.49	2180	1024	.53	2177	550	213	.38	34	365	.38	34

The results are quite surprising. Matlab is by far the fastest. For the small 3×3 pattern (the normal case for applications), RelView is faster than Mathematica, but it is the opposite for the large 100×100 pattern. For both Matlab and Mathematica, the times for dilation and erosion are similar, while those of RelView for erosion are worse than for dilation. The main difference between the expression for dilation (35) and that for erosion (39) is that two compositions become residuals. So, the implementation of residuals seem to be less efficient than that of composition.

An advantage of RelView is that it is easy to modify the programs to explore other geometries. For instance, choosing for the successor S_{d1} one that does modulo addition results in images wrapping around the grid horizontally, giving a cylinder. Doing the same also for S_{c1} yields a torus.

By (24), any two-dimensional black and white image can be represented by a vector. Using an obvious generalisation of (32) with n-ary parallel products, this is also true of n-dimensional structures. One can then calculate morphological operations on vectors. For instance, for dilation,

$$\mathsf{vec}(R_1 \rhd R_2) = \mathsf{parsucc}_1(R_2)\,;\mathsf{vec}(R_1), \tag{42}$$

where

$$\mathsf{parsucc}_1(P) = (\bigcup i, j \mid iPj : [S_{\mathsf{d}1}^{i\mathsf{T}},\ S_{\mathsf{c}1}^{j\mathsf{T}}]). \tag{43}$$

Thus the vector representation of the image, $\mathsf{vec}(R_1)$, is transformed into the vector representation of the dilation, $\mathsf{vec}(R_1 \rhd R_2)$. The drawback is that the transformer $\mathsf{parsucc}_1(P)$ may be large, due to the parallel product of the successor relations for the image.

5 Graph Morphology

Let Q and R be relations. The notation $\mathsf{ih}(\sigma, Q, R)$ means that σ is an injective homomorphism from Q to R, i.e.,

$$\mathsf{ih}(\sigma, Q, R) \;\Leftrightarrow\; \sigma\,;\sigma^{\mathsf{T}} = \mathsf{I} \wedge \sigma^{\mathsf{T}}\,;\sigma \subseteq \mathsf{I} \wedge \sigma^{\mathsf{T}}\,;Q\,;\sigma \subseteq R. \tag{44}$$

Since σ is a mapping, the condition $\sigma^{\mathsf{T}}\,;Q\,;\sigma \subseteq R$ is equivalent to $Q \subseteq \sigma\,;R\,;\sigma^{\mathsf{T}}$, by (11).

For our purpose, a graph G is a 4-tuple (V, R, r, b), where $R : V \leftrightarrow V$ is a homogenous relation and $r, b\colon V$ are vectors representing the *roots* and the *buds* of G [6]. In the sequel, G_i denotes the graph (V_i, R_i, r_i, b_i), for any index i.

The *bud-complementation* and *root-bud-exchange* unary operations are defined as follows:

- $G^{-} = (V, R, r, \overline{b})$ complements the bud vector b;
- $G^{\leftrightarrow} = (V, R, b, r)$ exchanges the roots and the buds.

Note that in both cases the relation R stays the same.

Define a partial ordering $\sqsubseteq$ on graphs by

$$\begin{aligned}(V_1, R_1, r_1, b_1) &\sqsubseteq (V_2, R_2, r_2, b_2)\\ &\Leftrightarrow\; V_1 = V_2 \wedge R_1 = R_2 \wedge r_1 = r_2 \wedge b_1 \subseteq b_2.\end{aligned} \tag{45}$$

The *dilation* $G_1 \rhd G_2$ of G_1 by G_2 is defined by

$$G_1 \rhd G_2 = (V_1,\ R_1,\ r_1,\ (\bigcup \sigma \mid \mathsf{ih}(\sigma, R_2, R_1) \wedge \sigma^{\mathsf{T}}\,; r_2 \cap b_1 \neq \mathsf{O} : \sigma^{\mathsf{T}}\,; b_2)). \tag{46}$$

This definition of dilation corresponds to that of [6], restricted to the binary case (rather than the grey level case), but generalised to directed graphs (rather than undirected graphs). The terminology *roots* and *buds* is that of Heijmans et al., who use it for the structuring graph (G_2 in (46)). To keep things simple, we use the same terminology for the source graph (G_1 in (46)); this is possible because we consider only the binary case (in [6], the nodes of the source graph are weighted by grey level values). Images are a special case of graphs: for an image, the underlying discrete grid is a regular undirected graph, the origin is a root and a coordinate point that belongs to the image is a bud.

Thus $G_1 \triangleright G_2$ is the same graph as G_1, except for the buds. This definition can be understood as follows: if the roots of the embedding of G_2 by σ have a nonempty intersection with the buds of G_1, then the buds of the embedding of G_2 are buds of $G_1 \triangleright G_2$. We now transform this intuitive expression to a form that will lead to an easy calculation of the residuals.

$$
\begin{array}{ll}
& (\bigcup\sigma \mid \mathsf{ih}(\sigma, R_2, R_1) \wedge \sigma^\mathsf{T} ; r_2 \cap b_1 \neq \mathsf{O} : \sigma^\mathsf{T} ; b_2) \\
= & \qquad \langle \text{Tarski rule (7) \& Boolean algebra} \rangle \\
& (\bigcup\sigma \mid \mathsf{ih}(\sigma, R_2, R_1) : \mathsf{L} ; (\sigma^\mathsf{T} ; r_2 \cap b_1) ; \mathsf{L} \cap \sigma^\mathsf{T} ; b_2) \\
= & \qquad \langle r_2 \text{ and } b_1 \text{ are vectors \& (4) \& (3)} \rangle \\
& (\bigcup\sigma \mid \mathsf{ih}(\sigma, R_2, R_1) : \mathsf{L} ; (\sigma^\mathsf{T} ; r_2 \cap b_1) \cap \sigma^\mathsf{T} ; b_2) \\
= & \qquad \langle r_2 \text{ is a vector \& (5) \& (6)} \rangle \\
& (\bigcup\sigma \mid \mathsf{ih}(\sigma, R_2, R_1) : r_2^\mathsf{T} ; \sigma ; b_1 \cap \sigma^\mathsf{T} ; b_2) \\
= & \qquad \langle r_2 \text{ and } b_2 \text{ are vectors \& (3)} \rangle \\
& (\bigcup\sigma \mid \mathsf{ih}(\sigma, R_2, R_1) : \sigma^\mathsf{T} ; b_2 ; r_2^\mathsf{T} ; \sigma ; b_1)
\end{array}
$$

By this derivation and (46),

$$
G_1 \triangleright G_2 = (V_1,\, R_1,\, r_1,\, (\bigcup\sigma \mid \mathsf{ih}(\sigma, R_2, R_1) : \sigma^\mathsf{T} ; b_2 ; r_2^\mathsf{T} ; \sigma ; b_1)). \qquad (47)
$$

The subexpression $r_2^\mathsf{T};\sigma;b_1$ is equal to O if the roots of the embedded G_2 have an empty intersection with the buds of G_1; otherwise, by the Tarski rule (7) and the fact that r_2 and b_1 are vectors, it is equal to L and then $\sigma^\mathsf{T};b_2;r_2^\mathsf{T};\sigma;b_1 = \sigma^\mathsf{T};b_2$, so things are like with (46).

Then,

$$
\begin{array}{ll}
& G_1 \triangleright G_2 \sqsubseteq G_3 \\
\Leftrightarrow & \qquad \langle (47) \text{ \& } (45) \rangle \\
& V_1 = V_3 \wedge R_1 = R_3 \wedge r_1 = r_3 \wedge (\bigcup\sigma \mid \mathsf{ih}(\sigma, R_2, R_1) : \sigma^\mathsf{T};b_2;r_2^\mathsf{T};\sigma;b_1) \subseteq b_3.
\end{array}
$$

We continue with the expression for buds only and first aim at the left residual $G_3 \not\triangleright G_2$.

$$
\begin{array}{ll}
& (\bigcup\sigma \mid \mathsf{ih}(\sigma, R_2, R_1) : \sigma^\mathsf{T} ; b_2 ; r_2^\mathsf{T} ; \sigma ; b_1) \subseteq b_3 \\
\Leftrightarrow & \qquad \langle \text{Definition of } \cup \rangle \\
& (\forall\sigma \mid \mathsf{ih}(\sigma, R_2, R_3) : \sigma^\mathsf{T} ; b_2 ; r_2^\mathsf{T} ; \sigma ; b_1 \subseteq b_3) \\
\Leftrightarrow & \qquad \langle (8) \rangle
\end{array}
$$

$$
\begin{aligned}
& (\forall\sigma \mid \mathsf{ih}(\sigma, R_2, R_3) : b_1 \subseteq (\sigma^\mathsf{T} ; b_2 ; r_2^\mathsf{T} ; \sigma)\backslash b_3) \\
\Leftrightarrow \quad & \qquad \langle \text{Definition of } \cap \rangle \\
& b_1 \subseteq (\textstyle\bigcap\sigma \mid \mathsf{ih}(\sigma, R_2, R_3) : (\sigma^\mathsf{T} ; b_2 ; r_2^\mathsf{T} ; \sigma)\backslash b_3) \\
\Leftrightarrow \quad & \qquad \langle (9) \ \ \& \ \ (6) \rangle \\
& b_1 \subseteq (\textstyle\bigcap\sigma \mid \mathsf{ih}(\sigma, R_2, R_3) : \overline{\sigma^\mathsf{T} ; r_2 ; b_2^\mathsf{T} ; \sigma ; \overline{b_3}}) \\
\Leftrightarrow \quad & \qquad \langle \text{Boolean algebra} \rangle \\
& b_1 \subseteq \overline{(\textstyle\bigcup\sigma \mid \mathsf{ih}(\sigma, R_2, R_3) : \sigma^\mathsf{T} ; r_2 ; b_2^\mathsf{T} ; \sigma ; \overline{b_3})}
\end{aligned}
$$

Since by definition $G_1 \rhd G_2 \sqsubseteq G_3 \Leftrightarrow G_1 \sqsubseteq G_3 \not\rhd G_2$, we obtain from the last two derivations the *left erosion* as the left residual

$$
\begin{aligned}
G_3 \not\rhd G_2 &= (V_3, R_3, r_3, \overline{(\textstyle\bigcup\sigma \mid \mathsf{ih}(\sigma, R_2, R_3) : \sigma^\mathsf{T} ; r_2 ; b_2^\mathsf{T} ; \sigma ; \overline{b_3})}) \\
&= (G_3^- \rhd G_2^{\leftrightarrow})^-,
\end{aligned}
\tag{48}
$$

where (47) is used for the last transformation. Note the similarity with (9).

The calculation of right erosion $G_2 = G_1 \rhd\!\!\!\!\backslash\, G_3$ yields

$$
b_2 \subseteq (\textstyle\bigcap\sigma \mid \mathsf{ih}(\sigma, R_2, R_1) : \sigma ; b_3 \cup \overline{\mathsf{L} ; b_1^\mathsf{T} ; \sigma^\mathsf{T} ; r_2}).
$$

So, the situation is the same as for images, i.e., the buds b_2 of G_2 depend on R_2 and r_2. Comments like those after (41) can be made here. One can fix the geometry by giving V_2, R_2 and r_2, and then calculate the buds b_2, which correspond to the images in the previous section.

Now back to dilation (47). An algorithm for calculating dilation based on (47) may proceed by enumerating all injective homomorphisms σ while taking the join of the $\sigma^\mathsf{T} ; b_2 ; r_2^\mathsf{T} ; \sigma ; b_1$. This is very costly, as the number of combinations grows quickly with the size of the graphs. What follows shows that the search can be made a bit more efficient.

Expressing the vectors b_1 and r_2 as the union of the points they contain, one can rewrite the expression for the buds of $G_1 \rhd G_2$ in (46) as follows:

$$
\begin{aligned}
& (\textstyle\bigcup\sigma \mid \mathsf{ih}(\sigma, R_2, R_1) \wedge \sigma^\mathsf{T} ; r_2 \cap b_1 \neq \mathsf{O} : \sigma^\mathsf{T} ; b_2) \\
={} & (\textstyle\bigcup b_1', r_2' \mid b_1', r_2' \text{ points } \ \wedge b_1' \subseteq b_1 \wedge r_2' \subseteq r_2 : \\
& \qquad (\textstyle\bigcup\sigma \mid \mathsf{ih}(\sigma, R_2, R_1) \wedge \sigma^\mathsf{T} ; r_2' \cap b_1' \neq \mathsf{O} : \sigma^\mathsf{T} ; b_2)).
\end{aligned}
\tag{49}
$$

Now consider the following derivation.

$$
\begin{aligned}
& \mathsf{ih}(\sigma, R_2, R_1) \ \wedge \ \sigma^\mathsf{T} ; r_2' \cap b_1' \neq \mathsf{O} \\
\Leftrightarrow \quad & \qquad \langle (44) \rangle \\
& \sigma ; \sigma^\mathsf{T} = \mathsf{I} \ \wedge \ \sigma^\mathsf{T} ; \sigma \subseteq \mathsf{I} \ \wedge \ \sigma^\mathsf{T} ; R_2 ; \sigma \subseteq R_1 \wedge \sigma^\mathsf{T} ; r_2' \cap b_1' \neq \mathsf{O} \\
\Rightarrow \quad & \qquad \langle (11) \ \ \& \ \ \sigma^\mathsf{T} ; r_2' \text{ is a point by } (12) \ \ \& \ \ (13) \rangle \\
& R_2 ; \sigma \subseteq \sigma ; R_1 \ \wedge \ \sigma^\mathsf{T} ; R_2 \subseteq R_1 ; \sigma^\mathsf{T} \ \wedge \ \sigma^\mathsf{T} ; r_2' \subseteq b_1'
\end{aligned}
$$

$$\Leftrightarrow \qquad \langle (6) \rangle$$

$$R_2 ; \sigma \subseteq \sigma ; R_1 \wedge R_2^{\mathsf{T}} ; \sigma \subseteq \sigma ; R_1^{\mathsf{T}} \wedge r_2'^{\mathsf{T}} ; \sigma \subseteq b_1'^{\mathsf{T}}$$

$$\Leftrightarrow \qquad \langle (8 \text{ \& Boolean algebra} \rangle$$

$$\sigma \subseteq R_2 \backslash (\sigma ; R_1) \cap R_2^{\mathsf{T}} \backslash (\sigma ; R_1^{\mathsf{T}}) \cap r_2'^{\mathsf{T}} \backslash b_1'^{\mathsf{T}}$$

Hence $\mathsf{ih}(\sigma, R_2, R_1) \wedge \sigma^{\mathsf{T}} ; r_2' \cap b_1' \neq \mathsf{O}$ implies that σ is both a forward and a backward simulation, and that the bud b_1' simulates the root r_2'. The largest such simulation, which is thus a superset of all injective homomorphisms σ, can easily be computed as the greatest fixed point of the monotonic function $f(X) := R_2 \backslash (X ; R_1) \cap R_2^{\mathsf{T}} \backslash (X ; R_1^{\mathsf{T}}) \cap r_2'^{\mathsf{T}} \backslash b_1'^{\mathsf{T}}$. From this largest simulation, the injective homomorphisms can then be extracted by looking at all possibilities compatible with their inclusion in the simulation. This is what the RelView algorithm that we have implemented does. Similar comments can be made for erosion.

The execution times for computing the dilation of a sample of graphs are given in Table 2. The size of the graphs is the number of vertices. The times quickly increase with the size of the graphs, which is to be expected for an NP-complete problem (finding an injective homomorphism is the same as finding an isomorphic subgraph) [3]. It remains to be seen whether a better algorithm can be found for RelView.

Table 2. Dilation of G_1 by G_2: CPU times with RelView

	Size of $G_2 = \|V_2\| = 3$							
Size of $R_1 = \|V_1\|$	50	100	150	200	250	300	350	400
Time	0.3	2.7	12	31	66	105	195	244
	Size of $G_2 = \|V_2\| = 4$							
Size of $R_1 = \|V_1\|$	10	20	30	40	50	100	110	120
Time	0.02	0.3	1.4	2.5	10	242	439	656
	Size of $G_2 = \|V_2\| = 5$							
Size of $R_1 = \|V_1\|$	10	20	30	35	40	45	50	55
Time	0.03	4.1	44	71	106	306	504	1085

6 Conclusion

Like other articles, for instance [1], this one shows that RelView is a nice tool for programming functions that can be expressed with relation algebra. It also shows that the computation times are competitive, at least when compared to the well-known commercial environment Mathematica for the computation of dilation and erosion.

There are many possible extensions to this work.

1. With respect to the material of Sect. 4, we intend to explore other geometries, by using different origin and successor relations, and we plan to program more morphological operations with RelView, like closing and opening. These are obtained by composing dilation with erosion, but it may be possible to find optimisations based on relational transformations.
2. The operation of right erosion deserves more investigation, both for images and graphs.
3. The constraint of homomorphism imposed to the graph pattern G_2 in Sect. 5 is very strong. On the other hand, the corresponding constraint for images is very weak, being essentially based on a notion of proximity in a regular grid. There are intermediate possibilities. One that comes to mind after the use of simulations in Sect. 5 is to simply require that a simulation relation exists between the two graphs, rather than a homomorphism.
4. The graphs of Sect. 5, with their roots and buds, are close to automata, except for the fact that their edges are not labelled. We plan to investigate morphological operations applied to automata, with the possibility of extending them to the corresponding languages and to Kleene algebra with tests [9] or Kleene algebra with domain [2].

Acknowledgements. We gratefully acknowledge the input of the anonymous referees and the financial support of NSERC (Natural Sciences and Engineering Research Council of Canada).

References

1. Berghammer, R.: Computing minimal extending sets by relation-algebraic modeling and development. J. Log. Algebr. Methods Program. **83**, 103–119 (2014)
2. Desharnais, J., Möller, B., Struth, G.: Kleene algebra with domain. ACM Trans. Comput. Log. (TOCL) **7**(4), 798–833 (2006)
3. Garey, M.R., Johnson, D.S.: Computers and Intractability: A Guide to the Theory of NP-Completeness. W.H. Freeman and Company, New York (1979)
4. Gries, D., Schneider, F.B.: A Logical Approach to Discrete Math. Springer, New York (1993)
5. Heijmans, H.J.A.M., Ronse, C.: The algebraic basis of mathematical morphology I. Dilations and erosions. Comput. Vis. Graph. Image Process. **50**, 245–295 (1990)
6. Heijmans, H.J.A.M., Nacken, P., Toet, A., Vincent, L.: Graph morphology. J. Visual Commun. Image Represent. **3**(1), 24–38 (1992)
7. Heijmans, H.J.A.M., Vincent, L.: Graph morphology in image analysis. In: Dougherty, E. (ed.) Mathematical Morphology in Image Processing, pp. 171–203. Marcel-Dekker, New York (1992)
8. Klette, R., Rosenfeld, A.: Digital Geometry - Geometric Methods for Digital Picture Analysis. Elsevier, Amsterdam (2004)
9. Kozen, D.: Kleene algebra with tests. ACM Trans. Program. Lang. Syst. **19**(3), 427–443 (1997)
10. Schmidt, G.: Relational Mathematics. Encyclopedia of Mathematics and Its Applications, vol. 132. Cambridge University Press, Cambridge (2010)

11. Schmidt, G., Ströhlein, T.: Relations and Graphs. Springer, New York (1988)
12. Serra, J.: Image Analysis and Mathematical Morphology. Academic Press, London (1982)
13. Stell, J.G.: Relations in mathematical morphology with applications to graphs and rough sets. In: Winter, S., Duckham, M., Kulik, L., Kuipers, B. (eds.) COSIT 2007. LNCS, vol. 4736, pp. 438–454. Springer, Heidelberg (2007). doi:10.1007/978-3-540-74788-8_27
14. Stell, J.G.: Relations on hypergraphs. In: Kahl, W., Griffin, T.G. (eds.) RAMICS 2012. LNCS, vol. 7560, pp. 326–341. Springer, Heidelberg (2012). doi:10.1007/978-3-642-33314-9_22
15. Data. http://www2.ift.ulaval.ca/~desharnais/RAMiCS2017/RAMiCS2017.zip
16. Matlab homepage. https://www.mathworks.com/products/matlab.html
17. Mathematica homepage. https://www.wolfram.com/mathematica/
18. RelView homepage. http://www.informatik.uni-kiel.de/~progsys/relview/

Tool-Based Relational Investigation of Closure-Interior Relatives for Finite Topological Spaces

Rudolf Berghammer(✉)

Institut für Informatik, Universität Kiel, 24098 Kiel, Germany
rub@informatik.uni-kiel.de

Abstract. In a topological space $(X, \mathcal{T})$ at most 7 distinct sets can be constructed from a set $A \in 2^X$ by successive applications of the closure and interior operation in any order. If sets so constructed are called closure-interior relatives of A, then for each topological space $(X, \mathcal{T})$ with $|X| \geq 7$ there exists a set with 7 closure-interior relatives; for $|X| < 7$, however, 7 closure-interior relatives of a set cannot co-exist. Using relation algebra and the RelView tool we compute all closure-interior relatives for all topological spaces with less than 7 points. From these results we obtain that for all finite topological spaces $(X, \mathcal{T})$ the maximum number of closure-interior relatives of a set is $|X|$, with one exception: For the indiscrete topology $\mathcal{T} = \{\emptyset, X\}$ on a set X with $|X| = 2$ there exist two sets which possess $|X| + 1$ closure-interior relatives.

1 Introduction

Systematic experiments are an accepted means for doing science and they become increasingly important as one proceeds in investigations. Meanwhile they have also become important in formal sciences, like mathematics and theoretical computer science. In this context tool support is indispensable. Typical applications of tools are theorem proving, (numerical or algebraic) computations, the (random) generation of inputs for experiments, the visualisation of the computed results and the construction of appropriate graphical displays that illustrate decisive principles and concepts underlying a problem solution. Use is made of general computer algebra systems, but also of systems for specific applications.

RelView (cf. [5,20]) is such a *specific purpose computer algebra system* for the manipulation and visualisation of (binary) relations, relational prototyping and relational programming. Computational tasks can be described by short and concise programs, which frequently consist of only a few lines that present the relation-algebraic expressions or formulae of the notions in question. Such programs are easy to alter in case of slightly changed specifications. Combining this with RelView's possibilities for visualisation, animation and the random generation of relations makes the tool very useful for systematic experiments and scientific research. Another advantage of the tool is its very efficient implementation of relations via BDDs (binary decision diagrams). It leads to an amazing

P. Höfner et al. (Eds.): RAMiCS 2017, LNCS 10226, pp. 60–76, 2017.
DOI: 10.1007/978-3-319-57418-9_4

computational power as, for example, demonstrated in [4,5,7], and allows to experiment also with very large relations.

We have successfully combined relation algebra and RelView for many years and in rather different areas. In this paper we continue the relational treatment of topology, i.e., the work of [8,17,18], with a new application. Using relation algebra and the RelView tool we compute for all topological spaces $(X, \mathcal{T})$ with less than 7 points and all sets $A \in 2^X$ the set $CI(A)$ of the so-called closure-interior relatives of A, where the latter sets are obtained by starting with A and applying the operations of closure and interior in any order and any number of times. The computation of the sets $CI(A)$ is motivated by a variant of the well-known Kuratowski closure-complement theorem (cf. [14]). This variant is stated in [9] and says that for all topological spaces $(X, \mathcal{T})$ and all sets $A \in 2^X$ it holds $|CI(A)| \leq 7$. A further motivation are results of [12,15] concerning this variant. In [12] a finite topological space with 7 points is presented that allows 7 closure-interior relatives of a set and from [12,15] it follows that in case of less than 7 points 7 closure-interior relatives of a set cannot co-exist. So, it is natural to ask in case of less than 7 points for the maximum number of closure-interior relatives a set can produce. The answer to this question is given by the numerical data we will present in Sect. 2. In combination with the results of [9,12,15] our data imply that for all finite and non-empty sets X, all topologies $\mathcal{T}$ on X and all sets $A \in 2^X$ it holds $|CI(A)| \leq |X|$, apart from one case. Precisely for $|X| = 2$ and the indiscrete topology $\{\emptyset, X\}$ there exist two sets $A \in 2^X$ with $|CI(A)| = |X| + 1$. The part of the paper following Sect. 2 consists in the relation-algebraic preliminaries (Sect. 3), a short description of RelView and the preparation of some basic code (Sect. 4), the development of the RelView-program for the computation of the numerical data (Sect. 5), applications of modifications of this program (Sect. 6) and some concluding remarks (Sect. 7).

2 Problem Background and the Numerical Results

We assume the reader to be familiar with the basic notions and facts of topology. Otherwise we refer to [13], for example. If topological spaces are defined via open sets, then a subset $\mathcal{T}$ of the powerset 2^X of a non-empty set X is called an (open sets) *topology* on X and $(X, \mathcal{T})$ is called a *topological space* if $\emptyset \in \mathcal{T}$, $X \in \mathcal{T}$, any union $\bigcup \mathcal{X}$ of an arbitrary subset $\mathcal{X}$ of $\mathcal{T}$ is in $\mathcal{T}$ and any intersection $\bigcap \mathcal{X}$ of a finite, non-empty subset $\mathcal{X}$ of $\mathcal{T}$ is in $\mathcal{T}$. The sets of the topology $\mathcal{T}$ are called *open* and a set $C \in 2^X$ is called *closed* if its *complement* $C^c := X \setminus C$ is open.

To formulate the *closure-complement theorem* of [14] and the variant of [9] (the *closure-interior theorem*), besides the complement A^c we need the *interior* $A^\circ := \bigcup\{O \in \mathcal{T} \mid O \subseteq A\}$ and *closure* $A^- := \bigcap\{C \in 2^X \mid A \subseteq C \wedge C \text{ closed}\}$ of a set $A \in 2^X$ with respect to a topological space $(X, \mathcal{T})$. Furthermore, we need for a set $A \in 2^X$ the subset $CI(A)$ of 2^X, inductively defined by the rules (i) $A \in CI(A)$ and (ii) $B^- \in CI(A)$ and $B^\circ \in CI(A)$ for all $B \in CI(A)$, and the subset $CK(A)$ of 2^X, inductively defined by the rules (i) $A \in CK(A)$ and (ii) $B^- \in CK(A)$ and $B^c \in CK(A)$ for all $B \in CK(A)$. The sets in $CI(A)$ are called the *closure-interior relatives* of A and those in $CK(A)$ are called the *closure-complement relatives* of A – all with respect to $(X, \mathcal{T})$. From the well-known rules $A^{\circ\circ} = A^{\circ}$, $A^{--} = A^{-}$, $A^{\circ-\circ-} = A^{\circ-}$ and $A^{-\circ-\circ} = A^{-\circ}$ we get

$$CI(A) = \{A, A^{\circ}, A^{-}, A^{-\circ}, A^{\circ-}, A^{\circ-\circ}, A^{-\circ-}\} \tag{1}$$

and (1) shows the closure-interior theorem of [9], i.e., the following result:

Theorem 2.1 [9]. *For all topological spaces $(X, \mathcal{T})$ and all sets $A \in 2^X$ it holds $|CI(A)| \leq 7$.*

Using $A^{cc} = A$ and the well-known rules $A^{\circ} = A^{c-c}$, $A^{\circ c} = A^{c-}$ and $A^{-c} = A^{c\circ}$, which connect complements, closures and interiors, we, furthermore, obtain

$$CK(A) = CI(A) \cup \{B^c \mid B \in CI(A)\} \tag{2}$$

and (2) and Theorem 2.1 imply the closure-complement theorem of [14], i.e.:

Theorem 2.2 [14]. *For all topological spaces $(X, \mathcal{T})$ and all sets $A \in 2^X$ it holds $|CK(A)| \leq 14$.*

The bounds of both theorems are sharp. For Theorem 2.2 this is already shown in [14] by means of the standard topology on the set of real numbers. In [12] a finite topology $\mathcal{T}$ on a set X with $|X| = 7$ and a 3-point subset A of X are presented such that $|CI(A)| = 7$ and $|CK(A)| = 14$. In this paper it is also shown that for all finite topological spaces $(X, \mathcal{T})$ and all sets $A \in 2^X$ it holds

$$|X| < 7 \implies |CK(A)| < 14. \tag{3}$$

That for all (i.e., also possibly infinite) topological spaces $(X, \mathcal{T})$ and all sets $A \in 2^X$ it holds

$$|CI(A)| = 7 \iff |CK(A)| = 14 \tag{4}$$

is mentioned in [3] and proved in [15]. After these preparations we are able to prove the following analogon of (3) for closure-interior relatives.

Theorem 2.3. *For all finite topological spaces $(X, \mathcal{T})$ with $|X| < 7$ and all sets $A \in 2^X$ it holds $|CI(A)| < 7$.*

Proof. We use contradiction and assume to the contrary that there exist a finite topological space $(X, \mathcal{T})$ with $|X| < 7$ and a set $A \in 2^X$ with $|CI(A)| \geq 7$. Then Theorem 2.1 shows $|CI(A)| = 7$ such that $|CK(A)| = 14$ due to (4). But $|X| < 7$ and $|CK(A)| = 14$ contradict (3). □

Property (3) and Theorem 2.3 give raise to the question how large the two sets $CK(A)$ and $CI(A)$ actually may get in dependence on the number of points. For the sets $CK(A)$ the following answer is given in [2]: If $(X, \mathcal{T})$ is a finite topological space with $|X| \leq 7$, then for all sets $A \in 2^X$ it holds $|CK(A)| \leq 2|X|$. To answer the question also for the sets $CI(A)$, we have used the RelView tool and checked for all finite topological spaces $(X, \mathcal{T})$ with $1 \leq |X| \leq 6$ whether there exists a set $A \in 2^X$ with $|CI(A)| = k$, where k ranges from 1 to 7. The development of the program (which also shows its correctness) is postponed to Sect. 5, but the computed data are already presented in the following table. Here the entry for $|X| = m$ and $k = n$ equals the number of topologies $\mathcal{T}$ on an m-point set X such that – relative to the topological space $(X, \mathcal{T})$ – there exists a set $A \in 2^X$ with $|CI(A)| = n$. According to the notions used in [3,11], we call such a set A an *n-CI-set with respect to* $(X, \mathcal{T})$. The numbers in the column marked with $k = 1$ coincide with the numbers of all possible topologies on X (cf. e.g., with the data of [10]). This follows from the fact that $\emptyset^- = \emptyset^\circ = \emptyset$ and $X^- = X^\circ = X$ for all topological spaces $(X, \mathcal{T})$, which leads to $\emptyset$ and X as 1-CI-sets.

	$k = 1$	2	3	4	5	6	7
$\lvert X\rvert = 1$	1	0	0	0	0	0	0
2	4	2	**1**	0	0	0	0
3	29	24	16	0	0	0	0
4	355	340	286	84	0	0	0
5	6 942	6 890	6 581	3 385	420	0	0
6	209 527	209 324	207 594	138 090	35 490	1 440	0

We also have computed the number of topologies on a 7-point set X such that there exists a 7-CI-set. According to RelView, for 10 080 of the 9 535 241 topologies on X a 7-CI-set exists. With the help of the numerical data of the table we now can prove the following analogon of the result of [2] for the sets $CI(A)$.

Theorem 2.4. *For all topological spaces $(X, \mathcal{T})$ we have (i) $|CI(A)| \leq |X| + 1$ for all sets $A \in 2^X$ and (ii) that there exists a set $A \in 2^X$ with $|CI(A)| = |X| + 1$ if and only if $|X| = 2$ and $\mathcal{T}$ is the indiscrete topology $\{\emptyset, X\}$.*

Proof. For $|X| < 7$ the first claim follows from the data of the above table and for $|X| \geq 7$ it follows from Theorem 2.1.

To prove the direction "$\Rightarrow$" of the second claim, assume that there exists a set $A \in 2^X$ with $|CI(A)| = |X| + 1$. Then Theorem 2.1 yields $|X| \leq 6$. The data of the above table show that there exists exactly one topological space with less than 7 points such that the assumption holds. From the boldface entry we get $|X| = 2$. It is easy to check that in this case the only topology on X for which there exists a 3-CI-set is the indiscrete topology $\{\emptyset, X\}$. Here both singleton sets of points produce 3 closure-interior relatives (whereas $CI(\emptyset) = \{\emptyset\}$ and $CI(X) = \{X\}$). From this, also direction "$\Leftarrow$" of the second claim follows. □

If we call a set $A \in 2^X$ such that $|CK(A)| = n$ an *n-CK-set with respect to* the topological space $(X, \mathcal{T})$, then (4) says that, given any topological space, for $n := 7$ a set is a n-CI-set if and only if it is a $2n$-CK-set. Theorem 2.4 shows that the result does not generalise to all n from $\{1, \dots, 7\}$.

3 Relation-Algebraic Preliminaries

Given sets X and Y, we denote the set of all relations with *source* X and *target* Y (the powerset $2^{X \times Y}$) by $[X \leftrightarrow Y]$ and write $R : X \leftrightarrow Y$ instead of $R \in [X \leftrightarrow Y]$. If the sets X and Y of the *type* $X \leftrightarrow Y$ of R are finite (and linearly ordered), we may consider R as a Boolean $|X| \times |Y|$ matrix. Such an interpretation is also used as one of the graphical representations of relations by RelView. Thus, in the following we often use Boolean matrix terminology and notation. In particular, we speak of 0- and 1-entries, of rows and columns and we write $R_{x,y}$ instead of $x\,R\,y$ or $(x, y) \in R$ to express relationships. We will employ the following five basic operations on relations: $\overline{R}$ (*complement*), $R \cup S$ (*union*), $R \cap S$ (*intersection*), R^{T} (*transposition*) and $R;S$ (*composition*). We also will use the constants O (*empty relation*), L (*universal relation*) and I (*identity relation*). Here we overload the symbols, i.e., avoid the binding of types to them. Furthermore, we will use the tests $R \subseteq S$ (*inclusion*) and $R = S$ (*equality*). We assume the reader to be familiar with these concepts; otherwise we refer to [16]. For more details on the notions we introduce in the following, we also refer to [16].

By $syq(R, S) := \overline{R^{\mathsf{T}};\overline{S}} \cap \overline{\overline{R^{\mathsf{T}}};S}$ the *symmetric quotient* of $R : X \leftrightarrow Y$ and $S : X \leftrightarrow Z$ is defined. From this specification we get the typing $syq(R, S) : Y \leftrightarrow Z$ and, given $y \in Y$ and $z \in Z$, also that $syq(R, S)_{y,z}$ if and only if for all $x \in X$ it holds $R_{x,y}$ if and only if $S_{x,z}$. Since the latter description uses relationships between elements, it is called an *element-wise description*.

Restricting the common notion of a relational vector (as used e.g., in [16]) slightly, in the present paper a *vector* is a relation v with the specific set $\mathbf{1} := \{\bot\}$ as target. Since in a relationship $v_{x,\bot}$ the second index $\bot$ is irrelevant, we write in the following v_x instead of $v_{x,\bot}$. In the matrix model vectors correspond to Boolean column vectors. We say that $v : X \leftrightarrow \mathbf{1}$ *models* (or is the *vector-model* of) the subset V of X if $x \in V$ and v_x are equivalent, for all $x \in X$. In such a case $inj(v) : V \leftrightarrow X$ denotes the *embedding relation* of V into X induced by v, that is, the identity function from V to X regarded as a relation of type $V \leftrightarrow X$. Using an element-wise description we, therefore, get $inj(v)_{y,x}$ if and only if $y = x$, for all $y \in V$ and $x \in X$. A (relational) *point* is a vector $p : X \leftrightarrow \mathbf{1}$ such that $p \neq \mathsf{O}$ and $p;p^{\mathsf{T}} \subseteq \mathsf{I}$ (i.e., p is a non-empty and injective vector). It is easy to see that then p models a singleton subset of X and it corresponds to a Boolean column vector with exactly one 1-entry. If $\{x\}$ is modeled by p, then we say that *p models the element $x \in X$*. Hence, the point $p : X \leftrightarrow \mathbf{1}$ models the element $x \in X$ if and only if for all $y \in X$ it holds p_y if and only if $x = y$. To model subsets of powersets, we also will use *membership relations* $\mathsf{M} : X \leftrightarrow 2^X$, element-wisely described by $\mathsf{M}_{x,Y}$ if and only if $x \in Y$, for all $x \in X$ and $Y \in 2^X$. If $v : 2^X \leftrightarrow \mathbf{1}$ is the vector-model of a subset $\mathcal{V}$ of 2^X and we define $R : X \leftrightarrow \mathcal{V}$ by $R := \mathsf{M};inj(v)^{\mathsf{T}}$, then we get $R_{x,V}$ if and only if $x \in V$, for all $x \in X$ and $V \in \mathcal{V}$. By reason of this element-wise description, a relation $R : X \leftrightarrow \mathcal{V}$ is called the *membership-model* of a subset $\mathcal{V}$ of 2^X if for all $x \in X$ and $V \in \mathcal{V}$ it holds $R_{x,V}$ if and only if $x \in V$. In Boolean matrix terminology this means that each set of $\mathcal{V}$ is modeled by exactly one column of R. To go back from the membership-model $R : X \leftrightarrow \mathcal{V}$ of $\mathcal{V}$ to the vector-model $v : 2^X \leftrightarrow \mathbf{1}$ of $\mathcal{V}$, we use the symmetric quotient construction, since its element-wise description yields $v = \mathit{syq}(\mathsf{M}, R);\mathsf{L}$, where $\mathsf{L} : \mathcal{V} \leftrightarrow \mathbf{1}$.

For each direct product of sets there exist the two projection functions which decompose a pair $u = (u_1, u_2)$ into its first component u_1 and its second component u_2. When working in a relational context, it is useful to consider instead of these functions the corresponding two *projection relations* $\pi : X \times Y \leftrightarrow X$ and $\rho : X \times Y \leftrightarrow Y$, element-wisely described by $\pi_{(u_1,u_2),x}$ if and only if $u_1 = x$ and $\rho_{(u_1,u_2),y}$ if and only if $u_2 = y$, for all $(u_1, u_2) \in X \times Y$, $x \in X$ and $y \in Y$. The projection relations enable us to specify the well-known pairing operation of functional programming in two versions. In this paper we only need one of them. The *left-pairing* $[\![R, S]\!] : X \times Y \leftrightarrow Z$ of the relations $R : X \leftrightarrow Z$ and $S : Y \leftrightarrow Z$ is given by $[\![R, S]\!] := \pi;R \cap \rho;S$, where $\pi : X \times Y \leftrightarrow X$ and $\rho : X \times Y \leftrightarrow Y$ are as above. Element-wisely this definition means that $[\![R, S]\!]_{(u_1,u_2),z}$ if and only if $R_{u_1,z}$ and $S_{u_2,z}$, for all $(u_1, u_2) \in X \times Y$ and $z \in Z$. Projection relations and left-pairing also allow us to define a Boolean algebra isomorphism

$$\mathit{vec} : [X \leftrightarrow Y] \to [X \times Y \leftrightarrow \mathbf{1}] \qquad \mathit{Rel} : [X \times Y \leftrightarrow \mathbf{1}] \to [X \leftrightarrow Y] \tag{5}$$

between $([X \leftrightarrow Y], \cup, \cap, \overline{})$ and $([X \times Y \leftrightarrow \mathbf{1}], \cup, \cap, \overline{})$ by $\mathit{vec}(R) = [\![R, \mathsf{I}]\!];\mathsf{L}$, where $\mathsf{I} : Y \leftrightarrow Y$ and $\mathsf{L} : Y \leftrightarrow \mathbf{1}$, and $\mathit{Rel}(v) = \pi^{\mathsf{T}};(\rho \cap v;\mathsf{L})$, where $\pi : X \times Y \leftrightarrow X$ and $\rho : X \times Y \leftrightarrow Y$ are the projection relations of the direct product $X \times Y$ and $\mathsf{L} : \mathbf{1} \leftrightarrow Y$. Then the element-wise description of the left-pairing operation yields $\mathit{vec}(R)_{(x,y)}$ if and only if $R_{x,y}$, for all $R : X \leftrightarrow Y$, $x \in X$ and $y \in Y$. For the inverse function *Rel* the element-wise descriptions of the projection relations show $\mathit{Rel}(v)_{x,y}$ if and only if $v_{(x,y)}$, for all $v : X \times Y \leftrightarrow \mathbf{1}$, $x \in X$ and $y \in Y$.

The disjoint union of sets leads to the two *injection relations* $\imath : X \leftrightarrow X \uplus Y$ and $\kappa : Y \leftrightarrow X \uplus Y$, which allow the definition of two relational sums. In RelView only one is implemented. It takes $R : X \leftrightarrow Z$ and $S : Y \leftrightarrow Z$ and yields the relation $R + S := \imath^{\mathsf{T}};R \cup \kappa^{\mathsf{T}};S : X \uplus Y \leftrightarrow Z$. As the tool enumerates a disjoint union $X \uplus Y$ of two (finite) sets by listing the elements of X in front of those of Y, the Boolean matrix of $R + S$ is obtained by putting the matrix of R on top of the matrix of S. Hence, in case of a relation $R : X \leftrightarrow Y$ and a vector $v : Y \leftrightarrow \mathbf{1}$ the concatenation operation $\mathit{conc}(R, v) := (R^{\mathsf{T}} + v^{\mathsf{T}})^{\mathsf{T}}$ adds the Boolean vector at the right of the last column of R. In Sect. 5 we will use an obvious generalisation that concatenates 7 vectors to a single relation with 7 columns.

4 The RelView Tool and Some Basic Code

Continuing work at the University of the German Forces, since 1993 we develop at Kiel University a specific purpose computer algebra system for the manipulation and visualisation of relations, relational prototyping and relational programming, called RelView. The tool is written in the C programming language, uses reduced ordered BDDs for implementing relations and makes full use of a graphical user interface. Details can be found, e.g., in [4,5,20]. Via [20] also the newest version of the tool is available (Version 8.2, released January 2016).

The main purpose of the RelView tool is the evaluation of relation-algebraic expressions. These are constructed from the relations of its workspace using pre-defined operations and tests (including those presented in Sect. 3), user-defined relational functions and user-defined relational programs.

In the programming language of RelView a relational function is defined as it is customary in mathematics, i.e., as $F(R_1, \ldots, R_n) = E$, where F is the function name, $R_1, \ldots, R_n$ are the parameters (standing for the input relations) and E is a relation-algebraic expression that specifies how the result is computed from the input. E.g., the definition of the function $vec : [X \leftrightarrow Y] \to [X \times Y \leftrightarrow \mathbf{1}]$ of (5) immediately leads to the following relational function:

```
vec(R) = dom([|R,I(R^*R)]).
```

In this RelView code *dom* and I are pre-defined RelView operations. Assuming $R : X \leftrightarrow Y$, a comparison of the specification $vec(R)$ in Sect. 3 and the above relational function *vec* shows that a call of *dom* composes the argument from the right with a universal vector of appropriate type and a call of I computes an identity relation with the same type as the argument. In RelView the symbol "$*$" denotes the composition operation and the symbol "^" denotes the transposition operation. So, the argument of I is the relation $R^{\mathsf{T}};R$, which implies that the second argument of the left-pairing of the relational function is the identity relation of type $Y \leftrightarrow Y$, exactly as in the specification $vec(R)$.

A relational program in RelView is much like a function procedure in conventional programming languages, except that it uses only relations as data type and is not able to modify the workspace (i.e., applications are free of side-effects). It starts with a head line containing the program name and the list of parameters, which again stand for relations. Then the declarations of the local relational domains and functions and of the variables follow. Domain declarations for direct products can be used to introduce projection relations and those for disjoint unions can be used to introduce injection relations. The third part of a relational program is the body, a while-program over relations. As a program computes a value, finally, its last part consists of a return-clause, which is a relation-algebraic expression whose value after the execution of the body is the result.

In contrast with the definition of the function *vec* of (5), in the definition of its inverse, the function $Rel : [X \times Y \leftrightarrow \mathbf{1}] \to [X \leftrightarrow Y]$ of (5), the projection relations $\pi : X \times Y \leftrightarrow X$ and $\rho : X \times Y \leftrightarrow Y$ of the direct product $X \times Y$ are explicitly used. Therefore, a RelView-version of *Rel* requires the use of a relational program in which a product domain for $X \times Y$ has to be declared. From the latter the projection relations π and ρ may then be obtained with the help of the pre-defined RelView operations *p-1* and *p-2* and two assignments that store the results of their applications in local variables, as shown in the following relational program *Rel* (where the symbol "&" denotes RelView's intersection operation):

```
Rel(v,S)
  DECL XY = PROD(S*S^,S^*S);
       pi, rho
  BEG  pi = p-1(XY); rho = p-2(XY)
       RETURN pi^*(rho & v*L1n(S))
  END.
```

RelView is not able to derive from the type $X \times Y \leftrightarrow \mathbf{1}$ of the input v the two sets X and Y of the direct product $X \times Y$. These, however, are necessary for the declaration of the product domain XY for $X \times Y$ and have to be provided in the form of two relations of type $X \leftrightarrow X$ resp. $Y \leftrightarrow Y$. In the program *Rel* the auxiliary input $S : X \leftrightarrow Y$ allows to construct such relations as $S;S^{\mathsf{T}} : X \leftrightarrow X$ and $S^{\mathsf{T}};S : Y \leftrightarrow Y$. With the help of S and the pre-defined RelView operation *L1n* also the transposed vector $\mathsf{L} : \mathbf{1} \leftrightarrow Y$ of the specification *Rel*(v) is obtained.

In view of the relational program we will develop in Sect. 5, now we develop a specification *ToRel*(p) $: X \leftrightarrow X$, that computes for a point $p : [X \leftrightarrow X] \leftrightarrow \mathbf{1}$ the relation modeled by p. Assume $R : X \leftrightarrow X$ to be the latter. With the help of the membership relation $\mathsf{M} : X^2 \leftrightarrow [X \leftrightarrow X]$ we then obtain for all $x, y \in X$:

$$
\begin{aligned}
\mathit{ToRel}(p)_{x,y} &\iff R_{x,y} && \text{assumption} \\
&\iff \mathsf{M}_{(x,y),R} && \text{element-wise descr. } \mathsf{M} \\
&\iff \exists S \in [X \leftrightarrow Y] : \mathsf{M}_{(x,y),S} \wedge R = S \\
&\iff \exists S \in [X \leftrightarrow Y] : \mathsf{M}_{(x,y),S} \wedge p_S && p \text{ models } R \\
&\iff (\mathsf{M};p)_{(x,y)} \\
&\iff \mathit{Rel}(\mathsf{M};p)_{x,y} && \text{element-wise descr. } \mathit{Rel}
\end{aligned}
$$

From this result we get the specification *ToRel*(p) $=$ *Rel*(M;p) and a translation into a relational program *ToRel* looks as follows:

```
ToRel(p,S)
  DECL XX = PROD(S,S);
       pi, M
  BEG  pi = p-1(XX); M = epsi(pi)
       RETURN Rel(M*p,S)
  END.
```

Here the pre-defined RelView operation *epsi* computes for an arbitrary relation of type $Y \leftrightarrow Z$ the membership relation $\mathsf{M} : Y \leftrightarrow 2^Y$. The second (auxiliary) input $S : X \leftrightarrow X$ of the program *ToRel* is used at two places. In the declaration of the product domain XX for X^2 it provides the carrier set X. The first projection π of the direct product X^2 then is used to obtain via its source and the operation *epsi* the membership relation $\mathsf{M} : X^2 \leftrightarrow [X \leftrightarrow X]$. Furthermore, in the call of the relational program *Rel* the input S provides the type for the result.

In Sect. 5 we will also use the following relational program *clReps*, where the input $R : X \leftrightarrow X$ is assumed to be an equivalence relation:

```
clReps(R)
  DECL v
  BEG  v = point(dom(R));
       WHILE -eq(R*v,dom(R)) DO
         v = v | point(-(R*v) & dom(R)) OD
       RETURN v
  END.
```

Here the symbol "-" denotes the complement operation, the symbol "|" denotes the union operation, the pre-defined RelView operation *eq* tests the equality of relations (such that *-eq* tests non-equality) and the pre-defined RelView operation *point* yields for a non-empty vector a point that is contained in the vector, i.e., a point that models an element of the set modeled by the vector. By combining relation algebra and assertion-based program verification, in [6] it is shown that the output vector $v : X \leftrightarrow \mathbf{1}$ of *clReps* models a complete set of representatives for all equivalence classes of R, i.e., a set which contains for each equivalence class of R exactly one representative.

5 Computation of the Numerical Data via RelView

The development (including the correctness proof) of the relational program for computing the numerical data of Sect. 2, that we will present in this section, is based upon the membership-models for topologies. Such a modeling easily allows to compute interiors and closures with relation-algebraic means. Given $T : X \leftrightarrow \mathcal{T}$ as the membership-model of a topology $\mathcal{T}$ on the set X and $a : X \leftrightarrow \mathbf{1}$ as the vector-model of the set $A \in 2^X$, we can calculate the equivalence

$$\begin{aligned}
x \in A^\circ &\iff x \in \bigcup\{O \in \mathcal{T} \mid O \subseteq A\} && \text{definition interior}\\
&\iff \exists O \in \mathcal{T} : O \subseteq A \wedge x \in O\\
&\iff \exists O \in \mathcal{T} : x \in O \wedge \forall y \in O : y \in A\\
&\iff \exists O \in \mathcal{T} : x \in O \wedge \forall y \in X : y \in O \Rightarrow a_y && a \text{ models } A\\
&\iff \exists O \in \mathcal{T} : T_{x,O} \wedge \neg\exists y \in X : T^{\mathsf{T}}_{O,y} \wedge \overline{a}_y && T \text{ membership-model}\\
&\iff \exists O \in \mathcal{T} : T_{x,O} \wedge \overline{T^{\mathsf{T}};\overline{a}}_O\\
&\iff (T;\overline{T^{\mathsf{T}};\overline{a}})_x,
\end{aligned}$$

for all $x \in X$. As a consequence, we obtain that by

$$\mathit{Int}(T,a) := T;\overline{T^{\mathsf{T}};\overline{a}} : X \leftrightarrow \mathbf{1} \tag{6}$$

the vector-model of the interior A° of A is specified. From this, the well-known rule $A^- = A^{coc}$ and the fact that the complement $\overline{v}$ of a vector v models the complement B^c of the set B modeled by v, we obtain

$$\mathit{Clos}(T,a) := \overline{\mathit{Int}(T,\overline{a})} = \overline{T;\overline{T^{\mathsf{T}};a}} : X \leftrightarrow \mathbf{1} \tag{7}$$

as specification of the vector-model of the closure A^- of A. A translation of (6) and (7) into the programming language of RelView is trivial. It leads to the following relational functions *Int* and *Clos*:

```
Int(T,a) = T*-(T^*-a).        Clos(T,a) = -(T*-(T^*a)).
```

A specific property of the right-hand sides of the specifications of $Int(T, a)$ and $Clos(T, a)$ in (6) resp. (7) is that these are constructed from the parameter a using only the Boolean operations on relations and composition from the left with a relation-algebraic expression free of a. This implies that they are *column-wise extendible* with respect to a in the sense of [7] and, hence, a replacement of the vector $a : X \leftrightarrow \mathbf{1}$ by the membership-model $M : X \leftrightarrow \mathcal{M}$ of a subset $\mathcal{M}$ of 2^X leads to specifications $Int(T, M) : X \leftrightarrow \mathcal{M}$ and $Clos(T, M) : X \leftrightarrow \mathcal{M}$ with the property that for all $x \in X$ and sets $A \in \mathcal{M}$ the relationships $M_{x,A}$, $Int(T, M)_{x,A^\circ}$ and $Clos(T, M)_{x,A^-}$ are equivalent. Using matrix terminology this means that, if A is modeled by the nth column of M, where $1 \leq n \leq |\mathcal{M}|$, then A° is modeled by the nth column of $Int(T, M)$ and A^- is modeled by the nth column of $Clos(T, M)$. Especially for M as membership relation $\mathsf{M} : X \leftrightarrow 2^X$ and the 6 relations $I, C, E, F, G, H : X \leftrightarrow 2^X$ defined by

$$\begin{array}{lll} I := Int(T, \mathsf{M}) & E := Int(T, C) & G := Int(T, F) \\ C := Clos(T, \mathsf{M}) & F := Clos(T, I) & H := Clos(T, E) \end{array} \qquad (8)$$

we get for all $n \in \{1, \dots, 2^{|X|}\}$ and sets $A \in 2^X$ the following property: If the nth column of M models the set A, then the nth column of I resp. C, E, F, G and H models the set A° resp. $A^-, A^{-\circ}, A^{\circ-}, A^{\circ-\circ}$ and $A^{-\circ-}$. Together with Eq. (1) this shows for all $k \in \{1, \dots, 7\}$ that A is a k-CI-set if and only if exactly k of the nth columns of $\mathsf{M}, I, C, E, F, G$ and H are pair-wise different. This is the decisive idea behind the program we want to develop.

To give an example, we consider the two-point set $X := \{a, b\}$ and the topology $\mathcal{T}_1 := \{\emptyset, \{a\}, X\}$ on X. We use RelView to compute the membership relation and the 6 relations of (8). The next 7 pictures show, from left to right, the Boolean matrix representations of the relations $\mathsf{M}, I, C, E, F, G$ and H, where a black square means a 1-entry and a white square means a 0-entry.

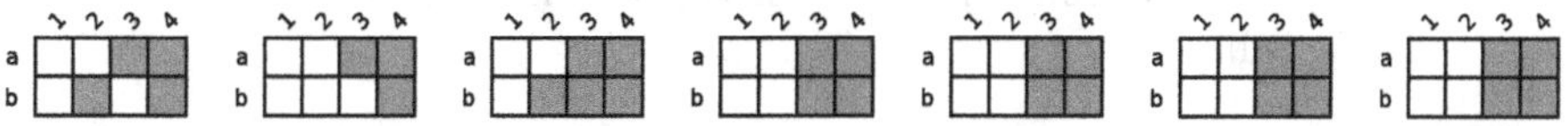

Hence, $I \subset E = F = G = H \subset C$. From the first columns of these 7 Boolean matrices we get $CI(\emptyset) = \{\emptyset\}$, from the second ones $CI(\{b\}) = \{\emptyset, \{b\}\}$, from the third ones $CI(\{a\}) = \{\{a\}, X\}$ and from the last ones $CI(X) = \{X\}$. If we take the indiscrete topology $\mathcal{T}_2 := \{\emptyset, X\}$ on X, then RelView yields the following Boolean matrices for $\mathsf{M}, I, C, E, F, G$ and H:

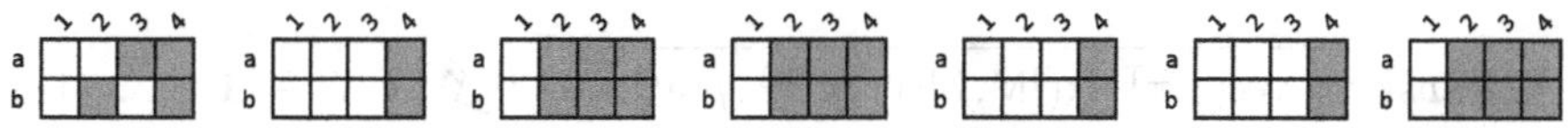

Here we have $I = F = G \subset C = E = H$. Now a column-wise comparison of these 7 Boolean matrices shows $CI(\emptyset) = \{\emptyset\}$, $CI(\{b\}) = \{\emptyset, \{b\}, X\}$, $CI(\{a\}) =$

$\{\emptyset, \{a\}, X\}$ and $CI(X) = \{X\}$. In particular, the singleton sets $\{a\}$ and $\{b\}$ lead to $3 = |X| + 1$ closure-interior relatives, as stated in the proof of Theorem 2.4.

In order to apply the above mentioned idea to all possible topologies on a given finite set X, where $1 \leq |X| \leq 6$, we have to generate for each topology $\mathcal{T}$ on X the respective membership-model $T : X \leftrightarrow \mathcal{T}$. For this purpose we use the well-known fact that (because topologies on finite sets are so-called Alexandroff-topologies) there exists a 1-to-1 correspondence between the set $\mathfrak{P}_X$ of all preorder relations on X and the set $\mathfrak{T}_X$ of all topologies on X; see e.g., [1]. The direction we are interested in is the translation from an arbitrary preorder relation $Q : X \leftrightarrow X$ to the corresponding topology $\mathcal{T}_Q$. Again it is known that $\mathcal{T}_Q$ consists of the lower sets of the preordered set (X, Q). Based on this description, the following calculation shows that $\overline{\mathsf{L};(Q;\mathsf{M} \cap \overline{\mathsf{M}})}^{\mathsf{T}} : 2^X \leftrightarrow \mathbf{1}$ relation-algebraically specifies the vector-model of $\mathcal{T}_Q$, where $A \in 2^X$ is an arbitrary set, $\mathsf{L} : \mathbf{1} \leftrightarrow X$ is a transposed universal vector and $\mathsf{M} : X \leftrightarrow 2^X$ is a membership relation:

$$\begin{aligned}
\overline{\mathsf{L};(Q;\mathsf{M} \cap \overline{\mathsf{M}})}_{\perp,A} &\iff \neg\exists x \in X : \mathsf{L}_{\perp,x} \wedge (Q;\mathsf{M} \cap \overline{\mathsf{M}})_{x,A} \\
&\iff \neg\exists x \in X : (Q;\mathsf{M})_{x,A} \wedge \overline{\mathsf{M}}_{x,A} \\
&\iff \neg\exists x \in X : (\exists y \in X : Q_{x,y} \wedge \mathsf{M}_{y,A}) \wedge \overline{\mathsf{M}}_{x,A} \\
&\iff \neg\exists x, y \in X : Q_{x,y} \wedge y \in A \wedge x \notin A \\
&\iff \forall x, y \in X : y \in A \wedge Q_{x,y} \Rightarrow x \in A
\end{aligned}$$

In the fourth step the element-wise description of M is applied. With the technique of Sect. 3 the vector-model of $\mathcal{T}_Q$ leads to the specification

$$\mathit{ToTop}(Q) := \mathsf{M};\mathit{inj}(\overline{\mathsf{L};(Q;\mathsf{M} \cap \overline{\mathsf{M}})}^{\mathsf{T}})^{\mathsf{T}} : X \leftrightarrow \mathcal{T}_Q \tag{9}$$

of the membership-model of $\mathcal{T}_Q$. The RelView-version of (9) looks as follows:

```
ToTop(Q)
  DECL M
  BEG  M = epsi(Q)
       RETURN M*inj(-(L1n(Q)*(Q*M & -M))^)^
  END.
```

Being able to go from a preorder relation on X to the corresponding Alexandroff-topology, the next task is to model the set $\mathfrak{P}_X$ by a vector. Here we use that

$$\mathit{reflR} := \overline{\mathit{vec}(\mathsf{I})^{\mathsf{T}};\overline{\mathsf{M}}}^{\mathsf{T}} : [X \leftrightarrow X] \leftrightarrow \mathbf{1} \tag{10}$$

specifies the vector-model of the set of all reflexive relations on X and

$$\mathit{transR} := \overline{\mathit{vec}(\rho;\pi^{\mathsf{T}})^{\mathsf{T}};([\![\mathsf{M}, \mathsf{M}]\!] \cap [\![\pi;\pi^{\mathsf{T}}, \rho;\rho^{\mathsf{T}}]\!];\overline{\mathsf{M}})}^{\mathsf{T}} : [X \leftrightarrow X] \leftrightarrow \mathbf{1}, \tag{11}$$

specifies the vector-model of the set of all transitive relations on X, such that $\mathit{reflR} \cap \mathit{transR}$ specifies the vector-model of the set $\mathfrak{P}_X$, i.e., is the specification we are looking for. Both specifications (10) and (11) use the membership relation $\mathsf{M} : X^2 \leftrightarrow [X \leftrightarrow X]$. In (10) additionally $\mathsf{I} : X \leftrightarrow X$ is used and in (11) additionally the projection relations $\pi, \rho : X^2 \leftrightarrow X$ of the direct product X^2 are applied. To prove the specifications (10) and (11) as correct, we assume an arbitrary relation $R : X \leftrightarrow X$. In case of (10) we then calculate as follows:

$$\begin{aligned} \mathit{reflR}_R &\iff \overline{\mathit{vec}(\mathsf{I})^{\mathsf{T}};\overline{\mathsf{M}}}_{\perp,R} \\ &\iff \neg\exists\, u \in X^2 : \mathit{vec}(\mathsf{I})_u \wedge \overline{\mathsf{M}}_{u,R} \\ &\iff \forall\, u \in X^2 : \mathit{vec}(\mathsf{I})_u \Rightarrow \mathsf{M}_{u,R} \\ &\iff \forall\, u \in X^2 : \mathsf{I}_{u_1,u_2} \Rightarrow R_{u_1,u_2} \\ &\iff \forall\, x \in X : R_{x,x} \end{aligned}$$

Here we assume $u = (u_1, u_2)$ and use in the fourth step the element-wise descriptions of M and $\mathit{vec}(\mathsf{I})$. The next calculation proves the correctness of (11):

$$\begin{aligned} \mathit{transR}_R &\iff \overline{\mathit{vec}(\rho;\pi^{\mathsf{T}})^{\mathsf{T}};([\![\mathsf{M},\mathsf{M}]\!] \cap [\![\pi;\pi^{\mathsf{T}},\rho;\rho^{\mathsf{T}}]\!];\overline{\mathsf{M}})}_{\perp,R} \\ &\iff \neg\exists\, u,v \in X^2 : \mathit{vec}(\rho;\pi^{\mathsf{T}})_{(u,v)} \wedge ([\![\mathsf{M},\mathsf{M}]\!] \cap [\![\pi;\pi^{\mathsf{T}},\rho;\rho^{\mathsf{T}}]\!];\overline{\mathsf{M}})_{(u,v),R} \\ &\iff \neg\exists\, u,v \in X^2 : \\ &\qquad \mathit{vec}(\rho;\pi^{\mathsf{T}})_{(u,v)} \wedge [\![\mathsf{M},\mathsf{M}]\!]_{(u,v),R} \wedge ([\![\pi;\pi^{\mathsf{T}},\rho;\rho^{\mathsf{T}}]\!];\overline{\mathsf{M}})_{(u,v),R} \\ &\iff \neg\exists\, u,v \in X^2 : (\rho;\pi^{\mathsf{T}})_{u,v} \wedge \mathsf{M}_{u,R} \wedge \mathsf{M}_{v,R} \wedge \\ &\qquad\qquad \exists\, w \in X^2 : [\![\pi;\pi^{\mathsf{T}},\rho;\rho^{\mathsf{T}}]\!]_{(u,v),w} \wedge \overline{\mathsf{M}}_{w,R} \\ &\iff \neg\exists\, u,v \in X^2 : u_2 = v_1 \wedge R_{u_1,u_2} \wedge R_{v_1,v_2} \wedge \\ &\qquad\qquad \exists\, w \in X^2 : (\pi;\pi^{\mathsf{T}})_{u,w} \wedge (\rho;\rho^{\mathsf{T}})_{v,w} \wedge \neg R_{w_1,w_2} \\ &\iff \neg\exists\, u,v,w \in X^2 : \\ &\qquad u_2 = v_1 \wedge R_{u_1,u_2} \wedge R_{v_1,v_2} \wedge u_1 = w_1 \wedge v_2 = w_2 \wedge \neg R_{w_1,w_2} \\ &\iff \neg\exists\, u,v \in X^2 : u_2 = v_1 \wedge R_{u_1,u_2} \wedge R_{v_1,v_2} \wedge \neg R_{u_1,v_2} \\ &\iff \forall\, u,v \in X^2 : R_{u_1,u_2} \wedge R_{v_1,v_2} \wedge u_2 = v_1 \Rightarrow R_{u_1,v_2} \\ &\iff \forall\, x,y,z \in X : R_{x,y} \wedge R_{y,z} \Rightarrow R_{x,z} \end{aligned}$$

Here we assume $u = (u_1, u_2)$, $v = (v_1, v_2)$ and $w = (w_1, w_2)$, use the element-wise description of $\mathit{vec}(\rho;\pi^{\mathsf{T}})$ in the fourth step, that of left-pairings in the fourth and fifth step, that of the membership relation M in the fifth step and that of the projection relations π and ρ in the fifth and sixth step.

Again it is straightforward to translate (10) and (11) into RelView code. In case of the specification (10) we obtain the following relational program *reflR*:

```
reflR(v)
  DECL XX = PROD(v*v^,v*v^);
       pi, M
  BEG  pi = p-1(XX); M = epsi(pi)
       RETURN -(vec(I(v*v^))^*-M)^
  END.
```

Compared with (10), the program *reflR* uses a parameter $v : X \leftrightarrow \mathbf{1}$ that makes the assumed set X available. Also the next relational program *transR* for implementing (11) uses this parameter for providing X:

```
transR(v)
  DECL XX = PROD(v*v^,v*v^);
       pi, rho, M
  BEG  pi = p-1(XX); rho = p-2(XX); M = epsi(pi)
       RETURN -(vec(rho*pi^)^*([|M,M] & [|pi*pi^,rho*rho^]*-M))^
  END.
```

Having the relational programs *reflR* and *transR* and a vector $v : X \leftrightarrow \mathbf{1}$ at hand, a run through all points contained in the vector *reflR*$(v) \cap$ *transR*(v) allows to generate the membership-models of all topologies on X. Namely, if by means of the assignment $p = point(reflR(v) \cap transR(v))$ such a point $p : [X \leftrightarrow X] \leftrightarrow \mathbf{1}$ is selected, then the call $ToRel(p, v;v^{\mathsf{T}})$ computes the preorder relation Q modeled by p and, therefore, the evaluation of the expression $ToTop(ToRel(p, v;v^{\mathsf{T}}))$ returns the membership-model of the corresponding topology $\mathcal{T}_Q$.

After these preparations we are able to formulate the following relational program *genCI* for the computation of the numerical data of Sect. 2:

```
genCI(v,Lk)
  DECL conc(m,i,c,e,f,g,h) = (m^+i^+c^+e^+f^+g^+h^)^;
       po, res, p, M, T, I, C, E, F, G, H, w, q, R
  BEG  po = reflR(v) & transR(v); res = O(po);
       WHILE -empty(po) DO
         p = point(po);
         M = epsi(v); T = ToTop(ToRel(p,v*v^));
         I = Int(T,M); C = Clos(T,M);
         E = Int(T,C); F = Clos(T,I);
         G = Int(T,F); H = Clos(T,E);
         w = L1n(M)^;
         WHILE -empty(w) DO
           q = point(w);
           R = conc(M*q,I*q,C*q,E*q,F*q,G*q,H*q);
           IF cardeq(clReps(syq(R,R)),Lk) THEN res = res | p FI;
           w = w & -q OD;
         po = po & -p OD
       RETURN res
  END.
```

This program expects as input a vector $v : X \leftrightarrow \mathbf{1}$, that provides the set X, and a universal vector L^k with a k-element source, that provides the number $k \in \mathbb{N}$. The result of *genCI*(v, L^k) is a vector *res*: $[X \leftrightarrow X] \leftrightarrow \mathbf{1}$. It models the set of all preorder relations Q on X such that there exists a k-CI-set $A \in 2^X$ with respect to $(X, \mathcal{T}_Q)$ (with $\mathcal{T}_Q$ as topology corresponding to Q). In the table of Sect. 2 we have collected for each result of *genCI*(v, L^k), where $1 \leq |X| \leq 6$ and $1 \leq k \leq 7$, the number of 1-entries (i.e., the cardinality of the set it models). RelView delivers this information when it depicts a relation in its relation window.

To facilitate the subsequent explanation of the program, we have marked the lines of its body. When executing *genCI*(v, L^k), first, the vector-model of the set $\mathfrak{P}_X$ is computed and stored in the local variable *po*; then the result variable *res* is defined as empty (line 1). After this initialisation the program runs through all points $p \subseteq po$ (lines 2–14) and performs successively three actions:

1. Selection of p and computation of the membership relation $\mathsf{M} : X \leftrightarrow 2^X$ and of the membership-model $T : X \leftrightarrow \mathcal{T}$ of the topology $\mathcal{T}$ that corresponds to the preorder relation modeled by p (line 3–4).
2. Computation of the 6 relations $I, C, E, F, G, H : X \leftrightarrow 2^X$ of (8) for the topology $\mathcal{T}$ (lines 5–7).
3. Test, whether there exists an n in $\{1, \ldots, 2^{|X|}\}$ such that precisely k of the nth columns of the relations $\mathsf{M}, I, C, E, F, G$ and H are pair-wise different, and addition of the preorder relation to the result (expressed by the union of *res* and p) in case of a positive answer (lines 8–13).

The test of action 3 is realised by a loop through all points q contained in the universal vector $\mathsf{L} : 2^X \leftrightarrow \mathbf{1}$. In line 11 the columns of $\mathsf{M}, I, C, E, F, G, H$ designated by q are computed as vectors $\mathsf{M};q$, $I;q$, $C;q$, $E;q$, $F;q$, $G;q$, $H;q : X \leftrightarrow \mathbf{1}$ and then concatenated to a single relation R with source X and 7 columns via the technique described at the end of Sect. 3 and the local relational function *conc*. So, precisely k of the 7 vectors are pair-wise different if and only if precisely k of the 7 columns of R are pair-wise different. From the element-wise description of the symmetric quotient it follows that $syq(R, R)_{m,n}$ if and only if the mth column of R equals the nth column of R. Hence, exactly k of the 7 columns of R are pair-wise different if and only if the number of equivalence classes of the equivalence relation $syq(R, R)$ is k. In line 12 the latter property is tested by comparing the number of 1-entries of *clReps*$(syq(R, R))$ (i.e., the number of equivalence classes of $syq(R, R)$) with the number of 1-entries of the universal vector L^k (i.e., the number k) using the pre-defined RelView-operation *cardeq* for testing the equality of the cardinality of relations.

Next, we present an application. We consider again the two-point set $X := \{a, b\}$, since already 3 points lead to matrices and vectors which are too large to be presented here. The following RelView-pictures show, from top to bottom, the membership relation $\mathsf{M} : X^2 \leftrightarrow [X \leftrightarrow X]$ and then the transposed results $res_k^{\mathsf{T}} : \mathbf{1} \leftrightarrow [X \leftrightarrow X]$ of the calls *genCI*(v, L^k), where $v : X \leftrightarrow \mathbf{1}$ and $k \in \{1, 2, 3\}$:

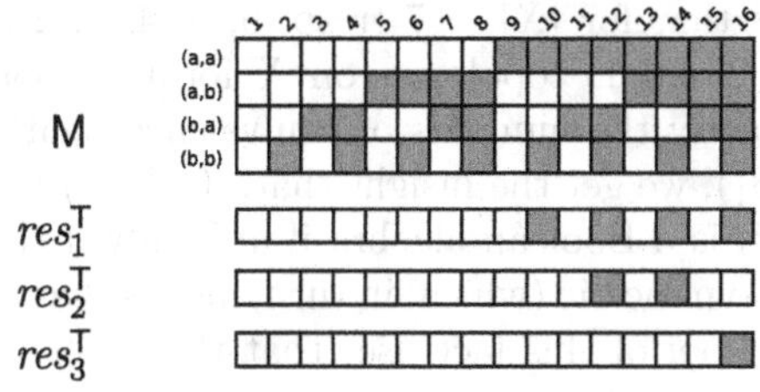

Since each topology leads to a 1-CI-set (cf. Sect. 2), from M and res_1^{T} it follows that there exist precisely 4 preorder relations on the set X, viz. the identity relation (column 10), corresponding to the discrete topology 2^X, the relation $\{(a,a),(b,a),(b,b)\}$ (column 12), corresponding to the topology $\{\emptyset,\{b\},X\}$, the relation $\{(a,a),(a,b),(b,b)\}$ (column 14), corresponding to the topology $\{\emptyset,\{a\},X\}$, and the universal relation (column 16), corresponding to the indiscrete topology $\{\emptyset,X\}$. From M and res_2^{T} we get that exactly the second and the third topology allow 2-CI-sets and M and res_3^{T} show that the indiscrete topology is the only topology on X that allows 3-CI-sets.

6 Modifications of the Program with Applications

In Sect. 1 we have mentioned that RelView-programs are easy to alter in case of slightly changed specifications. Following, we present examples for this.

First, we insert the assignment $R = (R\hat{} + -R\hat{})\hat{}$ after line 11 of the body of the program *genCI* and change the name to *genCK*. Because of (2), for $v : X \leftrightarrow \mathbf{1}$ then the call $genCK(v, \mathsf{L}^k)$ computes a vector that models the set of all preorder relations on X such that the corresponding topologies lead to k-CK-sets. From (4) we obtain $genCI(\mathsf{L}^7,\mathsf{L}^7) = genCK(\mathsf{L}^7,\mathsf{L}^{14})$. Generalising this property, RelView yields $genCI(\mathsf{L}^k,\mathsf{L}^k) = genCK(\mathsf{L}^k,\mathsf{L}^{2k})$ for $k \in \{1,4,5,6,7\}$. Hence, apart from the cases $|X| = 2$ and $|X| = 3$ each finite topological space $(X,\mathcal{T})$ allows a $|X|$-CI-set if and only if it allows a $2|X|$-CK-set. Concerning the first exception and assuming $X := \{a,b\}$, from the above pictures we already know that the vector $genCI(\mathsf{L}^2,\mathsf{L}^2)$ models the set $\{\{(a,a),(b,b),(a,b)\},\{(a,a),(b,b),(b,a)\}\}$. RelView computes that the set modeled by the vector $genCK(\mathsf{L}^2,\mathsf{L}^4)$ additionally contains the universal relation on X. When $|X| = 3$, we obtain from the numerical data of Sect. 2 that the vector $genCI(\mathsf{L}^3,\mathsf{L}^3)$ models a set of 16 preorder relations. RelView shows $genCI(\mathsf{L}^3,\mathsf{L}^3) \supset genCK(\mathsf{L}^3,\mathsf{L}^6)$ and that the vector $genCK(\mathsf{L}^3,\mathsf{L}^6)$ models a set with 12 preorder relations only.

In [11] a good deal of work is done in the connections of the sets of $CI(A)$ and, based on this, in the connections of compositions of the interior- and closure-operation and their use in characterising the underlying topological space $(X,\mathcal{T})$. Translated into our context, e.g., one result of [11] says that $(\mathcal{T},\cup,\cap,^c)$ is a Boolean algebra if and only if $C = E$, with the relations C and E as defined in (8). Another result of [11] says that the closure of any open set is open if and only if $F = G$, again with F and G from (8). So, RelView can be used to mechanise the characterisations of [11]. E.g., an obvious modification of the program *genCI* shows that for $|X| = 7$ (resp. 6, 5, 4, 3, 2 and 1) exactly 877 (resp. 203, 52, 15, 5, 2 and 1) topologies on X form a Boolean algebra. Since these numbers are exactly the numbers of equivalence relations on X (i.e., the *Bell numbers* B_7 to B_1), we get the insight that, if $(X,\mathcal{T})$ is a finite topological space, then $(\mathcal{T},\cup,\cap,^c)$ is a Boolean algebra if and only if the preorder relation corresponding to $\mathcal{T}$ is symmetric (which, in turn, means that $(X,\mathcal{T})$ is a so-called R_0-space). A formal proof of this fact uses that the topology $\mathcal{T}$ equals the set

$\mathcal{C}_\mathcal{T}$ of closed sets of $(X, \mathcal{T})$ and that the preorder relation corresponding to the topology $\mathcal{C}_\mathcal{T}$ on X is the transpose of the preorder relation corresponding to $\mathcal{T}$.

7 Concluding Remarks

Although Theorem 2.4 is a novel analogon of the result of [2] with a remarkable characterisation of indiscrete topologies on two-point sets, we do not regard it as the main contribution of the paper. In our view more important are the developments of the relational program *genCI* and of the auxiliary relational functions and programs, especially those of *reflR* and *transR*, and the application of *genCI* and its modifications. This again demonstrates the considerable potential of relation algebra and RelView in modeling, specification and problem solving. In our opinion more important than Theorem 2.4 is also the use of the tool for computational mathematics and mathematical experiments. In [8] experiments with RelView lead to new topological results, too, viz. that, given a finite topological space $(X, \mathcal{T})$ with $\subseteq$-least base $\mathcal{B}_*$ for $\mathcal{T}$, (i) the number of bases for $\mathcal{T}$ is $2^{|\mathcal{T}|-|\mathcal{B}_*|}$, (ii) the $\cap$-irreducible sets of $\mathcal{B}_*$ form a $\subseteq$-minimal subbase for $\mathcal{T}$ and (iii) this specific subbase is even the $\subseteq$-least element of the set of those subbases for $\mathcal{T}$ which are contained in the $\subseteq$-least base $\mathcal{B}_*$ for $\mathcal{T}$. For the future we plan to extend our investigations to variations of the Kuratowski problem as, e.g., discussed in [19] by the addition of intersection and/or union. Following the case-distinction approach of [2], we also want to prove Theorem 2.4 as usual in mathematics, i.e., without the use of the RelView tool.

References

1. Alexandroff, P.: Diskrete Räume. Mat. Sb. NS **2**, 501–518 (1937)
2. Anusiak, J., Shum, K.P.: Remarks on finite topological spaces. Colloq. Math. **23**, 217–223 (1971)
3. Aull, C.E.: Classification of topological speces. Bull. Acad. Polon. Sci. Ser. Math. Astro. Phys. **15**, 773–778 (1967)
4. Berghammer, R., Leoniuk, B., Milanese, U.: Implementation of relational algebra using binary decision diagrams. In: de Swart, H.C.M. (ed.) RelMiCS 2001. LNCS, vol. 2561, pp. 241–257. Springer, Heidelberg (2002). doi:10.1007/3-540-36280-0_17
5. Berghammer, R., Neumann, F.: RelView – an OBDD-based computer algebra system for relations. In: Ganzha, V.G., Mayr, E.W., Vorozhtsov, E.V. (eds.) CASC 2005. LNCS, vol. 3718, pp. 40–51. Springer, Heidelberg (2005). doi:10.1007/11555964_4
6. Berghammer, R., Winter, M.: Embedding mappings and splittings with applications. Acta Inform. **47**, 77–110 (2010)
7. Berghammer, R.: Column-wise extendible vector expressions and the relational computation of sets of sets. In: Hinze, R., Voigtländer, J. (eds.) MPC 2015. LNCS, vol. 9129, pp. 238–256. Springer, Cham (2015). doi:10.1007/978-3-319-19797-5_12
8. Berghammer, R., Winter, M.: Solving computational tasks on finite topologies by means of relation algebra and the RelView tool. J. Log. Algebr. Methods Program. **88**, 1–25 (2017)

9. Chapman, T.A.: A further note on closure and interior operations. Am. Math. Mon. **68**, 524–529 (1962)
10. Erné, M., Stege, K.: Counting finite posets and topologies. Order **8**, 247–265 (1991)
11. Gardner, B.J., Jackson, M.: The Kuratowski closure-complement theorem. N. Z. J. Math. **38**, 9–44 (2008)
12. Herda, H.H., Metzler, R.C.: Closure and interior in finite topological spaces. Colloq. Math. **15**, 211–216 (1966)
13. Kelley, J.L.: General Topology. Springer, New York (1975)
14. Kuratowski, C.: Sur l'opération $\overline{A}$ de l'analysis situs. Fund. Math. **3**, 182–199 (1922)
15. Langford, E.: Characterization of Kuratowski 14-sets. Am. Math. Mon. **78**, 362–367 (1971)
16. Schmidt, G.: Relational Mathematics. Cambridge University Press, Cambridge (2010)
17. Schmidt, G., Berghammer, R.: Contact, closure, topology, and the linking of rows and column types. J. Log. Algebr. Program. **80**, 339–361 (2011)
18. Schmidt, G.: A point-free relation-algebraic approach to general topology. In: Höfner, P., Jipsen, P., Kahl, W., Müller, M.E. (eds.) RAMICS 2014. LNCS, vol. 8428, pp. 226–241. Springer, Cham (2014). doi:10.1007/978-3-319-06251-8_14
19. Sheman, D.: Variations on Kuratowski's 14-set theorem. Am. Math. Mon. **117**, 113–123 (2010)
20. RelView-homepage: http://www.informatik.uni-kiel.de/~progsys/relview/

Varieties of Cubical Sets

Ulrik Buchholtz[1,2P (✉)] and Edward Morehouse[3,4P]

[1] Department of Philosophy, Carnegie Mellon University, Pittsburgh, PA 15213, USA

[2] Fachbereich Mathematik, Technische Universität Darmstadt, Schlossgartenstraße 7, 64289 Darmstadt, Germany

buchholtz@mathematik.tu-darmstadt.de

[3] Carnegie Mellon School of Computer Science, 5000 Forbes Avenue, Pittsburgh, PA 15213, USA

emorehouse@wesleyan.edu

[4] Department of Mathematics and Computer Science, Wesleyan University, 265 Church Street, Middletown, CT 06459, USA

Abstract. We define a variety of notions of cubical sets, based on sites organized using substructural algebraic theories presenting PRO(P)s or Lawvere theories. We prove that all our sites are test categories in the sense of Grothendieck, meaning that the corresponding presheaf categories of cubical sets model classical homotopy theory. We delineate exactly which ones are even strict test categories, meaning that products of cubical sets correspond to products of homotopy types.

1 Introduction

There has been substantial interest recently in the use of cubical structure as a basis for higher-dimensional type theories (Angiuli et al. 2017; Awodey 2016; Bezem et al. 2014; Cohen et al. 2016). A question that quickly arises is, what sort of cubical structure, because there are several plausible candidates to choose from.

The suitability of a given cubical structure as a basis for a higher-dimensional type theory is dependent on at least two considerations: its proof-theoretic characteristics (e.g. completeness with respect to a given class of models, decidability of equality of terms, existence of canonical forms) and its homotopy-theoretic characteristics, most importantly, that the synthetic homotopy theory to which it gives rise should agree with the standard homotopy theory for topological spaces.

Contributions. In this paper we organize various notions of cubical sets along several axes, using substructural algebraic theories as a guiding principle (Mauri 2005). We define a range of *cube categories* (or *cubical sites*), categories $\mathbb{C}$ for which the corresponding presheaf category $\widehat{\mathbb{C}}$ can be thought of as a category of cubical sets. We then consider each of these from the perspective of *test categories* (Grothendieck 1983), which relates presheaf categories and homotopy theory. We give a full analysis of our cubical sites as test categories.

P. Höfner et al. (Eds.): RAMiCS 2017, LNCS 10226, pp. 77–92, 2017.
DOI: 10.1007/978-3-319-57418-9_5

1.1 Cube Categories

Our cube categories are presented as monoidal categories with a single generating object, X, representing an abstract dimension. To give such a presentation, then, is to give a collection of morphism generators, $f : X^{\otimes n} \to X^{\otimes m}$, and a collection of equations between parallel morphisms. Such a presentation is known in the literature as a "PRO". Two closely-related notions are those of "PROP" and of "Lawvere theory", the difference being that a PROP assumes that the monoidal category is *symmetric*, and a Lawvere theory further assumes that it is *cartesian*. These additional assumptions manifest themselves in the proof theory as the admissibility of certain structural rules: in a PROP, the structural rule of *exchange* is admissible, and in a Lawvere theory *weakening* and *contraction* are admissible as well.

The distinguishing property of a theory making it "cubical", rather than of some other "shape", is the presence of two generating morphisms, $d^0, d^1 : 1 \to X$ representing the face maps (where "1" is the monoidal unit). This is because combinatorially, an n-dimensional cube has $2n$ many $(n-1)$-cube faces, namely, two in each dimension.

1.2 Test Categories

In his epistolary research diary, *Pursuing Stacks*, Grothendieck set out a program for the study of abstract homotopy. The homotopy theory of topological spaces can be described as that of (weak) higher-dimensional groupoids. Higher-dimensional groupoids also arise from combinatorially-presented structures, such as simplicial or, indeed, cubical sets, which can thus be seen as alternative presentations of the classical homotopy category, **Hot**. That is, **Hot** arises either from the category of topological spaces, or from the category of simplicial sets, by inverting a class of morphisms (the weak equivalences).

Recall that for any small category A and any cocomplete category $\mathcal{E}$, there is an equivalence between functors $i : A \to \mathcal{E}$ and adjunctions

$$\widehat{A} \underset{i_!}{\overset{i^*}{\underset{\longrightarrow}{\overset{\longleftarrow}{\top}}}} \mathcal{E} \tag{1}$$

where $i_!$ is the left Kan extension of i, and $i^*(X)(a) = \mathrm{Hom}_{\mathcal{E}}(i(a), X)$.[1] Here $\widehat{A}$ is the category of functors $A^{\mathrm{op}} \to \mathbf{Set}$, also known as the presheaf topos of A. The category of simplicial sets is defined as the presheaf category $\mathbf{sSet} = \widehat{\Delta}$, where Δ is the category of non-empty finite totally ordered sets and order-preserving maps.

The category of small categories, **Cat**, likewise presents the homotopy category, by inverting the functors that become weak equivalences of simplicial sets after applying the nerve functor, which is the right adjoint functor $N = i^*$ coming from the inclusion $\Delta \hookrightarrow \mathbf{Cat}$. Grothendieck realized that this allows us to

[1] In this paper, all small categories are considered as strict categories, and **Cat** denotes the 1-category of these.

compare the presheaf category $\widehat{A}$, for any small category A, with the homotopy category in a canonical way, via the functor $i_A : A \to \mathbf{Cat}$ given by $i_A(a) = A_{/a}$. The left Kan extension is in this case also denoted i_A and we have $i_A(X) = A_{/X}$ for X in $\widehat{A}$. He identified the notion of a *test category*, which is one for which the homotopy category of its category of presheaves is equivalent to the homotopy category via the right adjoint i_A^*, after localization at the weak equivalences. In other words, presheaves over a test category are models for homotopy types of spaces/∞-groupoids. In this way, the simplex category Δ is a test category, and the classical cube category of Serre and Kan, $\mathbb{C}_{(\mathrm{w},\cdot)}$ in our notation, is also a test category.

One perceived benefit of the simplex category Δ is that the induced functor $\mathbf{sSet} \to \mathbf{Hot}$ preserves products, whereas the functor $\widehat{\mathbb{C}_{(\mathrm{w},\cdot)}} \to \mathbf{Hot}$ does not. A test category A for which the functor $\widehat{A} \to \mathbf{Hot}$ preserves products is called a *strict* test category. We shall show that most natural cube categories are in fact strict test categories.

2 Cube Categories

In this paper, we consider as base categories the syntactic categories for a range of monoidal theories, all capturing some aspect of the notion of an interval. We call the resulting categories "cube categories", because the monoidal powers of the interval then correspond to cubes[2].

The generating morphisms for these cube categories can be classified either as *structural*, natural families corresponding to structural rules of a proof theory, or as *algebraic*, distinguished by the property of having coarity one. Our cube categories vary along three dimensions: the structural rules present, the signature of algebraic function symbols, and the equational theory.

This section is organized as follows: in Subsect. 2.1 we recall the basics of algebraic theories in monoidal categories, and in Subsect. 2.2 we discuss how monoidal languages are interpreted in monoidal categories. Then we introduce in Subsect. 2.3 the monoidal languages underlying our cubical theories, so that in Subsect. 2.4 we can introduce the standard interpretations of these and our cubical theories. In Subsect. 2.5 we give a tour of the resulting cube categories.

2.1 Monoidal Algebraic Theories

We assume the reader is familiar with ordinary algebraic theories and their categorical incarnations as Lawvere theories. The idea of monoidal algebraic theories as described by Mauri (2005) is to generalize this to cases where only a subset of the structural rules (weakening, exchange and contraction) are needed to describe the axioms. Think for example of the theory of monoids over a signature of a neutral element 1 and a binary operation. The axioms state that

[2] Here we say "cube" for short instead of "hypercube" for an arbitrary dimensional power of an interval. A prefix will indicate the dimension, as in 0-cube, 1-cube, etc.

$1x = x = x1$ (in the context of one variable x) and $x(yz) = (xy)z$ (in the context of the variables x, y and z). Note that the terms in these equations each contain all the variables of the context and in the same order. Thus, no structural rules are needed to form the equations, and hence the notion of a monoid makes sense in any monoidal category.

Let us now make this more precise. For the structural rules, we follow Mauri (*ibid.*) and consider any subset of $\{\mathrm{w}, \mathrm{e}, \mathrm{c}\}$ (w for weakening, e for exchange, and c for contraction), except that whenever c is present, so is e. Thus we consider the following lattice of subsets of structural rules:

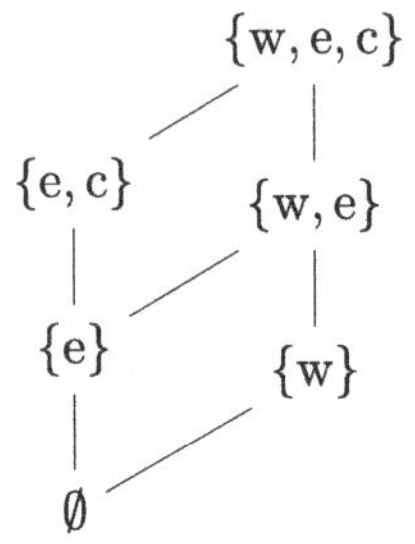

(2)

It would be possible to consider monoidal theories over operations that have any number of incoming and outgoing edges, representing morphisms $f : A_1 \otimes \cdots \otimes A_n \to B_1 \otimes \cdots \otimes B_m$ in a monoidal category, where the A_i and B_j are sorts. However, it simplifies matters considerably, and suffices for our purposes, to consider only operations of the usual kind in algebra, with any arity of incoming edges and exactly one outgoing edge, as in $f : A_1 \otimes \cdots \otimes A_n \to B$.

Furthermore, we shall consider only single-sorted theories, defined over a signature Σ given by a set of function symbols with arities. For a single-sorted signature, the types can of course be identified with the natural numbers.

A (single-sorted, algebraic) *monoidal language* consists of a pair of a set of structural rules together with an algebraic signature. A term t in a context of n free variables $x_1, \ldots, x_n$ is built up from the function symbols in such a way that when we list the free variables in t from left to right:

- every variables occurs unless w (weakening) is a structural rule;
- the variables occur in order unless e (exchange) is a structural rule; and
- there are no duplicated variables unless c (contraction) is a structural rule.

For the precise rules regarding term formation and the proof theory of equalities of terms, we refer to Mauri (2005).

2.2 Interpretations

An interpretation of a monoidal language in a monoidal category $(\mathcal{E}, 1, \otimes)$ will consist of an object X representing the single sort, together with morphisms representing the structural rules and morphisms representing the function symbols (including constants).

The structural rules w, e, c are interpreted respectively by morphisms

$$\varepsilon : X \to 1$$
$$\tau : X \otimes X \to X \otimes X$$
$$\delta : X \to X \otimes X$$

satisfying certain laws (Mauri 2005, (27–36)). These laws specify also the interaction between these morphisms and the morphisms corresponding to the function symbols. When $\mathcal{E}$ is symmetric monoidal, we interpret τ by the braiding of $\mathcal{E}$, and when $\mathcal{E}$ is cartesian monoidal, we interpret everything using the cartesian structure. A function symbol f of arity n is interpreted by a morphism $|f| : X^{\otimes n} \to X$. This morphism and the structural morphisms are required to interact nicely, e.g., $\varepsilon \circ |f| = \varepsilon^{\otimes n}$ (cf. *loc. cit.*).

In a syntactic category for the empty theory (of which the syntactic category for a non-empty theory is a quotient), every morphism factors uniquely as a structural morphism followed by a functional one (*op. cit.*, Proposition 5.1). When we impose a theory, we may of course lose uniqueness, but we still have existence.

2.3 Cubical Monoidal Languages

All of our cubical signatures will include the two endpoints of the interval as nullary function symbols, 0 and 1. For the rest, we consider the two "connections" $\vee$ and $\wedge$, as well as the reversal, indicated by a prime, $'$. This gives us the following lattice of 6 signatures:

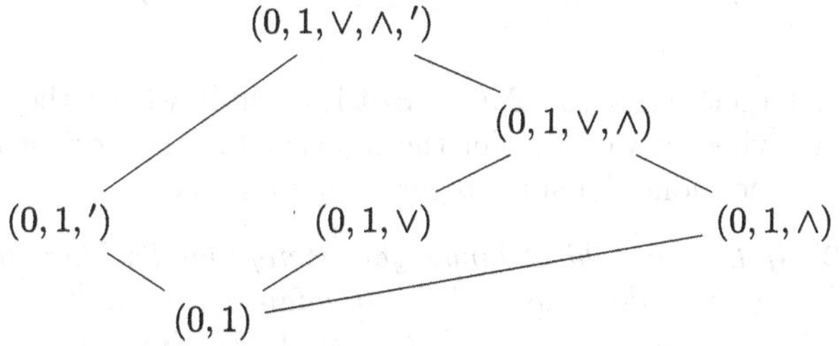

(3)

Combined with the 6 possible combinations of structural rules, we are thus dealing with 36 distinct languages.

Definition 1. *Let $L_{(a,b)}$, where a is one of the six substrings of "wec" corresponding to* (2) *and b one of the six substrings of "$\vee\wedge'$" corresponding to* (3), *denote the language with structural rules from a and signature obtained from* $(0,1)$ *by expansion with the elements of b.*

We shall, however, mostly consider the 18 languages with weakening present (otherwise we are dealing with *semi-cubical sets*).

The third dimension of variation for our cube categories is the algebraic theory for a given language. Here we shall mostly pick the theory of a particular standard structure, so let us pause our discussion of cube categories to introduce our standard structures.

2.4 The Canonical Cube Categories

We consider the interval structures of the following objects in the cartesian monoidal categories **Top** and **Set** (so all structural rules are supported, interpreted by the global structure):

- the standard topological interval, $I = [0, 1]$ in **Top**;
- the standard 2-element set, $I = \mathbb{2} = \{0, 1\}$ in **Set**.

For the topological interval, we take $x \vee y = \max\{x, y\}$, $x \wedge y = \min\{x, y\}$ and $x' = 1 - x$. These formulas also apply to the 2-element set $\mathbb{2}$.

We can also consider the 3-element Kleene algebra (cf. Subsect. 2.5) $\mathbb{3} = \{0, u, 1\}$ with $u' = u$ and the 4-element de Morgan algebra $\mathbb{D} = \{0, u, v, 1\}$ with $u' = u$ and $v' = v$ in **Set** (called the *diamond*), as further structures. It is interesting to note here that $\mathbb{3}$ has the same theory as $[0, 1]$ in the full language, namely the theory of Kleene algebras (Gehrke et al. 2003). Of course, $\mathbb{2}$ gives us the theory of Boolean algebras, while $\mathbb{D}$ gives the theory of de Morgan algebras.

Definition 2. *With (a, b) as in Definition 1, and T a theory in the language $L_{(a,b)}$, let $\mathbb{C}_{(a,b)}(T)$ denote the syntactic category of the theory T. We write $\mathbb{C}_{(a,b)}$ for short for $\mathbb{C}_{(a,b)}(\mathrm{Th}([0, 1]))$ for the syntactic category of the topological interval with respect to the monoidal language $L_{(a,b)}$.*

We say that $\mathbb{C}_{(a,b)}$ is the *canonical* cube category for the language $L_{(a,b)}$.

Proposition 1. *The canonical cube category $\mathbb{C}_{(a,b)}$ is isomorphic to the monoidal subcategory of* **Top** *generated by $[0, 1]$ with respect to the language $L_{(a,b)}$.*

In fact, since the forgetful functor **Top** $\to$ **Set** is faithful, we get the same theory in every language whether we consider the interval $[0, 1]$ in **Top** or in **Set**, and it generates the same monoidal subcategory in either case.

Proposition 2. *If L is a cubical language strictly smaller than the maximal language, $L_{(\mathrm{wec},\vee\wedge')}$, then the theory of the standard structure $[0, 1]$ is the same as the theory of the standard structure $\mathbb{2} = \{0, 1\}$ in* **Set**. *For $L_{(\mathrm{wec},\vee\wedge')}$ the theory of $[0, 1]$ equals the theory of the three-element Kleene algebra, $\mathbb{3}$, and is in fact the theory of Kleene algebras.*

Proof. If L does not have reversal, then we can find for any points $a < b$ in $[0, 1]$ a homomorphism $f : [0, 1] \to \mathbb{2}$ in **Set** with $f(a) < f(b)$ (cut between a and b), so $[0, 1]$ and $\mathbb{2}$ have the same theory.

If L does not have contraction, each term is monotone in each variable. By pushing reversals towards the variables and applying the absorption laws, we can find for each resulting term $s[x, y_1, \ldots y_n]$ instantiations $b_1, \ldots, b_n \in \mathbb{2}$ with $s[0, b_1, \ldots, b_n] \neq s[1, b_1, \ldots, b_n]$. It follows that if two terms agree on instantiations in $\mathbb{2}$, then they use each variable the same way: either with or without a reversal. Hence we can ignore the reversals and use the previous case.

The last assertion is proved by Gehrke et al. (2003). $\square$

Corollary 1. *For each canonical cube category $\mathbb{C}_{(a,b)}$ we have decidable equality of terms.*

Proof. This follows because equality is decidable in the theory of Kleene algebras, as this theory is characterized by the object $\mathbb{3}$. Hence we can decide equality by the method of "truth-tables", however, relative to the elements of $\mathbb{3}$. □

Of course, any algorithm for equality in the theory of Kleene algebras can be used, including the one based on disjunctive normal forms (*op. cit.*).

We can in the presence of weakening quite simply give explicit axiomatizations for each of the canonical cube categories $\mathbb{C}_{(a,b)}$: take the subset of axioms in Table 1 that make sense in the language $L_{(a,b)}$. Without weakening, this will not suffice, as we should need infinitely many axioms to ensure for instance that $s \wedge 0 = t \wedge 0$ where s, t are terms in the same variable context.

Table 1. Cubical axioms

Axiom	Lang. req.	Name
$x \vee (y \vee z) = (x \vee y) \vee z$	$(\cdot, \vee)$	$\vee$-associativity
$0 \vee x = x = x \vee 0$	$(\cdot, \vee)$	$\vee$-unit
$1 \vee x = 1 = x \vee 1$	$(\mathrm{w}, \vee)$	$\vee$-absorption
$x \vee y = y \vee x$	$(\mathrm{e}, \vee)$	$\vee$-symmetry
$x \vee x = x$	$(\mathrm{ec}, \vee)$	$\vee$-idempotence
$x \wedge (y \wedge z) = (x \wedge y) \wedge z$	$(\cdot, \wedge)$	$\wedge$-associativity
$1 \wedge x = x = x \wedge 1$	$(\cdot, \wedge)$	$\wedge$-unit
$0 \wedge x = 0 = x \wedge 0$	$(\mathrm{w}, \wedge)$	$\wedge$-absorption
$x \wedge y = y \wedge x$	$(\mathrm{e}, \wedge)$	$\wedge$-symmetry
$x \wedge x = x$	$(\mathrm{ec}, \wedge)$	$\wedge$-idempotence
$x'' = x$	$(\cdot, ')$	$'$-involution
$0' = 1$	$(\cdot, ')$	$'$-computation
$x \wedge (y \vee z) = (x \wedge y) \vee (x \wedge z)$	$(\mathrm{ec}, \vee\wedge)$	Distributive law 1
$x \vee (y \wedge z) = (x \vee y) \wedge (x \vee z)$	$(\mathrm{ec}, \vee\wedge)$	Distributive law 2
$x = x \vee (x \wedge y) = x \wedge (x \vee y)$	$(\mathrm{wec}, \vee\wedge)$	Lattice-absorption
$(x \vee y)' = x' \wedge y'$	$(\cdot, \vee\wedge')$	de Morgan's law
$x \wedge x' \leq y \vee y'$	$(\mathrm{wec}, \vee\wedge')$	Kleene's law

When we compare our canonical cube categories to the categories of Grandis and Mauri (2003), we find $\mathbb{I} = \mathbb{C}_{(\mathrm{w},\cdot)}$, $\mathbb{J} = \mathbb{C}_{(\mathrm{w},\vee\wedge)}$, $\mathbb{K} = \mathbb{C}_{(\mathrm{we},\vee\wedge)}$, and $!\mathbb{K} = \mathbb{C}_{(\mathrm{we},\vee\wedge')}$.

For the full language, $L_{(\mathrm{wec},\vee\wedge')}$, there are two additional interesting cube categories: $\mathbb{C}_{\mathrm{dM}}$ (de Morgan algebras) and $\mathbb{C}_{\mathrm{BA}}$ (boolean algebras), where de Morgan algebras satisfy all the laws of Table 1 except Kleene's law, and boolean algebras of course satisfy additionally the law $x \vee x' = 1$. The case of de Morgan algebras is noteworthy as the basis of the cubical model of type theory of Cohen et al. (2016).

Definition 3. *For each of our cube categories $\mathbb{C}$, we write $[n]$ for the object representing a context of n variables, and we write $\square^n$ for the image of $[n]$ under the Yoneda embedding* $\mathrm{y} : \mathbb{C} \hookrightarrow \widehat{\mathbb{C}}$.

2.5 A Tour of the Menagerie

Plain Cubes. In the canonical cube category for the language $L_{(\mathrm{w},\cdot)}$ the monoidal unit $[0]$ is terminal and the interpretation of weakening is generated by the unique degeneracy map $\varepsilon : [1] \to [0]$. The points 0 and 1 are interpreted by face maps $\eta_0, \eta_1 : [0] \to [1]$. In the following we let $\{i, j\}$ range over $\{0, 1\}$ with the assumption that $i \neq j$.

Because $[0]$ is terminal we have the *face-degeneracy laws*, $\varepsilon \circ \eta_i = \mathrm{Id}_{[0]}$, which we represent string-diagrammatically as:

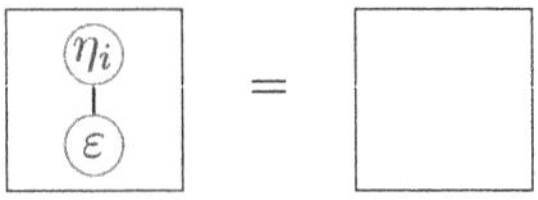

This law is invisible in the algebraic notation.

In a presheaf T, the face maps give rise to reindexing functions between the fibers, $\partial^i_k : T[n+1] \to T[n]$, where $1 \leq k \leq n+1$. In particular, when n is 0 these pick out the respective boundary points of an interval. Similarly, the degeneracy map gives rise to reindexing functions between the fibers, $*_k : T[n] \to T[n+1]$, where $1 \leq k \leq n+1$. In particular, when n is 0 this determines a degenerate interval $*(a) \in T[1]$ on a point $a \in T[0]$, and the face-degeneracy laws tell us its boundary points: $\partial^0(*(a)) = \partial^1(*(a)) = a$.

Adding the structural law of exchange to the language adds a natural isomorphism $\tau : [2] \to [2]$ to the syntactic category. In a presheaf, this lets us permute any adjacent pair of a cube's dimensions by reflecting it across the corresponding diagonal (hyper)plane. In particular, $T(\tau)$ reflects a square across its main diagonal.

Cubes with Diagonals. In the canonical cube category for the language $L_{(\mathrm{wec},\cdot)}$ the monoidal product is cartesian and the interpretation of contraction is generated by the diagonal map $\delta : [1] \to [2]$. In this case, the pair (δ, ε) forms a cocommutative comonoid.

Because δ is natural we have the *face-diagonal laws*, $\delta \circ \eta_i = \eta_i \otimes \eta_i$, which we represent string-diagrammatically as:

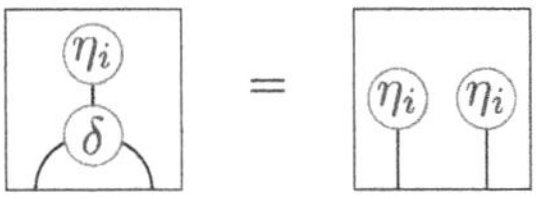

In a presheaf T, the diagonal map gives rise to reindexing functions between the fibers, $d_k : T[n+2] \to T[n+1]$, where $1 \leq k \leq n+1$. In particular, when n is 0 this picks out the main diagonal of a square, and the face-diagonal laws tell us its boundary points.

In the cubical semantics, the fact that ε is a comonoid counit for δ tells us that the diagonal of a square formed by degenerating an interval is just that interval. Likewise, the fact that δ is coassociative tells us that the main diagonal interval of a higher-dimensional cube is well-defined.

Cubes with Reversals. In the canonical cube category for the language $L_{(\cdot,')}$ we have an involutive reversal map $\rho : [1] \to [1]$. A reversal acts by swapping the face maps, giving the *face-reversal laws*, $\rho \circ \eta_i = \eta_j$, which we represent string-diagrammatically as:

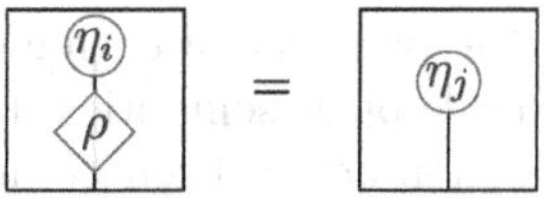

This embodies the equations $0' = 1$ and $1' = 0$.

In a presheaf T, the reversal map gives rise to endomorphisms on the fibers, $!_k : T[n+1] \to T[n+1]$, where $1 \leq k \leq n+1$. In particular, when n is 0 this reverses an interval, and the face-reversal laws tell us its boundary points.

If the signature contains weakening, then we have the *reversal-degeneracy law*, $\varepsilon \circ \rho = \varepsilon$. In the cubical interpretation, this says that a degenerate interval is invariant under reversal. If the signature contains contraction, then we have the *reversal-diagonal law*, $\delta \circ \rho = (\rho \otimes \rho) \circ \delta$. Cubically, this says that the reversal of a square's diagonal is the diagonal of the square resulting from reversing in each dimension.

Cubes with Connections. In the canonical cube category for the language $L_{(\mathrm{w},\vee\wedge)}$, the connectives are interpreted by connection maps $\mu_0, \mu_1 : [2] \to [1]$. Each pair (μ_i, η_i) forms a monoid. Furthermore, the unit for each monoid is an absorbing element for the other: $\mu_j \circ (\eta_i \otimes \mathrm{Id}) = \eta_i \circ \varepsilon = \mu_j \circ (\mathrm{Id} \otimes \eta_i)$. Grandis and Mauri (2003) call such structures "dioids", so we refer to these as the *dioid absorption laws*, and represent them string-diagrammatically as:

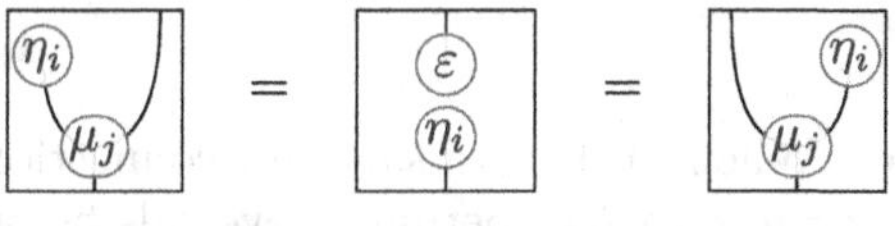

Equationally, these amount to the $\vee$- and $\wedge$-absorption laws of Table 1.

In the cubical semantics, connections can be seen as a variant form of degeneracy, identifying adjacent, rather than opposite, faces of a cube. Of course in order for this to make sense, a cube must have at least two dimensions. In a presheaf T, the connection maps give rise to reindexing functions between the fibers, $\ulcorner_k, \lrcorner_k : T[n+1] \to T[n+2]$, where $1 \leq k \leq n+1$. In particular, when n is 0 these act as follows:

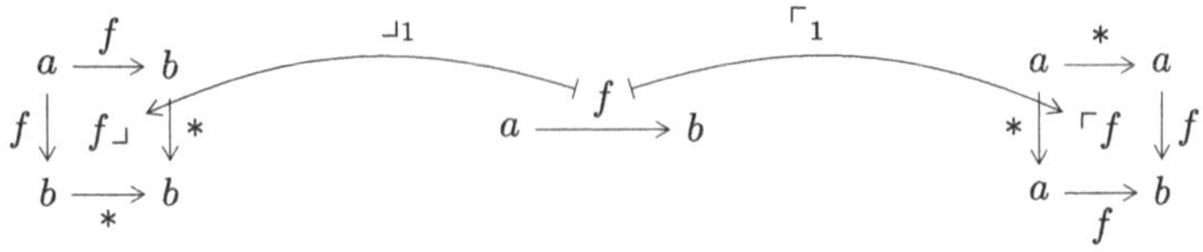

It may help to think of the interval f as a folded paper fan, with its "hinge" at the domain end in the case of $\lrcorner$, and at the codomain end in the case of $\ulcorner$. The respective connected squares are then obtained by "opening the fan". The monoid unit laws give us the (generally) non-degenerate faces of a connected square, while the dioid absorption laws give us the (necessarily) degenerate ones. Monoid associativity says that multiply-connected higher-dimensional cubes are well-defined.

Because [0] is terminal we also have the *connection-degeneracy laws*, $\varepsilon \circ \mu_i = \varepsilon \otimes \varepsilon$, which say that a connected square arising from a degenerate interval is a doubly-degenerate square. In the presence of exchange, we further assume that the μ_i are commutative. In the cubical semantics, this implies that connected cubes are invariant under reflection across their connected dimensions.

When adding contraction to the signature, we must give a law for rewriting a diagonal following a connection such that the structural maps come first. This is done by the *connection-diagonal laws*, $\delta \circ \mu_i = (\mu_i \otimes \mu_i) \circ (\mathrm{Id} \otimes \tau \otimes \mathrm{Id}) \circ (\delta \otimes \delta)$, which we represent string-diagrammatically as:

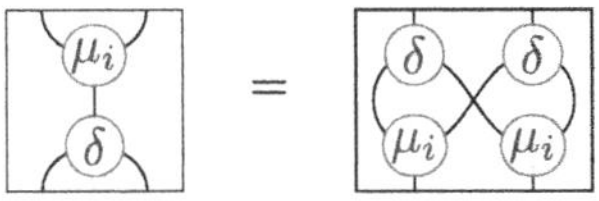

Algebraically, this says that two copies of a conjunction consists of a pair of conjunctions, each on copies of the respective terms; cubically, it gives the connected square on the diagonal interval of another square in terms of a product of diagonals in the 4-cube resulting from connecting each dimension separately. The whole structure, then, is a pair of bicommutative bimonoids, $(\delta, \varepsilon, \mu_i, \eta_i)$ related by the dioid absorption laws. We refer to these as *linked bimonoids.*

Additionally, we may impose the laws of bounded, modular, or distributive lattices, each of which implies its predecessors, and all of which imply the *diagonal-connection laws*, $\mu_i \circ \delta = \mathrm{Id}[1]$, known in the literature (rather generically) as "special" laws. These correspond to the $\vee$- and $\wedge$-idempotence laws of Table 1.

The Full Signature. When both reversals and connections are present, the *de Morgan laws*, $\rho \circ \mu_i = \mu_j \circ (\rho \otimes \rho)$ permute reversals before connections:

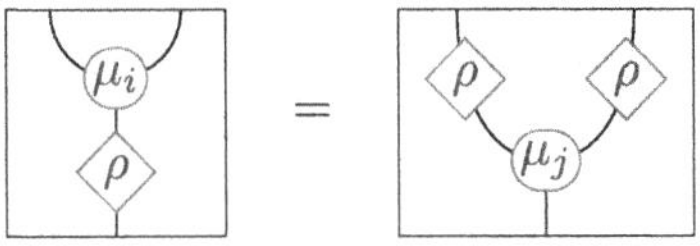

These laws imply each other as well as their nullary versions, the face-reversal laws.

Using the algebraic characterization of order in a lattice, $x \wedge y = x \iff x \leq y \iff x \vee y = y$, we can express the *Kleene law* as $\mu_i \circ (\mu_i \otimes \mu_j) \circ (\mathrm{Id} \otimes \rho \otimes \mathrm{Id} \otimes \rho) \circ (\delta \otimes \delta) = \mu_i \circ (\mathrm{Id} \otimes \rho) \circ (\delta \otimes \varepsilon)$:

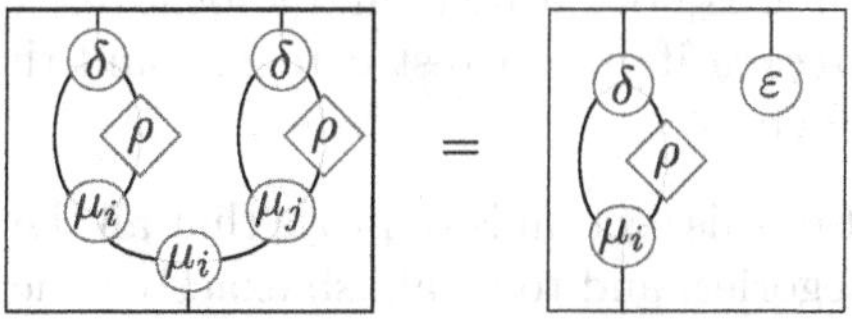

Finally, we arrive at the structure of a boolean algebra by assuming the *Hopf laws*, $\mu_i \circ (\mathrm{Id} \otimes \rho) \circ \delta = \eta_j \circ \varepsilon$:

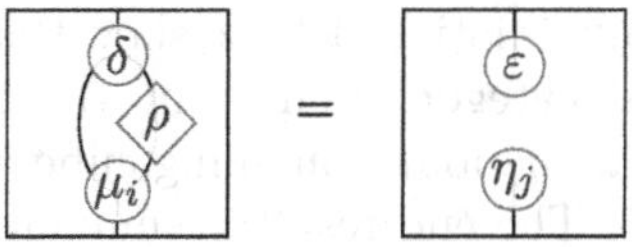

Cubically, these say that the anti-diagonal of a connected square is degenerate.

3 Test Categories

As mentioned in the introduction, a test category is a small category A such that the presheaf category $\widehat{A}$ can model the homotopy category after a canonical localization. Here we recall the precise definitions (cf. Maltsiniotis (2005)).

The inclusion functor $i : \Delta \to \mathbf{Cat}$ induces via (1) an adjunction $i_! : \widehat{\Delta} \leftrightarrows \mathbf{Cat} : N$ where the right adjoint N is the *nerve* functor. This allows us to transfer the homotopy theory of simplicial sets to the setting of (strict) small categories, in that we define a functor $f : A \to B$ to be a *weak equivalence* if $N(f)$ is a weak equivalence of simplicial sets. A small category A is called *aspheric* (or *weakly contractible*) if the canonical functor $A \to 1$ is a weak equivalence. Since any natural transformation of functors induces a homotopy, it follows that any category with a natural transformation between the identity functor and a constant endofunctor is contractible, and hence aspheric. In particular, a category with an initial or terminal object is aspheric. It follows from Quillen's Theorem A (Quillen 1973) that a functor $f : A \to B$ is a weak equivalence, if it is *aspheric*, meaning that all the slice categories $A_{/b}$ are aspheric, for $b \in B$.[3] By duality, so is any *coaspheric* functor $f : A \to B$, one for which the coslice categories ${}_{b\backslash}A$, for $b \in B$, are all aspheric. Finally, we say that a presheaf X in $\widehat{A}$ is *aspheric*, if $A_{/X}$ is (note that $A_{/X}$ is the category of elements of X).

For any small category A, we can use the adjunction induced by the functor $i_A : A \to \mathbf{Cat}$, $i_A(a) = A_{/a}$ (as mentioned in Subsect. 1.2) to define the class of weak equivalences in $\widehat{A}$, $\mathcal{W}_A$; namely, $f : X \to Y$ in $\widehat{A}$ is in $\mathcal{W}_A$ if $i_A(f) : A_{/X} \to A_{/Y}$ is a weak equivalence of categories. Following Grothendieck, we make the following definitions:

[3] We follow Grothendieck and let the slice $A_{/b}$ denote the comma category $(f \downarrow b)$ of the functors $f : A \to B$ and $b : 1 \to B$. Similarly, the coslice ${}_{b\backslash}A$ denotes $(b \downarrow f)$.

- A is a *weak test category* if the induced functor $\overline{i_A} : (\mathcal{W}_A)^{-1}\widehat{A} \to \mathbf{Hot}$ is an equivalence of categories (the inverse is then $\overline{i_A^*}$).
- A is a *local test category* if all the slices $A_{/a}$ are weak test categories.
- A is a *test category* if A is both a weak and a local test category.
- A is a *strict test category* if A is a test category and the functor $\widehat{A} \to \mathbf{Hot}$ preserves finite products.

The goal of the rest of this section is to prove that all the cube categories *with weakening* are test categories, and to establish exactly which ones are strict test categories. First we introduce Grothendieck interval objects. These allow us to show that any cartesian cube category is a strict test category (Corollary 2) as well as to show that all the cube categories are test categories (Corollary 3). Then we adapt an argument of Maltsiniotis (2009) to show that any cube category with a connection is a strict test category (Theorem 3). Finally, we adapt another argument of his to show that the four remaining cube categories fail to be strict test categories (Theorem 4). The end result is summarized in Table 2.

Table 2. Which canonical cube categories $\mathbb{C}_{(a,b)}$ are test (t) or even strict test (st) categories. The bottom-right corner refers to the cube categories corresponding to de Morgan, Kleene and boolean algebras.

a∖b	·	′	∨	∧	∨∧	∨∧′
w	t	t	st	st	st	st
we	t	t	st	st	st	st
wec	st	st	st	st	st	st/st/st

In order to study test categories, Grothendieck (1983) introduced the notion of an *interval* (*segment* in the terminology of Maltsiniotis (2005)) in a presheaf category $\widehat{A}$. This is an object I equipped with two global elements d^0, d^1. As such, it is just a structure for the initial cubical language, $L_{(\cdot,\cdot)}$ in the cartesian monoidal category $\widehat{A}$, and $\square^1$ is thus canonically a Grothendieck interval in all our categories of cubical sets, $\widehat{\mathbb{C}}$.

An interval is *separated* if the equalizer of d^0 and d^1 is the initial presheaf, $\varnothing$. For cubical sets, this is the case if and only if $0 = 1$ is not derivable in the base category, in any variable context.

The following theorem is due to Grothendieck (1983, 44(c)), cf. Theorem 2.6 of Maltsiniotis (2009). We say that A is *totally aspheric* if A is non-empty and all the products $\mathrm{y}(a) \times \mathrm{y}(b)$ are aspheric, where $\mathrm{y} : A \to \widehat{A}$ is the Yoneda embedding.

Theorem 1 (Grothendieck). *If A is a small category that is totally aspheric and has a separated aspheric interval, then A is a strict test category.*

Corollary 2. *If $\mathbb{C}$ is any canonical cube category over the full set of structural rules, $\mathbb{C}_{(\mathrm{wec},b)}$, or the cartesian cube category of de Morgan algebras or that of boolean algebras, then $\mathbb{C}$ is a strict test category.*

Proof. Since $\mathbb{C}$ has finite products it is totally aspheric, as the Yoneda embedding preserves finite products. The Grothendieck interval corresponding to the 1-cube is representable and hence aspheric. This interval is separated as $0 = 1$ is not derivable in any context. □

The following theorem is from Grothendieck (1983, 44(d), Proposition on p. 86):

Theorem 2 (Grothendieck). *If A is a small aspheric category with a separated aspheric interval (I, d^0, d^1) in $\widehat{A}$, and $i : A \to \mathbf{Cat}$ is a functor such that for any a in A, $i(a)$ has a final object, and there is a map of intervals $i_!(I) \to 2$ in* **Cat**, *then A is a test category.*

In this case, i is in fact a *weak test functor*, meaning that $i^* : \mathbf{Cat} \to \widehat{A}$ induces an equivalence $\mathbf{Hot} \to (\mathcal{W}_A)^{-1}\widehat{A}$.

Corollary 3. *Any canonical cube category $\mathbb{C}$ is a test category.*

Proof. The conditions of the theorem hold trivially for any cube category without reversal, taking i to be the canonical functor sending I to 2. If we have reversals, then we can define i by sending I to the category with three objects, $\{0\}, \{1\}, \{0,1\}$ and arrows $\{0\}, \{1\} \to \{0,1\}$. Since this is a Kleene algebra, this functor is well-defined in all cases. □

The cube categories $\mathbb{C}_{\mathrm{dM}}$ and $\mathbb{C}_{\mathrm{BA}}$ of de Morgan and boolean algebras are test categories by Corollary 2.

We now turn to the question of which non-cartesian cube categories are strict test categories. The following theorem was proved by Maltsiniotis (2009, Proposition 3.3) for the case of $\mathbb{C}_{(\mathrm{w},\vee)}$. The same proof works more generally, so we obtain:

Theorem 3. *Any canonical cube category $\mathbb{C}_{(a,b)}$ where b includes one of the connections $\vee$, $\wedge$ is a strict test category.*

That leaves four cases: $(\mathrm{w}, \cdot)$, $(\mathrm{w}, ')$, $(\mathrm{we}, \cdot)$, $(\mathrm{we}, ')$. The first of these, the "classical" cube category, is not a strict test category by the argument of Maltsiniotis (2009, Sect. 5).

Note the unique factorizations we have for these categories: every morphism $f : [m] \to [n]$ factors as degeneracies, followed by (possibly) symmetries, followed by (possibly) reversals, followed by face maps. The isomorphisms are exactly the compositions of reversals and symmetries (if any).

Next, we use variations of the argument of Maltsiniotis (*loc. cit.*) to show that none of these four sites are strict test categories by analyzing the homotopy type of the slice category $\mathbb{C}_{/\Box^1 \times \Box^1}$ in each case.

Theorem 4. *The canonical cube categories $\mathbb{C}_{(\mathrm{w},\cdot)}$, $\mathbb{C}_{(\mathrm{w},')}$, $\mathbb{C}_{(\mathrm{we},\cdot)}$ and $\mathbb{C}_{(\mathrm{we},')}$ are not strict test categories.*

Proof. Let $\mathbb{C}$ be one of these categories. We shall find a full subcategory A of $\mathbb{C}_{/\square^1\times\square^1}$ that is not aspheric, and such that the inclusion is a weak equivalence. Hence $\mathbb{C}_{/\square^1\times\square^1}$ is not aspheric, and $\mathbb{C}$ cannot be a strict test category.

An object of the slice category $\mathbb{C}_{/\square^1\times\square^1}$ is given by a dimension and two terms corresponding to that dimension. We can thus represent it by a variable context $x_1\cdots x_n$ (for an n-cube $[n]$) and two terms s,t in that context.

For the classical cube category case $(\mathrm{w},\cdot)$, we let A contain the following objects, cf. Maltsiniotis (2009, Proposition 5.2):

- 4 zero-dimensional objects $(\cdot,(i,j))$, $i,j\in\{0,1\}$.
- 5 one-dimensional objects; 4 corresponding to the sides of a square, $(x,(i,x))$ and $(x,(x,i))$ with $i\in\{0,1\}$, and 1 corresponding to its diagonal, $(x,(x,x))$.
- 2 two-dimensional objects: $(xy,(x,y))$ and $(xy,(y,x))$ (let us call them, respectively, the northern and southern hemispheres).

In the presence of the exchange rule, we do not need both of the two-dimensional objects (so we retain, say, the northern hemisphere), and in the presence of reversal, we need additionally the anti-diagonal, $(x,(x,x'))$.

Note that in each case A is a partially ordered set (there is at most one morphism between any two objects), and we illustrate the incidence relations between the objects in Fig. 1. The figure also illustrates how to construct functors $F:A\to\mathbf{Top}$, which are cofibrant with respect to the Reedy model structure on this functor category, where A itself is considered a directed Reedy category relative to the obvious dimension assignment (points of dimension zero, lines of dimension one, and the hemispheres of dimension two). We conclude that the homotopy colimit of F (in **Top**) is weakly equivalent to the ordinary colimit of F, which is seen to be equivalent to $S^2\vee S^1$, $S^2\vee S^1\vee S^1$, S^1 and $S^1\vee S^1$, respectively for the cases $(\mathrm{w},\cdot)$, $(\mathrm{w},')$, $(\mathrm{we},\cdot)$ and $(\mathrm{we},')$. Since F takes values in contractible spaces, this homotopy colimit represents in each case the homotopy type of the nerve of A. It follows that A is not aspheric.

It remains to see that the inclusion $A\hookrightarrow\mathbb{C}_{/\square^1\times\square^1}$ is in each case a weak equivalence. For this we use the dual of Quillen's Theorem A, and show that for each object $(x_1\cdots x_n,(s,t))$ (written (s,t) for short) of $\mathbb{C}_{/\square^1\times\square^1}$, the coslice category $B_{(s,t)} = {}_{(s,t)\backslash}A$ has an initial object, and is hence aspheric.

As in Maltsiniotis (*loc. cit.*), we make case distinctions on (s,t). Note that for the languages under consideration, a term can have at most one free variable.

- $(s,t)=(x('),y('))$ for distinct variables x,y. The initial object is a hemisphere (the northern one in the presence of exchange, the southern one without exchange and in case x,y appear in reversed order in the variable context).
- $(s,t)=(x,x)$ or (x',x'). The initial object is then the diagonal in A.
- $(s,t)=(x,x')$ or (x',x). The initial object is here the anti-diagonal.
- $(s,t)=(i,x('))$ or $(x('),i)$. The initial object is the corresponding side of the square.
- $(s,t)=(i,j)$. The initial object is then the corresponding vertex of the square. □

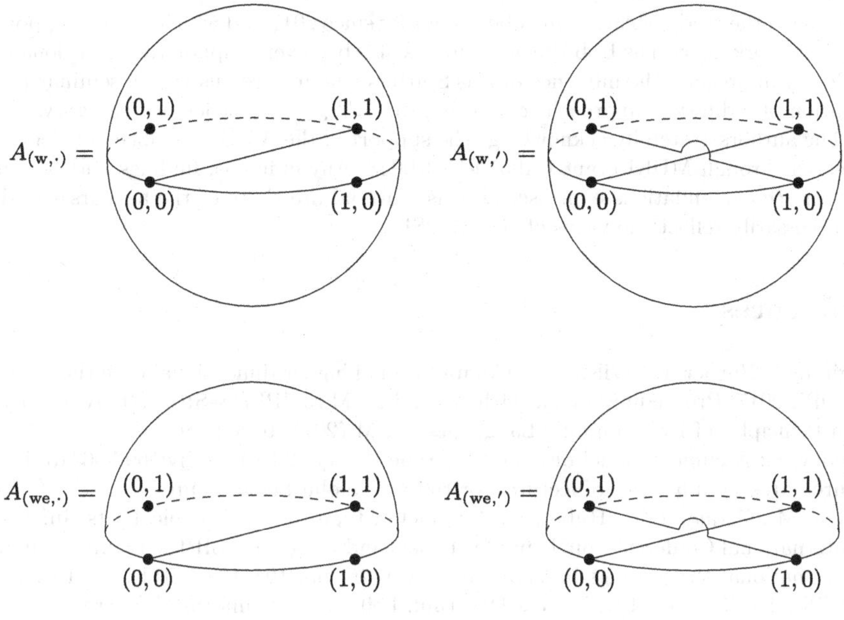

Fig. 1. The partially ordered sets $A = A_{(a,b)}$ in their topological realizations.

We remark that this proof method fails in the presence of a connection, where terms can now refer to multiple variables. And in the presence of contraction, the diagonal (x, x) is incident on the northern hemisphere, which is then a maximal element in A, and hence A is contractible. These observations explain why it is only those four cases that can fail to give strict test categories.

4 Conclusion

We have espoused a systematic algebraic framework for describing notions of cubical sets, and we have shown that all reasonable cubical sites are test categories. We have shown that a cubical site is a strict test category precisely when it is cartesian monoidal or includes one of the connections.

To improve our understanding of the homotopy theory of each cubical site, we need to investigate the induced Quillen equivalence between the corresponding category of cubical sets $\widehat{\mathbb{C}}$ with the Cisinski model structure and the category of simplicial sets $\widehat{\Delta}$ with the Kan model structure. For instance, we can ask whether the fibrations are those satisfying the cubical Kan filling conditions. Furthermore, to model type theory, one should probably require that the model structures lift to algebraic model structures. We leave all this for future work.

Acknowledgements. We wish to thank the members of the HoTT group at Carnegie Mellon University for many fruitful discussions, in particular Steve Awodey who has encouraged the study of cartesian cube categories since 2013 and who has been supportive of our work, as well as Bob Harper who has also been very supportive. Additionally, we deeply appreciate the influence of Bas Spitters who inspired us with a seminar presentation of a different approach to showing that $\mathbb{C}_{(\mathrm{wec},\cdot)}$ is a strict test category.

The authors gratefully acknowledge the support of the Air Force Office of Scientific Research through MURI grant FA9550-15-1-0053. Any opinions, findings and conclusions or recommendations expressed in this material are those of the authors and do not necessarily reflect the views of the AFOSR.

References

Angiuli, C., Harper, R., Wilson, T.: Computational higher-dimensional type theory. In: POPL 2017: Proceedings of the 44th Annual ACM SIGPLAN-SIGACT Symposium on Principles of Programming Languages. ACM (2017, to appear)

Awodey, S.: A cubical model of homotopy type theory (2016). arXiv:1607.06413. Lecture notes from a series of lectures for the Stockholm Logic group

Bezem, M., Coquand, T., Huber, S.: A model of type theory in cubical sets. In: 19th International Conference on Types for Proofs and Programs, LIPIcs, Leibniz Leibniz International Proceedings in Informatics, vol. 26, pp. 107–128 (2014). doi:10.4230/LIPIcs.TYPES.2013.107. Schloss Dagstuhl. Leibniz-Zent. Inform., Wadern

Cohen, C., Coquand, T., Huber, S., Mörtberg, A.: Cubical type theory: a constructive interpretation of the univalence axiom. In: 21st International Conference on Types for Proofs and Programs, LIPIcs, Leibniz International Proceedings in Informatics (2016, to appear). Schloss Dagstuhl. Leibniz-Zent. Inform., Wadern

Gehrke, M., Walker, C.L., Walker, E.A.: Normal forms and truth tables for fuzzy logics. Fuzzy Sets Syst. **138**(1), 25–51 (2003). doi:10.1016/S0165-0114(02)00566-3. Selected papers from the 21st Linz Seminar on Fuzzy Set Theory (2000)

Grandis, M., Mauri, L.: Cubical sets and their site. Theory Appl. Categ. **11**(8), 185–211 (2003). http://www.tac.mta.ca/tac/volumes/11/8/11-08abs.html

Grothendieck, A.: Pursuing stacks. Manuscript (1983). http://thescrivener.github.io/PursuingStacks/

Maltsiniotis, G.: La théorie de l'homotopie de Grothendieck. Astérisque, vol. 301 (2005). https://webusers.imj-prg.fr/~georges.maltsiniotis/ps/prstnew.pdf

Maltsiniotis, G.: La catégorie cubique avec connexions est une catégorie test stricte. Homology Homotopy Appl. **11**(2), 309–326 (2009). doi:10.4310/HHA.2009.v11.n2.a15

Mauri, L.: Algebraic theories in monoidal categories. Unpublished preprint (2005)

Quillen, D.: Higher algebraic K-theory: I. In: Bass, H. (ed.) Higher K-Theories. LNM, vol. 341, pp. 85–147. Springer, Berlin (1973). doi:10.1007/BFb0067053

Non-associative Kleene Algebra and Temporal Logics

Jules Desharnais[1] and Bernhard Möller[2(✉)]

[1] Université Laval, Quebec City, QC, Canada
jules.desharnais@ift.ulaval.ca
[2] Institut für Informatik, Universität Augsburg, Augsburg, Germany
bernhard.moeller@informatik.uni-augsburg.de

Abstract. We introduce new variants of Kleene star and omega iteration for the case where the iterated operator is neither associative nor has a neutral element. The associated *repetition algebras* are used to give closed semantic expressions for the *Until* and *While* operators of the temporal logic CTL* and its sublogics CTL and LTL. Moreover, the relation between the semantics of these logics can be expressed by homomorphisms between repetition algebras, which is a more systematic and compact approach than the ones taken in earlier papers.

Keywords: Temporal logics · Semantics · Kleene algebra · Repetition algebra

1 Introduction

The temporal logic CTL* and its sublogics CTL and LTL (see [7] for an excellent survey) are prominent tools in the analysis of concurrent and reactive systems. Although they are well understood, one still rarely finds algebraic treatments of their semantics which provide a better understanding and yield simpler (and, at the same time, completely formal) proofs of the semantic properties.

In the present paper we take up the approach of [15] and refine it in several ways. First, we present a variant of Kleene algebra where the underlying multiplication is not assumed to be associative. Such an operator arises, e.g., in the semantics of the until operator of CTL* and its relatives. Therefore we present a general investigation of the star and omega for such operators in what we call repetition algebras. The relation between various semantics for CTL* and CTL can then be expressed by homomorphisms between repetition algebras; in particular, several tedious ad-hoc applications of the principle of least/greatest fixed point fusion that occurred in [15] are now replaced by a single proof for general repetition algebras. Another new feature is a much cleaner separation between finite and infinite traces than in the predecessor paper. Also a number of new results concerning the universal trace quantifier A and the globality operator G arise. For lack of space we omit all proofs; they are found in the report [4].

P. Höfner et al. (Eds.): RAMiCS 2017, LNCS 10226, pp. 93–108, 2017.
DOI: 10.1007/978-3-319-57418-9_6

2 Modelling CTL*

To make the paper self-contained we recall some basic facts about CTL*; in this we largely follow [7]. Formulas in CTL* characterise sets of traces, where a trace is a finite or infinite sequence of program states. A set Φ of *atomic propositions* is used to distinguish sets of states. The syntax of the language Ψ of CTL* *formulas* over Φ is given by the grammar

$$\Psi ::= \bot \mid \Phi \mid \Psi \to \Psi \mid \mathsf{E}\Psi \mid \mathsf{X}\Psi \mid \Psi \mathbin{\mathsf{U}} \Psi,$$

where $\bot$ denotes falsity, $\to$ is logical implication, E is the existential quantifier on traces, and X and U are the next-time and until operators.

We briefly recall the informal semantics. A trace is said to satisfy an atomic formula iff its first state does. A trace σ satisfies $\mathsf{E}\varphi$ iff there is some trace τ that satisfies φ and has the same first state as σ. The formula $\mathsf{X}\varphi$ holds for a trace σ if φ holds for the remainder of σ after one step. A trace σ satisfies $\varphi \mathbin{\mathsf{U}} \psi$ iff after a finite number (including zero) of X steps within σ the remaining trace satisfies ψ and all intermediate trace pieces for which ψ does not yet hold satisfy φ.

The logical connectives $\neg, \wedge, \vee, \mathsf{A}$ are defined, as usual, by $\neg\varphi =_{df} \varphi \to \bot$, $\top =_{df} \neg\bot$, $\varphi \wedge \psi =_{df} \neg(\varphi \to \neg\psi)$, $\varphi \vee \psi =_{df} \neg\varphi \to \psi$ and $\mathsf{A}\varphi =_{df} \neg\mathsf{E}\neg\varphi$. Moreover, the "finally" operator F and the "globally" operator G are defined by

$$\mathsf{F}\psi =_{df} \top \mathsf{U} \psi \quad \text{and} \quad \mathsf{G}\psi =_{df} \neg\mathsf{F}\neg\psi.$$

Informally, $\mathsf{F}\psi$ holds if after a finite number of steps the remainder of the trace satisfies ψ, while $\mathsf{G}\psi$ holds if after every finite number of steps ψ still holds.

The sublanguages Ξ of *state formulas*[1] that denote sets of states and Π of *trace formulas*[2] that denote sets of computation traces are given by

$$\Xi ::= \bot \mid \Phi \mid \Xi \to \Xi \mid \mathsf{E}\Pi,$$
$$\Pi ::= \Xi \mid \Pi \to \Pi \mid \mathsf{X}\Pi \mid \Pi \mathbin{\mathsf{U}} \Pi.$$

To motivate our algebraic semantics, we briefly recapitulate the standard CTL* semantics of formulas. Its basic objects are traces σ from Σ^ω, the set of infinite words over some set Σ of states. The i-th element of σ (indices starting with 0) is denoted σ_i, and σ^i is the trace that results from σ by removing its first i elements. Hence $\sigma^0 = \sigma$.

Each atomic proposition $\pi \in \Phi$ is associated with the set $\Sigma_\pi \subseteq \Sigma$ of states for which π holds. The relation $\sigma \models \varphi$ of *satisfaction* of a formula φ by a trace σ is defined inductively (see e.g. [7]) by

$$\begin{array}{llll}
\sigma \not\models \bot, & & \sigma \models \mathsf{E}\varphi & \text{iff } \exists\, \tau \in \Sigma^\omega : \tau_0 = \sigma_0 \text{ and } \tau \models \varphi, \\
\sigma \models \pi & \text{iff } \sigma_0 \in \Sigma_\pi, & \sigma \models \mathsf{X}\varphi & \text{iff } \sigma^1 \models \varphi, \\
\sigma \models \varphi \to \psi & \text{iff } \models \varphi \text{ implies } \sigma \models \psi, & \sigma \models \varphi \mathbin{\mathsf{U}} \psi & \text{iff } \exists\, j \geq 0 : \ \sigma^j \models \psi \text{ and} \\
 & & & \phantom{\text{iff }} \forall\, k < j : \ \sigma^k \models \varphi.
\end{array}$$

In particular, $\sigma \models \neg\varphi$ iff $\sigma \not\models \varphi$.

[1] In the literature this set is usually called Σ. We avoid this, since throughout the paper we use Σ for sets of states.

[2] In the literature these are mostly called *path formulas*.

We quickly repeat the proof of validity of the CTL^* axiom

$$\neg \mathsf{X}\varphi \leftrightarrow \mathsf{X}\neg\varphi, \tag{1}$$

since this will be crucial for the algebraic representation of X in Sect. 6:

$$\sigma \models \neg\mathsf{X}\varphi \Leftrightarrow \sigma \not\models \mathsf{X}\varphi \Leftrightarrow \sigma^1 \not\models \varphi \Leftrightarrow \sigma^1 \models \neg\varphi \Leftrightarrow \sigma \models \mathsf{X}\neg\varphi.$$

3 Semirings, Quantales and Iteration

We formulate our more abstract developments in terms of algebraic structures. The elements of these structures may, for instance, stand for sets of traces.

Definition 3.1

1. An *idempotent left semiring*, briefly IL-semiring, is a structure $(A, +, \cdot, 0, 1)$ such that $(A, +, 0)$ is a commutative monoid with idempotent addition, that is, $(A, \cdot, 1)$ is a monoid, multiplication distributes from the right over addition and 0 is a *left annihilator* for multiplication, that is, $0 \cdot a = 0$ for all $a \in A$. An IL-semiring is *left-distributive* if multiplication distributes over addition also from the left.
2. Every IL-semiring can be partially ordered by setting $a \leq b \Leftrightarrow_{df} a + b = b$. Then $+$ and $\cdot$ are isotone w.r.t. $\leq$ and 0 is the least element. Moreover, $a + b$ is the supremum of $a, b \in A$. An IL-semiring is *bounded* if it has a greatest element $\top$.
3. An IL-semiring is called a *left quantale* [12] if $\leq$ induces a complete lattice and multiplication distributes over arbitrary suprema from the right. The infimum and the supremum of a subset $B \subseteq A$ are denoted by $\bigsqcap B$ and $\bigsqcup B$, respectively. Their binary variants are $a \sqcap b$ and $a \sqcup b$ (the latter coinciding with $a + b$).
4. In left quantales finite and infinite iteration can be defined as least and greatest fixed points, namely $a^* =_{df} \mu x \,.\, 1 + a \cdot x$ and $a^\omega =_{df} \nu x \,.\, a \cdot x$. For details and properties see [12].
5. The IL-semiring/left quantale is *Boolean* if $(A, \leq)$ induces a Boolean algebra.

Quantales (or *standard Kleene algebras* [2]) have been used in many contexts other than that of program semantics (cf. the general reference [16]). They have the advantage that the general fixpoint calculus is available there. A number of our proofs need the principle of fixpoint fusion which is a second-order principle; in the first-order setting of conventional Kleene algebras [9] only special cases of it, like the induction and co-induction rules, can be used as axioms.

Example 3.2. We want to use an algebra of sets of traces. We set $\Sigma^\infty =_{df} \Sigma^+ \cup \Sigma^\omega$, where Σ^+ is the set of non-empty finite traces over Σ. The operator . denotes concatenation of traces. First we define the partial operation of the *fusion product* that glues traces together at a common point, if any. For $\sigma, \tau \in \Sigma^\infty$,

$$\sigma \bowtie \tau = \begin{cases} \sigma & \text{if } \sigma \in \Sigma^\omega, \\ \sigma'.x.\tau' & \text{if } \sigma \in \Sigma^+, \sigma = \sigma'.x, \tau = x.\tau' \text{ for some } x \in \Sigma, \\ \text{undefined otherwise.} \end{cases}$$

The *purely infinite* and *purely finite* parts of a set V of traces are $\mathsf{inf}\,V =_{df} V \cap \Sigma^\omega$ and $\mathsf{fin}\,V =_{df} V - \mathsf{inf}\,V$. With this we extend $\bowtie$ to trace sets V, W as

$$V \bowtie W =_{df} \mathsf{inf}\,V \cup \{s \bowtie t : s \in \mathsf{fin}\,V \wedge t \in W\}.$$

This operation has the set Σ, viewed as a set of one-element traces, as its neutral element. Moreover, $V \bowtie \emptyset = \mathsf{inf}\,V$ and hence $V \bowtie \emptyset = \emptyset$ iff $\mathsf{inf}\,V = \emptyset$. This will be generalised in Sect. 7.

Now we define the Boolean left quantale $\mathsf{TRC}(\Sigma)$ of sets of finite and infinite traces by $\mathsf{TRC}(\Sigma) =_{df} (\mathcal{P}(\Sigma^\infty), \cup, \bowtie, \emptyset, \Sigma)$. This quantale has the greatest element $\top = \Sigma^\infty$ and is even left-distributive. A transition relation over a state set Σ can be modelled in $\mathsf{TRC}(\Sigma)$ as a set R of words of length 2. The powers R^i of R consist of traces of length $i+1$ that are generated by R-transitions. In particular, we instantiate R to $\Sigma^2 =_{df} \Sigma.\Sigma$, the set of all two-letter words and hence the most general next-step transition relation. Then $\mathsf{TRC}(\Sigma)$ is generated by Σ^2 as $\mathsf{TRC}(\Sigma) = (\Sigma^2)^* \cup (\Sigma^2)^\omega$. This is generalised in Sect. 8. □

Next to an abstract representation of sets of traces we will also need one for sets of states. This is achieved by the notion of tests [10].

Definition 3.3. A *test* in an IL-semiring is an element p that has a complement $\neg p$ relative to the multiplicative unit 1, namely $p + \neg p = 1$ and $p \cdot \neg p = 0 = \neg p \cdot p$. The set of all tests in A is denoted by $\mathsf{test}(A)$.

The element $\neg p$ is uniquely determined by these axioms if it exists. In a Boolean IL-semiring every element $p \leq 1$ is a test with $\neg p = \overline{p} \sqcap 1$, where $\overline{}$ is the general complement operator (that need not exist in non-Boolean IL-semirings). The expressions $p \cdot a$ and $a \cdot p$ abstractly represent restriction of the traces in a to the ones that start and end in p-states, resp.

In $\mathsf{TRC}(\Sigma)$ the multiplicative identity Σ has exactly the subsets of Σ as its sub-objects, hence there the tests faithfully represent sets of states.

Using tests we can also define a domain operator and the modal operators diamond and box (cf. [5]). Due to the existence of $\top$ we can use a slightly different but equivalent axiomatisation than given there; the equivalence is established by Lemma 9.1 of that paper.

Definition 3.4. A bounded IL-semiring A is called a *domain IL-semiring* if it has a *domain operator* $\ulcorner : A \to \mathsf{test}(A)$ axiomatised, for $a, b \in A, q \in \mathsf{test}(A)$, by the Galois connection $\ulcorner a \leq q \Leftrightarrow a \leq q \cdot \top$ together with the axiom of *locality*

$$\ulcorner(a \cdot b) = \ulcorner(a \cdot \ulcorner b). \tag{2}$$

Then we set $|a\rangle q =_{df} \ulcorner(a \cdot q)$ and $|a]q =_{df} \neg|a\rangle\neg q$.

The locality property means that the domain of a composition does not depend on the inner structure of the second operand, but only on its starting states.

In $\mathsf{TRC}(\Sigma)$, for trace set V the domain $\ulcorner V$ consists of all starting letters of traces in V. Moreover $|V\rangle P$ for some set $P \subseteq \Sigma$ is the set of all starting states of traces in V that end in some state in P, hence a kind of inverse image of P under V. Dually, $|V]P$ consists of those states x for which all traces in V starting in x have their final states, if any, in P.

We recall a few basic properties; see [5] for more details.

Lemma 3.5. *Let A be a domain IL-semiring, $a, b \in A$ and $p, q \in \mathsf{test}(A)$.*

1. $a = {}^{\ulcorner}a \cdot a$ *and* ${}^{\ulcorner}(p \cdot a) \leq p$.
2. ${}^{\ulcorner}(p \cdot \top) = p$.
3. $p \leq q \Leftrightarrow p \cdot \top \leq q \cdot \top$.
4. *If $a \sqcap b$ exists then* $p \cdot (a \sqcap b) = p \cdot a \sqcap b = a \sqcap p \cdot b$. *Hence if $b \leq a$ then* $p \cdot a \sqcap b = p \cdot b$. *In particular,* $\underline{p \cdot \top} \sqcap b = p \cdot b$.
5. *If A is Boolean then* $\neg p \cdot \top = \overline{p \cdot \top}$.
6. $|a \cdot b\rangle q = |a\rangle(|b\rangle q)$ *and* $|a \cdot b]q = |a](|b]q)$.
7. $p \cdot |b\rangle q = |p \cdot b\rangle q$ *(import/export).*
8. $p \leq q \cdot |a]p \Rightarrow p \leq |a^*]q$ *(box induction).*

By these properties we can represent the set of all possible traces that start with some state in set p by the *test ideal* $p \cdot \top$. By Part 3 the set of test ideals is isomorphic to the set of tests.

4 General Algebraic Semantics of CTL*

We now give our algebraic interpretation of CTL^* over a Boolean left domain quantale A. As a preparation we transform the semantics from Sect. 2 into a set-based one by assigning to each formula φ the set $[\![\varphi]\!] =_{df} \{\sigma \mid \sigma \models \varphi\}$ of traces that satisfy it.

$$
\begin{aligned}
[\![\bot]\!] &= \emptyset, & [\![\mathsf{E}\varphi]\!] &= {}^{\ulcorner}[\![\varphi]\!] \bowtie \Sigma^\omega, \\
[\![\pi]\!] &= \Sigma_\pi \bowtie \Sigma^\omega, & [\![\mathsf{X}\,\varphi]\!] &= \Sigma^2 \bowtie [\![\varphi]\!], \\
[\![\varphi \to \psi]\!] &= \overline{[\![\varphi]\!]} \cup [\![\psi]\!], & [\![\varphi\,\mathsf{U}\,\psi]\!] &= \bigcup_{j \geq 0} ((\Sigma^2)^j \bowtie [\![\psi]\!] \cap \bigcap_{k<j} (\Sigma^2)^k \bowtie [\![\varphi]\!]).
\end{aligned}
$$

Note that in Σ^2 the power is taken w.r.t. the concatenation operator . whereas j and k denote powers w.r.t. $\bowtie$.

As in this set-based semantics, every atomic proposition $\pi \in \Phi$ is algebraically associated with a set $\Sigma_\pi \subseteq \Sigma$ of states, i.e., with an element of $\mathsf{test}(\mathsf{TRC}(\Sigma))$. Therefore, to save some notation, in the algebraic semantics we simply set $\Phi = \mathsf{test}(A)$. Moreover, we fix an element x (where x stands for "next" and corresponds to Σ^2) that represents the transition system underlying the logic. The precise requirements for x will be discussed in Sect. 6. Then the concrete semantics above generalises to a function $[\![\_]\!] : \Psi \to A$, where $[\![\varphi]\!]$ abstractly represents the set of traces satisfying formula φ.

Definition 4.1. The *general algebraic semantics* $[\![\varphi]\!]$ of CTL^* formula φ is defined inductively over the structure of φ. This results from the set-based semantics by a straightforward translation of the concrete operators of $\mathsf{TRC}(\Sigma)$ into the corresponding quantale operators:

$$\begin{aligned} [\![\bot]\!] &= 0, & [\![\mathsf{E}\varphi]\!] &= \ulcorner[\![\varphi]\!]\cdot\top, \\ [\![p]\!] &= p\cdot\top, & [\![\mathsf{X}\,\varphi]\!] &= \mathsf{x}\cdot[\![\varphi]\!], \\ [\![\varphi\to\psi]\!] &= \overline{[\![\varphi]\!]}+[\![\psi]\!], & [\![\varphi\,\mathsf{U}\,\psi]\!] &= \bigsqcup_{j\geq 0}(\mathsf{x}^j\cdot[\![\psi]\!]\sqcap\bigsqcap_{k<j}\mathsf{x}^k\cdot[\![\varphi]\!]). \end{aligned}$$

As a word of warning, the definition $[\![p]\!]=p\cdot\top$ does not correspond exactly to the TRC semantics, where $[\![\pi]\!]=\Sigma_\pi\bowtie\Sigma^\omega$ and $\Sigma^\omega\neq\top$. This problem will be taken up in Sect. 7.

Using the above definitions, it is easy to check that

$$[\![\varphi\vee\psi]\!]=[\![\varphi]\!]+[\![\psi]\!],\ [\![\varphi\wedge\psi]\!]=[\![\varphi]\!]\sqcap[\![\psi]\!],\ [\![\neg\varphi]\!]=\overline{[\![\varphi]\!]},\ [\![\top]\!]=\top. \tag{3}$$

Then the above semantics coincides with that of Sect. 2, as far as infinite streams are concerned. This is discussed in detail in Sects. 6 and 7.

To exemplify our semantics we state a number of properties of the trace quantifiers. In particular, we work out a more explicit form of the A semantics.

Corollary 4.2. $[\![\mathsf{EE}\psi]\!]=[\![\mathsf{E}\psi]\!]$ *and* $[\![\mathsf{AA}\psi]\!]=[\![\mathsf{A}\psi]\!]$ *and* $[\![\mathsf{A}\psi]\!]=\neg\ulcorner\overline{[\![\psi]\!]}\cdot\top$.

Moreover, for the CTL^* axiom $\mathsf{EX}\top$ [7] we obtain the following result.

Lemma 4.3. $[\![\mathsf{EX}\top]\!]=\top\Leftrightarrow\ulcorner\mathsf{x}=1$.

In a relational setting the property $\ulcorner\mathsf{x}=1$ means that x is a left-total transition relation.

5 Modified Iteration and the Semantics of Until

We now deal with the semantics of the until operator. To bring the corresponding expression in Definition 4.1 into more palatable shape we introduce a bit of notation. For elements $a,b\in A$ and $j\in\mathbb{N}$ we set

$$a\,\boxed{j}\,b=_{df}\mathsf{x}^j\cdot b\sqcap\bigsqcap_{k<j}\mathsf{x}^k\cdot a, \tag{4}$$

which is the expression occurring in the right hand side of the semantic equation for $[\![\varphi\,\mathsf{U}\,\psi]\!]$ when $a=[\![\varphi]\!]$ and $b=[\![\psi]\!]$. It states that φ holds j times and then ψ holds. The idea is now to find an inductive formulation of $\boxed{j}$ driven by j. For the induction base we calculate, using the definitions of $\boxed{0}$ and powers, neutrality of 1 and lattice algebra, $a\,\boxed{0}\,b=\mathsf{x}^0\cdot b\sqcap\bigsqcap_{k<0}\mathsf{x}^k\cdot a=b\sqcap\top=b$. To proceed with the induction step we need an assumption about x that is closely related to (1), as is discussed in detail in Sect. 6. This condition reads

$$\forall\,a,b\in A:\mathsf{x}\cdot(a\sqcap b)=\mathsf{x}\cdot a\sqcap\mathsf{x}\cdot b. \tag{LDM}$$

It means that left multiplication by x distributes through binary and hence non-empty finite meets. With that we calculate as follows. By definition, splitting the $\sqcap$ expression, definition of powers and neutrality of 1, commutativity of $\sqcap$, index shift, (LDM), definition of $\boxed{j}$ and the definition below,

$$\begin{aligned} a \boxed{j+1} b &= \mathsf{x}^{j+1} \cdot b \sqcap \bigsqcap_{k<j+1} \mathsf{x}^k \cdot a \;=\; \mathsf{x}^{j+1} \cdot b \sqcap \mathsf{x}^0 \cdot a \sqcap \bigsqcap_{k=1}^{j} \mathsf{x}^k \cdot a \\ &= a \sqcap \mathsf{x} \cdot \mathsf{x}^j \cdot b \sqcap \bigsqcap_{k=1}^{j} \mathsf{x} \cdot \mathsf{x}^{k-1} \cdot a \;=\; a \sqcap \mathsf{x} \cdot \mathsf{x}^j \cdot b \sqcap \bigsqcap_{l<j} \mathsf{x} \cdot \mathsf{x}^l \cdot a \\ &= a \sqcap \mathsf{x} \cdot (\mathsf{x}^j \cdot b \sqcap \bigsqcap_{l<j} \mathsf{x}^l \cdot a) \;=\; a \sqcap \mathsf{x} \cdot (a \boxed{j} b) \;=\; a \,\Box\, (a \boxed{j} b), \end{aligned}$$

where

$$c \,\Box\, d =_{df} c \boxed{1} d = c \sqcap \mathsf{x} \cdot d. \tag{5}$$

The inductive clause for $\boxed{j}$ will be the basis for an inductive (or recursive) formulation of the until semantics.

We can now formulate the semantics of until more compactly as

$$[\![\varphi \,\mathsf{U}\, \psi]\!] = \bigsqcup_{j \geq 0} [\![\varphi]\!] \boxed{j} [\![\psi]\!]. \tag{6}$$

Below we will relate this to a fixed point equation for U.

The operator $\Box$ enjoys a number of pleasant properties, as will be seen below. However, in general it is neither associative nor does it have a neutral element. Nevertheless it gives rise to an analogue of the Kleene star which will even allow us to bring the semantics of the until operator into closed form.

To do this, we abstract from the concrete definitions above.

Definition 5.1. Consider a set S and an arbitrary, possibly non-associative operator $\Box : S \times S \to S$.

1. We define the iterations $\boxed{j}$ of $\Box$ as above by

$$a \boxed{0} b =_{df} b, \qquad a \boxed{j+1} b =_{df} a \,\Box\, (a \boxed{j} b).$$

2. The structure $(S, \Box)$ is called a *repetition algebra*[3] if S is a complete lattice with order $\leq$, least element 0 and binary supremum operator $+$, and $\Box$ is isotone in both arguments.
3. In a repetition algebra we define variants of the star and omega operators:

$$a \boxed{*} b =_{df} \mu f_{a,b} \text{ where } f_{a,b}(x) =_{df} a \,\Box\, x + b \;,\; a^{\boxed{\omega}} =_{df} \nu x \,.\, a \,\Box\, x. \tag{7}$$

In fact, $\boxed{*}$ corresponds to Kleene's original definition of $*$ as an infix operator in [8]. Not surprisingly, $\boxed{*}$ and $^{\boxed{\omega}}$ enjoy properties analogous to those of * and $^\omega$. We recall that an endofunction on a complete lattice is *(co-)continuous* if it preserves all joins (meets) of non-empty chains.

[3] We would have preferred the term *iteration algebra* which, however, is already used in [1] and follow-up papers with a different meaning.

Lemma 5.2. *Consider a repetition algebra $(S, \Box)$.*

1. *The operators $\boxed{*}$ and $\boxed{\omega}$ are isotone.*
2. $a \boxed{i+j} a = a \boxed{i} (a \boxed{j} b)$.
3. *If $\Box$ is right-strict, i.e., if $a \Box 0 = 0$ for all a, and distributes through arbitrary joins and binary meets in its right argument then $f_{a,b}$ from (7) is continuous and $a \boxed{*} b = \bigsqcup_{j \geq 0} a \boxed{j} b$.*
4. $b \leq a \boxed{*} b$.
5. $a \Box b \leq a \boxed{*} b$.
6. $a \boxed{*} (a \Box b) \leq a \Box (a \boxed{*} b)$.
7. $a \boxed{*} (a \boxed{*} b) = a \boxed{*} b$.
8. *If $\Box$ is left-strict, i.e., if $0 \Box a = 0$ for all a, then $0 \boxed{*} b = b$ and $0^{\boxed{\omega}} = 0$.*
9. *If $a \Box 0 = 0$ then $a \boxed{*} 0 = 0$.*
10. *If $\Box$ is left-distributive then $a \boxed{*} (b + c) = a \boxed{*} b + a \boxed{*} c$.*
11. $a \boxed{*} a^{\boxed{\omega}} = a^{\boxed{\omega}}$.
12. *If S is a universally distributive complete lattice then $\nu f_{a,b} = \mu f_{a,b} + a^{\boxed{\omega}} = a \boxed{*} b + a^{\boxed{\omega}}$.*

A main tool used in the subsequent sections is that of projections from one repetition algebra to another.

Definition 5.3. Let $(S_i, \Box_i)_{i=1,2}$ be repetition algebras. A *homomorphism* between them is a function $h : S_1 \to S_2$ that is continuous and strict and preserves $+$ and $\Box$ in that $h(a +_1 b) = h(a) +_2 h(b)$ and $h(a \Box_1 b) = h(a) \Box_2 h(b)$ for all $a, b \in S_1$.

Lemma 5.4. *Let $(S_i, \Box_i)_{i=1,2}$ be repetition algebras with a* homomorphism *$h : S_1 \to S_2$. Then h preserves $\boxed{*}$ as well, i.e., $h(a \boxed{*}_1 b) = h(a) \boxed{*}_2 h(b)$ for all $a, b \in S_1$. If h is co-continuous and* co-strict, *i.e., satisfies $h(\top) = \top$, then it also preserves $\boxed{\omega}$, i.e., $h(a^{\boxed{\omega}_1}) = h(a)^{\boxed{\omega}_2}$ for all $a \in S_1$.*

We now return to the concrete instance of $\Box$ defined in (5). To make use of Lemma 5.2 we need to ensure that $\Box$ has the necessary properties. Fortunately, this is achieved by stipulating besides (LDM) a second requirement on the semantic element x, motivated by the semantics of X as follows. In $\mathsf{TRC}(\Sigma)$, for arbitrary formula φ and its semantics $V = [\![\varphi]\!]$ we want

$$[\![\mathsf{X}\varphi]\!] = \mathsf{x} \bowtie V = \mathsf{x} \bowtie \bigcup_{v \in V} \{v\} = \bigcup_{v \in V} \mathsf{x} \bowtie \{v\}.$$

Therefore, we require that in the abstract quantale semantics left multiplication by x distributes through arbitrary joins.

Definition 5.5. In a left quantale A we call $\mathsf{x} \in A$ a *step* if left multiplication by x distributes through arbitrary joins and binary meets. In particular, $\mathsf{x} \cdot 0 = 0$.

Now Lemma 5.2 applies and yields the following theorem that provides an important check of the adequacy of our definitions.

Theorem 5.6. *Assume a Boolean left domain quantale with a step* x. *Then*

$$[\![\varphi \mathsf{U} \psi]\!] = [\![\varphi]\!] \circledast [\![\psi]\!].$$

This yields the following simpler closed representation of F from Sect. 2:

Corollary 5.7. $[\![\mathsf{F}\psi]\!] = \mathsf{x}^* \cdot [\![\psi]\!]$. *In particular,* $[\![\mathsf{F}\top]\!] = \top$.

The operator G and its relation with the $\boxed{\omega}$ operator are treated in Sect. 7.

6 The Next-Time Operator

We now discuss the connection between (1) and (LDM) in the algebraic setting. To satisfy (1), we need to have for all formulas φ and their semantic values $a =_{df} [\![\varphi]\!]$ that $\overline{\mathsf{x} \cdot a} = [\![\neg \mathsf{X}\varphi]\!] = [\![\mathsf{X}\neg\varphi]\!] = \mathsf{x} \cdot \overline{a}$. This semantic property can equivalently be characterised as follows (Parts 1 and 2 were already shown in [3]).

Lemma 6.1. *Consider a Boolean IL-semiring* A *and* $\mathsf{x} \in A$.

1. *If* x *is left-distributive, i.e.,* $\mathsf{x} \cdot (a + b) = \mathsf{x} \cdot a + \mathsf{x} \cdot b$ *for all* a, b, *and satisfies* $\forall a \in A : \mathsf{x} \cdot \overline{a} \leq \overline{\mathsf{x} \cdot a}$ *then* (LDM) *and* $\mathsf{x} \cdot 0 = 0$ *hold.*
2. *If* (LDM) *and* $\mathsf{x} \cdot 0 = 0$ *hold then so does* $\forall a \in A : \mathsf{x} \cdot \overline{a} \leq \overline{\mathsf{x} \cdot a}$.
3. *If* x *is left-distributive then* $\forall a \in A : \overline{\mathsf{x} \cdot a} \leq \mathsf{x} \cdot \overline{a} \Leftrightarrow \mathsf{x} \cdot \top = \top \Leftrightarrow \mathsf{x}^\omega = \top$.
4. *If* x *satisfies* (LDM) *and* $\forall a : \mathsf{x} \cdot \overline{a} = \overline{\mathsf{x} \cdot a}$ *then* x *is left-distributive.*

In relation algebra, the special case $\mathsf{x} \cdot \overline{1} \leq \overline{\mathsf{x}}$ of the property in Part 1 characterises x as a partial function and is equivalent to the full property $\forall a : \mathsf{x} \cdot \overline{a} \leq \overline{\mathsf{x} \cdot a}$ [17]. But in general quantales the special and the full case are not equivalent [3]. Moreover, again from [3], we know that in quantales such as TRC left multiplication by an element x distributes over meet iff x is prefix-free, i.e., if no member of x is a prefix of another member. This holds in particular if all words in x have equal length, which is the case if x models a transition relation and hence consists only of words of length 2. The equivalent condition $\forall a : \mathsf{x} \cdot a \sqcap \mathsf{x} \cdot \overline{a} = 0$ was used in the computation calculus of R.M. Dijkstra [6].

But what about Lemma 6.1.3? Only rarely will a quantale be "generated" by x in the sense that $\mathsf{x}^\omega = \top$. We deal with this problem in Sects. 7 and 8.

7 Infinitary Semantics of CTL*

Before we tackle a general algebraic solution to the problem mentioned at the end of the previous section, let us look at the concrete quantale $A = \mathsf{TRC}(\Sigma)$. There we definitely do *not* have $\mathsf{x}^\omega = \top$ for $\mathsf{x} = \Sigma^2$, since $\mathsf{x}^\omega = \Sigma^\omega = \mathsf{inf}\, A$, where the inf operator was introduced in Example 3.2.

We will show that restricting the semantics given in Sect. 4 to infinite words remedies this problem, while at the same time faithfully reflecting the original semantics of CTL*, which was given in terms of infinite sequences of states anyway.

To obtain an abstract algebraic version of this, we need some additional notions. The key is the observation in Example 3.2 that $V \bowtie \emptyset = \mathsf{inf}\, V$ and hence $V \bowtie \emptyset = \emptyset$ iff $\mathsf{inf}\, V = \emptyset$.

This motivates the following definition.

Definition 7.1. Assume a bounded IL-semiring A.

1. The *purely infinite part* of $a \in A$ is $\mathsf{inf}\, a =_{df} a \cdot 0$. We call a *purely infinite* or *non-terminating* if $a = \mathsf{inf}\, a$. We set $\mathsf{N} =_{df} \mathsf{inf}\, \top$; hence N is the greatest *n*onterminating element. The set of all purely infinite elements is denoted by $\mathsf{infel}(A)$.
2. Dually, we call a *purely finite* if $\mathsf{inf}\, a = a \cdot 0 \leq 0$, i.e., if its purely infinite part is trivial. The right hand side is equivalent to $a \cdot 0 = 0$.
3. If A is Boolean we can define the *purely finite part* of $a \in A$ analogously as in $\mathsf{TRC}(\Sigma)$ by $\mathsf{fin}\, a =_{df} a - \mathsf{inf}\, a$.

We state some simple consequences of the definition; for more details see [12].

Lemma 7.2. *Consider arbitrary* $a, b \in A$.

1. *If* b *is purely infinite then so is* $a \cdot b$.
2. $\mathsf{inf}\,(a \cdot b) = a \cdot \mathsf{inf}\, b$. *In particular,* inf *commutes with left restriction, i.e., for* $p \in \mathsf{test}(A)$, $\mathsf{inf}\,(p \cdot b) = p \cdot \mathsf{inf}\, b$.
3. $a \cdot \mathsf{N} \leq \mathsf{N}$.
4. *The operator* inf *is a kernel operator, i.e., it is contractive (*$\mathsf{inf}\, a \leq a$*), isotone and idempotent (*$\mathsf{inf}\,(\mathsf{inf}\, a) = \mathsf{inf}\, a$*). By the latter fact the functionality of the operator can be made precise as* $\mathsf{inf} : A \to \mathsf{infel}(A)$.

Now we can give our modified semantics for CTL^*.

Definition 7.3. The *infinitary semantics* $[\![\varphi]\!]_i$ of a CTL^* formula φ over a Boolean left domain quantale is defined as follows:

- $[\![\mathsf{E}\varphi]\!]_i =_{df} \ulcorner[\![\varphi]\!]_i \cdot \mathsf{N}$.
- For all other formulas φ we set $[\![\varphi]\!]_i =_{df} \mathsf{inf}\, [\![\varphi]\!]$.

As an auxiliary we define complementation relative to N as $\neg_i\, a =_{df} \mathsf{N} - a$. This satisfies the following properties.

Theorem 7.4. *Assume a Boolean left quantale* A *with a step* x.

1. *The pair* $(\mathsf{infel}(A), \Box_i)$, *where* $\Box_i$ *is the restriction of* $\Box$ *to* $\mathsf{infel}(A)$, *is a repetition algebra and* inf *is a homomorphism from* $(A, \Box)$ *to* $(\mathsf{infel}(A), \Box_i)$.
2. $[\![\neg\varphi]\!]_i = \neg_i\, [\![\varphi]\!]_i$ *and* $\neg_i \neg_i\, a = \mathsf{inf}\, a$.
3. *The semantics* $[\![\,]\!]_i$ *propagates inductively:*

$$[\![\bot]\!]_i = 0, \qquad [\![\mathsf{X}\,\varphi]\!]_i = \mathsf{x} \cdot [\![\varphi]\!]_i,$$
$$[\![p]\!]_i = p \cdot \mathsf{N}, \qquad [\![\varphi\, \mathsf{U}\, \psi]\!]_i = [\![\varphi]\!]_i \boxast_i [\![\psi]\!]_i,$$
$$[\![\varphi \to \psi]\!]_i = \neg_i\, [\![\varphi]\!]_i + [\![\psi]\!]_i.$$

In addition,

$$[\![\varphi \vee \psi]\!]_i = [\![\varphi]\!]_i + [\![\psi]\!]_i,\ [\![\varphi \wedge \psi]\!]_i = [\![\varphi]\!]_i \sqcap [\![\psi]\!]_i,\ [\![\mathsf{A}\varphi]\!]_i = \neg\ulcorner(\neg_i\, [\![\varphi]\!]_i) \cdot \mathsf{N}.$$

4. If $\mathsf{N} \leq \mathsf{x} \cdot \mathsf{N}$ *(and hence* $\mathsf{N} \leq \mathsf{x}^\omega$*) then for all* $a \in A$ *we have* $\mathsf{inf}\,(\mathsf{x} \cdot \overline{a}) = \mathsf{inf}\,\overline{\mathsf{x} \cdot a}$. *In particular,* $[\![\mathsf{X}\,\neg\varphi]\!]_i = [\![\neg\mathsf{X}\,\varphi]\!]_i$*. Furthermore, for all* $a \in A$ *we have* $\neg_{\sqcap_i}(\mathsf{x} \cdot \mathsf{inf}\,\overline{a}) = \mathsf{x} \cdot \mathsf{inf}\,a$.
5. If $\mathsf{N} \leq \mathsf{x} \cdot \mathsf{N}$ *then* $[\![\mathsf{F}\psi]\!]_i = \mathsf{x}^* \cdot [\![\psi]\!]_i$ *and* $[\![\mathsf{G}\psi]\!]_i = [\![\psi]\!]_i^{\boxed{\omega}_i}$.

This means that we have now obtained a semantics which faithfully mirrors the original CTL^* semantics.

We combine the results of this theorem with our results on the until operator.

Corollary 7.5. *Assume again* $\mathsf{N} \leq \mathsf{x} \cdot \mathsf{N}$ *and define, for formulas* φ *and* ψ *the abbreviation* $\varphi \mathsf{W} \psi \Leftrightarrow_{df} \mathsf{G}\varphi \vee (\varphi \mathsf{U} \psi)$. *Then* $[\![\varphi \mathsf{W} \psi]\!]_i = \nu y \,.\, [\![\psi]\!]_i + ([\![\varphi]\!]_i \,\square_i\, y)$.

In the literature the operator W is known as *weak until* or *while*. It expresses that φ holds forever or else ψ will eventually hold with φ holding all the time before that.

8 Generated Quantales

In view of Theorem 7.4.4 we introduce a new notion.

Definition 8.1. Assume a Boolean quantale A with a step $\mathsf{x} \in A$. Then A is called x-*generated* if $\top = \nu x \,.\, 1 + \mathsf{x} \cdot x = \mathsf{x}^* + \mathsf{x}^\omega$ and $\mathsf{x}^\omega \leq \mathsf{N}$. If additionally $\ulcorner\mathsf{N} = 1$ then A is *strongly* x*-generated.*

The definition means that all elements of A can be obtained by finite or infinite iteration of x. The constraint $\mathsf{x}^\omega \leq \mathsf{N}$ serves to exclude "pseudo-infinite" iterations of x. Strong generation means that all starting states can be extended into infinite computations.

Example 8.2. The quantale $\mathsf{TRC}(\Sigma)$ (Example 3.2) is strongly Σ^2-generated, while its reduct to finite traces is not.

The definition of generatedness has important structural consequences. For any IL-semiring let

$$\mathsf{rtest}(A) =_{df} \{p \cdot \mathsf{N} \mid p \in \mathsf{test}(A)\} \tag{8}$$

be the set of *relative test ideals* of A; each of them characterises the set of infinite traces with starting states in a state set p.

Lemma 8.3. *Consider an* x*-generated quantale* A.

1. $\mathsf{N} = \mathsf{x}^\omega$ *and* $\mathsf{N} \sqcap \mathsf{x}^* = 0$. *Hence* x^ω *and* x^* *are complements of each other.*
2. $\mathsf{N} = \mathsf{x} \cdot \mathsf{N}$.
3. $\mathsf{x}^\omega = \mathsf{inf}\,(\mathsf{x}^\omega)$ *and hence* $\mathsf{x}^\omega \cdot \mathsf{x}^\omega = \mathsf{x}^\omega = (\mathsf{x}^\omega)^\omega$.

Consider now the concrete operator $c \,\square\, d =_{df} c \sqcap \mathsf{x} \cdot d$ *from* (5).

4. For all $a \in A$ *we have* $a^{\boxed{\omega}} \leq \mathsf{N}$.

5. $\mathsf{x}^{\boxed{\omega}} = 0$.
6. *If* $a \in A$ *is purely infinite then* $a^{\boxed{\omega}} = \bigsqcap_{k\in\mathbb{N}} \mathsf{x}^k \cdot a$.

Assume now that A *is strongly* x*-generated.*

7. $\ulcorner\mathsf{x} = 1$.
8. *The sets* $\mathsf{test}(A)$ *and* $\mathsf{rtest}(A)$ *are order-isomorphic.*

We can extend Lemma 8.3.5 a bit further. Together with Lemma 8.3.6 we obtain $[\![\mathsf{G}\psi]\!]_i = \bigsqcap_{i\in\mathbb{N}} \mathsf{x}^i \cdot [\![\psi]\!]_i$. Hence, in a $*$-continuous quantale [9], i.e., a quantale with $a \cdot b^* \cdot c = \bigsqcup \{a \cdot b^n \cdot c \mid n \in \mathbb{N}\}$ for all a, b, c, we therefore have the pleasantly symmetric formulations $[\![\mathsf{F}\psi]\!]_i = \bigsqcup_{i\in\mathbb{N}} \mathsf{x}^i \cdot [\![\psi]\!]_i$ and $[\![\mathsf{G}\psi]\!]_i = \bigsqcap_{i\in\mathbb{N}} \mathsf{x}^i \cdot [\![\psi]\!]_i$.

9 Towards CTL: The Semantics of State Formulas

In this section we show, among other properties, that the semantics of each state formula has the special form of a test ideal and hence directly corresponds to a test, i.e., an abstract representation of a set of states. This will be the key to the simplified CTL semantics in Sect. 10. Throughout this section we assume an x-generated quantale.

Theorem 9.1. *Let* φ *be a state formula of* CTL^*.

1. $[\![\varphi]\!]$ *is a test ideal, and hence, by Lemma 3.5.2,* $[\![\varphi]\!] = \ulcorner[\![\varphi]\!] \cdot \top$.
2. $[\![\varphi]\!]_i$ *is a relative test ideal, i.e.,* $[\![\varphi]\!]_i = \ulcorner[\![\varphi]\!] \cdot \mathsf{N}$.
3. $[\![\mathsf{E}\varphi]\!] = [\![\varphi]\!]$.
4. $[\![\mathsf{A}\varphi]\!] = [\![\varphi]\!]$.

Parts 3 and 4 show that state formulas are closed under E and A. In addition we have the following result.

Lemma 9.2. *State formulas are closed under* $\neg$, $\wedge$ *and* $\vee$.

Next, we state some properties of U and its relatives for state formulas.

Lemma 9.3. *Let* φ, ψ *be state formulas of* CTL^* *with* $[\![\varphi]\!] = p \cdot \top$ *and* $[\![\psi]\!] = q \cdot \top$ *for suitable tests* p, q.

1. $[\![\varphi \mathsf{U} \psi]\!] = (p \cdot \mathsf{x})^* \cdot q \cdot \top = ([\![\varphi]\!] \sqcap \mathsf{x})^* \cdot [\![\psi]\!]$.
2. $[\![\mathsf{G}\psi]\!]_i = (q \cdot \mathsf{x})^\omega$. *Hence we have the "shunting rule"* $(q \cdot \mathsf{x})^\omega = \neg_i (\mathsf{x}^* \cdot \neg q \cdot \mathsf{N})$.

Now we deal with EX.

Lemma 9.4. *For a state formula* φ *we have* $[\![\mathsf{EX}\varphi]\!] = [\![\mathsf{EXE}\varphi]\!]$ *and hence* $[\![\mathsf{EX}\varphi]\!]_i = [\![\mathsf{EXE}\varphi]\!]_i$.

We conclude this section by noting that in the infinitary semantics EX and AX are De Morgan duals; again the proof is a straightforward calculation.

Lemma 9.5. $[\![\mathsf{AX}\varphi]\!]_i = [\![\neg\mathsf{EX}\neg\varphi]\!]_i$.

From this and Lemma 9.4 we obtain the last result of this section.

Corollary 9.6. $[\![\mathsf{AX}\varphi]\!]_i = [\![\mathsf{AXA}\varphi]\!]_i$.

10 From CTL* to CTL

For a number of applications the sublogic CTL of CTL* suffices. We will see that it can be modelled in plain Kleene algebra. Syntactically, CTL consists of the CTL* state formulas that use trace formulas of the restricted form

$$\Pi ::= \mathsf{X}\,\Xi \mid \Xi\,\mathsf{U}\,\Xi. \tag{9}$$

From the previous section we already know that the semantics of every CTL formula is a test ideal t, from which, by Theorem 9.1.1, we can extract the corresponding test (or state set) as $\ulcorner t$. This is reflected by the simplified semantics[4] $[\![\varphi]\!]_d =_{df} \ulcorner([\![\varphi]\!]_i)$ which enables us to calculate solely with tests. Throughout this section we assume $\ulcorner\mathsf{N} = 1$, so that by locality (2) $\ulcorner(a \cdot \mathsf{N}) = \ulcorner a$ for all a.

First we state another homomorphic property.

Lemma 10.1. *Over a complete Boolean semiring A the structure* $(\mathsf{test}(A), \Box_d)$ *with* $p \Box_d q =_{df} |p \cdot \mathsf{x}\rangle q$ *is a repetition algebra and* $\ulcorner : \mathsf{rtest}(A) \to \mathsf{test}(A)$ *is a homomorphism from* $(\mathsf{rtest}(A), \Box_i)$ *to* $(\mathsf{test}(A), \Box_d)$. *Moreover,* $p \boxast_d q = |(p\cdot\mathsf{x})^*\rangle q$.

For the Boolean connectives we obtain by disjunctivity of domain and Lemma 3.5 together with Theorem 7.4.3 and standard domain properties,

$$[\![\varphi \vee \psi]\!]_d = [\![\varphi]\!]_d + [\![\psi]\!]_d, \quad [\![\varphi \wedge \psi]\!]_d = [\![\varphi]\!]_d \cdot [\![\psi]\!]_d, \quad [\![\neg\varphi]\!]_d = \neg[\![\varphi]\!]_d. \tag{10}$$

Next, we state some laws for A.

Lemma 10.2. *For atomic proposition* $p \in \mathsf{test}(A)$,

$$[\![\mathsf{A}\bot]\!]_d = 0, \qquad [\![\mathsf{A}\top]\!]_d = 1,$$
$$[\![\mathsf{A}(p \vee \varphi)]\!]_d = p + [\![\mathsf{A}\varphi]\!]_d, \qquad [\![\mathsf{A}(p \wedge \varphi)]\!]_d = p \cdot [\![\mathsf{A}\varphi]\!]_d.$$

Now we can calculate $[\![\_]\!]_d$ for all CTL formulas by induction on their syntactic structure, cf. the grammar in (9). We use implication $\to$ between tests, defined as $p \to q =_{df} \neg p + q$.

Theorem 10.3

(1) $[\![\bot]\!]_d = 0$, (2) $[\![p]\!]_d = p$,
(3) $[\![\varphi \to \psi]\!]_d = [\![\varphi]\!]_d \to [\![\psi]\!]_d$, (4) $[\![\mathsf{EX}\varphi]\!]_d = |\mathsf{x}\rangle[\![\varphi]\!]_d$,
(5) $[\![\mathsf{AX}\varphi]\!]_d = |\mathsf{x}][\![\varphi]\!]_d = [\![\mathsf{AXA}\varphi]\!]_d$, (6) $[\![\mathsf{E}(\varphi\mathsf{U}\psi)]\!]_d = |([\![\varphi]\!]_d \cdot \mathsf{x})^*\rangle[\![\psi]\!]_d$,
(7) $[\![\mathsf{A}(\varphi\mathsf{U}\psi)]\!]_d = \neg\,\ulcorner\overline{(\mathsf{x}^* \cdot [\![\psi]\!]_d \cdot \mathsf{N})} \cdot |(\neg[\![\psi]\!]_d \cdot \mathsf{x})^*]([\![\varphi]\!]_d + [\![\psi]\!]_d)$.

Parts (4) and (5) mean that the existential and universal quantifiers of CTL are semantically reflected as the existential and universal modal operators diamond and box. Part (6) means that the starting states of the traces in $[\![\mathsf{E}(\varphi\mathsf{U}\psi)]\!]_d$ are precisely those from which after finitely many X steps through φ states a ψ state can be reached. Part (7) characterises $[\![\mathsf{A}(\varphi\mathsf{U}\psi)]\!]_d$ as the set of those states from which eventually a ψ state must be reached and for which iteration through non-ψ states must lead to a φ or a ψ state.

[4] The subscript $_d$ stands for "domain".

11 From CTL* to LTL

The logic LTL is the fragment of CTL* in which only A may occur, once and outermost only, as trace quantifier. More precisely, LTL has no state formulas apart from those of the form $\mathsf{A}\varphi$ and the trace formulas are given by

$$\Pi ::= \Phi \mid \bot \mid \Pi \to \Pi \mid \mathsf{X}\,\Pi \mid \Pi \,\mathsf{U}\, \Pi.$$

Over an x-generated semiring, the LTL semantics is embedded into the CTL* one by assigning to $\varphi \in \Pi$ the semantic value $[\![\mathsf{A}\varphi]\!]_i$.

The reason for this is the following. An arbitrary CTL* formula φ may be called *valid* if its semantics is the set of all traces, abstractly, if $[\![\varphi]\!]_i = \mathsf{N}$. This is related to the A quantifier:

Lemma 11.1. $[\![\varphi]\!]_i = \mathsf{N} \Leftrightarrow [\![\mathsf{A}\varphi]\!]_i = \mathsf{N}$.

Although the infinitary semantics adequately reflects the standard LTL semantics, we present another view of the concrete case $A = \mathsf{TRC}(\Sigma)$ for some set Σ of states (cf. Example 3.2). Since we want to set up a similar connection to modal operators as in the CTL case (Theorem 10.3), we embed the carrier set $\mathcal{P}(\Sigma^\infty)$ of $\mathsf{TRC}(\Sigma)$ into the relational semiring $\mathsf{REL}(\Sigma^\infty)$ by encoding every subset $V \subseteq \Sigma^\infty$ as the relational test $h(V) =_{df} \{(\sigma,\sigma) \mid \sigma \in V\}$.

Based on this we define another semantic mapping $[\![\,]\!]_\mathsf{L}$ as

$$[\![\varphi]\!]_\mathsf{L} =_{df} h([\![\varphi]\!]_i). \tag{11}$$

Next, we mimic the semantic element x relationally. In $\mathsf{TRC}(\Sigma)$ we had $\mathsf{x} = \Sigma^2$, which was used to "glue" transitions to the front of traces. In $\mathsf{REL}(\Sigma^\infty)$ we replace this by the relation $N =_{df} \{(\sigma, \sigma^1) \mid \sigma \in \Sigma^\omega\}$, where, as in Sect. 2, σ^1 is σ with its first state removed. Now for a subset $V \subseteq \Sigma^\infty$,

$$h(\mathsf{x} \bowtie V) = |N\rangle h(V). \tag{12}$$

This allows the construction of yet another semantic homomorphism.

Lemma 11.2. *The structure* $(\mathsf{test}(\mathsf{REL}(\Sigma^\infty)), \Box_\mathsf{L})$ *with* $P \Box_\mathsf{L} Q =_{df} P \,;\, |\mathsf{N}\rangle Q$ *is a repetition algebra and* h *from* (11) *is a homomorphism from* $(\mathcal{P}(A^\omega), \Box_i)$ *to* $(\mathsf{test}(\mathsf{REL}(\Sigma^\infty)), \Box_\mathsf{L})$. *Here* ; *denotes relational composition.*

From this, Theorem 7.4 and Lemma 5.4 we obtain, with $\cdot = \bowtie$, $\mathsf{N} = A^\omega$ and $P \to Q = \neg P + Q$ (P, Q relational tests),

$$[\![\bot]\!]_\mathsf{L} = \emptyset, \qquad [\![\mathsf{X}\varphi]\!]_\mathsf{L} = |N\rangle [\![\varphi]\!]_\mathsf{L},$$
$$[\![p]\!]_\mathsf{L} = h(p \cdot \mathsf{N}), \qquad [\![\varphi \,\mathsf{U}\, \psi]\!]_\mathsf{L} = [\![\psi]\!]_\mathsf{L} \circledast_\mathsf{L} [\![\varphi]\!]_\mathsf{L},$$
$$[\![\varphi \to \psi]\!]_\mathsf{L} = [\![\varphi]\!]_\mathsf{L} \to [\![\psi]\!]_\mathsf{L}.$$

From the last equation we obtain

$$[\![\neg\varphi]\!]_\mathsf{L} = \neg[\![\varphi]\!]_\mathsf{L}, \qquad [\![\top]\!]_\mathsf{L} = h(\mathsf{N}),$$
$$[\![\varphi \vee \psi]\!]_\mathsf{L} = [\![\varphi]\!]_\mathsf{L} + [\![\psi]\!]_\mathsf{L} \qquad [\![\varphi \wedge \psi]\!]_\mathsf{L} = [\![\varphi]\!]_\mathsf{L} \,;\, [\![\psi]\!]_\mathsf{L}.$$

Moreover, we can simplify the U operator. Let $P =_{df} [\![\varphi]\!]_{\mathsf{L}}$ and $Q =_{df} [\![\psi]\!]_{\mathsf{L}}$. By Lemma 11.2 with Lemma 5.4, definition of diamond with the import/export law from Lemmas 3.5.7 and 8,

$$P \boxed{*}_{\mathsf{L}} Q = \mu Y \,.\, Q + P \,;\, |N\rangle Y \;= \mu Y \,.\, Q + |P \,;\, N\rangle Y \;= |(P \,;\, N)^*\rangle Q.$$

From this we obtain

$$[\![\varphi \,\mathsf{U}\, \psi]\!]_{\mathsf{L}} = |([\![\varphi]\!]_{\mathsf{L}} \,;\, N)^*\rangle [\![\psi]\!]_{\mathsf{L}}, \quad [\![\mathsf{F}\psi]\!]_{\mathsf{L}} = |N^*\rangle [\![\psi]\!]_{\mathsf{L}}, \quad [\![\mathsf{G}\psi]\!]_{\mathsf{L}} = |N^*][\![\psi]\!]_{\mathsf{L}}.$$

This shows that for LTL we can weaken the requirements on the underlying semantic algebra even further, viz. to that of a modal Kleene algebra.

Finally we briefly resume the discussion on axiom (1) in this interpretation.

$$[\![\mathsf{X}\neg\varphi]\!]_{\mathsf{L}} = \neg[\![\mathsf{X}\varphi]\!]_{\mathsf{L}} \;\Leftrightarrow\; |N\rangle \neg [\![\varphi]\!]_{\mathsf{L}} = \neg |N\rangle [\![\varphi]\!]_{\mathsf{L}} \;\Leftrightarrow\; |N][\![\varphi]\!]_{\mathsf{L}} = |N\rangle [\![\varphi]\!]_{\mathsf{L}}$$

for all φ. This means that N has to be a total and deterministic relation, which is the case if the function $\lambda x \,.\, \mathsf{x} \cdot x$ is surjective and injective, i.e., a bijection. These properties hold for the element Σ^2 that generates Σ^ω.

Note that the condition $|N] = |N\rangle$ does not propagate to $|N^*]$ and $|N^*\rangle$, since these correspond to iterated conjunction and disjunction, resp.

12 Conclusion

We have provided a compact algebraic semantics for full CTL* in the framework of modal quantales and shown that for the two sublogics CTL and LTL the semantics can be mapped to closed expressions using modal operators as well as Kleene star and ω iteration. Compared with representations of CTL* in the modal μ-calculus the compactness is achieved, since in quantales the modal operators are defined for ω-regular expressions (and even more generally), not only for atomic actions. Moreover, we have shown that for CTL and LTL the requirements on the semantic algebra can be relaxed to that of an omega (Sects. 9 and 10) or even just a Kleene algebra (Sect. 11).

As a non-trivial application, the article [14] shows that the algebraic semantics developed in this paper can be transferred to the setting of Concurrent Kleene Algebras and hence allow temporal reasoning about sequential subthreads there.

Future research will concern use of the algebraic semantics for concrete calculations in case studies as well the extension from the current propositional case to the first-order one; for this Tarskian frames as introduced in [11] seem promising candidates.

Acknowledgement. We are grateful to Roland Glück and to the anonymous referees for valuable comments.

References

1. Bloom, S., Ésik, Z.: Iteration algebras. Int. J. Found. Comput. Sci. **3**(3), 245–302 (1992)
2. Conway, J.: Regular Algebra and Finite Machines. Chapman & Hall, Boca Raton (1971)
3. Desharnais, J., Möller, B.: Characterizing determinacy in Kleene algebras. Inf. Sci. **139**, 253–273 (2001)
4. Desharnais, J., Möller, B.: Non-associative Kleene algebra and temporal logics. https://www.informatik.uni-augsburg.de/de/lehrstuehle/dbis/pmi/publications/all_pmi_tech-reports/tr-RAMICS16/
5. Desharnais, J., Möller, B., Struth, G.: Kleene algebra with domain. ACM Trans. Comput. Logic **7**, 798–833 (2006)
6. Dijkstra, R.M.: Computation calculus bridging a formalisation gap. Sci. Comput. Program. **37**, 3–36 (2000)
7. Emerson, E.A.: Temporal and modal logic. In: van Leeuwen, J. (ed.) Handbook of Theoretical Computer Science. Vol. B: Formal Models and Semantics, pp. 995–1072. Elsevier, Amsterdam (1991)
8. Kleene, S.: Representation of events in nerve nets and finite automata. In: Shannon, C., McCarthy, J. (eds.) Automata Studies, pp. 3–41. Princeton University Press, Princeton (1956)
9. Kozen, D.: A completeness theorem for Kleene algebras and the algebra of regular events. Inf. Comput. **110**, 366–390 (1994)
10. Kozen, D.: Kleene algebras with tests. ACM Trans. Program. Lang. Syst. **19**, 427–443 (1997)
11. Kozen, D.: Some results in dynamic model theory. Sci. Comput. Program. **51**, 3–22 (2004)
12. Möller, B.: Lazy Kleene algebra. In: Kozen, D. (ed.) MPC 2004. LNCS, vol. 3125, pp. 252–273. Springer, Heidelberg (2004). doi:10.1007/978-3-540-27764-4_14. Revised version in [13]
13. Möller, B.: Kleene getting lazy. Sci. Comput. Program. **65**, 195–214 (2007)
14. Möller, B., Hoare, T.: Exploring an interface model for CKA. In: Hinze, R., Voigtländer, J. (eds.) MPC 2015. LNCS, vol. 9129, pp. 1–29. Springer, Cham (2015). doi:10.1007/978-3-319-19797-5_1
15. Möller, B., Höfner, P., Struth, G.: Quantales and temporal logics. In: Johnson, M., Vene, V. (eds.) AMAST 2006. LNCS, vol. 4019, pp. 263–277. Springer, Heidelberg (2006). doi:10.1007/11784180_21
16. Rosenthal, K.: Quantales and their applications. Pitman Research Notes in Math. No. 234 Longman Scientific and Technical (1990)
17. Schmidt, G., Ströhlein, T.: Relations and Graphs: Discrete Mathematics for Computer Scientists. EATCS Monographs on Theoretical Computer Science. Springer, Heidelberg (1993)

Algebraic Investigation of Connected Components

Roland Glück(✉)

Deutsches Zentrum für Luft- und Raumfahrt,
Am Technologiezentrum 4, 86159 Augsburg, Germany
roland.glueck@dlr.de

Abstract. This paper characterizes connected components of both directed and undirected graphs as atomic fixpoints. As algebraic structure for our investigations we combine complete Boolean algebras with the well-known theory of Kleene Algebra with domain. Using diamond operators as an algebraic generalization of relational image and preimage we show how connected components can be modeled as atomic fixpoints of functions operating on tests and prove some advanced theorems concerning connected components.

1 Introduction

Algebraic reasoning about relations, graphs and graph algorithms has become a rising area of research in the past years. [BSW15,FK12] use Dedekind categories as algebraic tool whereas [BHS15] relies on classical relation algebra. Approaches considering also cardinality functions appear in [BDHS16,BPS16]. Another idea which is suitable also for edge-weighted graphs are fuzzy relations which are treated in [KF99,Kaw06]. [Kah14] uses ideas from category theory for modeling graph transformations. The considered problems include bipartitions as in [BSW15] or network flows as in [Kaw06] whereas there is no deeper examination of (stronlgy) connected components by algebraic means. Some known work as in [SS93] characterizes strongly connected components by means of equivalence relations but does not tackle algorithmic issues as in [Sha81]; a gap which this paper aims to narrow.

The approach here combines two known structures. As first component, we use complete distributive lattices with complement which allows us to reason about join, meet and complement of graphs. The second ingredient of our combined structure are Kleene algebras for modeling additionally composition and iteration, corresponding to reflexive-transitive hull and reachability. This allows the reuse of plenty already proven theorems in both areas and puts the results in a more general context than a pure relation algebraic approach. We assume basic knowledge of lattice theory, graph theory and relation algebra and refer the reader to [Bir67,JR92] for lattice theory, to [Jun05] for graph theory and to [SS93] for relation algebra.

P. Höfner et al. (Eds.): RAMiCS 2017, LNCS 10226, pp. 109–126, 2017.
DOI: 10.1007/978-3-319-57418-9_7

A perseverative task in algebraic reasoning about graphs and relations is the formalization of single nodes and elements. Usually, this challenge is tackled by point relations or point axioms as in [FK12,Kaw06]. Here we use atomic tests for this purpose which motivates the detailed investigation of atomicity.

In Sect. 2 we investigate some lattice-theoretic aspects of atomicity. Section 3 introduces the concept of graph algebras, a combination of a quantale and Kleene algebra with domain. We use the results of these sections in Sect. 4 to develop an algebraic characterization of connected components. The last Sect. 5 summarizes our results and gives an outlook to future work.

2 Full Atomic Lattices

The main structure we investigate in this section will be a *full atomic lattice.* As usual, a complete Boolean algebra is a structure $\mathcal{M} = (M, \sqsubseteq, \bigsqcup, \bigsqcap, \bot, \top, ^{-})$ where $(M, \sqsubseteq)$ is an ordered set with least and greatest elements $\bot$ and $\top$, resp., $\bigsqcup$ and $\bigsqcap$ are supremum and infimum with respect to $\sqsubseteq$, supremum distributes over arbitrary infima and vice versa, and $^{-}$ is the complement satisfying the de Morgan's laws $\overline{\bigsqcup M'} = \bigsqcap\{\overline{m'} \mid m' \in M'\}$ and $\overline{\bigsqcap M'} = \bigsqcup\{\overline{m'} \mid m' \in M'\}$ for all $M' \subseteq M$. $\sqcup$ and $\sqcap$ serve as abbreviations for binary supremum and infimum, resp. We define the symbols $\sqsupseteq$, $\sqsubset$, $\not\sqsubseteq$ and $\not\sqsupseteq$ by $m \sqsupseteq n \Leftrightarrow_{df} n \sqsubseteq m$, $m \sqsubset n \Leftrightarrow_{df} m \sqsubseteq n \wedge m \neq n$, $m \not\sqsubseteq n \Leftrightarrow_{df} \neg m \sqsubseteq n$ and $m \not\sqsupseteq n \Leftrightarrow_{df} \neg m \sqsupseteq n$. An element $m \in M$ is called *non-bottom* if $m \neq \bot$. In this setting we make the following definition:

Definition 2.1. *Let $\mathcal{M} = (M, \sqsubseteq, \bigsqcup, \bigsqcap, \bot, \top, ^{-})$ be a complete Boolean algebra. A non-bottom element $m^a \in M$ is called* atomic *if for all non-bottom $m \in M$ the implication $m \sqsubseteq m^a \Rightarrow m = m^a$ holds. The set of all atomic elements of $\mathcal{M}$ is denoted by* $\mathsf{atom}(\mathcal{M})$. *$\mathcal{M}$ is a* full atomic lattice *if $m = \bigsqcup\{m^a \in \mathsf{atom}(\mathcal{M}) \mid m^a \sqsubseteq m\}$ holds for all $m \in M$.*

As a convention, we will denote atomic elements always with a superscript a. In the sequel we list some simple but useful properties of full atomic lattices (the proofs are omitted here for brevity but can be found under [Glüa]):

Lemma 2.2. *Let $\mathcal{M} = (M, \sqsubseteq, \bigsqcup, \bigsqcap, \bot, \top, ^{-})$ be a full atomic lattice, and consider arbitrary atoms $m^a, n^a \in \mathsf{atom}(\mathcal{M})$, arbitrary $m, n \in M$ and an arbitrary $M' \subseteq M$. Then the following properties hold:*

1. $m^a \sqcap m \neq \bot \Leftrightarrow m^a \sqcap m = m^a$
2. $m^a \sqcap n^a = \bot \Leftrightarrow m^a \neq n^a$
3. $m^a \sqsubseteq m \Leftrightarrow m^a \sqcap m \neq \bot$
4. $m^a \sqsubseteq \bigsqcup M' \Leftrightarrow \exists m' \in M' : m^a \sqsubseteq m'$
5. $m^a \sqsubseteq m \Leftrightarrow m^a \not\sqsubseteq \overline{m}$
6. $m \sqsubseteq n \wedge m^a \sqsubseteq n \Rightarrow m^a \sqsubseteq m \vee m^a \sqsubseteq n \sqcap \overline{m}$
7. $m \sqcap n = \bot \wedge n^a \sqsubseteq n \Rightarrow m \sqcup n^a \sqsupset m$
8. $m \not\sqsubseteq n \Rightarrow \exists o^a \in \mathsf{atom}(\mathcal{M}) : o^a \sqsubseteq m \wedge o^a \not\sqsubseteq n$

Given an undirected graph, the set-valued reachability function (mapping a set of nodes to the set of therefrom reachable nodes) is a closure operator with the special property that the complement of a fixpoint is a fixpoint again. This motivates the following definition in a more general context:

Definition 2.3. *A* complementary strict distributive closure *or* csd closure *is a function f on the carrier set M of a full atomic lattice $\mathcal{M}$ with the following properties:*

- $f(\bigsqcup M') = \bigsqcup f(M')$ *for all* $M' \subset M$ (distributivity)
- $m \sqsubseteq f(m)$ *for all* $m \in M$ (extensivity)
- $f(f(m)) = f(m)$ *for all* $m \in M$ (idempotence)
- $f(\overline{f(m)}) = \overline{f(m)}$ *for all* $m \in M$ (complementary idempotence)

An elementary implication of distributivity is isotony of f, i.e., we have $m \sqsubseteq n \Rightarrow f(m) \sqsubseteq f(n)$, so the set fix_f of fixpoints of f is also a complete sublattice due to the Knaster-Tarski theorem (see [Tar55]). Similarly, idempotence and complementary idempotence imply the closeness of fix_f under complementation. By lattice theory and distributivity we have $f(\bot) = f(\bigsqcup \emptyset) = \bigsqcup \emptyset = \bot$, so $\bot$ and $\top$ (by closeness of fix_f under complementation) are fixpoints of f. Together this means that $\mathcal{FIX}_f =_{df} (\mathsf{fix}_f, \sqsubseteq, \bigsqcup, \bigsqcap, \bot, \top, \bar{\ })$ is a complete Boolean algebra. The following lemmata will help to show that $\mathcal{FIX}_f$ is even a full atomic lattice.

Lemma 2.4. *Let f be a csd closure on a full atomic lattice $\mathcal{M}$ and consider an arbitrary $m^a \in \mathsf{atom}(\mathcal{M})$. Then $f(m^a)$ is an atom in $\mathcal{FIX}_f$.*

Proof. Consider an arbitrary atom $m^a \in \mathsf{atom}(\mathcal{M})$ and assume there is a non-bottom $m \in \mathsf{fix}_f$ with $m \sqsubset f(m^a)$ (note that due to idempotence we have $f(m^a) \in \mathsf{fix}_f$). According to Lemma 2.2.6 there are two cases:

1. $m^a \sqsubseteq m$: here we have $f(m) = m \sqsubset f(m^a)$ which contradicts isotony of f.
2. $m^a \sqsubseteq f(m^a) \sqcap \overline{m}$: in this case, $f(m^a) \sqcap \overline{m}$ is also a fixpoint of f (remember the closeness of fix_f under infima and complementation). Moreover, we have $f(m^a) \sqcap \overline{m} \sqsubset f(m^a)$ by lattice theory. Putting this together, we obtain $m^a \sqsubseteq f(m^a) \sqcap \overline{m}$ and $f(f(m^a) \sqcap \overline{m}) \sqsubset f(m^a)$ which contradicts the isotony of f. ■

Together with the atomicity of $\mathcal{M}$ this shows the following lemma:

Lemma 2.5. *Let f be a complementary distributive closure on a full atomic lattice $\mathcal{M} = (M, \sqsubseteq, \bigsqcup, \bigsqcap, \bot, \top, \bar{\ })$ and consider an arbitrary $x \in \mathsf{fix}_f$. Then we have the identity $x = \bigsqcup\{f(m^a) \mid m^a \in \mathsf{atom}(\mathcal{M}) \wedge m^a \sqsubseteq x\}$.*

The next lemma asserts also the reverse direction of Lemma 2.4:

Lemma 2.6. *Let f be a csd closure on a full atomic lattice $\mathcal{M}$ and consider an arbitrary $x^a \in \mathsf{atom}(\mathcal{FIX}_f)$. Then every atom $m^a \in \mathsf{atom}(\mathcal{M})$ with $m^a \sqsubseteq x^a$ fulfills $f(m^a) = x^a$.*

Proof. Due to Lemma 2.5 we have $f(m^a) \sqsubseteq x^a$, and strictness of f and atomicity of m^a yield $f(m^a) \sqsupset \bot$. Atomicity of x^a implies now $f(m^a) = x^a$. ■

Atomicity and distributivity of a csd closure imply the following corollary:

Corollary 2.7. *Let f be a csd closure and consider an arbitrary $x^a \in \mathsf{atom}(\mathcal{FIX}_f)$. Then for every non-bottom $x' \sqsubseteq x^a$ we have $f(x') = x^a$.*

Putting these results together, we obtain the following theorem:

Theorem 2.8. *Let f be a csd closure on a full atomic lattice $\mathcal{M}$. Then $\mathcal{FIX}_f$ is atomic with $\mathsf{atom}(\mathcal{FIX}_f) = f(\mathsf{atom}(\mathcal{M}))$.*

This theorem generalizes Theorem 3.13 in [GMS09] where the role of f is played by a generalization of the image function of equivalence relations.

3 Graph Algebras

Clearly, the set of endorelations over a set forms a complete distributive lattice with the subset relation as order, union and meet as supremum and infimum, and the full and empty relation as top and bottom element, resp. In order to reason about relations and graphs we need some additional ingredients. The first one is an abstract model of relational composition:

Definition 3.1. *A structure $Q = (M, \sqsubseteq, \bigsqcup, \bigsqcap, \bot, \top, 1, \cdot)$ is called a* quantale *if $(M, \sqsubseteq, \bigsqcup, \bigsqcap, \bot, \top)$ is a complete distributive lattice and $\cdot : M \times M \to M$ is an associative function (called* multiplication*) which distributes from both sides over arbitrary suprema and has 1 as both left and right neutral element. Moreover, Q obeys the* Tarski rule *$m \neq \bot \Leftrightarrow \top \cdot m \cdot \top = \top$ for all $m \in M$.*

This definition implies isotony of multiplication in both arguments and $\bot \cdot m = \bot = m \cdot \bot$ for all $m \in M$. For brevity we may often write mn instead of $m \cdot n$. Multiplication is sub-conjunctive, i.e., we have $m \cdot \bigsqcap N \cdot o \sqsubseteq \bigsqcap\{mno \mid n \in N\}$. The Tarski rule is not included in the standard definition of a quantale, however, it holds in the relational quantale which is the object of our investigations.

Quantales model the interplay between relational composition and union but lack the possibility of reasoning about subsets of the carrier set of a relation. This gap can be filled by tests, defined as follows (see e.g. [DMS04]):

Definition 3.2. *Given a quantale $Q = (M, \sqsubseteq, \bigsqcup, \bigsqcap, \bot, \top, 1, \cdot)$ we call an element $p \in M$ a* test *if there is an element $\neg p \in M$ with $p \sqcup \neg p = 1$ and $p \cdot \neg p = \bot = \neg p \cdot p$. The set of all tests of Q is denoted by $\mathsf{test}(Q)$.*

In the relation lattice the set of tests equals the set of subidentities and corresponds in an obvious way with subsets of the carrier set. On tests, multiplication is idempotent and coincides with infimum. Clearly, $\bot$ and 1 are tests; for further properties see again [DMS04]. [MHS06] shows that multiplication by tests distributes even over infima, so we have $p(m \sqcap n)q = pmq \sqcap pnq$ for arbitrary $m, n \in M$ and $p, q \in \mathsf{test}(Q)$. We call a test p^a *atomic* if $q \sqsubseteq p^a \Leftrightarrow q = p^a$ holds for all non-bottom tests q and denote the set of atomic tests by $\mathsf{att}(Q)$. If $\mathcal{TEST}_Q =_{df} (\mathsf{test}(Q), \sqsubseteq, \sqcup, \sqcap, \bot, 1, \neg)$ is a full atomic lattice we call Q *test-atomic.*

Atomic tests serve to model single elements of the carrier set of a relation since they correspond to singleton subsets of the identity relation. In the sequel we denote tests by p, q, r, …, decorated on demand by indices, primes and similar ornaments.

Using tests we can define the *forward diamond* $|\rangle$ and *backward diamond* $\langle|$ as functions of the type $M \to \mathsf{test}(Q) \to \mathsf{test}(Q)$ by the equivalences $|m\rangle p \sqsubseteq q \Leftrightarrow \neg q m p \sqsubseteq \bot \Leftrightarrow \langle m|\neg q \sqsubseteq \neg p$ for all $m \in M$ and $p, q \in \mathsf{test}(Q)$. Furthermore, we require *modality*, i.e., $\langle mn|p = \langle n|\langle m|p$ and $|mn\rangle p = |m\rangle|n\rangle p$. Both diamond operators distribute in every argument over arbitrary suprema and are hence strict and isotone in every argument. More details can be found e.g. in [DMS04]. In the relational lattice, forward and backward diamond correspond to preimage and image operations, resp.

The last refinement enables us to reason about iteration, the reflexive-transitive hull and hence reachability and connectivity (see e.g. [Koz]):

Definition 3.3. *A structure* $\mathcal{GA} = (M, \sqsubseteq, \bigsqcup, \bigsqcap, \bot, \top, \cdot, 1, {}^*, {}^\circ, |\rangle, \langle|)$ *is called a* graph algebra *if* $(M, \sqsubseteq, \bigsqcup, \bigsqcap, \bot, \top, 1, \cdot)$ *is a test-atomic quantale with forward and backward diamond operations* $|\rangle$ *and* $\langle|$*, and the* star *operation* ${}^* : M \to M$ *fulfills the following properties for all* $m, n, o \in M$*:*

- $1 \sqcup mm^* \sqsubseteq m^*$ *and* $1 \sqcup m^*m \sqsubseteq m$ (star unfold)
- $n \sqcup mo \sqsubseteq o \Rightarrow m^*n \sqsubseteq o$ *and* $n \sqcup om \sqsubseteq o \Rightarrow nm^* \sqsubseteq o$ (star induction)

The converse *operator* ${}^\circ : M \to M$ *enjoys the property* $|m^\circ\rangle p = \langle m|p$ *for all* $m \in M$ *and tests* p*. Finally,* $\mathcal{GA}$ *fulfills the* all-or-nothing *property stating* $p^a m q^a = \bot \Leftrightarrow p^a m q^a \neq p^a \top q^a$ *for all* $m \in M$ *and atomic tests* p^a *and* q^a *(implying* $\bot \neq p^a \top q^a$*).*

Clearly, for all $m \in M$ we have $1 \sqsubseteq m^*$ due to star unfold. Moreover, the star operation is isotone. For further properties see again [DMS04,Koz]. The intention of the definition of the converse operator is to model the relational converse by swapping image and preimage. If we interpret $p^a g q^a = \bot$ as the non-existence of an edge between the nodes corresponding to p^a and q^a we have to ensure that for two algebraic graphs g_1 and g_2 with $p^a g_1 q^a \neq \bot \neq p^a g_2 q^a$ the meet $p^a(g_1 \sqcap g_2)q^a$ is also non-bottom. In our definition, this motivates the introduction of the all-or-nothing property.

We refer to the elements of the carrier set of a graph algebra as *algebraic graphs*. An algebraic graph g is called *reflexive* if $1 \sqsubseteq g$, and *transitive* if $gg \sqsubseteq g$ holds. g is *symmetric* if $\langle g^\circ|p = \langle g|p$ holds for all tests p. An *algebraic equivalence* is a reflexive, transitive and symmetric algebraic graph. For an algebraic equivalence g we have $g = 1 \cdot g \sqsubseteq gg$ due to neutrality of 1 and isotony of multiplication and hence even the equality $g = gg$.

From now on we will reason only about graph algebras and carry over the namings (atomic) test and the notions of diamonds in an obvious manner. First we list some technical lemmata which will be useful in the further course.

Lemma 3.4. *Let* p^a *be an atomic test. Then the identity* $p^a \top p^a = p^a$ *holds.*

Proof. Idempotence of multiplication on tests and multiplicative neutrality of 1 yield $p^a = p^a 1 p^a$. By all-or-nothing we have either $p^a 1 p^a = \bot$ or $p^a 1 p^a = p^a \top p^a$, however, the first case contradicts the atomicity of p^a. ■

Lemma 3.5. *For atomic tests p^a and q^a and arbitrary m we have the equivalences $p^a \sqsubseteq |m\rangle q^a \Leftrightarrow p^a m q^a = p^a \top q^a \Leftrightarrow q^a \sqsubseteq \langle m| p^a \Leftrightarrow p^a m q^a \neq \bot$.*

Proof. Consider atomic tests p^a and q^a and an arbitrary m with $p^a \sqsubseteq |m\rangle q^a$. By Lemma 2.2.3 this is equivalent to $p^a \sqcap |m\rangle q^a \neq \bot$, and because on tests multiplication and infimum coincide we can rewrite this condition as $p^a \cdot |m\rangle q^a \neq \bot$. By shunting we obtain $|m\rangle q^a \not\sqsubseteq \neg p^a$ which is by contraposition of the diamond's definition the same as $p^a m q^a \not\sqsubseteq \bot$. Because $\bot$ is the least element this is equivalent to $p^a m q^a \neq \bot$, and the all-or-nothing axiom yields the equivalence of this condition and $p^a m q^a = p^a \top q^a$. Symmetrically we obtain the equivalence of $q^a \sqsubseteq \langle m| p^a$ and $p^a m q^a \neq \bot$. The last equivalence is a consequence of all-or-nothing. ■

Lemma 3.6. *For all non-bottom tests p the equalities $|\top\rangle p = 1 = \langle \top | p$ hold.*

Proof. We fix an arbitrary atomic test q^a and another atomic test $p^a \sqsubseteq p$. Then we have $q^a \sqsubseteq \langle \top | p^a$ by Lemma 3.5 and hence $1 \sqsubseteq \langle \top | p^a$ by atomicity of $\mathsf{test}(\mathcal{GA})$ and supremum properties. Because the diamond operation is isotone this implies $1 \sqsubseteq \langle \top | p$, however, 1 is the greatest test so we even have the equality $1 = \langle \top | p$. The other equality can be shown analogously. ■

The next lemma deals with the interplay between converse and diamond operators. Intuitively, it states that the preimage function of a relation is the same as the image function of its converse.

Lemma 3.7. *For all m and tests p we have the equality $\langle m^\circ | p = |m\rangle p$.*

Proof. We reason as follows:

$$
\begin{array}{lr}
\langle m^\circ | p \sqsubseteq |m\rangle p \Leftrightarrow & \{ \text{ diamond property } \} \\
p \cdot m^\circ \cdot (\neg |m\rangle p) \sqsubseteq \bot \Leftrightarrow & \{ \text{ diamond property } \} \\
|m^\circ\rangle(\neg |m\rangle p) \sqsubseteq \neg p \Leftarrow & \{ \text{ definition of the converse } \} \\
\langle m | (\neg |m\rangle p) \sqsubseteq \neg p \Leftrightarrow & \{ \text{ diamond property } \} \\
(\neg |m\rangle p) \cdot m \cdot p \sqsubseteq \bot \Leftrightarrow & \{ \text{ diamond property } \} \\
|m\rangle p \sqsubseteq |m\rangle p \Leftrightarrow & \{ \text{ reflexivity of } \sqsubseteq \} \\
\mathsf{true} &
\end{array}
$$

In an analogous manner we show $|m\rangle p \sqsubseteq \langle m^\circ | p$ which finishes the proof. ■

This lemma and the definition of the converse imply the following lemma:

Lemma 3.8. *For all g and tests p we have $|(g^\circ)^\circ\rangle p = |g\rangle p$ and $\langle (g^\circ)^\circ | p = \langle g | p$.*

Now we can show that the star and the converse operation commute with each other:

Lemma 3.9. *For all m and tests p we have the equalities $|(m^*)^\circ\rangle p = |(m^\circ)^*\rangle p$ and $\langle(m^*)^\circ|p = \langle(m^\circ)^*|p$.*

Proof. Definition of the converse yields $|(m^*)^\circ\rangle p = \langle m^*|p$. [DMS06] shows that $\langle m^*|p$ is the least fixpoint of the function $f(p) =_{df} p \sqcup \langle m|p$ which by definition of the converse again equals $g(p) =_{df} p \sqcup |m^\circ\rangle p$. Again by [DMS06] the least fixpoint of g is $|(m^\circ)^*\rangle p$. The other equality follows analogously by Lemma 3.7. ■

Lemma 3.10. *For all m and tests p and q we have the equalities $pmq = \bigsqcup\{p^a mq^a \mid p^a, q^a \in \mathsf{att}(\mathcal{GA}), p^a \sqsubseteq p, q^a \sqsubseteq q\} = \bigsqcup\{p^a mq^a \mid p^a, q^a \in \mathsf{att}(\mathcal{GA}), p^a \sqsubseteq p, q^a \sqsubseteq q, p^a mq^a \neq \bot\}$.*

Proof. We reason as follows:

$$
\begin{array}{ll}
pmq = & \{\text{atomicity of } \mathsf{test}(\mathcal{GA})\} \\
\bigsqcup\{p^a \in \mathsf{att}(GA), p^a \sqsubseteq p\} \cdot m \cdot \bigsqcup\{q^a \in \mathsf{att}(GA), q^a \sqsubseteq q\} = & \{\text{ distributivity }\} \\
\bigsqcup\{p^a mq^a \mid p^a, q^a \in \mathsf{att}(GA), p^a \sqsubseteq p, q^a \sqsubseteq q\} = & \{\text{ neutrality of } \bot \text{ wrt. } \bigsqcup\ \} \\
\bigsqcup\{p^a mq^a \mid p^a, q^a \in \mathsf{att}(GA), p^a \sqsubseteq p, q^a \sqsubseteq q, p^a mq^a \neq \bot\} &
\end{array}
$$
■

Strictness of the supremum operation implies now the following corollary:

Corollary 3.11. *For all m and arbitrary tests p and q we have the equivalence $pmq \neq \bot \Leftrightarrow \exists p^a, q^a \in \mathsf{att}(\mathcal{GA}) : p^a \sqsubseteq p \wedge q^a \sqsubseteq q \wedge p^a mq^a \neq \bot$. In particular, for an atomic test p^a we have the equivalence $p^a \leq \langle m|q \Leftrightarrow \exists q^a \in \mathsf{att}(\mathcal{GA}) : q^a \leq q \wedge p^a \leq \langle m|q^a$.*

Two relations are equal iff their image functions coincide. In our setting, this is expressed by the following lemma:

Lemma 3.12. *For all m and n we have $m = n$ iff $\langle m|p = \langle n|p$ holds for all tests p.*

Proof. The direction from the left to right side is trivial so we show only the implication from the right side to the left. From the right side we conclude that $q^a \sqsubseteq \langle m|p^a \Leftrightarrow q^a \sqsubseteq \langle n|p^a$ holds for all atomic tests p^a and q^a. Lemma 3.5 and all-or-nothing imply $p^a mq^a \neq \bot \Leftrightarrow p^a nq^a \neq \bot$, and now the claim follows from Lemma 3.10 by setting $p = q = 1$. ■

Using Lemma 3.6 (together with $\langle\top|\bot = \bot$ by strictness of the diamond operator), Lemmas 3.8 and 3.9 we obtain the following lemma:

Lemma 3.13. *For all algebraic graphs g we have $\top = \top^\circ$, $g^{\circ\circ} = g$ and $(g^*)^\circ = (g^\circ)^*$.*

As in real relational life, converse distributes over infimum:

Lemma 3.14. *For all algebraic graphs m and n we have $(m \sqcap n)^\circ = m^\circ \sqcap n^\circ$.*

Proof. We show that $q^a \sqsubseteq \langle(m \sqcap n)^\circ|p^a \Leftrightarrow q^a \sqsubseteq \langle m^\circ \sqcap n^\circ|p^a$ holds for all atomic tests p^a and q^a; the rest follows analogously to the proof of Lemma 3.12. So we fix two arbitrary atomic tests p^a and q^a and reason as follows:

$$
\begin{array}{ll}
q^a \sqsubseteq \langle (m \sqcap n)^\circ | p^a \Leftrightarrow & \{ \text{Lemma 3.7} \} \\
q^a \sqsubseteq |(m \sqcap n)\rangle p^a \Leftrightarrow & \{ \text{Lemma 3.5} \} \\
q^a (m \sqcap n) p^a = q^a \top p^a \Leftrightarrow & \{ \text{distributivity of multiplication by tests} \} \\
q^a m p^a \sqcap q^a n p^a = q^a \top p^a \Leftrightarrow & \{ \text{lattice theory, all-or-nothing} \} \\
q^a m p^a = q^a \top p^a \wedge q^a n p^a = q^a \top p^a \Leftrightarrow & \{ \text{Lemma 3.5} \} \\
q^a \sqsubseteq |m\rangle p^a \wedge q^a \sqsubseteq |n\rangle p^a \Leftrightarrow & \{ \text{Lemma 3.7} \} \\
q^a \sqsubseteq \langle m^\circ | p^a \wedge q^a \sqsubseteq \langle n^\circ | p^a \Leftrightarrow & \{ \text{Lemma 3.5} \} \\
p^a m^\circ q^a = p^a \top q^a \wedge p^a n^\circ q^a = p^a \top q^a \Leftrightarrow & \{ \text{lattice theory, all-or-nothing} \} \\
p^a m^\circ q^a \sqcap p^a n^\circ q^a = p^a \top q^a \Leftrightarrow & \{ \text{distributivity of multiplication by tests} \} \\
p^a (m^\circ \sqcap n^\circ) q^a = p^a \top q^a \Leftrightarrow & \{ \text{Lemma 3.5} \} \\
q^a \sqsubseteq \langle m^\circ \sqcap n^\circ | p^a & \blacksquare
\end{array}
$$

The next lemma reflects the definition of relational composition:

Lemma 3.15. *Consider arbitrary algebraic graphs m and n and atomic tests p^a and q^a. Then $p^a mnq^a \neq \bot$ holds iff there is an atomic test r^a with $p^a mr^a \neq \bot \neq r^a nq^a$.*

Proof. "$\Rightarrow$": We have $p^a mnq^a = p^a m 1 nq^a = p^a m \bigsqcup \mathsf{att}(\mathcal{GA}) nq^a$ by multiplicative neutrality of 1 and atomicity of $\mathsf{test}(\mathcal{GA})$. By distributivity this transforms into $\bigsqcup \{p^a mr^a nq^a \mid r^a \in \mathsf{att}(\mathcal{GA})\}$, and due to idempotence of multiplication on tests this equals $\bigsqcup \{p^a mr^a \cdot r^a nq^a \mid r^a \in \mathsf{att}(\mathcal{GA})\}$. If for every atomic test r^a at least one of $p^a mr^a = \bot$ or $r^a nq^a = \bot$ holds the last supremum evaluates to $\bot$ which contradicts the assumption.
"$\Leftarrow$": First we calculate as follows (we assume tacitly that all tests with a superscript a are atomic):

$$
\begin{array}{ll}
p^a mnq^a = & \{ \text{Lemma 3.10, atomicity of } p^a \text{ and } q^a \} \\
\bigsqcup \{p^a mr^a_m \mid p^a mr^a_m \neq \bot\} \cdot \bigsqcup \{r^a_n nq^a \mid r^a_n nq^a \neq \bot\} = & \{ \text{distributivity} \} \\
\bigsqcup \{p^a mr^a_m r^a_n nq^a \mid p^a mr^a_m \neq \bot, r^a_n nq^a \neq \bot\} = & \{ \text{Lemma 2.2.2, } r^a_m \sqcap r^a_n = r^a_m r^a_n \} \\
\bigsqcup \{p^a mr^a nq^a \mid p^a mr^a \neq \bot, r^a nq^a \neq \bot\} &
\end{array}
$$

By assumption, the last set is nonempty, so together with all-or-nothing we have $p^a mnq^a \sqsupseteq p^a \top r^a \top q^a$. Due to the Tarski rule we have $\top r^a \top = \top$ which means $p^a mnq^a \sqsupseteq p^a \top q^a \sqsupset \bot$ by all-or-nothing. ■

By induction we obtain from this Lemma the following corollary:

Corollary 3.16. *Consider arbitrary algebraic graphs $m_1, m_2, \ldots, m_n$ and atomic tests p^a and q^a. Then $p^a m_1 m_2 \ldots m_n q^a \neq \bot$ holds iff there are atomic tests $r^a_1, r^a_2, \ldots, r^a_{n-1}$ such that $p^a m_1 r^a_1 \neq \bot \neq r^a_{n-1} m_n q^a$ and $r^a_i m_i r_{i+1} \neq \bot$ hold for all $1 \leq i \leq n-2$.*

Lemma 3.17. *Let p^a be an atomic test, q an arbitrary test and consider an algebraic graph g with $p^a \sqsubseteq \langle g | q$. Then there is an atomic test r^a with $r^a \sqsubseteq q$ and $p^a \sqsubseteq \langle g | r^a$.*

Proof. By atomicity of $\mathsf{test}(\mathcal{GA})$ we have $q = \bigsqcup\{q^a \mid q^a \in \mathsf{att}(\mathcal{GA}), q^a \sqsubseteq q\}$ which yields together with distributivity of the diamond the equality $\langle g|q = \bigsqcup\{\langle g|q^a \mid q^a \in \mathsf{att}(\mathcal{GA}), q^a \sqsubseteq q\}$. Lemma 2.2.4 implies the claim.

Lemma 3.18. *For arbitrary tests p, q, r and every transitive algebraic graph g we have the implication $p \sqsubseteq \langle g|q \wedge q \sqsubseteq \langle g|r \Rightarrow p \sqsubseteq \langle g|r$.*

Proof. First, we have $p \sqsubseteq \langle g|\langle g|r$ by isotony of the diamond. Modality yields $p \sqsubseteq \langle gg|r$, and transitivity of g does the remaining job. ■

Lemma 3.19. *Let g be a symmetric and transitive algebraic graph, p^a and q^a atomic tests and r an arbitrary test with $p^a \sqsubseteq \langle g|q^a$ and $p^a \sqsubseteq \langle g|r$. Then we have also $q^a \sqsubseteq \langle g|r$.*

Proof. We reason as follows:

$$
\begin{array}{lr}
p^a \sqsubseteq \langle g|q^a \wedge p^a \sqsubseteq \langle g|r \Rightarrow & \{\text{ symmetry of } g \text{ }\} \\
p^a \sqsubseteq |g\rangle q^a \wedge p^a \sqsubseteq \langle g|r \Rightarrow & \{\text{ Lemma 3.5 }\} \\
q^a \sqsubseteq \langle g|p^a \wedge p^a \sqsubseteq \langle g|r \Rightarrow & \{\text{ Lemma 3.18 }\} \\
q^a \sqsubseteq \langle g|r^a &
\end{array}
$$
■

4 Connected Components

4.1 Innately Connected Components

The connected components of both directed and undirected graphs induce equivalence relations on the nodes of the graph. For this reason we investigate first some properties of algebraic equivalences before dealing with connectivity.

Theorem 4.1. *Let g be an algebraic equivalence. Then the function $icc_g(p) =_{df} \langle g|p$ is a complementary strict distributive closure on $\mathcal{TEST}_{\mathcal{GA}}$.*

Proof. Distributivity of icc_g is a consequence of distributivity of the backward diamond. By reflexivity of g we have $1 \sqsubseteq g$ which yields extensivity of icc_g due to $\langle 1|p = p$ and isotony of the backward diamond. Moreover, we have $\langle g|\langle g|p = \langle gg|p$ by modality, and together with the algebraic equivalence property $gg = g$ this implies idempotence of icc_g.

For complementary idempotence we note first that $\langle g|\neg\langle g|p \sqsupseteq \langle 1|\neg\langle g|p = \neg\langle g|p$ holds due to reflexivity of g, isotony of the backward diamond and $\langle 1|q = q$. In order to prove the remaining inequality $\langle g|\neg\langle g|p \sqsubseteq \neg\langle g|p$ we assume $\langle g|\neg\langle g|p \not\sqsubseteq \neg\langle g|p$. Then there is by Lemma 2.2.8 an atomic test q^a with $q^a \sqsubseteq \langle g|\neg\langle g|p$ and $q^a \not\sqsubseteq \neg\langle g|p$. Now we reason as follows:

$$
\begin{array}{lr}
q^a \sqsubseteq \langle g|\neg\langle g|p \wedge q^a \not\sqsubseteq \neg\langle g|p \Rightarrow & \{\text{ Lemma 2.2.5 }\} \\
q^a \sqsubseteq \langle g|\neg\langle g|p \wedge q^a \sqsubseteq \langle g|p \Rightarrow & \{\text{ Lemma 3.17 }\} \\
\exists r^a \in \mathsf{att}(\mathcal{GA}) : r^a \sqsubseteq \neg\langle g|p \wedge q^a \sqsubseteq \langle g|r^a \wedge q^a \sqsubseteq \langle g|p \Rightarrow & \{\text{ Lemma 3.19 }\} \\
\exists r^a \in \mathsf{att}(\mathcal{GA}) : r^a \sqsubseteq \neg\langle g|p \wedge r^a \sqsubseteq \langle g|p \Rightarrow & \{\text{ lattice theory }\} \\
\exists r^a \in \mathsf{att}(\mathcal{GA}) : r^a \sqsubseteq \neg\langle g|p \sqcap \langle g|p \Rightarrow & \{\text{ lattice theory }\} \\
\exists r^a \in \mathsf{att}(\mathcal{GA}) : r^a \sqsubseteq \bot &
\end{array}
$$

However, this means $r^a = \bot$ which contradicts the atomicity of r^a. ■

Usually, a graph or subgraph is called connected if each of its nodes is reachable from every other of its nodes. Lifting this to sets of nodes this is equivalent that every subset of nodes is reachable from every nonempty subset of nodes. In this context, connectivity means reachability in an arbitrary number of steps. We will first investigate a more restricted version of connectivity which means reachability in exactly one step. Clearly, the connected components of the following definition correspond to equivalence classes, however, we chose the naming according to our further intentions.

Definition 4.2. *Given an algebraic equivalence g we call a test p* innately connected with respect to g *or* innately g-connected *for short if $p_2 \sqsubseteq \langle g|p_1$ holds for all non-bottom tests $p_1, p_2 \sqsubseteq p$. If p is innately g-connected and every test q with $q \sqsupset p$ is not innately g-connected then p is called an* innate connected component *of g. The set of all innate connected components of g is denoted by* $\mathsf{icc}(g)$.

It is easy to see that every atomic tests is innately connected so $\bot$ is no innate connected component.

Lemma 4.3. *Let g be an algebraic equivalence. Then a test p is innately g-connected iff for all atomic tests $p_1^a, p_2^a \sqsubseteq p$ the inequality $p_2^a \sqsubseteq \langle g|p_1^a$ holds.*

Proof. Clearly, the definition of innate g-connectivity implies $p_2^a \sqsubseteq \langle g|p_1^a$ for all atomic $p_1^a, p_2^a \sqsubseteq p$, so assume now that $p_2^a \sqsubseteq \langle g|p_1^a$ holds for all atomic tests $p_1^a, p_2^a \sqsubseteq p$. Given two arbitrary non-bottom tests $p_1, p_2 \sqsubseteq p$ we fix now an arbitrary atomic test $p_1^a \sqsubseteq p_1$ and reason as follows:

$$
\begin{array}{ll}
p_2 = & \{ \text{ atomicity of } \mathsf{test}(\mathcal{M}) \} \\
\bigsqcup\{p_2^a \mid p_2^a \in \mathsf{atom}(\mathcal{M}) \wedge p_2^a \sqsubseteq p_2\} \sqsubseteq & \{ \text{ assumption, lattice theory } \} \\
\bigsqcup\{\langle g|p_1^a\} \sqsubseteq & \{ p_1^a \sqsubseteq p_1, \text{ isotony of diamond } \} \\
\langle g|p_1 &
\end{array}
$$

■

Lemma 4.4. *Let g be an algebraic equivalence and pick an arbitrary $p \in \mathsf{icc}(g)$. Then p is contained in the set $\mathsf{fix}_{\langle g|}$ of fixpoints of $\langle g|$.*

Proof. We have $p \sqsubseteq p$ and hence by definition $p \sqsubseteq \langle g|p$ (chose $p_1 = p_2 = p$ in Definition 4.2) so it remains to show that $\langle g|p \sqsubseteq p$ holds. Therefore we assume that $\langle g|p \sqsubseteq p$ does not hold. Then we have $p \sqcup \langle g|p \sqsupset p$, and we can fix two arbitrary atomic tests $p_1^a, p_2^a \sqsubseteq p \sqcup \langle g|p \sqsupset p$ (recall the remark after Definition 4.2 which states $p \neq \bot$). We distinguish four cases:

1. $p_1^a \sqsubseteq p \wedge p_2^a \sqsubseteq p$: then we have $p_2^a \sqsubseteq \langle g|p_1^a$ because p is innately g-connected.
2. $p_1^a \sqsubseteq p \wedge p_2^a \sqsubseteq \langle g|p \sqcap \neg p$: by Corollary 3.11 there is an atomic test $p_3^a \sqsubseteq p$ with $p_2^a \sqsubseteq \langle g|p_3^a$, and due to innate g-connectivity of p we have also $p_3^a \sqsubseteq \langle g|p_1^a$, so we can reason as follows:

$$\begin{array}{ll} p_2^a \sqsubseteq & \{\text{ see above }\} \\ \langle g|p_3^a \sqsubseteq & \{p_3^a \sqsubseteq \langle g|p_1^a, \text{ isotony of diamond }\} \\ \langle g|\langle g|p_1^a = & \{\text{ modality and transitivity of } g\ \} \\ \langle g|p_1^a. & \end{array}$$

3. $p_1^a \sqsubseteq \langle g|p \sqcap \neg p \wedge p_2^a \sqsubseteq p$: due to symmetry of g and the previous case we have $p_2^a \sqsubseteq \langle g|p_1^a$.
4. $p_1^a \sqsubseteq \langle g|p \sqcap \neg p \wedge p_2^a \sqsubseteq \langle g|p \sqcap \neg p$: since p is non-bottom we can choose an arbitrary atomic $p_3^a \sqsubseteq p$. According to the two preceding cases, we have $p_2^a \sqsubseteq \langle g|p_3^a$ and $p_3^a \sqsubseteq \langle g|p_1^a$. A calculation analogous to Case 2 yields $p_2^a \sqsubseteq \langle g|p_1^a$.

In every case we have $p_2^a \sqsubseteq \langle g|p_1^a$, so by Lemma 4.3 $p \sqcup \langle g|p$ is g-connected which contradicts the maximality of p with respect to innate g-connectivity. ■

Lemma 4.5. *Let g be an algebraic equivalence and consider an arbitrary $p \in \mathsf{icc}(g)$. Then p is an atomic fixpoint of $\langle g|$.*

Proof. Due to Lemma 4.4, p is a fixpoint of $\langle g|$, so assume now there is a $q \in \mathsf{fix}_{\langle g|}$ with $0 \sqsubset q \sqsubset p$. Then we can chose two arbitrary atomic tests $p^a \sqsubseteq p \sqcap \neg q$ and $q^a \sqsubseteq q$ which implies also $p^a \sqsubseteq p$ and $q^a \sqsubseteq p$. Because p is innately g-connected we have $p^a \sqsubseteq \langle g|q^a$ and due to isotony of the diamond also $p^a \sqsubseteq \langle g|q$. However, q was assumed to be a fixpoint of $\langle g|$ which leads to $p^a \sqsubseteq q$, contradicting the choice of p^a. ■

Defining the lattice $\mathcal{FIX}_{\langle g|} =_{df} (\mathsf{fix}_{\langle g|}, \sqsubseteq, \bigsqcup, \bigsqcap, \bot, \top)$ (cf. Theorem 2.8) we can show that its atoms are innately connected components:

Lemma 4.6. *Let g be an algebraic equivalence and pick an arbitrary $p \in \mathsf{atom}(\mathcal{FIX}_{\langle g|})$. Then p is an innately connected component of g.*

Proof. First we show that an arbitrary $p \in \mathsf{atom}(\mathcal{FIX}_{\langle g|})$ is innately g-connected so we chose two non-bottom tests $q_1, q_2 \sqsubseteq p$. Due to Corollary 2.7 we have $p = \langle g|q_1$ which implies $q_2 \sqsubseteq \langle g|q_1$ by the assumption $q_2 \sqsubseteq p$.

Assume now there is an innately g-connected test q with $q \sqsupset p$. Then we have $q \sqsubseteq \langle g|p$ by definition of innate connectivity, and $p = \langle g|p$ because p is a fixpoint of $\langle g|$. Together this yields $q \sqsubseteq p$ which contradicts the assumption $q \sqsupset p$. ■

Theorem 4.7. *Let g be an algebraic equivalence. Then the set of fixpoints of $\langle g|$ forms a complete lattice with order $\sqsubseteq$ and least element $\bot$. Moreover, $\mathsf{icc}(g)$ is exactly the set of its atomic elements.*

Proof. Due to Theorem 4.1 $\langle g|$ is a csd closure on $\mathsf{test}(\mathcal{GA})$. The rest follows from Theorem 2.8 and Lemmas 4.5 and 4.6. ■

In particular, this means that the innately connected components of g are exactly of the form $\langle g|p^a$ with atomic tests p^a.

4.2 Algebraic Directed Acyclic Graphs

Given a directed graph G, the graph consisting of its strongly connected components as nodes and an edge between two components C_1 and C_2 iff there is an edge in G from some node in C_1 to some node in C_2 is acyclic (see e.g. [Jun05, Sha81]). Traditionally, a graph is said to be acyclic if it does not contain any cycles, i.e., if the meet of its transitive hull and the identity is empty. In the language of graph algebra, this leads to the following definition:

Definition 4.8. *An algebraic graph g is called an* algebraic directed acyclic graph *or* algebraic dag *for short if it fulfills the property* $g^+ \sqcap 1 = \bot$.

A sink in a directed acyclic graph is a node without outgoing edges. This motivates the following definition:

Definition 4.9. *Given an algebraic graph g, a non-bottom test s is called an* algebraic sink *of g if $\langle g|s = \bot$ holds.*

A finite directed acyclic graph contains always at least one sink. An analogous fact holds also for algebraic graphs, as stated in the next theorem:

Theorem 4.10. *Let g be an algebraic dag of a finite graph algebra $\mathcal{GA}$. Then there is an algebraic sink of g.*

Proof. We will show that there is an atomic test s^a with $\langle g|s^a = \bot$ which implies the claim obviously. In the sequel, we denote the cardinality of $\mathsf{att}(\mathcal{GA})$ by n_{at}. By finiteness of $\mathcal{GA}$ we have $g^* = \bigsqcup\{g^n \mid n \geq 0\}$ and hence $g^+ = \bigsqcup\{g^n \mid n \geq 1\}$. Assume now that for every atomic test s^a we have $\langle g|s^a \neq \bot$. Then we pick an arbitrary atomic test s^a_0 and construct a sequence of atomic tests $s^a_0, s^a_1, \ldots, s^a_{n_{at}}$ as follows: given s^a_i, we take an arbitrary atomic test s^a_{i+1} with $s^a_{i+1} \sqsubseteq \langle g|s^a_i$ (note that this is possible due to our assumption and atomicity of $\mathsf{test}(\mathcal{GA})$). By the pigeonhole principle, there are two indices n_1 and n_2 with $n_1 < n_2$ and $s^a_{n_1} = s^a_{n_2}$ which implies $s^a_{n_1} \sqsubseteq \langle g^{n_2-n_1}|s^a_{n_1}$ by $(n_2 - n_1)$-fold application of the backward diamond and modality. Herefrom we conclude $s^a_{n_1} \top s^a_{n_1} \sqsubseteq s^a_{n_1} g^{n_2-n_1} s^a_{n_1}$ by Lemma 3.5 and hence $s^a_{n_1} \top s^a_{n_1} \sqsubseteq g^+$ (remember that $n_1 < n_2$). So we can calculate:

$$
\begin{aligned}
& g^+ \sqcap 1 \sqsupseteq && \{ \text{ isotony of } \sqcap \} \\
& s^a_{n_1} \top s^a_{n_1} \sqcap 1 = && \{ \text{ Lemma 3.4 } \} \\
& s^a_{n_1} \sqcap 1 = && \{ \text{ atomicity of } \mathsf{test}(\mathcal{GA}) \} \\
& s^a_{n_1} \sqcap \bigsqcup \mathsf{att}(\mathcal{GA}) = && \{ \text{ distributivity } \} \\
& \bigsqcup\{s^a_{n_1} \sqcap p^a \mid p^a \in \mathsf{att}(\mathcal{GA})\} = && \{ \text{ Lemma 2.2.2, lattice theory } \} \\
& s^a_{n_1}
\end{aligned}
$$

Due to atomicity we have $s^a_{n_1} \neq \bot$, and we have the desired contradiction. ■

4.3 Undirected Graphs

An elementary approach to deal with an undirected graph $G = (V, E)$ is to define a directed graph $G' = (V', E')$ by $V' = V$ and $(v, w) \in E' \Leftrightarrow_{df} \{v, w\} \in E$. Clearly, G' is a symmetric directed graph with the same reachability properties as G. Hence, the connected components of G and the strongly connected components of G' coincide. Considerations as in Subsect. 4.1 lead to the following definition:

Definition 4.11. *Let g be a symmetric algebraic graph. A test p is said to be* connected with respect to g *or* g-connected *if $p_2 \sqsubseteq \langle g^*|p_1$ holds for all non-bottom tests $p_1, p_2 \sqsubseteq p$. If p is connected with respect to g and all tests q with $p \sqsubset q$ are not connected with respect to g then p is called a* connected component of g. *The set of all connected components of g is denoted by* $\mathsf{cc}(g)$.

Theorem 4.12. *Let g be a symmetric algebraic graph. Then the function $cc_g =_{df} \langle g^*|$ is a complementary distributive closure on* $\mathsf{test}(\mathcal{GA})$.

Proof. According to Theorem 4.1 it suffices to show that g^* is an algebraic equivalence. We have $1 \sqsubseteq g^*$ and $g^*g^* = g^*$ (hence $g^*g^* \sqsubseteq g^*$) by star properties and hence reflexivity and transitivity of g^*. For an arbitrary test p we have $\langle g^*|p = \langle (g^\circ)^*|p = \langle g^{*\circ}|p$ by symmetry of g and Lemma 3.13. ∎

Now, according to Theorem 4.7 and the subsequent remark a connected component of g can be described by $\langle g^*|p^a$ with an atomic test p^a. Algorithmically, this can be done by means of a BFS starting at p^a. Moreover, for two connected components c_1 and c_2 we have $c_1 \sqcap c_2 \neq \bot \Leftrightarrow c_1 = c_2$ by atomicity and Theorem 4.7.

The fact that there is no connection between two different connected components is stated in the next theorem:

Theorem 4.13. *Let g be a symmetric graph and consider two connected components c_1 and c_2 of g with $c_1 \neq c_2$. Then we have $c_1 g c_2 = \bot$.*

Proof. Assume that $c_1 g c_2 \neq \bot$. Then there are atomic tests $p_1^a \sqsubseteq c_1$ and $p_2^a \sqsubseteq c_2$ with $p_1^a g p_2^a \neq \bot$ by Corollary 3.11 and hence also $p_2^a \sqsubseteq \langle g|p_1^a$ by Lemma 3.5. Now we can calculate:

$$
\begin{aligned}
& c_1 = && \{ \text{ idempotence of supremum, } c_1 = \langle g^*|p_1^a \text{ by Lemma 2.6 } \} \\
& c_1 \sqcup \langle g^*|p_1^a \sqsupseteq && \{ \; g^* \sqsupseteq g, \text{ isotony of diamond operation } \} \\
& c_1 \sqcup \langle g|p_1^a \sqsupseteq && \{ \text{ choice of } p_2^a \} \\
& c_1 \sqcup p_2^a \sqsupset && \{ \; c_1 \sqcap c_2 = \bot, \text{ Lemma 2.2.7 } \} \\
& c_1
\end{aligned}
$$

This leads to the clear contradiction $c_1 \sqsupset c_1$. ∎

4.4 Directed Graphs

Analogously to Subsects. 4.1 and 4.3 we make the following definition:

Definition 4.14. *Let g be an arbitrary algebraic graph. A test p is said to be* strongly connected with respect to g *or* strongly g-connected *if $q_2 \sqsubseteq \langle g^*|q_1$ holds for all non-bottom tests $q_1, q_2 \sqsubseteq p$. If p is strongly connected with respect to g and all tests q with $q \sqsupset p$ are not strongly g-connected then p is called a* strongly connected component of g. *The set of all strongly connected components of g is denoted by* $\mathsf{scc}(g)$.

In general, g^* is no algebraic equivalence so we cannot immediately proceed as in Subsect. 4.3. The next lemmata serve to obtain an equivalent formulation of Definition 4.14 to which we can apply the results of Subsect. 4.1.

Lemma 4.15. *Let p be strongly g-connected. Then $pg^*p = p\top p$ holds.*

Proof. By definition of strong g-connectivity we have $p_2^a \sqsubseteq \langle g^*|p_1^a$ for all atomic tests $p_1^a, p_2^a \sqsubseteq p$. Due to Lemma 3.5 this is equivalent to $p_1^a g^* p_2^a \neq \bot$ so we calculate:

$$\begin{array}{lr}
pg^*p = & \{\text{ Lemma 3.10 }\} \\
\bigsqcup\{p_1^a g^* p_2^a \mid p_1^a, p_2^a \sqsubseteq p, p_1^a g^* p_2^a \neq \bot\} = & \{\text{ all-or-nothing }\} \\
\bigsqcup\{p_1^a \top p_2^a \mid p_1^a, p_2^a \sqsubseteq p\} = & \{\text{ distributivity, twice }\} \\
\bigsqcup\{p_1^a \mid p_1^a \sqsubseteq p\} \cdot \top \cdot \bigsqcup\{p_2^a \mid p_2^a \sqsubseteq p\} = & \{\text{ atomicity of } \mathsf{test}(\mathcal{GA}) \} \\
p\top p &
\end{array}$$

■

Lemma 4.16. *Let p be strongly g-connected. Then $pg^*p = p(g^\circ)^*p$ holds.*

Proof. We have $pg^*p = p\top p = p\top^\circ p = p(g^*)^\circ = p(g^\circ)^*p$ by Lemma 4.15, twice Lemmas 3.13 and 4.15 again. ■

Theorem 4.17. *For all g the function $scc_g =_{df} g^* \sqcap (g^\circ)^*$ is an algebraic equivalence.*

Proof. By star properties we have $1 \sqsubseteq g^*$ and $1 \sqsubseteq (g^\circ)^*$, so we have also $1 \sqsubseteq g^* \sqcap (g^\circ)^*$ due to the definition of the infimum which implies reflexivity of scc_g.

For transitivity we calculate as follows:

$$\begin{array}{lr}
(g^* \sqcap (g^\circ)^*) \cdot (g^* \sqcap (g^\circ)^*) \sqsubseteq & \{\text{ sub-conjunctivity of multiplication }\} \\
g^*g^* \sqcap g^*(g^\circ)^* \sqcap (g^\circ)^*g^* \sqcap (g^\circ)^*(g^\circ)^* \sqsubseteq & \{\text{ lattice theory }\} \\
g^*g^* \sqcap (g^\circ)^*(g^\circ)^* \sqsubseteq & \{\text{ star properties }\} \\
g^* \sqcap (g^\circ)^* &
\end{array}$$

Finally, we have $scc_g = g^* \sqcap (g^\circ)^* = (g^\circ)^* \sqcap g^* = (g^\circ)^* \sqcap ((g^*)^\circ)^\circ = (g^*)^\circ \sqcap ((g^\circ)^*)^\circ = (g^* \sqcap (g^\circ)^*)^\circ = {scc_g}^\circ$ by definition of scc_g, commutativity of $\sqcap$, multiple applications of Lemmas 3.13, 3.14 and definition of scc_g again which shows symmetry of scc_g. ■

This means that the strongly connected components of g have the form $\langle g^* \sqcap (g^\circ)^*|p^a$ with $p^a \in \mathsf{att}(\mathcal{GA})$. However, this characterization is not satisfying from an algorithmic point of view since the computation of the reflexive-transitive hull takes more than linear time in the general case. We will show now some properties which are used for a linear-time algorithm proposed in [Sha81].

Theorem 4.18. *Let g be an arbitrary algebraic graph. Then the* component graph $scg_g =_{df} \bigsqcup\{c_1 g c_2 \mid c_1, c_2 \in \mathsf{scc}(g), c_1 \neq c_2\}$ *is an algebraic dag.*

Proof. Assume that $scg_g^+ \sqcap 1 \neq \bot$. Then there is an atomic test p^a with $p^a scg_g^+ p^a \neq \bot$ which means $p^a \cdot \bigsqcup\{scg_g^n \mid n \geq 1\} \cdot p^a \neq \bot$. By distributivity this implies $\bigsqcup\{p^a \cdot scg_g^n \cdot p^a \mid n \geq 1\} \neq \bot$ so there is an $n \geq 1$ with $p^a \cdot scg_g^n \cdot p^a \neq \bot$, and hence there are strongly connected components $c_1, c_2, \ldots, c_{n+1}$ with $p^a c_1 g c_2 g c_3 \ldots c_n g c_{n+1} p^a \neq \bot$ (note that $c_i g c_{i+1} c_j g c_{j+1} \neq \bot$ implies $c_{i+1} = c_j$ due to Theorem 2.8, so we have $c_i g c_{i+1} c_j g c_{j+1} = c_i g c_j g c_{j+1}$). Applying Corollary 3.16 we conclude the existence of atomic tests $q_1^a, q_2^a, \ldots q_{2n}$ with $p^a c_1 q_1^a \neq \bot \neq q_{2n}^a c_{n+1} p^a$, $q_{2i-1}^a g q_{2i}^a \neq \bot$ for $1 \leq i \leq n$ and $q_{2i-2} c_i q_{2i-1} \neq \bot$ for $1 \leq i \leq n$. Now the terms of the form $q_{2i-2}^a c_i q_{2i-1}^a$ are non-bottom iff $q_{2i-2}^a = q_{2i-1}^a \sqsubseteq c_i$ holds and become in this case q_{2i-2}. Analogously, $p^a c_1$ and $c_{n+1} p^a$ are non-bottom iff $p^a \sqsubseteq c_1$ and $p^a \sqsubseteq c_{n+1}$ hold and become both p^a in every case. This implies the existence of atomic tests $r_1^a, r_2^a, \ldots, r_{n+1}^a$ with $r_1^a = r_{n+1}^a$, $r_i^a \sqsubseteq c_i$ and $r_i^a g r_{i+1}^a \neq \bot$. Again by Corollary 3.16 this implies $r_2^a g^{n-1} r_1^a \neq \bot$ and hence $r_2^a g^* r_1^a \neq \bot$, implying $r_1^a g^{*\circ} r_2^a \neq \bot$ by Lemma 3.5 and definition of the converse. On the other hand, we have $r_1^a g r_2^a \neq \bot$ from which we conclude $r_1^a g^* r_2^a \neq \bot$ and hence $r_1^a (g^* \sqcap g^{*\circ}) r_2^a \neq \bot$ by all-or-nothing, infimum properties and distributivity of multiplication by tests over infima. An immediate consequence thereof is $r_2^a \sqsubseteq \langle g^* \sqcap g^{*\circ}|r_1^a$ by Lemma 3.5, and on the other hand we have $c_1 = \langle g^* \sqcap g^{*\circ}|r_1^a$ by construction and hence $r_2^a \sqsubseteq c_1$. Therefrom we conclude $r_2^a \sqsubseteq c_1 \sqcap c_2$; however, by construction of scg_g we have $c_1 \neq c_2$ and hence $c_1 \sqcap c_2 = \bot$ by Theorem 2.8 which implies $r_2^a \sqsubseteq \bot$, contradicting the atomicity of r_2^a. ∎

Theorem 4.19. *Let g be a finite algebraic graph and consider a sink c_s of scg_g (see Theorem 4.18). Then for every atomic test $p_s^a \sqsubseteq c_s$ we have $\langle g^*|p_s^a = c_s$.*

Proof. Because of $c_s = \langle g^* \sqcap (g^\circ)^*|p_s^a$ and $g^* \sqcap (g^\circ)^* \sqsubseteq g^*$ we have $c_s \sqsubseteq \langle g^*|p_s^a$. In order to show $\langle g^*|p_s^a \sqsubseteq c_s$ we assume $\langle g^*|p_s^a \not\sqsubseteq c_s$ and fix an arbitrary atomic test q^a with $q^a \sqsubseteq \langle g^*|p_s^a$ and $q^a \not\sqsubseteq c_s$. Analogously to the proof of Theorem 4.10 there exist an $n \in \mathbb{N}$ and atomic tests $r_1^a, r_2^a, \ldots, r_n^a$ with $r_1^a = p_s^a$, $r_n^a = q^a$ and $r_{i+1} \sqsubseteq \langle g|r_i$. Because of $r_1^a \sqsubseteq c_s$ and $r_n^a \not\sqsubseteq c_s$ there is an index j such that $r_j^a \sqsubseteq c_s$ and $r_{j+1}^a \not\sqsubseteq c_s$. Then the strongly connected component $c_s' =_{df} \langle g^* \sqcap (g^\circ)^*|r_{j+1}$ fulfills $c_s \neq c_s'$. However, we have $r_j^a c_s = r_j^a$ and $r_{j+1}^a c_s' = r_{j+1}^a$ which leads to $r_j^a c_s g c_s' r_{j+1}^a \neq \bot$. This implies $\bot \sqsubset r_{j+1}^a \sqsubseteq \langle c_s g c_s'|r_j^a \sqsubseteq \langle c_s g c_s'|c_s$ which contradicts the fact that c_s is a sink of scg_g (remember $c_s \neq c_s'$ and the definition of scg_g). ∎

This last theorem is the base for the correctness of an algorithm for computing the strongly connected components of a graph G given in [Sha81]. It uses DFS in order to find a node in a sink of the graph consisting of the strongly connected components of G as nodes and an edge from component C_1 to component C_2 iff there is an edge in G from some node in C_1 to some in C_2. An algebraic formulation of this algorithm, especially finding the sink node, is beyond the scope of this paper.

5 Conclusion and Further Work

We demonstrated that connected components can be described in a one-sorted algebraic setting. The use of atomic tests together with parts not in first order logic are necessary to deal with graph properties concerning single nodes of a graph. This means that our approach is suitable for automated reasoning using proof assistants as COQ [Coq] or Isabelle [Isa] as demonstrated already e.g. in [GSW11]. In particular, correctness proofs for algorithms computing (strongly) connected components are a goal in near time. However, one crucial point is the algebraic description of DFS. While BFS is well understood in terms of Kleene Algebra (so e.g. in [DM11] reachability is handled basically via a BFS approach), DFS still awaits an algebraic treatment (the functional approach in [KL95] could serve as a first base therefore). Another natural generalization will be the application of the proposed methods to biconnected components.

There are also open questions of theoretical interest: how do the all-or-noting property and the Tarksi rule relate to each other? Mace4 (see [McC]) finds a counterexample (see [Glüb]) that the Tarski rule implies the all-or-nothing property, however, it does not find one for the other direction. Similarly, it is still open whether the Tarski rule is necessary for our results (it is used only once in the proof of Lemma 3.15). However, till now we cannot come up with a counterexample.

Due to the all-or-nothing property, our setting is tailored to unlabelled graphs. In order to deal with labelled graphs one possibility is to relax the all-or-nothing property in a way that infima of the form from the paragraph after Definition 3.3 become $\bot$ iff one of its operands equals $\bot$.

Acknowledgments. The author is grateful to Bernhard Möller and the anonymous reviewers for thorough proofreading and valuable hints and remarks which helped to improve the paper.

References

[BDHS16] Berghammer, R., Danilenko, N., Höfner, P., Stucke, I.: Cardinality of relations with applications. Discret. Math. **339**(12), 3089–3115 (2016)

[BHS15] Berghammer, R., Höfner, P., Stucke, I.: Tool-based verification of a relational vertex coloring program. In: Kahl, W., Winter, M., Oliveira, J.N. (eds.) RAMICS 2015. LNCS, vol. 9348, pp. 275–292. Springer, Cham (2015). doi:10.1007/978-3-319-24704-5_17

[Bir67] Birkhoff, G.: Lattice Theory, 3rd edn. American Mathematical Society, Providence (1967)

[BPS16] Brunet, P., Pous, D., Stucke, I.: Cardinalities of finite relations in Coq. In: Blanchette, J.C., Merz, S. (eds.) ITP 2016. LNCS, vol. 9807, pp. 466–474. Springer, Cham (2016). doi:10.1007/978-3-319-43144-4_29

[BSW15] Berghammer, R., Stucke, I., Winter, M.: Investigating and computing bipartitions with algebraic means. In: Kahl, W., Winter, M., Oliveira, J.N. (eds.) RAMICS 2015. LNCS, vol. 9348, pp. 257–274. Springer, Cham (2015). doi:10.1007/978-3-319-24704-5_16

[Coq] The Coq proof assistant. https://coq.inria.fr/

[DM11] Dang, H.-H., Möller, B.: Simplifying pointer Kleene algebra. In: Höfner, P., McIver, A., Struth, G. (eds.) Proceedings of 1st Workshop on Automated Theory Engineering, CEUR Workshop Proceedings, Wrocław, vol. 760, pp. 20–29. CEUR-WS.org (2011)

[DMS04] Desharnais, J., Möller, B., Struth, G.: Modal Kleene algebra and applications - a survey. J. Relat. Methods Comput. Sci. **1**, 93–131 (2004)

[DMS06] Desharnais, J., Möller, B., Struth, G.: Kleene algebra with domain. ACM Trans. Comput. Log. **7**, 798–833 (2006)

[FK12] Furusawa, H., Kawahara, Y.: Point axioms in dedekind categories. In: Kahl, W., Griffin, T.G. (eds.) RAMICS 2012. LNCS, vol. 7560, pp. 219–234. Springer, Heidelberg (2012). doi:10.1007/978-3-642-33314-9_15

[Glüa] Glück, R.: Atomic lattices. http://www.rolandglueck.de/Downloads/Atomiclattices.pdf

[Glüb] Glück, R.: Tarksi rule vs. all-or-nothing property. http://www.rolandglueck.de/Downloads/Tarski_all_or_nothing.in

[GMS09] Glück, R., Möller, B., Sintzoff, M.: A semiring approach to equivalences, bisimulations and control. In: Berghammer, R., Jaoua, A.M., Möller, B. (eds.) RelMiCS 2009. LNCS, vol. 5827, pp. 134–149. Springer, Heidelberg (2009). doi:10.1007/978-3-642-04639-1_10

[GSW11] Guttmann, W., Struth, G., Weber, T.: Automating algebraic methods in isabelle. In: Qin, S., Qiu, Z. (eds.) ICFEM 2011. LNCS, vol. 6991, pp. 617–632. Springer, Heidelberg (2011). doi:10.1007/978-3-642-24559-6_41

[Isa] Isabelle. https://isabelle.in.tum.de/

[JR92] Jipsen, P., Rose, H.: Varieties of Lattices, 1st edn. Springer, Heidelberg (1992)

[Jun05] Jungnickel, D.: Graphs, Networks and Algorithms, 2nd edn. Springer, Heidelberg (2005)

[Kah14] Kahl, W.: Graph transformation with symbolic attributes via monadic coalgebra homomorphisms. ECEASST **71** (2014)

[Kaw06] Kawahara, Y.: On the cardinality of relations. In: Schmidt, R.A. (ed.) RelMiCS 2006. LNCS, vol. 4136, pp. 251–265. Springer, Heidelberg (2006). doi:10.1007/11828563_17

[KF99] Kawahara, Y., Furusawa, H.: An algebraic formalization of fuzzy relations. Fuzzy Sets Syst. **101**(1), 125–135 (1999)

[KL95] King, D.J., Launchbury, J.: Structuring depth-first search algorithms in Haskell. In: Cytron, R.K., Lee, P. (eds.) Conference Record of POPL 1995, pp. 344–354. ACM Press (1995)

[Koz] Kozen, D.: A completeness theorem for Kleene algebras and the algebra of regular events. Inf. Comput. **110**(2), 366–390 (1994)

[McC] McCune, W.: Prover9 and Mace4. https://www.cs.unm.edu/mccune/mace4/

[MHS06] Möller, B., Höfner, P., Struth, G.: Quantales and temporal logics. In: Johnson, M., Vene, V. (eds.) AMAST 2006. LNCS, vol. 4019, pp. 263–277. Springer, Heidelberg (2006). doi:10.1007/11784180_21

[Sha81] Sharir, M.: A strong-connectivity algorithm and its applications in data flow analysis. Comput. Math. Appl. **7**(1), 67–72 (1981)

[SS93] Schmidt, G., Ströhlein, T.: Relations and Graphs: Discrete Mathematics for Computer Scientists. Springer, Heidelberg (1993)

[Tar55] Tarski, A.: A lattice-theoretical fixpoint theorem and its applications. Pac. J. Math. **5**(2), 285–309 (1955)

Stone Relation Algebras

Walter Guttmann(✉)

Department of Computer Science and Software Engineering,
University of Canterbury, Christchurch, New Zealand
walter.guttmann@canterbury.ac.nz

Abstract. We study a generalisation of relation algebras in which the underlying Boolean algebra structure is replaced with a Stone algebra. Many theorems of relation algebras generalise with no or small changes. Weighted graphs represented as matrices over extended real numbers form an instance. Relational concepts and methods can thus be applied to weighted graphs. All results are formally verified in Isabelle/HOL.

1 Introduction

Binary relations, which are the main instance of relation algebras, are essentially Boolean matrices. In graph theory they occur as the adjacency-matrix representation of unweighted graphs. It is therefore not surprising that relation-algebraic methods have been used to reason about graphs and to develop graph algorithms [8,9,11,45]. In this context weighted graphs are problematic simply because edge weights cannot be stored as entries of a Boolean matrix. Sometimes a workaround can be used, namely to represent weighted graphs by incidence matrices and weight functions [10]. However, keeping the direct representation of weighted graphs as matrices over numbers has benefits: it involves only one type of matrix, only a single matrix per graph, and only untyped (homogeneous) algebras which are better supported by theorem provers. Path problems and related algorithms have been treated successfully with this direct representation based on semirings with pre-orders and Kleene algebras [1,6,26,33]. Other graph problems, in particular the minimum spanning tree problem, seem to require more structure. Relation algebras provide additional structure, but need to be generalised to capture weighted graphs.

In order to verify Prim's algorithm for minimum spanning trees, we have proposed such a generalisation, Stone relation algebras, in [29]. Edge weights are typically numbers and form lattice and semiring structures (such as max-min and min-plus algebras). However, they do not form a Boolean algebra because a complement operation cannot be defined on the underlying linear order of numbers. The idea is to generalise the Boolean algebra structure just so much that edge weights can be represented while most of the structure is preserved. In particular, edge weights support a pseudocomplement operation and even form a Stone algebra. In Stone algebras, the involution property $\overline{\overline{x}} = x$ and the law of excluded middle $x \sqcup \overline{x} = \top$ are missing, but the weaker $\overline{x} \sqcup \overline{\overline{x}} = \top$ still holds, as do De Morgan's laws and $x \sqcap \overline{x} = \bot$. By forming matrices over

P. Höfner et al. (Eds.): RAMiCS 2017, LNCS 10226, pp. 127–143, 2017.
DOI: 10.1007/978-3-319-57418-9_8

Stone algebras we can hope to preserve much of the structure of relations. These matrices represent weighted graphs and we capture their algebraic properties by Stone relation algebras. The axioms of Stone relation algebras are based on the axioms of Tarski's relation algebras, which are modified to account for the weakening of the underlying lattice structure from Boolean algebras to Stone algebras.

Our previous paper gave the basic definitions and results with a focus on the verification of Prim's algorithm. In this paper, we study the properties of Stone relation algebras in more detail. Related work is discussed throughout the present paper. Its structure and contributions are as follows:

- In Sect. 2 we study pseudocomplemented algebras in general, and Stone algebras in particular. We also discuss the extended-real and matrix models of Stone algebras. Many results in this section are known from the literature; we contribute formally verified proofs of the algebraic properties and of the instantiation to the models.
- In Sect. 3 we study Stone relation algebras. Our contribution is to show that many results of relation algebras generalise to Stone relation algebras directly or, in some cases, with small changes. This includes algebraic properties that hold for all elements and ones that hold for specific classes of elements. Again, we formally prove our results including the instantiation to models.
- In Sect. 4 we study the weighted-graph model of Stone relation algebras. Our contribution is to characterise in logical terms the meaning of relation-algebraic properties when applied to weighted graphs. Also here, our results are formally stated and proved.
- In Sect. 5 we study Stone-Kleene relation algebras, which extend Stone relation algebras with the Kleene star operation. We contribute a number of algebraic properties, again formally proved.

All of our results are verified in Isabelle/HOL [42] using its integrated automated theorem provers and SMT solvers [14,43]. We omit the proofs, which can be found in the theory files available in the Archive of Formal Proofs [30,31] and at http://www.csse.canterbury.ac.nz/walter.guttmann/algebra/. The Archive currently stores the theories for Stone algebras and Stone relation algebras including the results presented in Sects. 2–4. The theories for Stone-Kleene relation algebras including the results of Sect. 5 are being prepared for it.

2 Pseudocomplemented Algebras

This section covers basic algebraic structures used in the present paper, including lattices, pseudocomplemented lattices and Stone algebras. These structures are further discussed in a number of textbooks [7,13,16,20,27]. Many results given in this section can be found in these textbooks. All results given in this section have been formally verified in Isabelle/HOL, mostly as part of a proof of Chen and Grätzer's construction theorem for Stone algebras [30].

Definition 1. A *bounded semilattice* is an algebraic structure $(S, \sqcup, \bot)$ where $\sqcup$ is associative, commutative and idempotent and has unit $\bot$:

$$x \sqcup (y \sqcup z) = (x \sqcup y) \sqcup z \qquad x \sqcup y = y \sqcup x \qquad x \sqcup x = x \qquad x \sqcup \bot = x$$

A *bounded lattice* is an algebraic structure $(S, \sqcup, \sqcap, \bot, \top)$ where $(S, \sqcup, \bot)$ and $(S, \sqcap, \top)$ are bounded semilattices and the following absorption axioms hold:

$$x \sqcup (x \sqcap y) = x \qquad x \sqcap (x \sqcup y) = x$$

A *bounded distributive lattice* is a bounded lattice where the following distributivity axioms hold (it is enough to postulate one of the two to obtain the other):

$$x \sqcup (y \sqcap z) = (x \sqcup y) \sqcap (x \sqcup z) \qquad x \sqcap (y \sqcup z) = (x \sqcap y) \sqcup (x \sqcap z)$$

The *lattice order* is given by

$$x \le y \Leftrightarrow x \sqcup y = y$$

A *(distributive) p-algebra* is an algebraic structure $(S, \sqcup, \sqcap, \overline{}, \bot, \top)$ such that $(S, \sqcup, \sqcap, \bot, \top)$ is a bounded (distributive) lattice and the pseudocomplement operation $\overline{}$ satisfies the equivalence

$$x \sqcap y = \bot \Leftrightarrow x \le \overline{y}$$

A *Stone algebra* is a distributive p-algebra satisfying the equation

$$\overline{x} \sqcup \overline{\overline{x}} = \top$$

An element $x \in S$ is *regular* if $\overline{\overline{x}} = x$ and *dense* if $\overline{x} = \bot$. A *Boolean algebra* is a Stone algebra whose elements are all regular.

Thus the pseudocomplement $\overline{y}$ of an element y is the $\le$-greatest element whose meet with y is $\bot$. The following result gives basic properties of pseudocomplements.

Theorem 2. *Let S be a p-algebra and let $x, y, z \in S$. Then the operation $\overline{}$ is $\le$-antitone, $x \sqcup \overline{x}$ is dense, and*

1. $\overline{\bot} = \top$
2. $\overline{\top} = \bot$
3. $x \le \overline{\overline{x}}$
4. $\overline{\overline{\overline{x}}} = \overline{x}$
5. $x \le \overline{y} \Leftrightarrow y \le \overline{x}$
6. $x \sqcap y = \bot \Leftrightarrow \overline{\overline{x}} \sqcap y = \bot$
7. $x \sqcap y \le \overline{z} \Leftrightarrow \overline{\overline{x}} \sqcap y \le \overline{z}$
8. $x \sqcap y \le \overline{z} \Leftrightarrow x \sqcap z \le \overline{y}$
9. $x \sqcap \overline{x} = \bot$
10. $\overline{x \sqcup y} = \overline{x} \sqcap \overline{y}$
11. $\overline{\overline{\overline{\overline{x}} \sqcup \overline{\overline{y}}}} = \overline{\overline{x \sqcup y}}$
12. $\overline{x \sqcap \overline{\overline{y}}} = \overline{x \sqcap y}$
13. $\overline{\overline{x \sqcap y}} = \overline{\overline{x}} \sqcap \overline{\overline{y}}$
14. $x \sqcap \overline{x \sqcap y} = x \sqcap \overline{y}$
15. $x \sqcap \overline{\overline{y}} \le \overline{\overline{x \sqcap y}}$
16. $x \sqcup \overline{\overline{y}} \le \overline{\overline{x \sqcup y}}$

In particular, the function $\lambda x.\overline{\overline{x}}$ is a closure operation, that is, idempotent, $\leq$-increasing and $\leq$-isotone. The image of the operation $^{-}$ is precisely the set of regular elements. They are closed under the operations $\sqcap$, $^{-}$, $\bot$ and $\top$. The dense elements of a p-algebra are precisely those mapped to $\top$ by the operation $\lambda x.\overline{\overline{x}}$. They are closed under the operations $\sqcup$, $\sqcap$, $\lambda x.\overline{\overline{x}}$ and $\top$. Equational axioms for p-algebras are obtained by adding Theorems 2.1, 2.2 and 2.14 to any set of equational axioms for bounded lattices.

In distributive p-algebras, we also obtain the following properties. By Theorem 3.1, every element x can be represented as the meet of a dense and a regular element.

Theorem 3. *Let S be a distributive p-algebra and let $x, y \in S$. Then*

1. $(x \sqcup \overline{x}) \sqcap \overline{\overline{x}} = x$
2. $x \sqcap y = \bot \wedge x \sqcup y = \top \Rightarrow \overline{x} = y$
3. $x \leq y \Leftrightarrow x \leq y \sqcup \overline{x}$
4. $x \leq y \Leftrightarrow x \sqcup \overline{x} \leq y \sqcup \overline{x}$

In a Stone algebra we obtain one of De Morgan's laws (the other is Theorem 2.10) and a number of weak shunting properties as the following result shows.

Theorem 4. *Let S be a Stone algebra and let $x, y, z \in S$. Then*

1. $\overline{x \sqcap y} = \overline{x} \sqcup \overline{y}$
2. $\overline{\overline{x \sqcup y}} = \overline{\overline{x}} \sqcup \overline{\overline{y}}$
3. $\overline{x \sqcup y} \sqcup \overline{x \sqcup \overline{y}} = \overline{x}$
4. $(x \sqcap \overline{y}) \sqcup (x \sqcap \overline{\overline{y}}) = x$
5. $\overline{\overline{x}} \sqcap y = \overline{\overline{x}} \sqcap z \wedge \overline{x} \sqcap y = \overline{x} \sqcap z \Rightarrow y = z$
6. $x \leq \overline{\overline{y}} \Leftrightarrow \top = \overline{x} \sqcup \overline{\overline{y}}$
7. $x \sqcap y \leq \overline{z} \Leftrightarrow x \leq \overline{z} \sqcup \overline{y}$
8. $x \sqcap \overline{y} \leq z \Leftrightarrow x \leq z \sqcup \overline{\overline{y}}$

The weak shunting property in Theorem 4.8 does not require the element z on the right-hand side to be regular. Another consequence is that the regular elements of a Stone algebra S are closed under the operation $\sqcup$, whence they form a Boolean subalgebra of S [27]. The dense elements of a Stone algebra form a distributive lattice with $\top$.

In the remainder of this section we look at instances of Stone algebras, notably extended real numbers and matrices over Stone algebras. Our considerations are motivated by weighted graphs. In this model we take edge weights from a Stone algebra and represent graphs by matrices containing edge weights.

For edge weights we use the extended real numbers $\mathbb{R}' = \mathbb{R} \cup \{\bot, \top\}$ with the operations max and min and the order $\leq$ extended so that $\bot$ is the $\leq$-least element and $\top$ is the $\leq$-greatest element. The resulting structure is a Stone algebra; the following result also shows the operation $\lambda x.\overline{\overline{x}}$ in this algebra.

Theorem 5. $(\mathbb{R}', \max, \min, ^{-}, \bot, \top)$ *is a Stone algebra with*

$$\overline{x} = \begin{cases} \top & \text{if } x = \bot \\ \bot & \text{if } x \neq \bot \end{cases} \qquad \overline{\overline{x}} = \begin{cases} \bot & \text{if } x = \bot \\ \top & \text{if } x \neq \bot \end{cases}$$

and the order $\leq$ on $\mathbb{R}'$ as the lattice order. The regular elements are $\bot$ and $\top$. All elements except $\bot$ are dense.

The operation $\lambda x.\overline{\overline{x}}$ checks whether its argument is different from $\bot$ and returns one of the Boolean elements $\bot$ or $\top$.

Weighted graphs are represented as matrices whose entries are edge weights. We therefore need to lift the Stone algebra structure to matrices according to the following result. Let $S^{A\times A}$ denote the set of square matrices with indices from a set A and entries from a set S. Such a matrix represents a directed graph with node set A and edge weights taken from S.

Theorem 6. *Let $(S, \sqcup, \sqcap, ^{-}, \bot, \top)$ be a Stone algebra and let A be a set. Then $(S^{A\times A}, \sqcup, \sqcap, ^{-}, \bot, \top)$ is a Stone algebra, where the operations $\sqcup, \sqcap, ^{-}, \bot, \top$ and the lattice order $\leq$ are lifted componentwise.*

Using pointwise liftings, the result holds more generally for the set S^X of all functions from X to S, for any set X.

It follows that the regular elements among the matrices over extended reals are the matrices over $\{\bot, \top\}$. They represent unweighted graphs: an entry $M_{ij} = \bot$ means that there is no edge from node i to node j in graph M, while $M_{ij} = \top$ means that there is an edge but no information about its weight is provided. Hence, on the matrix level the operation $\lambda x.\overline{\overline{x}}$ takes a weighted graph and produces an unweighted graph. The result $\overline{\overline{M}}$ represents the structure of the weighted graph M after forgetting the weights. Under this interpretation, the dense elements among the matrices correspond to complete graphs.

There are several approaches related to obtaining the structure of a weighted graph. In [47] an operation that gives the least 'crisp' relation containing a fuzzy relation is discussed. In [21] the 'shape' is a relation that represents a superset of the non-zero entries of a matrix of complex numbers; an operation that gives the non-zero entries is not considered. In [39] the 'support' is an operation on matrices over natural numbers that maps 0 to 0 and each non-zero entry to 1. In [36] weighted graphs are represented by matrices over commutative semirings and their structure is obtained by a 'flattening' operation that maps each entry x to the smallest multiplicatively idempotent element whose product with x is x. The multiplicatively idempotent elements in $\mathbb{R}$ are 0 and 1.

We conclude this section with a brief comparison of Stone algebras and Heyting algebras, which are bounded lattices where all relative pseudocomplements exist. The pseudocomplement of an element y relative to an element z is the $\leq$-greatest element whose meet with y is below z. By specialising $z = \bot$ it follows that Heyting algebras form distributive p-algebras. A counterexample generated by Nitpick [15] witnesses that, in general, Heyting algebras do not form Stone algebras this way. A counterexample given in [24, Example 4.6] shows that relative pseudocomplements need not exist in Stone algebras.

3 Stone Relation Algebras

In this section we further discuss the algebraic structure of matrices over Stone algebras. We have seen in Sect. 2 that such matrices form Stone algebras by lifting the operations componentwise. Moreover, they can be used to represent weighted graphs with edge weights taken from the extended reals. Finally, the subset of regular matrices obtained as the image of the closure operation $\lambda x.\overline{\overline{x}}$ represents unweighted graphs in this case. Because unweighted graphs correspond to relations, these observations suggest a generalisation of relation algebras to cover weighted graphs.

Definition 7. A *Stone relation algebra* $(S, \sqcup, \sqcap, \cdot, ^-, ^\top, \bot, \top, 1)$ is a Stone algebra $(S, \sqcup, \sqcap, ^-, \bot, \top)$ with a composition $\cdot$ and a converse $^\top$ and a constant 1 satisfying the following axioms (1)–(10). We abbreviate $x \cdot y$ as xy and let composition have higher precedence than the operators $\sqcup$ and $\sqcap$.

$$(xy)z = x(yz) \tag{1}$$
$$1x = x \tag{2}$$
$$(x \sqcup y)z = xz \sqcup yz \tag{3}$$
$$(xy)^\top = y^\top x^\top \tag{4}$$
$$(x \sqcup y)^\top = x^\top \sqcup y^\top \tag{5}$$
$$x^{\top\top} = x \tag{6}$$
$$\bot x = \bot \tag{7}$$
$$xy \sqcap z \leq x(y \sqcap x^\top z) \tag{8}$$
$$\overline{\overline{xy}} = \overline{\overline{x}}\,\overline{\overline{y}} \tag{9}$$
$$\overline{\overline{1}} = 1 \tag{10}$$

A *relation algebra* $(S, \sqcup, \sqcap, \cdot, ^-, ^\top, \bot, \top, 1)$ is a Stone relation algebra whose reduct $(S, \sqcup, \sqcap, ^-, \bot, \top)$ is a Boolean algebra.

An element $x \in S$ is a *vector* if $x\top = x$, a *co-vector* if $\top x = x$, *reflexive* if $1 \leq x$, *co-reflexive* if $x \leq 1$, *irreflexive* if $x \leq \overline{1}$, *symmetric* if $x = x^\top$, *asymmetric* if $x \sqcap x^\top = \bot$, *antisymmetric* if $x \sqcap x^\top \leq 1$, *transitive* if $xx \leq x$, *univalent* if $x^\top x \leq 1$, *injective* if $xx^\top \leq 1$, *total* if $1 \leq xx^\top$, *surjective* if $1 \leq x^\top x$, a *mapping* if x is univalent and total, *bijective* if x is injective and surjective, a *point* if x is a bijective vector, and an *atom* if both $x\top$ and $x^\top\top$ are bijective.

Tarski's relation algebras [46] require a Boolean algebra, axioms (1)–(6), and Theorem 8.20 below [40]. Axioms (7)–(10) follow from these properties. Another way to obtain relation algebras is by requiring a Boolean algebra and axioms (1)–(8) since axioms (9) and (10) immediately follow in Boolean algebras. There is a large body of research about Tarski's relation algebras; recent monographs are [32,41]. See [3] for an implementation of Tarski's relation algebras in Isabelle/HOL.

The Dedekind formula (8) or variants of it are known from [12,23,35,45]. In particular, Dedekind categories algebraically capture fuzzy relations, which are matrices over the real unit interval or complete distributive lattices used for modelling fuzzy systems [25,47]. In Dedekind categories composition is required to have a left residual and each Hom-set must be a complete distributive lattice and therefore a Heyting algebra [34]. Stone relation algebras maintain the signature of relation algebras. Algebras of relations with a smaller signature have been studied, for example, in [2,17]. In rough relation algebras [18] the lattice structure is required to be a double Stone algebra, which involves two dual pseudocomplement operations. A rough relation is a pair of upper and lower approximations of a relation with respect to a fixed indiscernibility relation [44].

Regular elements are closed under composition and its unit by axioms (9) and (10).

The following properties hold in Stone relation algebras.

Theorem 8. *Let S be a Stone relation algebra and let $w, x, y, z \in S$. Then*

1. $^\mathsf{T}$ *and* $\cdot$ *are* $\leq$*-isotone*
2. $\bot^\mathsf{T} = \bot$
3. $\top^\mathsf{T} = \top$
4. $1^\mathsf{T} = 1$
5. $(x \sqcap y)^\mathsf{T} = x^\mathsf{T} \sqcap y^\mathsf{T}$
6. $\overline{x}^\mathsf{T} = \overline{x^\mathsf{T}}$
7. $x\bot = \bot$
8. $x1 = 1$
9. $x \leq x\top$
10. $x \leq \top x$
11. $\top\top = \top$
12. $x \leq xx^\mathsf{T}x$
13. $x\top x\top = x\top$
14. $x(y \sqcup z) = xy \sqcup xz$
15. $x(y \sqcap z) \leq xy \sqcap xz$
16. $(x \sqcap y)z \leq xz \sqcap yz$
17. $x = (1 \sqcap xx^\mathsf{T})x = x(1 \sqcap x^\mathsf{T}x)$
18. $yx \sqcap z \leq (y \sqcap zx^\mathsf{T})x$
19. $xy \sqcap z = (x \sqcap zy^\mathsf{T})(y \sqcap x^\mathsf{T}z) \sqcap z$
20. $x^\mathsf{T}\overline{xy} \leq \overline{y}$
21. $xy \leq \overline{z} \Leftrightarrow x^\mathsf{T}z \leq \overline{y}$
22. $xy \leq \overline{z} \Leftrightarrow zy^\mathsf{T} \leq \overline{x}$
23. $xy \leq \overline{z} \Leftrightarrow yz^\mathsf{T} \leq \overline{x}^\mathsf{T}$
24. $xy \leq \overline{z} \Leftrightarrow z^\mathsf{T}x \leq \overline{y}^\mathsf{T}$
25. $xyz \leq \overline{w} \Leftrightarrow x^\mathsf{T}wz^\mathsf{T} \leq \overline{y}$
26. $xy \leq \overline{1} \Leftrightarrow yx \leq \overline{1}$
27. $x\overline{\overline{y}} \leq \overline{\overline{xy}}$
28. $\overline{x\overline{\overline{y}}} = \overline{xy}$
29. $\overline{\overline{x}}y \leq \overline{\overline{xy}}$
30. $\overline{\overline{\overline{x}}y} = \overline{xy}$
31. $xy \sqcap \overline{xz} = x(y \sqcap \overline{z}) \sqcap \overline{xz}$
32. $\overline{xy} \sqcup \overline{\overline{xz}} = \overline{x(y \sqcap \overline{z})} \sqcup \overline{\overline{xz}}$

Theorems 8.21–8.24 are weak versions of the Schröder equivalences of relation algebras: the elements on the right-hand sides of both inequalities must be regular. On the other hand, the conjugation property

$$xy \sqcap z = \bot \;\Leftrightarrow\; y \sqcap x^\mathsf{T}z = \bot \;\Leftrightarrow\; x \sqcap zy^\mathsf{T} = \bot$$

also holds in Stone relation algebras. Theorem 8.32 is another example how a property of relation algebras has been weakened, in this case by introducing double pseudocomplements. The original version is $\overline{xy} \sqcup xz = \overline{x(y \sqcap \overline{z})} \sqcup xz$ and appears as [40, Theorem 24(xxiv)].

Counterexamples generated by Nitpick witness that neither the Schröder equivalences of relation algebras nor [40, Theorem 24(xxiv)] hold in Stone relation algebras. Nevertheless, Theorem 8 shows that many properties of relation algebras already hold in Stone relation algebras.

We reuse the characterisations of vectors, co-reflexivity, injectivity and other properties known from relation algebras [45]. Consequences of these definitions are given by the following result. Once again, it shows that many properties generalise from relation algebras without changes.

Theorem 9. *Let S be a Stone relation algebra and let $w, x, y, z \in S$.*

1. *The regular elements of S are closed under the operation* $^\mathsf{T}$.
2. *The set of vectors of S is closed under the operations* $\sqcup, \sqcap, \cdot, \overline{}, \bot$ *and* $\top$.
3. *Every mapping, every bijective element and every atom is regular.*

If w and x are vectors, then

4. $(x \sqcap y)z = x \sqcap yz$
5. $y(z \sqcap x^\mathsf{T}) = yz \sqcap x^\mathsf{T}$
6. $(y \sqcap x^\mathsf{T})z = (y \sqcap x^\mathsf{T})(x \sqcap z)$
7. $(y \sqcap x^\mathsf{T})z = y(x \sqcap z)$
8. yx *is a vector*
9. $\overline{x}^\mathsf{T} x = \bot$
10. $xx^\mathsf{T}xx^\mathsf{T} \leq xx^\mathsf{T}$
11. $wx^\mathsf{T} = w \sqcap x^\mathsf{T}$

If w and x are co-reflexive, then

12. $x^\mathsf{T} = x$
13. $x\top \sqcap y = xy$
14. $x\top \sqcap 1 = x$
15. $\overline{x\top} \sqcap 1 = \overline{x} \sqcap 1$
16. $xx = x$
17. $xy \sqcap \overline{z} = xy \sqcap \overline{xz}$
18. $w \sqcap x = wx$
19. $wy \sqcap xy = (w \sqcap x)y$

If w is univalent and x is injective, then

20. $w(y \sqcap z) = wy \sqcap wz$
21. $w\overline{y} \leq \overline{wy}$
22. $(y \sqcap z)x = yx \sqcap zx$
23. $\overline{y}x \leq \overline{yx}$

If w is a mapping and x is bijective, then

24. $y \leq wz \Leftrightarrow w^\mathsf{T}y \leq z$
25. $w\overline{y} = \overline{wy}$
26. $y \leq zx \Leftrightarrow yx^\mathsf{T} \leq z$
27. $\overline{y}x = \overline{yx}$

Finally,

28. x *is a vector/univalent/total* $\Leftrightarrow$ x^T *is a co-vector/injective/surjective*
29. x *is total* $\Leftrightarrow x\top = \top$
30. x *is surjective* $\Leftrightarrow \top x = \top$

While vectors are closed under pseudocomplements in Stone relation algebras, a counterexample generated by Nitpick witnesses that x need not be a vector if $\overline{x}$ is a vector. In fact there are counterexamples in the weighted-graph model as shown below.

In order to instantiate Stone relation algebras by weighted graphs we proceed in two steps. First, we show how every Stone algebra gives rise to a Stone relation algebra by reusing some of the operations. Second, we lift the Stone relation algebra structure to matrices; this is similar to the lifting for Dedekind categories [47]. The following result also shows that every Stone relation algebra has a subalgebra that is a relation algebra. As a consequence we can work with weighted graphs in Stone relation algebras and use the full power of relation algebras for reasoning about their structure.

Theorem 10.

1. *The regular elements of a Stone relation algebra S form a relation algebra that is a subalgebra of S.*
2. *Let $(S, \sqcup, \sqcap, ^{-}, \bot, \top)$ be a Stone algebra. Then $(S, \sqcup, \sqcap, \sqcap, ^{-}, \lambda x.x, \bot, \top, \top)$ is a Stone relation algebra with meet as composition, $\top$ as its unit, and the identity function as converse.*
3. *Let $(S, \sqcup, \sqcap, \cdot, ^{-}, ^{\mathsf{T}}, \bot, \top, 1)$ be a Stone relation algebra and let A be a finite set. Then $(S^{A \times A}, \sqcup, \sqcap, \cdot, ^{-}, ^{\mathsf{T}}, \bot, \top, 1)$ is a Stone relation algebra, where the operations $\cdot$, ^{T} and 1 are defined by*

$$\begin{aligned} (M \cdot N)_{ij} &= \textstyle\bigsqcup_{k \in A} M_{ik} \cdot N_{kj} \\ (M^{\mathsf{T}})_{ij} &= (M_{ji})^{\mathsf{T}} \\ 1_{ij} &= \begin{cases} 1 & \text{if } i = j \\ \bot & \text{if } i \neq j \end{cases} \end{aligned}$$

The weighted-graph model is an instance of this construction because edge weights are taken from the Stone algebra of extended reals. Thus for a finite set A, the set of matrices $\mathbb{R}'^{A \times A}$ is a Stone relation algebra with the following operations:

$$\begin{aligned} (M \cdot N)_{ij} &= \max_{k \in A} \min\{M_{ik}, N_{kj}\} \\ (M^{\mathsf{T}})_{ij} &= M_{ji} \\ 1_{ij} &= \begin{cases} \top & \text{if } i = j \\ \bot & \text{if } i \neq j \end{cases} \end{aligned}$$

The remaining operations are lifted componentwise from the underlying Stone algebra.

Recall that the regular elements are the matrices over $\{\bot, \top\}$ in this case; they represent unweighted graphs. In particular, the graphs $\bot$, 1 and $\top$ are regular. Theorem 10.1 confirms that these matrices form a relation algebra.

We furthermore note that the way to obtain regular matrices (essentially relations) from weighted matrices by taking the image of $^{-}$ is similar to the way co-reflexive relations are obtained from relations by taking the image of the antidomain operation [22]. The operation $\lambda x.\overline{\overline{x}}$ corresponds to the domain operation and, if vectors are used instead of co-reflexives, to the operation $\lambda x.x\top$, which is a closure operation.

Next, we further discuss the difference between relation algebras and Stone relation algebras. The following list shows a number of properties of relation algebras that do not generally hold in Stone relation algebras. We give counterexamples found by Nitpick in the weighted-graph model of matrices over extended reals $\mathbb{R}'^{A \times A}$. Nitpick allows the user to set independent bounds for the size of matrices and the size of the set that approximates matrix entries in the search.

1. $xy \leq z \Leftrightarrow x^{\mathsf{T}}\overline{z} \leq \overline{y}$:
   This Schröder equivalence fails for $A = \{a\}$ and $x_{aa} = y_{aa} = 0$ and $z_{aa} = -1$.
2. $\overline{xy} \sqcup xz = \overline{x(y \sqcap \overline{z})} \sqcup xz$:
   This equation is [40, Theorem 24(xxiv)] and fails for $A = \{a\}$ and $x_{aa} = z_{aa} = 0$ and $y_{aa} = \top$.
3. $\overline{\overline{x}\top} \sqcap 1 \leq x$ for each vector x:
   This fails for $A = \{a\}$ and $x_{aa} = 0$.
4. $(\overline{\overline{x}\top} \sqcap 1)\top = x$ for each vector x:
   This fails for $A = \{a\}$ and $x_{aa} = 0$.
5. x is a vector if $\overline{x}$ is a vector:
   This holds for graphs with a single node but fails for $A = \{a, b\}$ and $x_{aa} = 0$ and $x_{ab} = x_{bb} = 1$ and $x_{ba} = \top$.
6. $\overline{x} \sqcap y$ is regular for each $x \neq \bot$:
   This holds for graphs with a single node but fails for $A = \{a, b\}$ and $x_{ba} = \bot$ and $y_{ba} = 1$ and all other entries of x and y set to $\top$.

Except in the last case, Nitpick indicated that the examples it found are potentially spurious, which might be due to the involved matrix products. All counterexamples have been verified manually.

Finally, we discuss two examples for proving properties of weighted graphs in Stone relation algebras. First, consider a graph G on a set of nodes A and a subset $B \subseteq A$ of the nodes. We work in the Stone relation algebra $S = \mathbb{R}'^{A \times A}$. The graph G is represented by an element $x \in S$ and the subset B of nodes is represented by a regular vector $v \in S$. The element vv^{T} describes the complete unweighted graph formed by the nodes in B. The meet $vv^{\mathsf{T}} \sqcap x$ restricts the edges of G to those that start and end in B; this is a weighted subgraph. By Theorem 9.10 we obtain

$$(vv^{\mathsf{T}} \sqcap x)(vv^{\mathsf{T}} \sqcap x) \leq vv^{\mathsf{T}}vv^{\mathsf{T}} \leq vv^{\mathsf{T}}$$

which shows that by following a sequence of two edges in the weighted subgraph we cannot leave the set of nodes in B. The claim extends to longer sequences of edges by using the Kleene star as in Sect. 5. Results like this are used for reasoning about Prim's minimum spanning tree algorithm, where in each step the constructed tree is a subgraph of the input and a spanning tree of the nodes that have already been visited.

Second, in the same setting let $e \in S$ such that $e \leq v\overline{v}^{\mathsf{T}}$. Such an e can represent a set of edges each of which goes from a node in B to a node outside of B. By Theorem 9.9 we obtain

$$ee \leq v\overline{v}^{\mathsf{T}}v\overline{v}^{\mathsf{T}} = \bot$$

which shows that it is not possible to follow two such edges in sequence. In Prim's algorithm, the edges considered for extending the spanning tree in each step satisfy this property. The obtained result is used for showing that the extended tree is acyclic.

4 Relational Properties of Weighted Graphs

In this section we study the weighted-graph model of Stone relation algebras. In particular, we discuss how relation-algebraic properties are interpreted in this instance. Throughout this section, a graph is an element of the Stone relation algebra $\mathbb{R}'^{A \times A}$ introduced in Sect. 3.

4.1 Mappings and Related Properties

We first look at univalent, injective, total, surjective and bijective matrices and at mappings.

Theorem 11. *Let $M \in \mathbb{R}'^{A \times A}$. Then M is*

1. *univalent $\Leftrightarrow$ in every row at most one entry is not $\bot$*
2. *injective $\Leftrightarrow$ in every column at most one entry is not $\bot$*
3. *total $\Leftrightarrow$ in every row at least one entry is $\top$*
4. *surjective $\Leftrightarrow$ in every column at least one entry is $\top$*
5. *a mapping $\Leftrightarrow$ in every row exactly one entry is $\top$ and the others are $\bot$*
6. *bijective $\Leftrightarrow$ in every column exactly one entry is $\top$ and the others are $\bot$*

Moreover,

7. *$\overline{M\top} = \bot \Leftrightarrow$ in every row at least one entry is not $\bot$*
8. *$\overline{\top M} = \bot \Leftrightarrow$ in every column at least one entry is not $\bot$*

Note that univalent, injective, total and surjective matrices may have entries which are neither $\bot$ nor $\top$. In the graph interpretation, univalent means that every node has at most one outgoing edge. To specify at least one outgoing edge, we can use the property in Theorem 11.7, which is equivalent to $\overline{\overline{M}}$ being total. Requiring M to be total is stronger: it means that at least one edge is labelled with $\top$. Therefore, to specify exactly one outgoing edge per node, the conjunction of univalent with the property in Theorem 11.7 has to be used. Requiring a mapping is more restrictive; in fact mappings are regular by Theorem 9.3. Similar remarks apply for injective, surjective and bijective matrices and the property in Theorem 11.8 with respect to the incoming edges of each node.

4.2 Vectors and Related Properties

We next look at vectors, co-vectors, points and atoms in the weighted-graph model of matrices over $\mathbb{R}'$.

Theorem 12. *Let $M \in \mathbb{R}'^{A \times A}$. Then M is*

1. *a vector $\Leftrightarrow$ in every row all entries are the same*
2. *a co-vector $\Leftrightarrow$ in every column all entries are the same*
3. *a point $\Leftrightarrow$ exactly one row is constant $\top$ and the others are constant $\bot$*
4. *an atom $\Leftrightarrow$ exactly one entry is $\top$ and the others are $\bot$*

Also vectors and co-vectors may have entries which are neither $\bot$ nor $\top$. As in the relational case, a matrix with just one column/row is sufficient to store the information contained in a vector/co-vector. Points and atoms are regular by Theorem 9.3. Their interpretation for graphs is the same as in the relational model: a point represents a node of the graph and an atom represents an edge. Weaker properties can again be obtained by replacing surjective with the property in Theorem 11.8 in the definitions of point and atom. In this case, all rows in a point would be $\bot$ except for one row, in which all entries would be the same, arbitrary non-$\bot$ value. Similarly, an atom would have exactly one non-$\bot$ value.

4.3 Orders and Related Properties

Finally, we look at reflexive, co-reflexive, irreflexive, symmetric, antisymmetric, asymmetric and transitive matrices over $\mathbb{R}'$.

Theorem 13. *Let $M \in \mathbb{R}'^{A \times A}$. Then M is*

1. *reflexive $\Leftrightarrow$ the diagonal is constant $\top$*
2. *co-reflexive $\Leftrightarrow$ all entries not on the diagonal are $\bot$*
3. *irreflexive $\Leftrightarrow$ the diagonal is constant $\bot$*
4. *symmetric $\Leftrightarrow$ $M_{ij} = M_{ji}$ for each $i, j \in A$*
5. *antisymmetric $\Leftrightarrow$ $M_{ij} = \bot$ or $M_{ji} = \bot$ for each $i \neq j \in A$*
6. *asymmetric $\Leftrightarrow$ $M_{ij} = \bot$ or $M_{ji} = \bot$ for each $i, j \in A$*
7. *transitive $\Leftrightarrow$ $M_{ik} \leq M_{ij}$ or $M_{kj} \leq M_{ij}$ for each $i, j, k \in A$*

Co-reflexive matrices share most properties of tests [28,38] except they do not form a Boolean algebra. Nevertheless, they form a Stone relation subalgebra in which composition and meet coincide and composition is idempotent. For example, the composition MN of a co-reflexive matrix M and an arbitrary matrix N restricts the elements of N in row i to at most M_{ii}. In particular, rows are filtered out if $M_{ii} = \bot$ and left unchanged if $M_{ii} = \top$. The composition NM has a similar effect on the columns of N.

Matrices that are reflexive, transitive and symmetric have a block-diagonal structure (that is, the base set A can be suitably partitioned by an equivalence relation). The entries in each block are different from $\bot$ but not necessarily $\top$.

Similarly, matrices that are reflexive, transitive and antisymmetric have the structure of a partial order. Again the non-$\bot$ entries may differ from $\top$. In the graph interpretation, antisymmetric means that there is at most one edge between any two different nodes. Asymmetric additionally requires that there are no loops; the latter property amounts to being irreflexive. Symmetric matrices can be used to represent undirected weighted graphs.

5 Stone-Kleene Relation Algebras

In this section we discuss iterated composition in Stone relation algebras. This works analogously to adding the Kleene star operation to relation algebras. In the graph model, this allows us to talk about reachability. We use the axioms of the Kleene star given in [37].

Definition 14. A *Stone-Kleene relation algebra* $(S, \sqcup, \sqcap, \cdot, \overline{}, ^\mathsf{T}, ^*, \bot, \top, 1)$ is a Stone relation algebra $(S, \sqcup, \sqcap, \cdot, \overline{}, ^\mathsf{T}, \bot, \top, 1)$ with an operation * satisfying the unfold and induction axioms

$$\begin{array}{ll} 1 \sqcup yy^* \leq y^* & z \sqcup yx \leq x \Rightarrow y^*z \leq x \\ 1 \sqcup y^*y \leq y^* & z \sqcup xy \leq x \Rightarrow zy^* \leq x \end{array}$$

and the axiom

$$\overline{\overline{x^*}} = \overline{\overline{x}}^* \tag{11}$$

An element $x \in S$ is *acyclic* if xx^* is irreflexive, and x is a *forest* if x is injective and acyclic.

Kleene algebras are based on idempotent semirings in [37], but we do not require more axioms than the above since all Stone relation algebras are idempotent semirings. Regular elements are closed under the Kleene star by axiom (11). The following properties hold in Stone-Kleene relation algebras.

Theorem 15. *Let S be a Stone-Kleene relation algebra and let $x, y \in S$.*

1. *The regular elements of S are closed under the operation *.*

Moreover

2. $x^{*\mathsf{T}} = x^{\mathsf{T}*}$
3. $x^\mathsf{T}(xx^\mathsf{T})^* \leq x^\mathsf{T}$ *if x is a vector*
4. $(xx^\mathsf{T})^* = 1 \sqcup xx^\mathsf{T}$ *if x is a vector*
5. $x^\mathsf{T}y^* = x^\mathsf{T}((x^\mathsf{T}y^*)^\mathsf{T}(x^\mathsf{T}y^*) \sqcap y)^*$ *if x is a vector*
6. $x^{\mathsf{T}*} \leq \overline{x}$ *if and only if x is acyclic*
7. *x is asymmetric if x is acyclic*
8. $x^*x^{\mathsf{T}*} \sqcap x^\mathsf{T}x \leq 1$ *if x is a forest*

As an example we discuss Theorem 15.5, which considers a graph y and a set of nodes x represented as a vector. Then $y^{\mathsf{T}*}x$ is a vector representing the set of nodes reachable from any node in x. The same set is represented by the left-hand side $x^\mathsf{T}y^*$ as a co-vector. The right-hand side uses the same construction except the graph y is restricted to those edges that start and end in this set of reachable nodes. Thus Theorem 15.5 states that to reach any of these nodes from x it suffices to take edges between these nodes. This property is used several times for proving the correctness of Prim's minimum spanning tree algorithm.

As another example, Theorem 15.6 has the following interpretation for an acyclic graph x. The left-hand side describes backward reachability in x. The inequality states that if a node q is reachable from a node p by going backward any number of steps in x, then there must not be an edge from p to q; otherwise we could combine it with the path from q to p to obtain a cycle in x. Moreover, this condition is equivalent to being acyclic.

In order to instantiate Stone-Kleene relation algebras by weighted graphs we extend the two-step process we used for Stone relation algebras in Sect. 3 by the

Kleene star operation. First, every Stone algebra gives rise to a Stone-Kleene relation algebra by setting $x^* = \top$. This is because the underlying bounded lattice forms a semiring where $1 = \top$. Second, we lift the Stone-Kleene relation algebra structure to matrices. Note that $x^* = \top$ does not generally hold in the matrix algebra; only the entries on the diagonal of x^* will be $\top$.

Theorem 16.

1. *Let* $(S, \sqcup, \sqcap, ^-, \bot, \top)$ *be a Stone algebra. Then, using the constant* $\top$ *function as the Kleene star,* $(S, \sqcup, \sqcap, \sqcap, ^-, \lambda x.x, \lambda x.\top, \bot, \top, \top)$ *is a Stone-Kleene relation algebra.*
2. *Let* $(S, \sqcup, \sqcap, \cdot, ^-, ^\mathsf{T}, ^*, \bot, \top, 1)$ *be a Stone-Kleene relation algebra and let* A *be a finite set. Then* $(S^{A \times A}, \sqcup, \sqcap, \cdot, ^-, ^\mathsf{T}, ^*, \bot, \top, 1)$ *is a Stone-Kleene relation algebra, where the operation* * *on matrices is defined using Conway's automata-based construction described in* [19].

The subalgebra of regular elements of a Stone-Kleene relation algebra is both a relation algebra and a Kleene algebra.

The proof of Theorem 16 formally verifies the correctness of Conway's construction for Kleene algebras. An implementation of the construction in Isabelle/HOL that extends [4] was given in [5] without a correctness proof.

6 Conclusion

In the present paper we have studied algebras for modelling weighted graphs. Stone relation algebras are designed to stay so close to relation algebras that relational methods and concepts can be reused, yet be general enough to capture weighted graphs. Like relation algebras, Stone relation algebras can be combined with Kleene algebras for reasoning about reachability. All of our results about these algebraic structures have been formally verified in Isabelle/HOL; this includes a proof that weighted graphs represented by matrices over extended reals form an instance.

We have applied these results in two case studies. The first is a formally verified proof of Chen and Grätzer's construction theorem for Stone algebras [30]. It involves extensive reasoning about algebraic structures in addition to reasoning in algebraic structures. The second case study is a formal verification of Prim's minimum spanning tree algorithm [29]. It uses Hoare logic and most of the proof can be carried out in Stone-Kleene relation algebras.

Section 4 interprets a number of relational properties for weighted graphs. Future work will consider further graph algorithms to understand the limits of what can be expressed algebraically in this model. The long-term goal of these efforts is a library for algebraic reasoning about weighted graphs and graph algorithms.

Acknowledgement. I thank Georg Struth and the anonymous referees for pointing out related work and for other helpful comments.

References

1. Aho, A.V., Hopcroft, J.E., Ullman, J.D.: The Design and Analysis of Computer Algorithms. Addison-Wesley Publishing Company, Reading (1974)
2. Andréka, H., Mikulás, Sz.: Axiomatizability of positive algebras of binary relations. Algebra Universalis **66**(1–2), 7–34 (2011)
3. Armstrong, A., Foster, S., Struth, G., Weber, T.: Relation algebra. Archive of Formal Proofs (2016, first version 2014)
4. Armstrong, A., Gomes, V.B.F., Struth, G., Weber, T.: Kleene algebra. Archive of Formal Proofs (2016, first version 2013)
5. Asplund, T.: Formalizing the Kleene star for square matrices. Bachelor thesis IT 14 002, Department of Information Technology, Uppsala Universitet (2014)
6. Backhouse, R.C., Carré, B.A.: Regular algebra applied to path-finding problems. J. Inst. Math. Appl. **15**(2), 161–186 (1975)
7. Balbes, R., Dwinger, P.: Distributive Lattices. University of Missouri Press, Columbia (1974)
8. Berghammer, R., Fischer, S.: Combining relation algebra and data refinement to develop rectangle-based functional programs for reflexive-transitive closures. J. Log. Algebr. Methods Program. **84**(3), 341–358 (2015)
9. Berghammer, R., von Karger, B.: Relational semantics of functional programs (Chap. 8). In: Brink, C., Kahl, W., Schmidt, G. (eds.) Relational Methods in Computer Science, pp. 115–130. Springer, Wien (1997). doi:10.1007/978-3-7091-6510-2_8
10. Berghammer, R., von Karger, B., Wolf, A.: Relation-algebraic derivation of spanning tree algorithms. In: Jeuring, J. (ed.) MPC 1998. LNCS, vol. 1422, pp. 23–43. Springer, Heidelberg (1998). doi:10.1007/BFb0054283
11. Berghammer, R., Rusinowska, A., de Swart, H.: Computing tournament solutions using relation algebra and RelView. Eur. J. Oper. Res. **226**(3), 636–645 (2013)
12. Bird, R., de Moor, O.: Algebra of Programming. Prentice Hall, Englewood Cliffs (1997)
13. Birkhoff, G.: Lattice Theory. Colloquium Publications, vol. XXV, 3rd edn. American Mathematical Society, Providence (1967)
14. Blanchette, J.C., Böhme, S., Paulson, L.C.: Extending Sledgehammer with SMT solvers. In: Bjørner, N., Sofronie-Stokkermans, V. (eds.) CADE 2011. LNCS (LNAI), vol. 6803, pp. 116–130. Springer, Heidelberg (2011). doi:10.1007/978-3-642-22438-6_11
15. Blanchette, J.C., Nipkow, T.: Nitpick: a counterexample generator for higher-order logic based on a relational model finder. In: Kaufmann, M., Paulson, L.C. (eds.) ITP 2010. LNCS, vol. 6172, pp. 131–146. Springer, Heidelberg (2010). doi:10.1007/978-3-642-14052-5_11
16. Blyth, T.S.: Lattices and Ordered Algebraic Structures. Springer, London (2005)
17. Bredihin, D.A., Schein, B.M.: Representations of ordered semigroups and lattices by binary relations. Colloq. Math. **39**(1), 1–12 (1978)
18. Comer, S.D.: On connections between information systems, rough sets and algebraic logic. In: Rauszer, C. (ed.) Algebraic Methods in Logic and in Computer Science. Banach Center Publications, vol. 28, pp. 117–124. Institute of Mathematics, Polish Academy of Sciences, Warsaw (1993)
19. Conway, J.H.: Regular Algebra and Finite Machines. Chapman and Hall, London (1971)

20. Curry, H.B.: Foundations of Mathematical Logic. Dover Publications, New York (1977)
21. Desharnais, J., Grinenko, A., Möller, B.: Relational style laws and constructs of linear algebra. J. Log. Algebr. Methods Program. **83**(2), 154–168 (2014)
22. Desharnais, J., Struth, G.: Internal axioms for domain semirings. Sci. Comput. Program. **76**(3), 181–203 (2011)
23. Freyd, P.J., Ščedrov, A.: Categories, Allegories. North-Holland Mathematical Library, vol. 39. Elsevier Science Publishers, Amsterdam (1990)
24. Fried, E., Hansoul, G.E., Schmidt, E.T., Varlet, J.C.: Perfect distributive lattices. In: Eigenthaler, G., Kaiser, H.K., Müller, W.B., Nöbauer, W. (eds.) Contributions to General Algebra, vol. 3, pp. 125–142. Hölder-Pichler-Tempsky, Wien (1985)
25. Goguen, J.A.: L-fuzzy sets. J. Math. Anal. Appl. **18**(1), 145–174 (1967)
26. Gondran, M., Minoux, M.: Graphs, Dioids and Semirings. Springer, New York (2008)
27. Grätzer, G.: Lattice Theory: First Concepts and Distributive Lattices. W. H. Freeman and Co., San Francisco (1971)
28. Guttmann, W.: Algebras for iteration and infinite computations. Acta Inf. **49**(5), 343–359 (2012)
29. Guttmann, W.: Relation-algebraic verification of Prim's minimum spanning tree algorithm. In: Sampaio, A., Wang, F. (eds.) ICTAC 2016. LNCS, vol. 9965, pp. 51–68. Springer, Cham (2016). doi:10.1007/978-3-319-46750-4_4
30. Guttmann, W.: Stone algebras. Archive of Formal Proofs (2016)
31. Guttmann, W.: Stone relation algebras. Archive of Formal Proofs (2017)
32. Hirsch, R., Hodkinson, I.: Relation Algebras by Games. Elsevier Science B.V., Amsterdam (2002)
33. Höfner, P., Möller, B.: Dijkstra, Floyd and Warshall meet Kleene. Form. Asp. Comput. **24**(4), 459–476 (2012)
34. Kawahara, Y., Furusawa, H.: Crispness in Dedekind categories. Bull. Inform. Cybern. **33**(1–2), 1–18 (2001)
35. Kawahara, Y., Furusawa, H., Mori, M.: Categorical representation theorems of fuzzy relations. Inf. Sci. **119**(3–4), 235–251 (1999)
36. Killingbeck, D., Teixeira, M.S., Winter, M.: Relations among matrices over a semiring. In: Kahl, W., Winter, M., Oliveira, J.N. (eds.) RAMiCS 2015. LNCS, vol. 9348, pp. 101–118. Springer, Cham (2015). doi:10.1007/978-3-319-24704-5_7
37. Kozen, D.: A completeness theorem for Kleene algebras and the algebra of regular events. Inf. Comput. **110**(2), 366–390 (1994)
38. Kozen, D.: Kleene algebra with tests. ACM Trans. Program. Lang. Syst. **19**(3), 427–443 (1997)
39. Macedo, H.D., Oliveira, J.N.: A linear algebra approach to OLAP. Form. Asp. Comput. **27**(2), 283–307 (2015)
40. Maddux, R.D.: Relation-algebraic semantics. Theoret. Comput. Sci. **160**(1–2), 1–85 (1996)
41. Maddux, R.D.: Relation Algebras. Elsevier B.V., Amsterdam (2006)
42. Nipkow, T., Paulson, L.C., Wenzel, M.: Isabelle/HOL: A Proof Assistant for Higher-Order Logic. LNCS, vol. 2283. Springer, Heidelberg (2002)
43. Paulson, L.C., Blanchette, J.C.: Three years of experience with Sledgehammer, a practical link between automatic and interactive theorem provers. In: Sutcliffe, G., Ternovska, E., Schulz, S. (eds.) Proceedings of 8th International Workshop on the Implementation of Logics, pp. 3–13 (2010)
44. Pawlak, Z.: Rough sets, rough relations and rough functions. Fundamenta Informaticae **27**(2–3), 103–108 (1996)

45. Schmidt, G., Ströhlein, T.: Relationen und Graphen. Springer, Heidelberg (1989)
46. Tarski, A.: On the calculus of relations. J. Symb. Log. **6**(3), 73–89 (1941)
47. Winter, M.: A new algebraic approach to L-fuzzy relations convenient to study crispness. Inf. Sci. **139**(3–4), 233–252 (2001)

Relation Algebras, Idempotent Semirings and Generalized Bunched Implication Algebras

Peter Jipsen(✉)

Chapman University, Orange, CA 92866, USA
jipsen@chapman.edu

Abstract. This paper investigates connections between algebraic structures that are common in theoretical computer science and algebraic logic. Idempotent semirings are the basis of Kleene algebras, relation algebras, residuated lattices and bunched implication algebras. Extending a result of Chajda and Länger, we show that involutive residuated lattices are determined by a pair of dually isomorphic idempotent semirings on the same set, and this result also applies to relation algebras. Generalized bunched implication algebras (GBI-algebras for short) are residuated lattices expanded with a Heyting implication. We construct bounded cyclic involutive GBI-algebras from so-called weakening relations, and prove that the class of weakening relation algebras is not finitely axiomatizable. These algebras play a role similar to representable relation algebras, and we identify a finitely-based variety of cyclic involutive GBI-algebras that includes all weakening relation algebras. We also show that algebras of down-closed sets of partially-ordered groupoids are bounded cyclic involutive GBI-algebras.

1 Introduction

Idempotent semirings, also known as dioids, play an important role in many applications in computer science, ranging from regular languages and Kleene algebras to shortest path algorithms using tropical semirings such as the max-plus semiring. They are also generalizations of distributive lattices, quantales, residuated lattices and relation algebras, each of which have been studied extensively in mathematics and logic. While it has been known for a long time that Boolean algebras, relation algebras and involutive residuated lattices have two isomorphic semiring reducts that are connected by an anti-isomorphism, the characterization of these algebras by coupled semirings has only recently been formalized in a result by Di Nola and Gerla [2] for MV-algebras and by Chajda and Länger [1] for bounded commutative integral involutive residuated lattices. In Sect. 2 we show that a more general result holds for arbitrary involutive residuated lattices, hence also for relation algebras. This leads to a shorter axiomatization for involutive residuated lattices using only two binary operations, two unary operations and a constant, which is useful for working with relation algebras and their generalizations in automated theorem provers and finite model finders.

P. Höfner et al. (Eds.): RAMiCS 2017, LNCS 10226, pp. 144–158, 2017.
DOI: 10.1007/978-3-319-57418-9_9

Residuated lattices are generalizations of relation algebras and of many other mathematical structures, including Heyting algebras, MV-algebras, basic logic algebras, lattice-ordered groups and quantales. They are also the algebraic semantics of many logical systems, such as intuitionistic logic, relevance logic, linear logic and other substructural logics. Even though they span so many algebraic and logical systems, residuated lattices have a simple definition and a surprisingly deep and elegant algebraic theory that is shared by all the special cases. In this paper we concentrate mostly on involutive residuated lattices expanded with a Heyting arrow. In the bounded and commutative case these algebras are known as bunched implication algebras, or BI-algebras, and have found significant applications in separation logic, a Hoare logic developed by Reynolds, O'Hearn, Pym and others for the verification of pointer data-structures, memory management algorithms and concurrent software. Most of the algebraic properties of BI-algebras hold also in the non-commutative setting of generalized bunched implication algebras, or GBI-algebras for short, so we take this more general approach. By definition a GBI-algebra is a residuated lattice with a Brouwerian algebra defined on the same lattice. In Sect. 3 we observe that relation algebras are a subvariety of bounded involutive GBI-algebras, so this provides an interesting connection between these classes of algebras. We investigate weakening relation algebras that are intuitionistic versions of representable relation algebras, and show that they are not finitely axiomatizable. In Sect. 4 we give partial-order semantics for these algebras and show that they are based on groupoids, i.e. small categories in which all morphisms are isomorphisms.

2 Coupled Semirings

A *semilattice* is of the form $(A, \vee)$ such that $\vee$ is a binary operation on the set A that is associative, commutative and idempotent ($x \vee x = x$), and in such an algebra $x \leq y \iff x \vee y = y$ defines a partial order. A *monoid* $(A, \cdot, 1)$ has an associative operation $\cdot$ such that $1x = x1 = x$. An *idempotent semiring* is an algebra $\mathbf{A} = (A, \vee, \cdot, 1)$ where $(A, \vee)$ is a semilattice, $(A, \cdot, 1)$ is a monoid and $\cdot$ distributes over $\vee$ in both arguments (i.e., $x(y \vee z) = xy \vee xz$ and $(x \vee y)z = xz \vee yz$). A *lattice* $(A, \wedge, \vee)$ is a pair of semilattices $(A, \vee)$ and $(A, \wedge)$ linked by the absorption laws $x \wedge (x \vee y) = x = x \vee (x \wedge y)$. A *residuated lattice* is of the form $\mathbf{A} = (A, \wedge, \vee, \cdot, 1, \backslash, /)$ where $(A, \wedge, \vee)$ is a lattice, $(A, \cdot, 1)$ is a monoid and $\backslash, /$ are the left and right residuals of $\cdot$, i.e., for all $x, y, z \in A$

$$xy \leq z \iff y \leq x\backslash z \iff x \leq z/y.$$

These equivalences imply that $(A, \vee, \cdot, 1)$ is an idempotent semiring since, e.g., $x(y \vee z) \leq w \iff y \vee z \leq x\backslash w \iff y, z \leq x\backslash w \iff xy \vee xz \leq w$. A residuated lattice is *bounded* if it has a bottom element $\bot$, *integral* if 1 is the top element and *commutative* if the identity $xy = yx$ holds. For an arbitrary constant 0 in a residuated lattice define the *linear negations* ${\sim}x = x\backslash 0$ and $-x = 0/x$. The constant 0 is *involutive* if ${\sim}{-}x = x = -{\sim}x$ for all elements x, and an *involutive residuated lattice* (also called an involutive FL-algebra) is a residuated lattice with an involutive 0. Such an algebra is *cyclic* if $\sim x = -x$. Note that a commutative involutive residuated lattice satisfies $x\backslash y = y/x$ and hence is always cyclic.

For example, a relation algebra $(A, \wedge, \vee, \neg, ;, ^{\smile}, 1)$ is a cyclic involutive residuated lattice if one defines $x\backslash y = \neg(x^{\smile}; \neg y)$, $x/y = \neg(\neg x; y^{\smile})$ and $0 = \neg 1$, and omits the operations $\neg, ^{\smile}$ from the signature. The cyclic linear negation is given by $\sim x = \neg(x^{\smile}) = (\neg x)^{\smile}$. An example that is a bounded commutative integral involutive residuated lattice is provided by the *standard MV-algebra* $([0,1], \min, \max, \cdot, 1, \backslash, /)$ where $xy = \max(x+y-1, 0)$ and $x\backslash y = y/x = \min(1-x+y, 1)$. The class of all MV-algebras is the variety generated by this unit-interval algebra (i.e. the smallest class closed under products, subalgebras and homomorphic images). We note that the variety of involutive residuated lattices has a decidable equational theory [3,17] while this is not the case for relation algebras.

In [2] Di Nola and Gerla showed that every MV-algebra is determined by a pair of coupled commutative semirings, and Chajda and Länger [1] generalized this construction to bounded commutative integral involutive residuated lattices. We show here that the result is actually valid in the general setting of involutive residuated lattices and hence includes all (reducts of) relation algebras.

For two algebras $\mathbf{A}, \mathbf{B}$ with the same signature, an *anti-isomorphism* $\alpha\colon \mathbf{A} \to \mathbf{B}$ is like an isomorphism, except that for all binary operations $*$ we have

$$\alpha(x *^{\mathbf{A}} y) = \alpha(y) *^{\mathbf{B}} \alpha(x)$$

(instead of $\alpha(x) *^{\mathbf{B}} \alpha(y)$).

A *generalized coupled semiring* is a triple $((A, \vee, \cdot, 1), (A, \wedge, +, 0), \alpha)$ such that

(i) $(A, \vee, \cdot, 1)$ and $(A, \wedge, +, 0)$ are idempotent semirings
(ii) $(A, \wedge, \vee)$ is a lattice (with order denoted by $\leq$)
(iii) α is an anti-isomorphism from $(A, \vee, \cdot, 1)$ to $(A, \wedge, +, 0)$
(iv) $x \leq y$ if and only if $1 \leq \alpha(x) + y$

Theorem 1. *Let* $\mathbf{A} = (A, \wedge, \vee, \cdot, 1, \backslash, /, 0)$ *be an involutive residuated lattice with linear negations* $\sim x = x\backslash 0$, $-x = 0/x$ *and define* $x + y = \sim((-y)\cdot(-x))$. *Then* $((A, \vee, \cdot, 1), (A, \wedge, +, 0), \sim)$ *is a generalized coupled semiring.*

Proof. In any residuated lattice $\cdot$ distributes over $\vee$ since $x(y \vee z) \leq w \iff y \vee z \leq x\backslash w \iff y, z \leq x\backslash w \iff xy, xz \leq w \iff xy \vee xz \leq w$, and likewise for $(x \vee y)z = xz \vee yz$. In an involutive residuated lattice the linear negations are order-reversing bijections, hence $\sim(x \vee y) = \sim y \wedge \sim x$. Replacing x, y by $\sim x, \sim y$ in the definition of $+$ shows that $\sim(xy) = \sim y + \sim x$, and $\sim 1 = 1\backslash 0 = 0$ since $x \leq 1\backslash 0 \iff x = 1x \leq 0$. Therefore (iii) is satisfied, and (i) follows since anti-isomorphisms preserve the structure of idempotent semirings. Obviously (ii) holds, so it remains to check (iv): $1 \leq \sim x + y \iff 1 \leq \sim((-y)x) \iff (-y)x \leq 0 \iff x \leq \sim - y = y$, where the middle equivalence holds because the linear negations are order-reversing and $-1 = 0$.

We now show that the converse also holds.

Theorem 2. *Let* $((A, \vee, \cdot, 1), (A, \wedge, +, 0), \alpha)$ *be a generalized coupled semiring and define* $x\backslash y = \alpha(\alpha^{-1}(y) \cdot x)$, $x/y = \alpha^{-1}(y \cdot \alpha(x))$. *Then* $\mathbf{A} = (A, \wedge, \vee, \cdot, 1, \backslash, /, 0)$ *is an involutive residuated lattice and* $\alpha(x) = {\sim}x$. *If* $\alpha = \alpha^{-1}$ *then* $\mathbf{A}$ *is cyclic, and if* 1 *is the top element of the first semiring then* $\mathbf{A}$ *is bounded and integral.*

Proof. By (i) and (ii) $(A, \wedge, \vee)$ is a lattice and $(A, \cdot, 1)$ is a monoid, so we need to show that $\backslash, /$ are residuals with 0 as involutive element. By (iv) we have

$$\begin{aligned} xy \le z &\iff 1 \le \alpha(xy) + z \\ &\iff 1 \le \alpha(y) + \alpha(x) + z \\ &\iff y \le \alpha(x) + z = \alpha(x) + \alpha(\alpha^{-1}(z)) = \alpha(\alpha^{-1}(z) \cdot x) = x\backslash z. \end{aligned}$$

To see that $xy \le z \iff x \le z/y$, first observe that (iv) is equivalent to $x \le y \iff 1 \le \alpha(\alpha^{-1}(y) \cdot x)$ and after replacing y by $\alpha(y)$ one obtains $x \le \alpha(y) \iff 1 \le \alpha(y \cdot x)$. From (iii) it follows that α and α^{-1} are order-reversing, so we compute

$$\begin{aligned} xy \le z &\iff \alpha(z) \le \alpha(xy) \\ &\iff 1 \le \alpha(x \cdot y \cdot \alpha(z)) \\ &\iff y \cdot \alpha(z) \le \alpha(x) \\ &\iff x \le \alpha^{-1}(y \cdot \alpha(z)) = z/y. \end{aligned}$$

Condition (iv) also implies that $\alpha(0) = 1$ since $1 \le 1 \implies 1 \le \alpha(\alpha^{-1}(1) \cdot 1) = \alpha(1 \cdot \alpha^{-1}(1)) \implies \alpha^{-1}(1) \le \alpha(1) = 0 \implies \alpha(0) \le 1$ and $0 \le 0 \implies 1 \le \alpha(1 \cdot 0) = \alpha(0)$. The element 0 is involutive since, ${\sim}x = x\backslash 0 = \alpha(\alpha^{-1}(0) \cdot x) = \alpha(1x) = \alpha(x)$, and $-x = 0/x = \alpha^{-1}(x \cdot \alpha(0)) = \alpha^{-1}(x1) = \alpha^{-1}(x)$. □

The preceding theorems show that all involutive residuated lattices are completely determined by their $\vee, \cdot$ structure and by an order-reversing bijection that satisfies property (iv). It also follows that the residuals are term-definable $x\backslash y = {\sim}((-y) \cdot x)$ and $x/y = -(y \cdot ({\sim}x))$, though this is a well-known result [6, p. 153].

As an application of the above result we obtain a fairly concise equational basis for the variety of involutive residuated lattices using the signature $\vee, \cdot, 1, {\sim}, -$ since the remaining operations are defined by $x \wedge y = {\sim}(-x \vee -y)$, $x\backslash y = {\sim}((-y) \cdot x)$, $x/y = -(y \cdot ({\sim}x))$ and $0 = {\sim}1$.

Theorem 3. *An algebra* $(A, \vee, \cdot, 1, {\sim}, -)$ *is (term equivalent to) an involutive residuated lattice if and only if the following 12 identities hold:*

- $(x \vee y) \vee z = x \vee (y \vee z)$, $x \vee y = y \vee x$ *(associativity and commutativity)*
- $x(y \vee z) = xy \vee xz$, $(x \vee y)z = xz \vee yz$ *(distributivity of* $\cdot$ *over* $\vee$*)*
- $(xy)z = x(yz)$, $x1 = x$ *(associativity and right identity of* $\cdot$*)*
- ${\sim}{-x} = x = -{\sim}x$ *(involution of linear negations)*

- $\sim(-(x \vee y) \vee -x) = x = \sim((-x) \vee -y) \vee x$ *(absorption laws)*
- $1 \leq \sim(x(\sim x))$, $x(\sim(yx)) \leq \sim y$ *(equivalent to $x \leq \sim y \iff 1 \leq \sim(yx)$).*

Proof. From Theorem 1 it follows that an involutive residuated lattice satisfies the above identities, where the last two are derived from condition (iv) of coupled semirings by $\sim y \leq \sim y \Rightarrow 1 \leq \sim(y(\sim y))$ and

$$\sim(yx) \leq \sim(yx) \Rightarrow 1 \leq \sim(yx(\sim(yx))) = \sim(y(x\sim(yx))) \Rightarrow x(\sim(yx)) \leq \sim y.$$

Conversely, assume $(A, \vee, \cdot, 1, \sim, -)$ is an algebra that satisfies the identities, and define $x \wedge y = \sim(-y \vee -x)$, $x + y = \sim((-y)(-x))$ and $0 = \sim 1$. It remains to show that $((A, \vee, \cdot, 1), (A, \wedge, +, 0), \sim)$ is a generalized coupled semiring. The absorption laws translate to the usual form $(x \vee y) \wedge x = x = (x \wedge y) \vee x$, and it is easy to see that $\wedge$ is associative and commutative. Since idempotence of $\wedge, \vee$ follow from the absorption laws, $(A, \wedge, \vee)$ is a lattice. The definition of $\wedge, +, 0$ and the involution identities show that $\sim$ is an anti-isomorphism from $(A, \vee, \cdot, 1)$ to $(A, \wedge, +, 0)$. Note that condition (iv) of coupled semirings is equivalent to $x \leq \sim y \iff 1 \leq \sim(yx)$. To see this holds we compute: $x \leq \sim y \Rightarrow y \leq -x$, hence $1 \leq \sim(y(\sim y)) \leq \sim(y(\sim -x)) = \sim(yx)$, and by distributivity over $\vee$, the operation $\cdot$ is order-preserving in each argument, so

$$1 \leq \sim(yx) \Rightarrow x \leq x(\sim(yx)) \leq \sim y.$$

Finally, since $-y = (-y)1$, we deduce $1x = x$ from the following equivalences: $x \leq y \iff 1 \leq \sim((-y)x) = \sim((-y)1x) \iff 1x \leq y$. Therefore $(A, \vee, \cdot, 1)$ is an idempotent semiring, and the anti-isomorphism $\sim$ shows the same holds for $(A, \wedge, +, 0)$. □

The standard equational basis for involutive residuated lattices has 15 identities and a signature with 5 binary operations. A short equational basis can be useful when searching for finite counterexamples or using automated theorem provers. It is not known if the given basis is irredundant, but it is interesting to note that it suffices to assume that 1 is a right-identity element.

It is easy to extend the preceding theorems to a categorical equivalence between the categories of involutive residuated lattices and generalized coupled semirings.

There is only one 2-element residuated lattice, namely the two-element lattice $\mathbf{2} = \{0, 1\}$ with $x \cdot y = x \wedge y$. Clearly this is a bounded commutative involutive lattice, and is in fact a Boolean algebra. There are three 3-element residuated lattices, the 3-element Gödel algebra $\mathbf{G}_3 = \{\bot, a, 1\}$ with $aa = a$ is integral but not involutive, the 3-element Łukasiewicz algebra $\mathbf{Ł}_3 = \{0, a, 1\}$ with $aa = 0$ which is both integral and cyclic involutive, and the Sugihara algebra $\mathbf{S}_3 = \{\bot, 1, \top\}$ which is cyclic involutive but not integral.

3 Distributive Residuated Lattices and Generalized Bunched Implication algebras

A *distributive residuated lattice* is a residuated lattice that satisfies the distributive law $x \wedge (y \vee z) = (x \wedge y) \vee (x \wedge z)$. A typical example of a distributive residuated lattice is given by a collection $\mathcal{R}$ of binary relations on a set X such that $\mathcal{R}$ is closed under intersection $\cap$, union $\cup$, composition $\circ$, residuals $\backslash, /$ and contains a relation E such that $E \circ R = R \circ E = R$ for all $R \in \mathcal{R}$. Here the operations $\backslash, /$ are defined by the usual expressions for residuals on binary relations: $R\backslash S = (R^{\smile} \circ S')'$ and $R/S = (R' \circ S^{\smile})'$, where $^{\smile}$ denotes the converse operation and $'$ is the operation of complementation with respect to the total relation $X^2 = X \times X$. Note that we are not assuming $\mathcal{R}$ is closed under $'$ or that it includes X^2.

A *Brouwerian algebra* is a residuated lattice that satisfies the identity $x \wedge y = xy$. Residuated operations always distribute over lattice joins, hence Brouwerian algebras satisfy the distributive law. Moreover, since $x \wedge y \leq x$ is equivalent to $y \leq x\backslash x$, it follows that Brouwerian algebras have a top element, denoted by the constant $\top$, which is also the identity element for $\wedge$. A *Heyting algebra* is a bounded Brouwerian algebra, i.e., it also has a bottom element $\bot$. The meet operation is commutative, hence $x\backslash y = y/x$ and this operation is usually called a Heyting implication and denoted $x \to y$. We now consider algebras that combine the signatures of residuated lattices and Brouwerian algebras.

A *generalized bunched implication algebra* (or GBI-algebra for short) $\mathbf{B} = (B, \wedge, \vee, \to, \top, \cdot, 1, \backslash, /)$ is a residuated lattice with an additional binary operation $\to$ that is a Heyting implication, i.e., for all $x, y, z \in B$

$$x \wedge y \leq z \iff y \leq x \to z.$$

A GBI-algebra is *bounded* if it also contains a bottom element $\bot$, and we consider $\bot$ to be a constant operation of the algebra. Hence bounded GBI-algebras have Heyting algebra reducts, while GBI-algebras have Brouwerian lattice reducts. In a bounded GBI-algebra, the *intuitionistic negation* is defined by $\neg x = x \to \bot$. An example of a GBI-algebra that is not bounded is given by the nonpositive integers $\mathbb{Z}^-$ with the operations $x \wedge y = \min(x, y)$, $x \vee y = \max(x, y)$, $xy = x + y$, $x\backslash y = y/x = \min(y - x, \top)$, $1 = \top = 0$ and

$$x \to y = \begin{cases} y & \text{if } x > y \\ \top & \text{otherwise.} \end{cases}$$

An *involutive* GBI-algebra is a (necessarily bounded) GBI-algebra that has an involutive constant 0, while a *Boolean* GBI-algebra is bounded GBI-algebra that satisfies the double negation identity $\neg\neg x = x$. For example, sequential algebras [9,10] are (term-equivalent to) a subvariety of Boolean GBI-algebras and relation algebras are (term-equivalent to) a subvariety of Boolean involutive GBI-algebras (see Theorem 6 below). Boolean GBI-algebras are also known as *residuated Boolean monoids* or *rm-algebras* [8,12].

A *bunched implication algebra* (or BI-algebra) is a bounded GBI-algebra that satisfies the identity $xy = yx$. These algebras are the algebraic semantics of separation logic, a programming logic for modeling mutable data structures and concurrent processes [15,16]. An advantage of the varieties of GBI-algebras and BI-algebras is that they have decidable equational theories [5,7], whereas the subvarieties of Boolean GBI-algebras and Boolean BI-algebras have undecidable equational theories [13].

In this section we study the algebraic structure of GBI-algebras and their connections with relation algebras and residuated lattices. Table 1 summarizes how many residuated lattices there are up to isomorphism on a set with n elements, and provides the same information for some of the subclasses introduced above.

Table 1. Number of algebras up to isomorphism on a set with n elements

Number of elements: $n =$	1	2	3	4	5	6	7	8
Residuated lattices	1	1	3	20	149	1488	18554	295292
GBI-algebras	1	1	3	20	115	899	7782	80468
Bunched implication algebras	1	1	3	16	70	399	2261	14358
Involutive residuated lattices	1	1	2	9	21	101	284	1464
Cyclic involutive resid. lattices	1	1	2	9	21	101	279	1433
Involutive GBI-algebras	1	1	2	9	8	43	49	282
Cyclic involutive GBI-algebras	1	1	2	9	8	43	48	281
Involutive BI-algebras	1	1	2	9	8	42	46	263
Boolean involutive BI-algebras	1	1	0	5	0	0	0	25
Relation algebras	1	1	0	3	0	0	0	13

Since (bounded) GBI-algebras have Brouwerian algebra reducts, they also satisfy the distributive law. A *relational GBI-algebra* is of the form $(\mathcal{R}, \cap, \cup, \rightarrow, \top, \circ, E, \backslash, /)$, where $\mathcal{R}$ is a collection of binary relations on a set X, and $\mathcal{R}$ is closed under these operations. Note that $\cup, \cap, \circ$ are the usual set-theoretic operations on binary operations, but $\top$ need only be transitive, $R \circ E = R = E \circ R$, and $\rightarrow, \backslash, /$ need only satisfy

$$R \cap S \subseteq T \iff S \subseteq R \rightarrow T \quad R \circ S \subseteq T \iff S \subseteq R \backslash T \iff R \subseteq T/S$$

for all $R, S, T \in \mathcal{R}$.

Natural examples of relational GBI-algebras are constructed as follows: Let $\mathbf{P} = (P, \sqsubseteq)$ be a partially ordered set, $Q \subseteq P^2$ an equivalence relation that contains $\sqsubseteq$, and define the set of *weakening relations* on $\mathbf{P}$ by $\mathrm{Wk}(\mathbf{P}, Q) = \{\sqsubseteq \circ R \circ \sqsubseteq \,:\, R \subseteq Q\}$. Since $\sqsubseteq$ is transitive and reflexive, this set can also be defined by $\{R \subseteq Q : \sqsubseteq \circ R \circ \sqsubseteq = R\}$. Theorem 5 below shows that $\mathrm{Wk}(\mathbf{P}, Q)$ is a bounded cyclic involutive relational GBI-algebra with Q as top element. If $Q = P \times P$, then we write $\mathrm{Wk}(\mathbf{P})$ instead of $\mathrm{Wk}(\mathbf{P}, Q)$ and call this algebra the *full weakening relation algebra*.

Weakening relations are the natural analogue of binary relations when the category **Set** of sets and functions is replaced by the category **Pos** of partially ordered sets and order-preserving functions. Since sets can be considered as discrete posets (i.e. ordered by the identity relation), **Pos** contains **Set** as a full subcategory, which implies that weakening relations are a substantial generalization of binary relations. They have applications in sequent calculi, proximity lattices/spaces, order-enriched categories, cartesian bicategories, bi-intuitionistic modal logic, mathematical morphology and program semantics, e.g. via separation logic.

Lemma 4. *Let* $\mathbf{P} = (P, \sqsubseteq)$ *be a poset,* Q *an equivalence relation that contains* $\sqsubseteq$*,* R *any binary relation on* P *and let* $R' = Q - R$*. Then*

1. $\sqsubseteq \circ R \circ \sqsubseteq = R$ *is equivalent to* $\sqsupseteq \circ R' \circ \sqsupseteq = R'$*, and*
2. $(\sqsupseteq \circ R \circ \sqsupseteq)'$ *is a weakening relation.*

Proof. 1. Assume $\sqsubseteq \circ R \circ \sqsubseteq = R$ and $(x, y) \in \sqsupseteq \circ R' \circ \sqsupseteq$. Then there exist $(u, v) \in Q$ such that $x \sqsupseteq u$, $(u, v) \notin R$ and $v \sqsupseteq y$. If $(x, y) \in R$ then $u \sqsubseteq x$ and $y \sqsubseteq v$ imply $(u, v) \in R$, which is a contradiction. Hence $(x, y) \in R'$ and therefore $\sqsupseteq \circ R' \circ \sqsupseteq = R'$. The converse is proved by a dual argument.
2. Let $(x, y) \in \sqsubseteq \circ (\sqsupseteq \circ R \circ \sqsupseteq)' \circ \sqsubseteq$. Then there exist $(u, v) \in Q$ such that $x \leq u$, $v \leq y$ and $(u, v) \notin \sqsupseteq \circ R \circ \sqsupseteq$. If $(x, y) \in \sqsupseteq \circ R \circ \sqsupseteq$ then there exist $(r, s) \in R$ such that $r \sqsubseteq x$ and $y \sqsubseteq s$. However, now transitivity implies $r \sqsubseteq u$ and $v \sqsubseteq s$, hence $(u, v) \in \sqsupseteq \circ R \circ \sqsupseteq$, a contradiction. Therefore $(x, y) \in (\sqsupseteq \circ R \circ \sqsupseteq)'$, and it follows that $\sqsubseteq \circ (\sqsupseteq \circ R \circ \sqsupseteq)' \circ \sqsubseteq \subseteq (\sqsupseteq \circ R \circ \sqsupseteq)'$. The reverse inclusion always holds by reflexivity. □

Theorem 5. *Let* $\mathbf{P} = (P, \sqsubseteq)$ *be a poset,* Q *an equivalence relation that contains* $\sqsubseteq$*, and for* $R, S \in \mathrm{Wk}(\mathbf{P}, Q)$ *define*

- $\top = Q$, $\bot = \emptyset$, $1 = \sqsubseteq$, $0 = \sqsupseteq'$,
- $R \to S = (\sqsupseteq \circ (R \cap S') \circ \sqsupseteq)'$ *where* $S' = Q - S$,
- $R \backslash S = (\sqsupseteq \circ R^{\smile} \circ S' \circ \sqsupseteq)'$ *and* $R/S = (\sqsupseteq \circ R' \circ S^{\smile} \circ \sqsupseteq)'$.

Then $\mathbf{Wk}(\mathbf{P}, Q) = (\mathrm{Wk}(\mathbf{P}, Q), \cap, \cup, \to, \top, \bot, \circ, 1, \backslash, /, 0)$ *is a bounded cyclic involutive relational GBI-algebra with involutive constant* 0*, and the linear negation is* ${\sim}R = R\backslash 0 = 0/R = R^{\smile\prime} = R^{\prime\smile}$.

Proof. Note that $\mathbf{Wk}(\mathbf{P}, Q)$ contains the empty set and is closed under $\circ$ and under (arbitrary) meets and joins. The operation $'$ is complementation with respect to Q, but it is not an operation on $\mathrm{Wk}(\mathbf{P}, Q)$. The relation $\sqsubseteq$ is an identity element for weakening relations since $\sqsubseteq \circ \sqsubseteq = \sqsubseteq$. The formula for $R \to S$ is justified by the lemma above and the following equivalences:

$$\begin{aligned}
R \cap S \subseteq T &\iff R \cap T' \cap S = \emptyset \\
&\iff R \cap T' \subseteq S' \\
&\iff R \cap T' \subseteq (\sqsupseteq \circ (R \cap T') \circ \sqsupseteq) \subseteq (\sqsupseteq \circ S' \circ \sqsupseteq) = S' \\
&\iff S \subseteq (\sqsupseteq \circ (R \cap T') \circ \sqsupseteq)' \\
&\iff S \subseteq R \to T.
\end{aligned}$$

For $R\backslash S$ the calculation is similar:

$$\begin{aligned} R\circ S\subseteq T &\iff R^\smile\circ T'\subseteq S' \qquad \text{(by relation algebra)}\\ &\iff R^\smile\circ T'\subseteq(\sqsupseteq\circ R^\smile\circ T'\circ\sqsupseteq)\subseteq(\sqsupseteq\circ S'\circ\sqsupseteq)=S'\\ &\iff S\subseteq(\sqsupseteq\circ R^\smile\circ T'\circ\sqsupseteq)'\\ &\iff S\subseteq R\backslash T \end{aligned}$$

and the argument for R/S is a mirror image.

Lemma 4 shows that $0=\sqsupseteq'$ is a weakening relation and the linear negations agree since

$$\sim x=x\backslash 0=(\sqsupseteq\circ x^\smile\circ\sqsupseteq''\circ\sqsupseteq)'=(\sqsupseteq\circ x^\smile\circ\sqsupseteq)'=(\sqsupseteq\circ\sqsupseteq''\circ x^\smile\circ\sqsupseteq)'=0/x=-x.$$

Hence $\sim R=(\sqsubseteq\circ R\circ\sqsubseteq)^{\smile\prime}=R^{\smile\prime}=R^{\prime\smile}$ for any weakening relation x, so $\sim\sim R=R^{\smile\prime\smile\prime}=R^{\smile\smile\prime\prime}=x$. □

In the previous proof we used the notation $^\smile$ for the converse operation on binary relations. In an abstract relation algebra, this operation is simply an order-preserving permutation that satisfies $x^{\smile\smile}=x$ and $(xy)^\smile=y^\smile x^\smile$, and it is definable by the composition of (cyclic) linear negation and complement: $x^\smile=\sim\neg x$ (where $\neg x=x\to\bot$). We extend this notation to bounded cyclic involutive GBI-algebras, but note that $x^{\smile\smile}=x$ only holds in the Boolean case, and adding $(xy)^\smile=y^\smile x^\smile$ gives an alternative definition of abstract relation algebras.

Theorem 6. *Boolean cyclic involutive GBI-algebras satisfy the identities* $\sim\neg x=\neg\sim x$, $(x\vee y)^\smile=x^\smile\vee y^\smile$, $(x\wedge y)^\smile=x^\smile\wedge y^\smile$ *and* $x^{\smile\smile}=x$. *They are relation algebras if and only if they also satisfy the identity* $(xy)^\smile=y^\smile x^\smile$.

Proof. By definition, Boolean GBI-algebras satisfy $\neg\neg x=x$, where $\neg x=x\to\bot$. The linear negations $\sim,-$ are anti-isomorphisms of the lattice structure, and since the complement $\neg$ is uniquely determined by the lattice structure, both $\sim$ and $-$ preserve complementation. Therefore $\sim\neg x=\neg\sim x$ and $-\neg x=\neg-x$ hold in any Boolean involutive residuated lattice. With $x^\smile$ defined as $\sim\neg x$ it follows that $x^{\smile\smile}=\sim\neg\sim\neg x=\sim\sim\neg\neg x=\sim\sim x$, and if the algebra is cyclic, then $x^{\smile\smile}=x$.

Now assume a Boolean cyclic involutive GBI-algebra satisfies the identity $(xy)^\smile=y^\smile x^\smile$. To see that it is a relation algebra, it suffices to show that De Morgan's Theorem K holds, i.e., $xy\wedge z=\bot\iff x^\smile z\wedge y=\bot$. This follows from the calculation below:

$$\begin{aligned} xy\wedge z=\bot\iff xy\le\neg z &\iff x\le(\neg z)/y=-(y\cdot(\sim\neg z))\\ &\iff x^\smile\le(-(y\cdot z^\smile))^\smile=-(z\cdot y^\smile)=\neg y/z\\ &\iff x^\smile z\le\neg y\iff x^\smile z\wedge y=\bot. \end{aligned}$$

The converse is obvious since relation algebras are Boolean cyclic involutive GBI-algebras and satisfy $x^{\smile\smile}=x$. □

We also note that the identities $\sim\neg x = \neg\sim x$ and $x^{\smile\smile} = x$ both fail in weakening relation algebras and in cyclic involutive GBI-algebras, e.g. in the 3-element Łukasiewicz algebra.

The smallest Boolean cyclic involutive GBI-algebra that fails the converse identity has 8 elements and was originally found in the context of residuated lattices with a De Morgan operation [4]. This algebra has atoms $1, a, b$ and satisfies $aa = a$, $bb = a \vee b$ and $ab = \top = ba$. The involutive element $0 = a \vee b = \neg 1$, and the linear negation satisfies $\sim 1 = 0$, $\sim a = 1 \vee a$ and $\sim b = 1 \vee b$. Hence $a^{\smile} = b$ and $b^{\smile} = a$, which implies that $(aa)^{\smile} = a^{\smile} = b \neq a^{\smile}a^{\smile} = bb = a \vee b$.

If $\mathbf{P}$ is a discrete poset then $\mathbf{Wk(P)}$ is the full representable relation algebra on the set P, so algebras of weakening relations play a role similar to representable relation algebras. Therefore we define the class WGBI of *weakening GBI-algebras* as all algebras that are embedded in a weakening relation algebra $\mathbf{Wk}(\mathbf{P}, Q)$ for some poset $\mathbf{P}$ and equivalence relation Q that contains $\sqsubseteq$. In fact the variety RRA of representable relation algebras is finitely axiomatized over WGBI.

Theorem 7. *1.* WGBI *is closed under subalgebras and products.*

2. RRA *is the subclass of algebras in* WGBI *that satisfy* $\neg\neg x = x$*, i.e., have Boolean algebra reducts.*

3. The class WGBI *is not finitely axiomatizable relative to the variety of all bounded cyclic involutive GBI-algebras.*

Proof. 1. Let $\{\mathbf{A}_i : i \in I\}$ be a family of algebras from WGBI. Then there exists a family of posets $\{\mathbf{P}_i : i \in I\}$ and equivalence relations $\{Q_i : i \in I\}$ such that $\mathbf{A}_i$ is embedded in $\mathbf{Wk}(\mathbf{P}_i, Q_i)$ for each $i \in I$. We can assume that the posets are disjoint, and define $\mathbf{P} = \bigcup_{i\in I} \mathbf{P}_i$, $Q = \bigcup_{i\in I} Q_i$. Then $\prod_{i\in I} \mathbf{Wk}(\mathbf{P}_i, Q_i) \cong \mathbf{Wk}(\mathbf{P}, Q)$ via the map that sends a tuple of disjoint weakening relations $(R_i : i \in I)$ to $\bigcup_{i\in I} R_i$. Since $\prod_{i\in I} \mathbf{A}_i$ is embedded in $\prod_{i\in I} \mathbf{Wk}(\mathbf{P}_i, Q_i)$, it follows that WGBI is closed under products. The closure under subalgebras holds because a composition of embeddings is again an embedding.

2. Let $\mathbf{A}$ be a member of WGBI that satisfies $\neg\neg x = x$. Then $\mathbf{A}$ is embedded in a weakening relation algebra $\mathbf{Wk}(\mathbf{P}, Q)$, so the identity element of $\mathbf{A}$ maps to the partial order $\sqsubseteq$ of the poset $\mathbf{P}$. Assume that $\sqsubseteq$ is not the identity relation on P, so there exist $p \neq q$ such that $p \sqsubseteq q$. Then $(q, p) \in \sqsupseteq\circ \sqsubseteq \circ\sqsupseteq$, hence it is not a member of $\neg\sqsubseteq = \sqsubseteq \to \bot = Q - \sqsupseteq\circ \sqsubseteq \circ\sqsupseteq$. It follows that $(q, p) \in \neg\neg\sqsubseteq$, which means that the identity $\neg\neg x = x$ fails for the identity element of $\mathbf{A}$, a contradiction. Therefore $\sqsubseteq$ is the identity relation, so $\mathbf{P}$ is a discrete poset, and $\mathbf{A}$ is a subalgebra of a representable relation algebra.

3. This is an immediate consequence of 2., since if WGBI were finitely axiomatizable, adding one more identity would give a finite axiomatization of RRA. However, Monk [14] proved that RRA is not finitely axiomatizable. □

Currently it has not been established whether WGBI is closed under homomorphic images, hence a variety, and whether it is a discriminator variety. Another interesting problem that arises is how to define a natural finitely-based variety that contains WGBI similar to Tarski's variety RA of (abstract) relation algebras relative to the variety RRA of all representable relation algebras. Clearly such a basis would include the axioms of bounded cyclic involutive GBI-algebras, but there are other simple identities that are satisfied by all weakening relations. In particular, one can define *domain* and *range* of a relation by the terms $d(x) = x\top \wedge 1$ and $r(x) = \top x \wedge 1$. In any lattice-ordered monoid with top element $\top$, $d(d(x)) = d(x)\top \wedge 1 \leq x\top\top = x\top$ and $d(x) \leq 1$, hence $d(d(x)) \leq d(x)$. Also $d(x) = d(x)1 \leq d(x)\top$, so $d(x) \leq d(d(x))$, and similarly $r(r(x)) = r(x)$.

Lemma 8. WGBI *satisfies the identities* $d(x)x = x$, $xr(x) = x$ *and* $\top x \top x \top = \top x \top$.

Proof. It suffices to show that the identities hold in any $\mathbf{Wk}(\mathbf{P}, Q)$. From $d(x) \leq 1$ it follows that $d(x)x \leq x$. For the reverse inclusion, let $(p, q) \in x$. Since $(q, p) \in Q$ and $(p, p) \in 1$, we have $(p, p) \in d(x)$, hence $(p, q) \in d(x)x$.

Clearly $\top x \top \leq \top$, so $\top x \top x \top \leq \top x \top = \top d(x) x \top \leq \top x \top x \top$. □

The smallest cyclic involutive GBI-algebra (or residuated lattice) where these identities fail is the 3-element Łukasiewicz algebra, with $0 < a < 1$ and satisfying $aa = 0$. Since $\top = 1$, we have $d(a) = a = r(a)$, but $aa \neq a$.

4 Partially-Ordered Groupoid Semantics for Some Cyclic Involutive GBI-Algebras

For an element a in a lattice $\mathbf{A} = (A, \wedge, \vee)$ the set $\{x \in A : x < a\}$ always has a least upper bound, which is either a or the largest element below a. In the latter case a is called *completely join-irreducible*, and a lattice is *join-perfect* if every element is a join of completely join-irreducible elements. *Completely meet-irreducible* elements and *meet-perfect* lattices are defined dually. A *perfect* lattice is defined to be both meet and join-perfect. Birkhoff showed that a finite distributive lattice $\mathbf{A}$ is determined by its poset $J(\mathbf{A})$ of completely join-irreducible elements (with the order induced by $\mathbf{A}$). The result also holds for complete perfect distributive lattices. Conversely, if $\mathbf{Q} = (Q, \leq)$ is a poset, then the set of downward closed subsets $D(\mathbf{Q})$ of $\mathbf{Q}$ forms a complete perfect distributive lattice under intersection and union. Moreover, $D(\mathbf{Q})$ is a Heyting algebra, with $U \to V = Q - {\uparrow}(U - V)$ for any $U, V \in D(\mathbf{Q})$.

For a poset $\mathbf{P}$ the weakening relation algebra $\mathrm{Wk}(\mathbf{P})$ is a complete and perfect GBI-algebra, and in this case the poset of completely join-irreducible elements is isomorphic to $\mathbf{Q} = \mathbf{P} \times \mathbf{P}^\partial$. The composition $\circ$ of $\mathbf{Wk}(\mathbf{P})$ is determined by its restriction to pairs of $\mathbf{Q}$, where it is a partial operation given by

$$(t, u) \circ (v, w) = \begin{cases} (t, w) & \text{if } u = v \\ \text{undefined} & \text{otherwise.} \end{cases}$$

For an arbitrary complete perfect GBI-algebra $\mathbf{A}$, the operation $\cdot$ is also determined by restricting to $J(\mathbf{A})$, but in general this requires a ternary relation to represent $\circ$. Here we consider the special case when the restriction of $\cdot$ to

$J(\mathbf{A})$ gives a partial operation on $J(\mathbf{A})$. The aim is to characterize the partially-ordered partial algebras that are the result of restricting from certain complete perfect bounded cyclic involutive GBI-algebras to their partially-ordered set of join-irreducibles.

For comparison, we first consider the classical case of relation algebras. A complete perfect relation algebra has a complete atomic Boolean algebra as reduct, and the set of join-irreducibles is the set of atoms. The operation of composition, restricted to atoms, is a partial operation precisely when the atoms form a (Brandt) groupoid [11, Sect. 5], or equivalently a small category with all morphism being invertible. In this case the relation algebra is in fact representable using the Cayley representation of the groupoid.

In the more general setting of cyclic involutive GBI-algebras we have a similar situation using partially-ordered groupoids. We first recall the definitions. A *groupoid* is defined as a partial algebra $\mathbf{G} = (G, \circ, ^{-1})$ such that $\circ$ is a partial binary operation and $^{-1}$ is a (total) unary operation on G that satisfy the following axioms:

1. $x \circ y, y \circ z \in G \implies (x \circ y) \circ z = x \circ (y \circ z)$,
2. $x \circ y \in G \iff x^{-1} \circ x = y \circ y^{-1}$,
3. $x \circ x^{-1} \circ x = x$ and $x^{-1-1} = x$.

These axioms imply $x \circ x^{-1} \in G$, as well as $x \circ y \in G \implies x \circ y \circ y^{-1} = x$ and $(x \circ y)^{-1} = y^{-1} \circ x^{-1}$. Typical examples of groupoids are disjoint unions of groups and the *pair-groupoid* $(X \times X, \circ, ^{\smile})$, where $(x, y)^{\smile} = (y, x)$ and $(x, y) \circ (z, w) = (x, w)$ if $y = z$ (undefined otherwise). A *partially-ordered groupoid* $(G, \leq, \circ, ^{-1})$, or *po-groupoid* for short, is a groupoid $(G, \circ, ^{-1})$ such that $(G, \leq)$ is a poset and

4. $x \leq y$ and $x \circ z, y \circ z \in G \implies x \circ z \leq y \circ z$,
5. $x \leq y \implies y^{-1} \leq x^{-1}$,
6. $x \circ y \leq z \circ z^{-1} \implies x \leq y^{-1}$.

Note that the implication $x \leq y$ and $z \circ x, z \circ y \in G \implies z \circ x \leq z \circ y$ follows from these axioms. If $\mathbf{P} = (P, \sqsubseteq)$ a poset with dual poset $\mathbf{P}^{\partial} = (P, \sqsupseteq)$ then $\mathbf{P} \times \mathbf{P}^{\partial} = (P \times P, \leq, \circ, ^{\smile})$ is a po-groupoid, called a *po-pair-groupoid*, with $(a, b) \leq (c, d) \iff a \sqsubseteq c$ and $b \sqsupseteq d$.

Theorem 9. *Let $\mathbf{G} = (G, \leq, \circ, ^{-1})$ be a partially-ordered groupoid. Then $D(\mathbf{G})$ is a bounded cyclic involutive GBI-algebra.*

Proof. The downsets of any poset form a complete perfect Heyting algebra under intersection and union. For downsets s, t the operation $\cdot$ is defined by $s \cdot t = {\downarrow}\{x \circ y : x \in s, y \in t\}$, and it is associative by Axiom 1. The identity of $D(\mathbf{G})$ is $1 = {\downarrow}\{x \circ x^{-1} : x \in G\}$, and cyclic involution is defined by ${\sim}s = G - \{x^{-1} : x \in s\}$. Hence $x \in {\sim}s \iff x^{-1} \notin s$. Axiom 5 ensures that ${\sim}s$ is again a downset, and since $x^{-1-1} = x$, it follows that ${\sim}{\sim}s = s$. It remains to check a version of the coupled semirings axiom: $s \subseteq {\sim}t \iff t \cdot s \subseteq 0 = {\sim}1$. Since every downset is a union of principal downsets, it suffices to consider $s = {\downarrow}x$ and $t = {\downarrow}y$ where $x, y \in G$. Now ${\downarrow}x \subseteq {\sim}{\downarrow}y \iff x^{-1} \notin {\downarrow}y \iff x^{-1} \not\leq y \iff x^{-1} \circ y^{-1} \not\leq z \circ z^{-1}$ for all

$z \in G$ using Axiom 6 in the forward direction, and using right-multiplication by $z^{-1} = y^{-1}$ in the reverse direction. This is equivalent to $(y \circ x)^{-1} \notin 1$, $\downarrow(y \circ x) \subseteq 0$ and finally $\downarrow y \cdot \downarrow x \subseteq 0$. □

In fact for a poset $\mathbf{P} = (P, \sqsubseteq)$ the weakening relation algebra $\mathbf{Wk}(\mathbf{P})$ is obtained from the po-pair-groupoid $\mathbf{G} = \mathbf{P} \times \mathbf{P}^\partial$, and for an equivalence relation $Q \subseteq P^2$, $\mathbf{Wk}(\mathbf{P}, Q)$ is obtained from the sub-po-groupoid $(Q, \leq, \circ, ^\smile)$. Hence every weakening relation algebra has po-groupoid semantics. For example, if one takes the 2-element chain $\mathbf{P} = \mathbf{C}_2 = (\{0,1\}, \sqsubseteq)$ with the usual order $0 \sqsubseteq 1$, then $P^2 = \{(0,0), (0,1), (1,0), (1,1)\}$ and

$$\mathrm{Wk}(\mathbf{C}_2) = \{\emptyset, \{(0,1)\}, \{(0,0),(0,1)\}, \{(0,1),(1,1)\}, \{(0,0),(0,1),(1,1)\}, P^2\}.$$

The linear negation $\sim$ dualizes this 6-element lattice and interchanges a, b. The 3-element chain $\mathbf{C}_3$ gives a 9-element po-groupoid, and $\mathbf{Wk}(\mathbf{C}_3)$ has 20 elements (see Fig. 1).

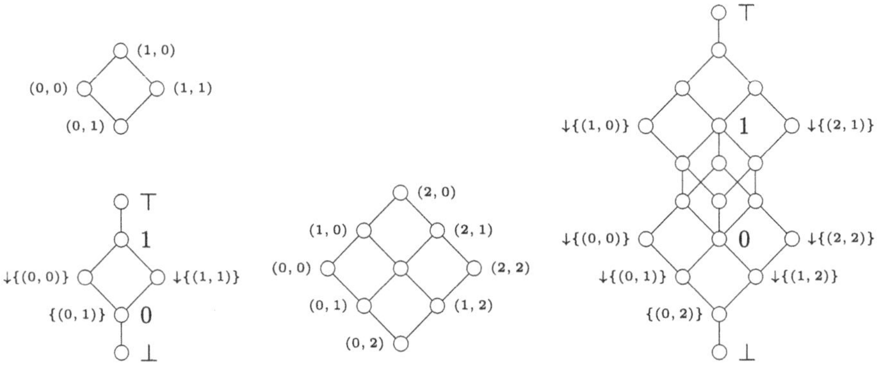

Fig. 1. Weakening relation algebras $\mathbf{Wk}(\mathbf{C}_2)$ and $\mathbf{Wk}(\mathbf{C}_3)$ and their po-pair-groupoids

However, there exist po-groupoids $\mathbf{G}$ such that $D(\mathbf{G})$ is not a weakening relation algebra. The smallest such po-groupoid is based on the pair-groupoid $\mathbf{G} = (\{0,1\}^2, \circ, ^\smile)$, but has only two pairs that are comparable: $(0,1) \leq (1,0)$, so $(0,0)$ and $(1,1)$ are not comparable to any other pairs. The cyclic involutive GBI-algebra $D(\mathbf{G})$ has 12 elements, which does not agree with the cardinality of any of the algebras $\mathbf{Wk}(\mathbf{P}, Q)$.

The last result shows that the cardinality of weakening relation algebras determined by a finite linear order is given by the central binomial series.

Theorem 10. *For an n-element chain* $\mathbf{C}_n$ *the weakening relation algebra* $\mathbf{Wk}(\mathbf{C}_n)$ *has cardinality* $\binom{2n}{n}$.

Proof. This follows from the observation that $D(\mathbf{C}_m \times \mathbf{C}_n)$ has cardinality $\binom{m+n}{n}$. For $n = 1$ this is immediate, since an m-element chain has $m+1$ down-closed sets. Assuming the result holds for n, note that $\mathbf{P} = \mathbf{C}_m \times \mathbf{C}_{n+1}$ is the disjoint union of $\mathbf{C}_m \times \mathbf{C}_n$ and $\mathbf{C}_m$, where we assume the additional m elements are not below any of the elements of $\mathbf{C}_m \times \mathbf{C}_n$. The number of downsets of $\mathbf{P}$ that contain an element a from the extra chain $\mathbf{C}_m$ as a maximal element is given by $\binom{k+n}{n}$ where k is the number of elements above a. Hence the total number of downsets of $\mathbf{P}$ is $\sum_{k=0}^{m} \binom{k+n}{n} = \binom{m+n+1}{n+1}$. □

5 Conclusion

The results in this paper provide connections between idempotent semirings, involutive residuated lattices, generalized bunched implication algebras and relation algebras. These ordered algebras have been extensively studied in algebraic logic and theoretical computer science, and they share many common features that allow techniques to transfer from one theory to the other. Weakening relation algebras extend representable relation algebras to nonclassical logic and are worthy of further investigation.

References

1. Chajda, I., Länger, H.: General coupled semirings of residuated lattices. Fuzzy Sets Syst. **303**, 128–135 (2016)
2. Di Nola, A., Gerla, B.: Algebras of Łukasiewicz's logic and their semiring reducts. Contemp. Math. **377**, 131–144 (2005)
3. Galatos, N., Jipsen, P.: Residuated frames with applications to decidability. Trans. AMS **365**, 2019–2049 (2013)
4. Galatos, N., Jipsen, P.: Relation algebras as expanded FL-algebras. Algebra Univers. **69**(1), 1–21 (2013)
5. Galatos, N., Jipsen, P.: Distributive residuated frames and generalized bunched implication algebras, to appear
6. Galatos, N., Jipsen, P., Kowalski, T., Ono, H.: Residuated Lattices: An Algebraic Glimpse at Substructural Logics. Studies in Logic and the Foundations of Mathematics, vol. 151. Elsevier, Amsterdam (2007)
7. Galmiche, D., Méry, D., Pym, D.J.: The semantics of BI and resource tableaux. Math. Struct. Comput. Sci. **15**(6), 1033–1088 (2005)
8. Jipsen, P.: Computer-aided investigations of relation algebras. Dissertation, Vanderbilt University (1992). http://www1.chapman.edu/~jipsen/dissertation/
9. Jipsen, P.: Representable sequential algebras and observation spaces. J. Relational Methods Comput. Sci. **1**, 235–250 (2004)
10. Jipsen, P., Maddux, R.D.: Nonrepresentable sequential algebras. Logic J. IPGL **5**(4), 565–574 (1997)
11. Jónsson, B., Tarski, A.: Boolean algebras with operators. Part II Am. J. Math. **74**, 127–162 (1952)
12. Jónsson, B., Tsinakis, C.: Relation algebras as residuated Boolean algebras. Algebra Univers. **30**(4), 469–478 (1993)
13. Kurucz, Á., Németi, I., Sain, I., Simon, A.: Decidable and undecidable logics with a binary modality. J. Logic Lang. Inf. **4**(3), 191–206 (1995)
14. Monk, D.: On representable relation algebras. Mich. Math. J. **11**, 207–210 (1964)

15. Pym, D.J.: The Semantics and Proof Theory of the Logic of Bunched Implications. Applied Logic Series, vol. 26. Kluwer Academic Publishers, Dordrecht (2002)
16. Reynolds, J.C., Logic, S.: A logic for shared mutable data structures. In: Proceedings of 17th IEEE Symposium on Logic in Computer Science (LICS 2002), Copenhagen, 22–25 July, pp. 55–74 (2002)
17. Wille, A.: A Gentzen system for involutive residuated lattices. Algebra Univers. **54**(4), 449–463 (2005)

Parsing and Printing of and with Triples

Sebastiaan J.C. Joosten(✉)

Computational Logic Group, UIBK Innsbruck, Innsbruck, Austria
Sebastiaan.Joosten@uibk.ac.at

Abstract. We introduce the tool Amperspiegel, which uses triple graphs for parsing, printing and manipulating data. We show how to conveniently encode parsers, graph manipulation-rules, and printers using several relations. As such, parsers, rules and printers are all encoded as graphs themselves. This allows us to parse, manipulate and print these parsers, rules and printers within the system. A parser for a context free grammar is graph-encoded with only four relations. The graph manipulation-rules turn out to be especially helpful when parsing. The printers strongly correspond to the parsers, being described using only five relations. The combination of parsers, rules and printers allows us to extract Ampersand source code from ArchiMate XML documents. Amperspiegel was originally developed to aid in the development of Ampersand.

1 Introduction

We introduce a framework for language transformations, called Amperspiegel. We see a language transformation as something that consists of three parts: a parser, a series of semantic transformations, and a printer. To describe these parts and their behaviour, we adopt the view that everything can be described in relations.

Languages are described by encoding a Context Free Grammar in four relations. Transformations are described using a set of declarative rules in a subset of relation algebra. The printing then occurs using the inverse of the parser.

Like the parser, also the transformation and the printer are expressed in relations. Consequently, the framework has some reflective capabilities. The name Amperspiegel stems from the framework's relation to Ampersand [4], while emphasising that it has reflection.[1] It is stand-alone software (http://github.com/sjcjoosten/Amperspiegel), so it can be used in projects other than Ampersand as well. Code specific to this paper can be found at: http://cl-informatik.uibk.ac.at/users/sjoosten/as/.

As an example, Sect. 7 creates a link between two tools: ArchiMate and Ampersand. We show how to parse files that describe a software architecture written in an ArchiMate XML file. The structure is transformed, and then

[1] Adding to Amperspiegel's reflection are the switches `collect` and `distribute`, which are not described in this paper.

P. Höfner et al. (Eds.): RAMiCS 2017, LNCS 10226, pp. 159–176, 2017.
DOI: 10.1007/978-3-319-57418-9_10

printed as a description of the same architecture as an Ampersand ADL file. This is done using Amperspiegel.

The focus of this paper is on the concepts behind Amperspiegel, seen as a stand-alone tool. Section 2 gives an overview of the tool and describes its use. We define a parser, a rule engine, Amperspiegel's embedding of a set of rules, and a printer, in Sects. 3, 4, 5 and 6 respectively.

Related Work. Several tools combine parsing and printing with transformations, including meta-programming languages such as Rascal [7] and Stratego [2], or programming language workbenches such as Spoofax [6]. Amperspiegel offers a fundamental approach to meta-programming, offering these features with a minimal implementation. Excluding a file that configures the initial state of Amperspiegel, it is under a thousand lines of Haskell code.

To achieve this, Amperspiegel borrows from several best practices. Using a Context Free Grammar for parsing and for printing is done before by Mark van den Brand [1]. Deriving new facts with rules, as Amperspiegel does, is similar to the declarative programming language datalog± [3]. Its restriction to triples, in a style like Amperspiegel, is described by Edward Robertson [9]. We have not seen a Context Free Grammar described through relations, and this allows to combine these concepts in a novel way. This makes building source-to-source transformations surprisingly easy and modular.

2 Overview of Amperspiegel

To transform languages, Amperspiegel can parse input, apply rules, produce output, and assemble these components in a single execution. This overview shows how components are assembled. Amperspiegel interprets command-line arguments as commands. They are executed from left to right.

The most important actions are '`apply`', '`parse`' and '`print`'. These actions are performed on structures that correspond to a kind of labelled graph. We refer to these structures as 'graph', and explain how they can be understood as a set of homogeneous relations. This interpretation is important, as we expect the Amperspiegel user to think of these structures as a description through several relations.

Initially, there are pre-defined graphs in Amperspiegel. Some of these graphs represent parsers. Using `parse`, a parser is used to parse an input file, creating another graph. Graphs can be manipulated by rules using `apply`, again creating a graph. A graph can be printed to stdout by `print`.

We illustrate Amperspiegel's command line interface by showing how to execute:

```
ds1 := parse data file1
ds2 := parse rule file2
res := apply ds2 ds1
print data res
```

This example uses built-in parsers and printers to read in some data (in `file1`), apply some transformation to it (given by `file2`) and print the result on stdout. It uses the same internal parser as a printer, called `data` to both read the data and print the result. The transformation is parsed using an internal parser called `rule`. For this code, Amperspiegel's command-line interface is used as follows:

```
amperspiegel -parse data file1 ds1 -parse rule file2 ds2 \
    -apply ds2 ds1 res -print data res
```

Since Amperspiegel is used to translate a variation of one language into another, a graph can be used in place of the default parser too:
`Amperspiegel -parse cfg path-to/parser mdp -parse mdp my-data ds1` uses the parser `path-to/parser`, described in `CFG` syntax, to parse the file `my-data` in the new syntax referred to as `mdp`.

Amperspiegel's graph-based notion of data is similar to that used for the semantic web. Another way to view such a graph is as a structure interpreting a set of binary-relation symbols.

Definition 1 (Graph). *A directed labeled graph* $G = (\mathcal{L}, V, E)$ *is given by a finite set of labels* L*, a set of vertices* V*, and a set of edges* $E \subseteq \mathcal{L} \times V \times V$*.*

In this paper we simply say *graph* when we mean a directed labeled graph. This notion of graph is useful when thinking about the implementation of Amperspiegel. From the perspective of an Amperspiegel user, however, it is more useful to think of this structure as a set of homogeneous binary relations. To help strengthen this way of thinking, we suggestively write $(v, w) \in_G r$ for $(r, v, w) \in E$. Indeed, when the label r occurs in an Amperspiegel script, it is natural to interpret it as a relation symbol. We say that a graph is *finite* if and only if its set of vertices is finite.

There is no way to access the structure of nodes in Amperspiegel, except through the edges in which they occur. Thus, the set of vertices is implicitly equal to those vertices that occur in an edge. In the following sections, we show how a finite graph can describe a parser, a printer and a data-transformation (set of rules).

3 Parsing

To specify parsers, we use Context Free Grammars. While a Context Free Grammar (CFG) is typically used to define a set of strings called 'language', we focus on how CFGs relate to graphs. This section relates CFGs to graphs in two ways: First, a CFG can be used to interpret a string as a parse graph. This allows the Amperspiegel user to read graphs from a file that has a certain file format. Second, a CFG can itself be encoded as a graph. This allows the Amperspiegel user to specify and use its own CFGs.

Definition 2 (Context Free Grammar). *A* CFG $g = (P, \Sigma, C, S)$ *is given by a relation* $C \subseteq P \times (P + \Sigma)^*$ *and a start symbol* $S \in P$*, where* P *denotes the* *finite set of non-terminals, and* Σ *denotes the set of terminals. A pair in* C *is called a production rule.*

We present a CFG by listing C. The set of terminals Σ is disjoint from P, and $S =$ '$\texttt{S}$'. See for instance Example 1. Strings in $(P+\Sigma)^*$ are given by separating elements in $P + \Sigma$ with spaces.

Example 1. $\texttt{S} \mapsto \texttt{0 L S}$ $\quad \texttt{S} \mapsto \varepsilon$ $\quad \texttt{L} \mapsto \texttt{S 1 L}$ $\quad \texttt{L} \mapsto \varepsilon$

It follows from convention that P is the two-element set containing $\texttt{S}$ and $\texttt{L}$, and that Σ contains $\texttt{0}$ and $\texttt{1}$.

3.1 Obtaining a Graph by Parsing a String

A CFG (P, Σ, C, S) gives rise to a parser graph $\mathbb{G}$ in which P are the labels, and $\Sigma^* \times P$ are the vertices. This graph is infinite, as it contains all possible parses. It is independent of the start nonterminal S. For a given string s, the parse graph is the subgraph of $\mathbb{G}$ of nodes and edges reachable from the node (s, S), which is guaranteed to be finite. We give an example before the definitions. The empty string is written as ε (Fig. 1).

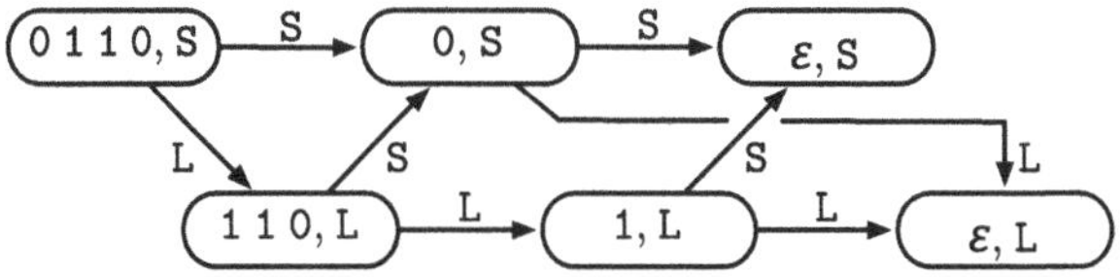

Fig. 1. The parse graph of Example 2

Example 2. For the CFG of Example 1, the parse graph of $\texttt{0 1 1 0}$ is given by:

$$((\texttt{0 1 1 0}, \texttt{S}), (\texttt{0}, \texttt{S})), ((\texttt{1 1 0}, \texttt{L}), (\texttt{0}, \texttt{S})), ((\texttt{1}, \texttt{L}), (\varepsilon, \texttt{S})), ((\texttt{0}, \texttt{S}), (\varepsilon, \texttt{S})) \in_G \texttt{S}$$
$$((\texttt{0 1 1 0}, \texttt{S}), (\texttt{1 1 0}, \texttt{L})), ((\texttt{1 1 0}, \texttt{L}), (\texttt{1}, \texttt{L})), ((\texttt{1}, \texttt{L}), (\varepsilon, \texttt{L})), ((\texttt{0}, \texttt{S}), (\varepsilon, \texttt{L})) \in_G \texttt{L}$$

In Example 2, each edge in the parse graph is of the form $((s_1, p), (s_2, p')) \in_{\mathbb{G}} p'$, indicating that s_1 parses as p via a production $(p, \cdots p' \cdots) \in C$, where the substring s_2 parses as p'. A parser graph captures all possible parse graphs, plus edges to terminal symbols that help in our definition of parser graph.

Definition 3 (Parser graph). *Given CFG* (P, Σ, C, S), *the graph* $\mathbb{G} = (P, \Sigma^* \times P, E)$ *is the parser graph of* (P, Σ, C, S), *in which* E *is the least set of edges such that for each* $(p, p_0 \cdots p_n) \in C$, *and for every* $s = s_0 \cdots s_n \in \Sigma^*$:

$$\left(\forall i \leq n. \begin{pmatrix} & (\exists x, p'. \ ((s_i, p_i), x) \in_{\mathbb{G}} p') \\ \vee & ((p_i, \varepsilon) \in C \wedge s_i = \varepsilon) \\ \vee & (p_i \in \Sigma \wedge s_i = p_i) \end{pmatrix}\right) \Rightarrow (\forall i \leq n. \ ((s, p), (s_i, p_i)) \in_{\mathbb{G}} p_i)$$

This formula states that if each of $p_0 \cdots p_n$ *can be parsed as a corresponding* $s_0 \cdots s_n$, *then* p *can be parsed as* $p_0 \cdots p_n$ *and corresponding edges exist in* $\mathbb{G}$.

Definition 4 (Parse graph). *A parse-graph for the string s and CFG (P, Σ, C, S) is the subgraph of the parser graph $\mathbb{G}$ that is reachable from (s, S) via edges in P.*

A parse graph of s is finite. It contains only vertices (s', v) in which s' is a substring of s and $v \in P$. There are at most $n(n+1)/2+1$ substrings in a string of length n, and P is finite. Therefore every parse graph is finite.

3.2 Describing a Context Free Grammar with a Graph

This section focuses on how CFGs have been implemented in Amperspiegel. We encode a CFG as a graph, allowing a light-weight implementation. This also allows us to express a CFG that can parse its own description and yield the CFG parser itself.

The CFG (P, Σ, C, S) is encoded as a graph $G = (\mathcal{L}, V, E)$, by making C explicit, and using a default element for S. The label `choice` $\in \mathcal{L}$ describes C. Amperspiegel does not have sums or lists as built-in types, so we reconstruct the type of vertices from the labels of edges. The structure of elements of $(P + \Sigma)^*$ is described using three labels: `recogniser`, `continuation` and `nonTerminal`. Amperspiegel uses `recogniser` and `continuation` rather than, say, `head` and `tail`. This choice is less likely to cause name clashes when combining graphs by taking their union, as we will do in Sect. 7. We combine the edges labelled `choice` with the ones that describe structure in a single graph, so $V = P+\Sigma+(P+\Sigma)^*$. In the sense of Sect. 4, vertices in $P+\Sigma$ act as constant symbols while vertices in $(P+\Sigma)^*$ act as variable symbols.

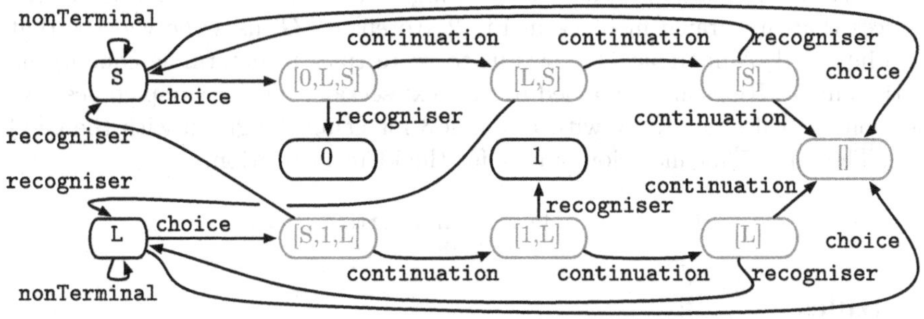

Fig. 2. The CFG of Example 1 drawn as a graph

For Example 1, the corresponding CFG is given as a graph in Fig. 2. Nodes that encode lists in $(P + \Sigma)^*$ are drawn in grey. The lists that make up these nodes are written in Haskell notation to emphasise difference between `S` $\in P$ and [`S`] $\in (P + \Sigma)^*$.

A CFG (P, Σ, C, S) corresponds to a graph G if:

$$(p, v) \in_G \texttt{choice} \Leftrightarrow (p, l(v)) \in C \qquad \left(\exists v'.\ (v', v) \in_G \texttt{nonTerminal}\right) \Leftrightarrow v \in P$$

$$l(v) = \begin{cases} v_1\, l(v_2) & \text{if} \quad (v, v_1) \in_G \texttt{recogniser} \quad \text{and} \quad (v, v_2) \in_G \texttt{continuation} \\ \varepsilon & \text{otherwise} \end{cases}$$

To ensure l is well-defined, the labels `recogniser` and `continuation` must describe univalent relations in G (`R` is *univalent* iff $(x, y), (x, z) \in_G \texttt{R}$ implies $y = z$).

Example 3. The following CFG describes the language for CFGs. It omits production rules for non-terminals P and Σ, as Amperspiegel has those production rules built-in. These built-in production rules are the only way to get constant symbols as vertices in the sense of Sect. 4. We write "$\mapsto$" for the terminal in Σ, to distinguish it from syntax.

$$\begin{aligned} \texttt{S} &\mapsto \varepsilon & \texttt{S} &\mapsto \texttt{S}\ P\ \texttt{"}\mapsto\texttt{"}\ \texttt{choice} \\ \texttt{nonTerminal} &\mapsto P & \texttt{choice} &\mapsto \texttt{continuation} \\ \texttt{continuation} &\mapsto \varepsilon & \texttt{continuation} &\mapsto \texttt{recogniser continuation} \\ \texttt{recogniser} &\mapsto \Sigma & \texttt{recogniser} &\mapsto \texttt{nonTerminal} \end{aligned}$$

The CFG in Example 3 describes the language of CFGs as used in this paper. It defines a parser yielding parse-graphs with the labels `choice`, `nonTerminal`, `recogniser` and `continuation`. So if G' is the parse graph of some string and the CFG in Example 3, then G' can be interpreted as a CFG in Amperspiegel.

In such G', some vertices are being interpreted as elements of Σ, and some are labels in the parser-graph corresponding to the CFG of G'. These are the vertices that are drawn in black in Fig. 2. To ensure G' uses the vertices that were intended, Amperspiegel allows us to write rules to determine equality on vertices in G'. Rules are explained in the next section, but for completeness, we mention the rules necessary with Example 3 for using the graph with the CFG here. They use $\sqsubseteq$ for inclusion, and $\mathbb{1}$ for the identity relation:

$$P \sqsubseteq \mathbb{1} \qquad \Sigma \sqsubseteq \mathbb{1} \qquad \texttt{nonTerminal} \sqsubseteq \mathbb{1}$$

4 Rules

To manipulate graphs in Amperspiegel, the programmer specifies rules. This is done in relation algebra to obtain a declarative, point-free language with attractive algebraic properties. Rules are evaluated with a deduction engine comparable to those for Datalog [3]. To this extent, Amperspiegel maintains a graph containing what it knows, and then makes it more specific by what it can prove. A typical use is to interpret a parse graph as initial knowledge, which is made specific by edges that can be deduced using the rules. This section introduces rules and shows how they are used.

Rules are formed over expressions. Expressions are built from relation symbols $\mathcal{L}$, a reserved symbol $\mathbb{1}$ which stands for the identity relation, and tuples (sets containing exactly one pair) written as $\langle a, b\rangle$ with a and b elements in a set of constants $\mathcal{K}$. We can also use the reserved symbol $\bot$, which stands for the empty relation. These are combined with the operations $\_ \sqcap \_$, $\_ ; \_$, and $\_^{\smile}$. The operations stand for intersection, relational composition, and relational converse, respectively. For a graph $G = (\mathcal{L}, \mathcal{K} + N, E)$, in which the vertices are *constant symbols* $\mathcal{K}$ or *variable symbols* N, the semantics of an expression $\mathcal{X}$, written as $[\![\mathcal{X}]\!]_G \subseteq (\mathcal{K} + N) \times (\mathcal{K} + N)$, is as in representable relation algebra. We assume $\mathcal{K}$ and N to be disjoint:

$$
\begin{aligned}
[\![l]\!]_G &= \{(x, y) \mid (x, y) \in_G l\} & [\![\mathbb{1}]\!]_G &= \{(v, v) \mid v \in (\mathcal{K} + N)\} \\
[\![\langle a, b\rangle]\!]_G &= \{(a, b)\} & [\![\bot]\!]_G &= \{\} \\
[\![L \sqcap R]\!]_G &= [\![L]\!]_G \cap [\![R]\!]_G & [\![L^{\smile}]\!]_G &= \{(y, x) \mid (x, y) \in [\![L]\!]_G\} \\
[\![L ; R]\!]_G &= \{(x, y) \mid \exists z.\ (x, z) \in [\![L]\!]_G \ \wedge\ (z, y) \in [\![R]\!]_G\}
\end{aligned}
$$

Definition 5 (Rule). *If L and R are expressions over sets of constant symbols $\mathcal{K}$ and labels $\mathcal{L}$, then $L \sqsubseteq R$ is a* rule. *We say that a graph G* satisfies *a set of rules $\mathcal{R}$, in symbols: $G \vDash \mathcal{R}$, iff for all $(L \sqsubseteq R) \in \mathcal{R}$ we have $[\![L]\!]_G \subseteq [\![R]\!]_G$. We say that a set of rules $\mathcal{R}$* implies *a rule r_0, in symbols: $\mathcal{R} \vDash r_0$, iff for all graphs G we have $(G \vDash \mathcal{R}) \Rightarrow (G \vDash \{r_0\})$.*

4.1 The Rule Engine by Example

We give a flavour of Amperspiegel's deduction engine, by showing how one can reason to construct a non-empty graph that satisfies a set of rules. Consider the example:

Example 4. These rules state that the label l stands for a total and self-inverse relation:

$$\mathbb{1} \sqsubseteq l ; l^{\smile} \tag{1}$$
$$\mathbb{1} \sqsubseteq l ; l \tag{2}$$
$$l ; l \sqsubseteq \mathbb{1} \tag{3}$$

Rule 1 states that l is total, and Rules 2 and 3 say that it is self-inverse.

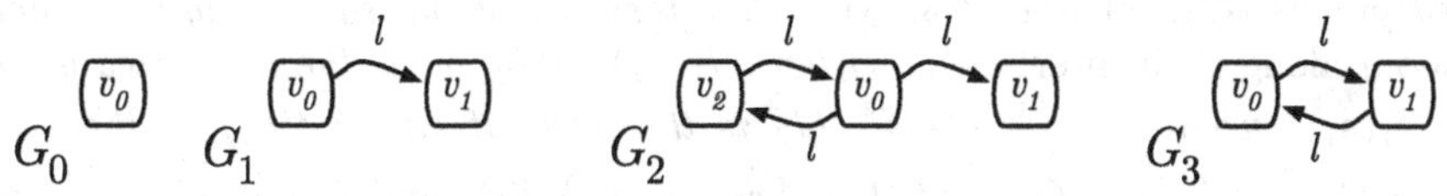

Fig. 3. Applying the rules of Example 4.

We construct a non-empty graph G that has no constant symbols, satisfying the rules of Example 4, to illustrate Amperspiegel's rule engine. See Fig. 3. Take $G_0 = (\{l\}, \{v_0\}, \{\})$ as initial non-empty graph. We identify a rule that does not hold on G_0, and a pair that shows why it does not. Rule 1 does not hold on G_0 as there must be some v_1 with $(v_0, v_1) \in_G l$. We therefore add the vertex v_1 to G_0, plus an edge from v_0 to v_1 with label l, which gives rise to G_1. On G_1, rule 2 states that some v_2 exists with $(v_0, v_2) \in_G l$ and $(v_2, v_0) \in_G l$. Changing G_1 to fix this adds two more edges and another vertex, giving G_2. Now rule 3 does not hold for $(v_2, v_1) \in [\![l \,;\, l]\!]_{G_2}$. Therefore, we identify v_1 and v_2 giving us G_3. This is a graph for which all rules hold.

4.2 Rule Engine Semantics

This section explains how Amperspiegel's rule engine is defined. We begin with some notions and notations. We overload a function $f : V_1 \to V_2$ to a function over sets: $f(V) = \{f(v) | v \in V\}$ for $V \subseteq V_1$, edges: $f(E) = \{(l, f(v_1), f(v_2)) \mid (l, v_1, v_2) \in E\}$, and graphs: $f((\mathcal{L}, V, E)) = (\mathcal{L}, f(V), f(E))$.

Our rule engine gradually changes a graph. We describe these changes in a categorical manner, inspired by Wolfram Kahl [5]. Such a change can be described by a homomorphism, which can be understood as a vertex map that preserves constant symbols and edge labels. This definition is used to describe all graph transformations.

Definition 6 (Graph homomorphism). *Take the graphs with shared sets of labels and constants* $G_1 = (\mathcal{L}, \mathcal{K} + N_1, E_1)$ *and* $G_2 = (\mathcal{L}, \mathcal{K} + N_2, E_2)$. *We say that a vertex map* $f : \mathcal{K} + N_1 \to \mathcal{K} + N_2$ *is a* graph homomorphism *iff* $\forall e \in E_1.\ f(e) \in E_2$, *and* $\forall k \in \mathcal{K}.\ f(k) = k$.

If there is a graph homomorphism $f : G_1 \to G_2$, we say that G_2 *is more specific than* G_1, or in symbols: $G_1 \leq G_2$. Graph homomorphisms between graphs with shared sets of labels and constants form a category in which graph homomorphisms are the morphisms. In the following, we assume fixed but arbitrary sets $\mathcal{L}$ of labels and $\mathcal{K}$ of constant symbols.

We use pushouts to combine two graphs. Note that due to the requirement that homomorphisms preserve constants, if $\mathcal{K}$ is non-empty, then the category of graph homomorphism does not have all colimits and not even all pushouts, since constants cannot be identified. For the pushouts that do exist, we introduce an abbreviating notation.

Definition 7 (Pushout along interfaces). *An* interfaced graph *is a pair* (G, s) *where* s *is a sequence of vertices of* G *called interface. Given two interfaced graphs* (G_1, s_1) *and* (G_2, s_2) *with interfaces of the same length* n, *their* pushout along their interfaces, *written* $(G_1, s_1) \sqcup (G_2, s_2)$ *is the interfaced graph* $(G_3, g_1(s_1))$ *where* $G_1 \overset{g_1}{\Rightarrow} G_3 \overset{g_2}{\Leftarrow} G_2$ *is the pushout (if existing) of the span* $G_1 \overset{f_1}{\Leftarrow} G_0 \overset{f_2}{\Rightarrow} G_2$ *over* $G_0 = (\mathcal{L}, \mathcal{K} + \{x_1, \ldots, x_n\}, \{\})$, *and* f_1 *and* f_2 *are graph homomorphisms defined by* $f_i(x_j) = s_i(j)$.

We aim to construct the least specific graph G such that $G \models \mathcal{R}$, called a least consequence graph. We define this to show correctness of our algorithm.

Definition 8 (Consequence graph). *Given a graph G_0 and a set of rules $\mathcal{R}$ over the same set of labels and set of constants. We say that G is a* consequence graph *of G_0 and $\mathcal{R}$, if $G \models \mathcal{R}$ and $G_0 \leq G$. Furthermore, G is a* least consequence graph *if for each consequence graph G' of G_0 and $\mathcal{R}$ we have $G \leq G'$.*

To construct a consequence graph, Amperspiegel repeatedly takes a rule that is not satisfied by a graph, and 'patches' this until there is nothing to repair. For a rule $L \sqsubseteq R$ with a pair in $[\![L]\!]_{G_0}$ that is not in $[\![R]\!]_{G_0}$, we do a *step*: A *patch* is created with the shape of R, which is combined into G_0 with a pushout.

Definition 9 (Patch). *The* patch *of an expression $\mathcal{X}$ over sets of labels $\mathcal{L}$ and constants $\mathcal{K}$, in symbols $(G, (v_1, v_2)) = \Delta(\mathcal{X})$, is a graph over $\mathcal{L}$ and a pair of vertices in that graph, inductively defined:*

$$\begin{aligned}
\Delta(\mathbb{1}) &= ((\mathcal{L}, \mathcal{K} + \{1\}, \{\}), (1, 1)) \\
\Delta(R \sqcap S) &= \Delta(R) \sqcup \Delta(S) \\
\Delta(R \,;\, S) &= (G', (v_1, v_4)) \\
&\quad \textit{where } (G', _) = (G_R, (v_2)) \sqcup (G_S, (v_3)) \\
&\quad \textit{and } (G_R, (v_1, v_2)) = \Delta(R) \textit{ and } (G_S, (v_3, v_4)) = \Delta(S) \\
\Delta(R^{\smile}) &= (G', (v_2, v_1)) \quad \textit{where} \quad (G', (v_1, v_2)) = \Delta(R) \\
\Delta(\langle a, b \rangle) &= ((\mathcal{L}, \mathcal{K}, \{\}), (a, b)) \\
\Delta(l) &= ((\mathcal{L}, \mathcal{K} + \{v_1, v_2\}, \{(l, v_1, v_2)\}), (v_1, v_2))
\end{aligned}$$

As with pushouts, $\Delta(\mathcal{X})$ may not be defined. Also, $\Delta(\bot)$ is intentionally left undefined.

Another example for an expression with undefined patch is $\mathbb{1} \sqcap \langle a, b \rangle$, if $a \neq b$, since the necessary pushout would have to identify the constant symbols a and b.

We use patches to work towards a consequence graph. This is done stepwise through $\mathcal{R}$-steps, that are given by the set of rules.

Definition 10 ($\mathcal{R}$-Step). *Let G be a graph. Let $(L \sqsubseteq R)$ be a rule, and let p be a pair of vertices in G such that:*

$$p \in [\![L]\!]_G \qquad\qquad p \notin [\![R]\!]_G$$

Then $G \xrightarrow[p]{L \sqsubseteq R} G'$ is a step *where $G' = \Delta(R) \sqcup (G, p)$ if defined, and $G' = \lightning$ otherwise. If $\mathcal{R}$ is a set of rules, then $G \xrightarrow{\mathcal{R}} G'$ is an* $\mathcal{R}$-step *if there exists a rule $r \in \mathcal{R}$ and a pair of vertices p in G such that $G \xrightarrow[p]{L \sqsubseteq R} G'$. If there is no $\mathcal{R}$-step for a graph G, then we say G is in $\mathcal{R}$-normal form. For notational convenience, $\xrightarrow{\mathcal{R}}$ is an endo-relation on the disjoint union of $\lightning$ with graphs, where $\lightning$ counts as an additional $\mathcal{R}$-normal form.*

Correctness of '$\mathcal{R}$-step' is understood as follows: If there is a terminating sequence $G_0 \xrightarrow{\mathcal{R}} \cdots \xrightarrow{\mathcal{R}} G_n \neq \lightning$, then G_n is a *least consequence graph* of G_0. This follows from observing that if $G_i \xrightarrow{\mathcal{R}} G_{i+1}$, then $G_i \leq G_{i+1}$. If G is a consequence graph of G_i and $\mathcal{R}$, then G is also a consequence graph of G_{i+1} and $\mathcal{R}$. This holds in particular if G is a least consequence graph. Finally, if G_n is a graph in $\mathcal{R}$-normal form, then G_n is a least consequence graph of G_n and $\mathcal{R}$. Furthermore, if $G \xrightarrow{\mathcal{R}} \lightning$, then there is no consequence graph of G and $\mathcal{R}$. This shows soundness of finding a consequence graph through a normalising sequence $G_0 \xrightarrow{\mathcal{R}} G_1 \cdots \xrightarrow{\mathcal{R}} G_n$ in which G_n is either $\lightning$ or a least consequence graph, which is what Amperspiegel's rule engine does.

Note that $\xrightarrow{\mathcal{R}}$ need not be weakly normalising or confluent, and the order in which we apply rules can determine whether we reach a normal form. It is possible to have an infinite sequence of $\mathcal{R}$-steps even though there are terminating sequences. To make this less likely, Amperspiegel ensures fairness: A sequence $G_0 \xrightarrow{\mathcal{R}} G_1 \cdots$ is fair if for all pairs p there are finitely many i such that $G_i \xrightarrow[p]{r} \_$. This condition is implemented by imposing a total order on the vertices, treating smallest vertices first, and making new vertices the largest elements in this order.

Amperspiegel's rule engine can terminate by finding the least consequence graph, or discovering that no such graph exists by reaching $\lightning$. The possibility of non-termination makes it that it is not a decision procedure. We leave the question whether Amperspiegel implements a semi-decision procedure as future work. We conjecture that the problem whether no least consequence graph exists is undecidable, yet semi-decidable, and that our procedure is a semi-decision procedure.

5 Amperspiegel's Embedding of the Rule Engine

This section shows how Amperspiegel uses the rule engine of the previous section to implement more general graph transformations, including destructive rules. We apply a rule system using the `apply` switch, which gets three arguments: a graph that encodes the rules $\mathcal{R}$, the name of a source graph $G_s = (\mathcal{L}_s, V_s, E_s)$, and the name for a target graph $G_t = (\mathcal{L}_t, V_t, E_t)$. The label set for rules is $\mathcal{L}_s + \mathcal{L}' + \mathcal{L}_t$. To ensure disjointness of these three sets of labels, `pre`, `during` and `post` are used as a prefix to labels respectively. The graph of which a least consequence graph is calculated is $G_0 = (\mathcal{L}_s + \mathcal{L}' + \mathcal{L}_t, V_s, E_s')$, in which E_s' contains the appropriately relabelled edges of E_s. The least consequence graph of G_0 and the rules $\mathcal{R}$ is then $G = (\mathcal{L}_s + \mathcal{L}' + \mathcal{L}_t, V_t, E)$. The target graph has the edges $E_t = \{(r, x, y) \mid (\mathsf{post}\, r, x, y) \in E\}$, where post is the rightmost constructor of the disjoint union $\mathcal{L}_s + \mathcal{L}' + \mathcal{L}_t$.

Consequently, the graph the procedure starts with only contains edges of G_s. The target graph will be overwritten. After obtaining the consequences by running the procedure, we only look at the edges that are in $\mathsf{post}(r)$ for some r and put those in G_t. For convenience, we allow labels of the form $\mathsf{during}(r)$, to allow labels for edges that do not end up in G_t, but are also guaranteed not to be used in G_s.

The user can use her own rules in Amperspiegel, as the rules are described as a graph. This follows the same pattern as describing a CFG with a graph. For an expression e, there is a pair (e, p) with p uniquely determined by e:

$$(e, p) \in_G \texttt{conjunct} \cup \texttt{compose} \cup \texttt{converse} \cup \texttt{pair} \cup \texttt{pre} \cup \texttt{during} \cup \texttt{post} \cup \texttt{id}$$

such that (e, p) occurs in exactly one of the relations mentioned, say l. If l is `conjunct` or `compose`, there are unique e_1 and e_2 such that $(p, e_1) \in_G$ `eFst` and $(p, e_2) \in_G$ `eSnd`. These e_1 and e_2 are, in turn, expressions again. If l is `pre`, `during` or `post`, p is a relation name (an unquoted string in $\mathcal{K}$). For `converse`, p is an expression. For `pair`, p is a pair of strings (quoted or unquoted) that can be accessed through the relations `pFst` and `pSnd`. If $l =$ `id`, p does not matter. A set of rules is a relation between expressions.

To take full advantage of rules as graphs, Amperspiegel allows a graph to contain both a grammar and rules, given by taking the union of the corresponding triples. We use these two together, by a switch called `-Parse` (note the capital `P`), that first parses and then applies the rules to the result. This makes many syntactical extensions straightforward to achieve. Take for instance the operation $\texttt{dom}(R)$, containing all pairs (x, x) for which x is in the domain of R, defined as follows:

$$\texttt{dom}(R) = (R\,;(R^{\smile})) \sqcap \mathbb{1}$$

We allow the relation `dom` to be used without changing Amperspiegel, by adding the following rule to the parser (for readability, we underline labels instead of writing `post`):

$$\texttt{pre dom} \sqsubseteq \underline{\texttt{conjunct}}\,;(\underline{\texttt{eFst}}\,;\underline{\texttt{compose}}\,;(\underline{\texttt{eFst}} \sqcap \underline{\texttt{eSnd}}\,;\underline{\texttt{converse}}) \sqcap \underline{\texttt{eSnd}}\,;\underline{\texttt{id}})$$

With this, we have seen an example of using rules in order to extend the syntax of rules. Section 7 contains another example where a syntax extention was useful.

6 Printing

We consider printer as a reverse operation to parsing. It is not always possible to reconstruct the original string. Consider for instance the following CFG, for lists with at least two words:

`Start` $\mapsto$ `Word Word`	`Start` $\mapsto$ `Word Start`
`Word` $\mapsto$ `e a t`	`Word` $\mapsto$ `t e a`

Printing of graphs that contain only univalent relations can be done unambiguously if for every non-terminal, each symbol occurs at most once on the right hand side of its production rules. We change the CFG to meet this condition, without changing the language it accepts:

$$\begin{array}{lll} \texttt{Start} \mapsto \texttt{Word}_1\ \ \texttt{Word}_2 & \texttt{Start} \mapsto \texttt{Word}\ \ \texttt{Start} & \texttt{e}' \mapsto \texttt{e} \\ \texttt{Word} \mapsto \texttt{e}'\ \texttt{a}'\ \texttt{t}' & \texttt{Word} \mapsto \texttt{t e a} & \texttt{a}' \mapsto \texttt{a} \\ \texttt{Word}_1 \mapsto \texttt{Word} & \texttt{Word}_2 \mapsto \texttt{Word} & \texttt{t}' \mapsto \texttt{t} \end{array}$$

When printing graphs that aren't a parse graph, we may encounter relations that are not univalent. For this purpose, we add the label `separator` to a graph describing a CFG, in addition to the four existing labels. The type of edges with this label can be thought of informally as $(P + \Sigma) \times \Sigma$, although Amperspiegel does not consider any structure on vertices.

The syntax for a printer closely follows that of a parser. The main difference is that we allow a relation to be named between square brackets, along with an optional separator string. This means that we can largely reuse the parser for a CFG as defined earlier. We drop the production-rule `recogniser` $\mapsto$ `nonTerminal` from Example 3, and replace it with:

$$\begin{array}{l} \texttt{recogniser} \mapsto \texttt{idNonTerminal} \\ \texttt{recogniser} \mapsto \texttt{"["\ recRelation\ "]"\ nonTerminal} \\ \texttt{recogniser} \mapsto \texttt{"["\ recRelation\ "SEPBY"\ separator\ "]"\ nonTerminal} \end{array}$$

One can think of `idNonTerminal` as a typed identity relation for those instances where we want to use the nonTerminal symbol as a label.

We recognise `recRelation` and `separator` as strings, and use the following rules:

$$\texttt{idNonTerminal} \sqsubseteq \mathbb{1} \qquad \texttt{recRelation} \sqsubseteq \mathbb{1}$$
$$\texttt{idNonTerminal} \sqsubseteq \texttt{nonTerminal}$$

7 Using Amperspiegel to Transform ArchiMate Files into Ampersand Code

In previous sections we discussed parsing, rules to evaluate, and printing. These are the necessary ingredients for transforming data structures. To demonstrate that Amperspiegel can do nontrivial work, it has been put to the test of practice. We picked a problem that was being solved at the time of writing in a software project in the Dutch government: to transform source code from ArchiMate [8] to Ampersand [4].

The specifics of the tools Ampersand and ArchiMate are not important to understand the transformation, but we give a little background: ArchiMate is a modeling tool to get an overview of a business, similar to UML yet more coarse grained. The tool helps users to build, visualise and modify architectures cooperatively, but does not feature a way to turn such architectures into code. For this purpose, we are interested in using another tool that describes architectures that does produce code, namely Ampersand. Ampersand can generate web-applications based on architectures, but often an architecture is already described in another language, in our case: ArchiMate.

To understand the transformation, it suffices to know that ArchiMate files are XML files describing 'elements'. Between these elements there are 'relations'. Elements are things like actors, business components, services, and infrastructure. A relation can be 'implements', describing which infrastructures implement which services.

The purpose of this section is to describe how one can create transformations with Amperspiegel. We define an XML parser, interpret the resulting graph as an ArchiMate model, and turn it into an Ampersand model. This section uses verbatim Amperspiegel syntax.

In the development of the XML parser, we keep the specification of syntax and rules in a single file. This changes the syntax for describing a CFG slightly: Each line should end with a dot, to keep the grammar unambiguous. We form rules, using `|-` as notation for $\sqsubseteq$, prefixed with `RULE`. We use `KEEP relationName` as syntax-sugar for:

```
RULE pre relationName |- post relationName
```

To achieve this, the parser for CFG's populates the relation `keep`, and the set of rules that is then applied to the result contains the rule:

```
RULE pre keep |- post rule;(post eFst;post pre /\ post eSnd;post post)
```

Similarly, `[expression -> elementName]` is a shorthand for the expression:

```
expression;<elementName,elementName>;expression~ /\ I
```

These short-hands are useful for the development of the XML parser and the transformation that follows it. We used them without changing Amperspiegel itself. We changed the Amperspiegel-scripts that define the parser for Amperspiegel-scripts instead. In the parser, `"[" pointExpression "->" pointElement "]"` is added in the right hand side of a production-rule for an expression. We also add these rules:

```
RULE pre pointExpression |- (post conjunct;(post eFst;(post compose;(post
        eFst /\ ((post eSnd;post compose);(post eSnd;post converse))))))
RULE pre pointElement |- (post conjunct;((post eFst;((((post compose;
       post eSnd);post compose);post eFst);pre pair)) /\ post eSnd))
```

7.1 Parsing XML

Building an XML parser lies outside of the scope what Amperspiegel was initially intended for: parsing Ampersand-like scripts. Consequently, Amperspiegel's lexer is not designed for parsing XML; it ignores comments and whitespace. Fortunately, we can get away with this by restricting ourselves to XML without text. This means tags, including attributes, are fine, but `<tag>text like this</tag>` is not. Such a tag would have to be replaced by an attribute-value, such as: `<tag value="text like this" />`.

An XML parser can then be defined as follows (`Start` is Amperspiegel's start symbol for a CFG):

```
Start > "<?xml" attributeList "?>" tagList.
Start > tagList.
tagList > tag tagList.
tagList > .
tag > "<" tagName attributeList ">" tagList "</" tagName ">".
tag > "<" tagName attributeList "/>".
tagName > UnquotedString .
attributeList > attribute attributeList.
attributeList > .
attribute > attributeName "=" attributeValue.
attributeValue > QuotedString.
attributeName > UnquotedString.

RULE pre UnquotedString |- I
RULE pre QuotedString |- I
RULE pre tagList |- I
RULE pre attributeList |- I
RULE (pre tagName) ~ ; pre tagName |- I     -- univalence of tagName

KEEP attributeName   KEEP attributeValue   KEEP attribute
KEEP tagName         KEEP tag
```

The first lines describe a CFG for XML. Note that the lines end with a dot, in order to distinguish KEEP statements from a continuation in which KEEP acts as recogniser. The rules for tagList and attributeList cause tag and attribute to be relations, rather than partial functions from the head of the list. We can forget the order-information of attributeList and tagList since for ArchiMate this order is irrelevant.

The rule for univalence of tagName requires a closing tag to match the opening tag, because the parser generates two tagName edges from the first tag rule to different tag names, which, after the contraction of UnquotedString edges, are string constant symbols. Parsing <openingtag></closingtag> will result in trying to identify two constants in $\mathcal{K}$ and produce the message:

```
Rules caused "openingtag" to be equal to "closingtag"
```

The XML we parse is well-formed, so these errors do not occur in practice.

7.2 Transforming a Graph

We parse XML such as the following. Figure 4 shows the first two lines parsed:

```
<element identifier="id-1311" xsi:type="BusinessProcess">
  <label xml:lang="en" value="Collect Premium"/></element>
<element identifier="id-1208" xsi:type="BusinessService">
  <label xml:lang="en" value="Premium Payment Service"/></element>
<relationship identifier="id-1329" source="id-1311"
  target="id-1208" xsi:type="RealisationRelationship" />
```

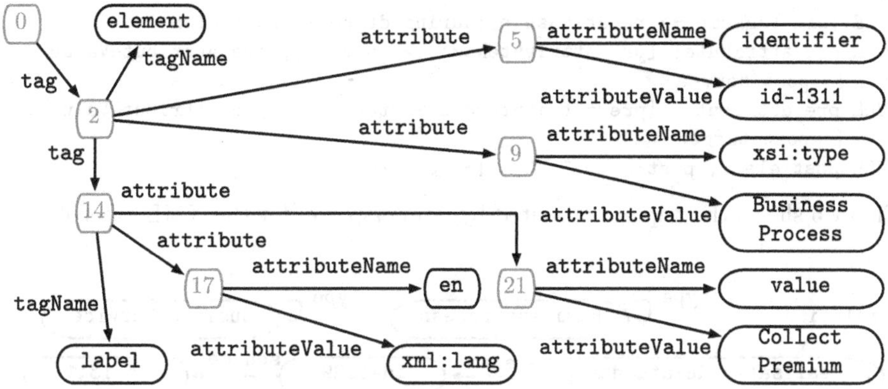

Fig. 4. Graph from applying parser and rules of Sect. 7.1 to two lines of XML.

The corresponding Ampersand code we will transform this XML into is:

```
CLASSIFY BusinessProcess ISA Element
CLASSIFY BusinessService ISA Element
RELATION RealisationRelationship :: Element * Element
POPULATION [ ( "Collect Premium" , "Premium Payment Service" ) ]
```

Here are some of the rules which we use to transform the parsed XML:

```
RULE pre attribute;[pre attributeName -> identifier]
     ; pre attributeValue |- I
RULE pre tag;[pre tagName -> label] |- during lab
RULE pre attribute; [pre attributeName -> value]
     ; pre attributeValue |- during value
RULE during lab; during value |- post label
```

The first rule states that identifiers are unique to elements, allowing us to use these as handlers. The second introduces a temporary abbreviation `lab` for `<label>` tags. The third introduces the abbreviation `value` for `value` attributes. The last creates the relation `label` from the value of pairs in `lab`.

To obtain all element types without duplicates, we use these rules:

```
RULE pre attribute; [pre attributeName -> xsi:type]
     ; pre attributeValue |- during dtype
RULE pre tag; [pre tagName -> element] |- during element
RULE during element; during dtype |- during X ; post type
RULE post type |- I
```

The first two rules create temporary shorthands: `dtype` and `element`. The third rule looks only at the element types, and creates a tuple in `type` with that target (and a fresh source). The fourth rule states that the source of that tuple should be equal to the target, removing duplicates. Finally, we obtain all relations and their triples:

```
RULE [pre tagName -> relationship];during dtype |- post elem ~
RULE pre attribute; [pre attributeName -> source] ; pre attributeValue
     |- post source
RULE pre attribute; [pre attributeName -> target] ; pre attributeValue
     |- post target
RULE post elem ; post elem ~ /\ I |- post relation
```

Figure 5 shows the triples computed by Amperspiegel for the XML excerpt.

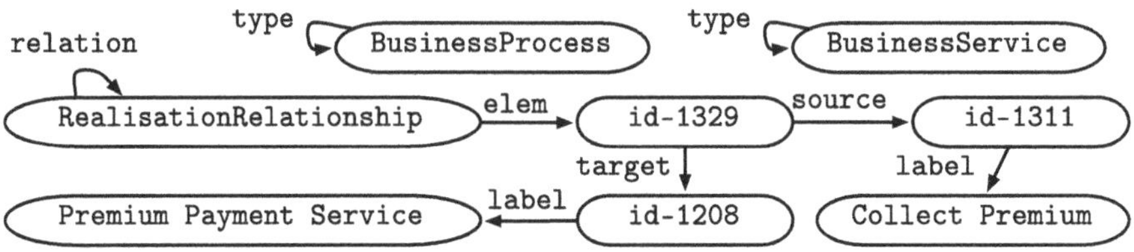

Fig. 5. The triples after applying the rules of Sect. 7.2

7.3 Printing a Graph

We define a printer such that there are no `identifier` values in the final output. Since the relations are not necessarily well typed in ArchiMate files, we create a type 'Element' to stand in for any type.

The printer is defined as follows:

```
Start > [I SEPBY "\n"] Statement.
Statement > "CLASSIFY" [type] UnquotedString "ISA Element".
Statement > "RELATION" [relation] UnquotedString
     ":: Element * Element\nPOPULATION [" [elem SEPBY "\n ,"] Pair "]".
Pair > "(" [source] Labeled "," [target] Labeled ")".
Labeled > [label] String.
```

The relation `I` is used in the first line of the printer. This determines which statements to print, and which not. For our purpose, we print all statements, by adding the rules:

```
RULE post type |- post I
RULE post relation |- post I
```

To summarise how we use Amperspiegel's tool-chain:

- Parse a CFG describing an XML parser in the file `xml.cfg`. To the result, apply the rules for CFGs. Put the result in the graph '`xml`'. On the commandline of Amperspiegel we write: `-Parse xml.cfg cfg xml`.
- Parse rules to convert the XML data specific to ArchiMate, and the corresponding printer specific to Ampersand. The corresponding file is `archi.cfg`. To Amperspiegel we pass: `-Parse archi.cfg cfg archi`

- Parse the ArchiMate xml file `Archisurance.xml` and apply the rules that go with the XML parser. This uses the graph '`xml`': `-Parse Archisurance.xml xml`. Since we omit the third argument, the result is put in the graph '`population`'.
- Apply the rules in the graph '`archi`' to `population`. Put the result in `population`: `-apply archi`.
- Print the graph '`population`' using the printer defined in '`archi`'. In Amperspiegel: `-print archi`.

We sequence the listed operations on the command line:

```
Amperspiegel -Parse xml.cfg cfg xml -Parse archi.cfg cfg archi \
          -Parse Archisurance.xml xml -apply archi -print archi
```

For the example XML code of Sect. 7.2, this produces exactly the mentioned Ampersand code. Parsing and printing a file of about 600 lines produces 209 lines in eleven seconds.

8 Discussion

Most parser implementations are a partial function from strings to finite tree structures. We use a standard parsing algorithm, and turn the result into a graph. Consequently, CFGs that generate infinite trees yet finite graphs remain future work.

Applying rules is slow: Amperspiegel traverses the right hand side expressions for every pair and applies the patch as it constructs it. Sharing work between applications of a rule may improve performance. We plan to use Amperspiegel to generate code out of a set of rules, hopefully boosting the performance of Amperspiegel. Ideally, we would also use Amperspiegel to generate code out of a CFG or a printer, making the core of Amperspiegel even simpler. As mentioned, Amperspiegel only consists of a thousand lines of Haskell code. We hope to further reduce this number in the process.

9 Conclusion

We introduced Amperspiegel, and used it for a source-to-source transformation, producing Ampersand code from ArchiMate code. To do so, the Amperspiegel syntax was extended in a convenient manner. This shows how triple graphs can be used to describe simple programs in a flexible, modular way.

Acknowledgements. I thank Wolfram Kahl for helping me greatly improve this paper's clarity in an intensive process of iterative feedback. I also thank the anonymous reviewers and Stef Joosten for their comments on an earlier version of this paper. Supported by the Austrian Science Fund (FWF) project Y757.

References

1. van den Brand, M., Visser, E.: Generation of formatters for context-free languages. ACM Trans. Softw. Eng. Methodol. (TOSEM) **5**(1), 1–41 (1996)
2. Bravenboer, M., Kalleberg, K.T., Vermaas, R., Visser, E.: Stratego/XT 0.17. a language and toolset for program transformation. Sci. Comput. Program. **72**(1), 52–70 (2008)
3. Gottlob, G., Lukasiewicz, T., Pieris, A.: Datalog+/-: questions and answers. In: Proceedings of the Fourteenth International Conference on Principles of Knowledge Representation and Reasoning (KR), pp. 682–685 (2014)
4. Joosten, S.: Software development in relation algebra with Ampersand. In: Pous, D., Struth, G., Höfner, P. (eds.) RAMiCS 2017. LNCS, vol. 10226, pp. 177–192. Springer, Cham (2017)
5. Kahl, W.: Algebraic graph derivations for graphical calculi. In: d'Amore, F., Franciosa, P.G., Marchetti-Spaccamela, A. (eds.) WG 1996. LNCS, vol. 1197, pp. 224–238. Springer, Heidelberg (1997). doi:10.1007/3-540-62559-3_19
6. Kats, L.C., Visser, E.: The Spoofax language workbench: rules for declarative specification of languages and ides. In: ACM SIGPLAN Conference on Object Oriented Programming, Systems, Languages and Applications (OOPSLA 2010), vol. 45, pp. 444–463. ACM (2010)
7. Klint, P., van der Storm, T., Vinju, J.: RASCAL: a domain specific language for source code analysis and manipulation. In: Proceedings of the 2009 9th IEEE International Working Conference on Source Code Analysis and Manipulation, pp. 168–177. SCAM 2009 (2009). http://dx.doi.org/10.1109/SCAM.2009.28
8. Lankhorst, M.M., Proper, H.A., Jonkers, H.: The architecture of the ArchiMate language. In: Halpin, T., Krogstie, J., Nurcan, S., Proper, E., Schmidt, R., Soffer, P., Ukor, R. (eds.) BPMDS/EMMSAD -2009. LNBIP, vol. 29, pp. 367–380. Springer, Heidelberg (2009). doi:10.1007/978-3-642-01862-6_30
9. Robertson, E.L.: Triadic Relations: An Algebra for the Semantic Web. In: Bussler, C., Tannen, V., Fundulaki, I. (eds.) SWDB 2004. LNCS, vol. 3372, pp. 91–108. Springer, Heidelberg (2005). doi:10.1007/978-3-540-31839-2_8

Software Development in Relation Algebra with Ampersand

Stef Joosten[1,2(✉)]

[1] Open Universiteit Nederland, Postbus 2960, 6401 DL Heerlen, The Netherlands
stef.joosten@ou.nl

[2] Ordina NV, Nieuwegein, The Netherlands

Abstract. Relation Algebra can be used as a programming language for building information systems. This paper presents a case study to demonstrate this principle. We have developed a database-application for legal reasoning as a case study, of which a small part is discussed in this paper to illustrate the mechanisms of programming in Relation Algebra. Beside being declarative, relation algebra comes with attractive promises for developing big software. The compiler that was used for this case study, Ampersand, is the result of an open source project. Ampersand has been tried and tested in practice and is available as free open source software.

Keywords: Relation algebra · Software development · Legal reasoning · Information systems design · Ampersand · MirrorMe · Big software

1 Introduction

This paper investigates how relation algebra can be used as a programming language for information systems. A compiler, Ampersand [21], is used to compile concepts, relations and rules into a working database-application. Ampersand is a syntactically sugared version of heterogeneous relation algebra [25]. We present a case study to demonstrate programming in relation algebra and its impact on the software development process. The case study takes the reader by the hand in the thinking process of a software developer who programs with relation algebra.

The use of relation algebra as a programming language is not new. It stands in the tradition of relation algebra [19], logic programming [18], database application programming [6], and formal specification of software. Ampersand uses these ideas to compile relation algebra to working software, an idea which is also found in RelView [4] by Berghammer (Univ. of Kiel). Some differences with earlier programming languages are discussed in Sect. 6, after presenting the case study.

The user may regard Ampersand as a programming language, especially suited for designing the back-end of information systems. The axioms of Tarski can be used to manipulate expressions in a way that preserves meaning [34]. This makes Ampersand a declarative language.

P. Höfner et al. (Eds.): RAMiCS 2017, LNCS 10226, pp. 177–192, 2017.
DOI: 10.1007/978-3-319-57418-9_11

Our case study shows an argument assistant for legal professionals, which was built as innovation project at Ordina. The purpose of this argument assistant is to support legal professionals in constructing a legal brief[1]. The challenge is to create a program that consists of relation algebra as much as possible. In doing so, we hope to learn more about software development in relation algebra.

Section 2 introduces Ampersand and its computational semantics. Section 3 introduces a theory of legal reasoning, which was developed for argument assistance. Section 4 discusses the programming mechanism in the application, and Sect. 5 visualizes that mechanism. Section 6 reflects on software development in Ampersand. It also provides an overview of the use of Ampersand in practice and an outlook to its further development.

2 Ampersand

In this section we explain the basics of Ampersand. The reader is expected to have sufficient background in relation algebra, in order to understand the remainder of this paper.

The core of an Ampersand-script is a tuple $\langle \mathbb{H}, \mathbb{R}, \mathbb{C}, \mathfrak{T} \rangle$, which consists of a set of rules $\mathbb{H}$, relations $\mathbb{R}$, concepts $\mathbb{C}$, and a type function $\mathfrak{T}$. Ampersand-scripts are interpreted by the compiler as an information system. The rules constitute a theory in heterogeneous relation algebra. They constrain a body of data that resides in a database. The Ampersand-compiler generates a database from relations in the script. A database-application[2] assists users to keep rules satisfied throughout the lifetime of the database. It is also generated by Ampersand.

A rule is an equality between two terms. Terms are built from relations. Ampersand interprets every relation as a finite set of pairs, which are stored in the database. The phase in which Ampersand takes a script, and turns it into a database, is what we will refer to as *compile-time*. The phase in which a user interacts with the database, is what we will refer to as *run-time*. At run-time, Ampersand can decide which rules are satisfied by querying the database. The compiler generates all software needed to maintain rules at run-time. If a rule is not satisfied as a result of data that has changed, that change is reverted (rolled back) to maintain a state in which all rules are satisfied. Changes to the database are not specified by the software developer, but generated by the compiler. Rules in Ampersand are maintained rather than executed directly.

Atoms are values that have no internal structure, meant to represent data elements in a database. From a business perspective, atoms are used to represent concrete items of the world, such as `Peter`, `1`, or `the king of France`. By

[1] A *brief* is a document that is meant to summarize a lawsuit for the judge and counterparty. It provides legal reasons for claims in a lawsuit based on regulations, precedents, and other legally acceptable sources. It shows how the reasoning applies to facts from the case.

[2] Ampersand generates an application that consists of a relational database and interface components. Currently this application runs server-side on a PHP/MySQL platform and on a web-browser on the client-side.

convention throughout the remainder of this paper, variables a, b, and c are used to represent *atoms*. The set of all atoms is called $\mathbb{A}$. Each atom is an instance of a *concept*.

Concepts (from set $\mathbb{C}$) are names we use to classify atoms in a meaningful way. For example, you might choose to classify `Peter` as a person, and `074238991` as a telephone number. We will use variables A, B, C, D to represent concepts. The term $\mathbb{I}_A$ represents the *identity relation* of concept A. The expression $a \in A$ means that atom a is an instance of concept A. In the syntax of Ampersand, concepts form a separate syntactic category, allowing a parser to recognize them as concepts. Ampersand also features specialization. Specialization is needed to allow statements such as: "An orange is a fruit that ....". Specialization is not relevant for the remainder of this paper.

Relations (from set $\mathbb{R}$) are used in information systems to store facts. A *fact* is a statement that is true in a business context. Facts are stored and kept as data in a computer. As data changes over time, so do the contents of these relations. In this paper relations are represented by variables r, s, and d. We represent the declaration of a relation r by $nm_{\langle A,B\rangle}$, in which nm is a name and A and B are concepts. We call A the source concept and B the target concept of the relation. The term $\mathbb{V}_{[A\times B]}$ represents the *universal relation* over concepts A and B.

The meaning of relations in Ampersand is defined by an interpretation function $\mathfrak{I}$. It maps each relation to a set of facts. Furthermore, it is a run-time requirement that the pairs in r are contained in its type:

$$\langle a,b\rangle \in \mathfrak{I}(nm_{\langle A,B\rangle}) \Rightarrow\ a \in A \wedge b \in B \tag{1}$$

Terms are used to combine relations using operators. The set of terms is called $\mathbb{T}$. It is defined by:

Definition 1 (terms).
The set of terms, $\mathbb{T}$, is the smallest set that satisfies, for all $r, s \in \mathbb{T}$, $d \in \mathbb{R}$ and $A, B \in \mathbb{C}$:

$$d \in \mathbb{T} \qquad (\textit{every relation is a term}) \tag{2}$$
$$(r \cap s) \in \mathbb{T} \qquad (\textit{intersection}) \tag{3}$$
$$(r - s) \in \mathbb{T} \qquad (\textit{difference}) \tag{4}$$
$$(r; s) \in \mathbb{T} \qquad (\textit{composition}) \tag{5}$$
$$r^{\smile} \in \mathbb{T} \qquad (\textit{converse}) \tag{6}$$
$$\mathbb{I}_A \in \mathbb{T} \qquad (\textit{identity}) \tag{7}$$
$$\mathbb{V}_{[A\times B]} \in \mathbb{T} \qquad (\textit{full set}) \tag{8}$$

Throughout the remainder of this paper, terms are represented by variables r, s, d, and t. The *type* of a term r is a pair of concepts given by $\mathfrak{T}(r)$. $\mathfrak{T}$ is a partial function that maps terms to types. If term r has a type, this term is called *type correct*. The Ampersand compiler requires all terms to be type correct, or else it will not generate any code. The type function and the restrictions it suffers are discussed in [16]. However, for the remainder of this paper this is irrelevant.

The meaning of terms in Ampersand is an extension of interpretation function $\mathfrak{I}$. Let A and B be finite sets of atoms, then $\mathfrak{I}$ maps each term to the set of pairs for which that term stands.

Definition 2 (interpretation of terms).
For every $A, B \in \mathbb{C}$ and $r, s \in \mathbb{T}$

$$\mathfrak{I}(r) = \{\langle a,b\rangle |\ a\ r\ b\} \tag{9}$$
$$\mathfrak{I}(r \cap s) = \{\langle a,b\rangle |\ \langle a,b\rangle \in \mathfrak{I}(r) \ \ and \ \ \langle a,b\rangle \in \mathfrak{I}(s)\} \tag{10}$$
$$\mathfrak{I}(r - s) = \{\langle a,b\rangle |\ \langle a,b\rangle \in \mathfrak{I}(r) \ \ and\ \langle a,b\rangle \notin \mathfrak{I}(s)\} \tag{11}$$
$$\mathfrak{I}(r;s) = \{\langle a,c\rangle |\ \ for\ some\ b,\ \langle a,b\rangle \in \mathfrak{I}(r) \ \ and\ \langle b,c\rangle \in \mathfrak{I}(s)\} \tag{12}$$
$$\mathfrak{I}(r^{\smile}) = \{\langle b,a\rangle |\ \langle a,b\rangle \in \mathfrak{I}(r)\} \tag{13}$$
$$\mathfrak{I}(\mathbb{I}_A) = \{\langle a,a\rangle |\ a \in A\} \tag{14}$$
$$\mathfrak{I}(\mathbb{V}_{[A\times B]}) = \{\langle a,b\rangle |\ a \in A, b \in B\} \tag{15}$$

Ampersand has more operators than the ones introduced in Definition 2: the complement (prefix unary $-$), Kleene closure operators (postfix $^+$ and *), left- and right residuals (infix $\backslash$ and $/$), relational addition (infix †), and product (infix $\times$). These are all expressible in the definitions above, so we have limited this exposition to the operators introduced above.

The complement operator is defined by means of the binary difference operator (Eq. 4).

$$\mathfrak{T}(r) = \langle A,B\rangle \ \Rightarrow\ \overline{r} \ = \ \mathbb{V}_{[A\times B]} - r \tag{16}$$

This definition is elaborated in [34].

A *rule* is a pair of terms $r, s \in \mathbb{T}$ with $\mathfrak{T}(r) = \mathfrak{T}(s)$, which is syntactically recognizable as a rule.

$$\text{RULE } r = s$$

This means $\mathfrak{I}(r) = \mathfrak{I}(s)$. In practice, many rules are written as:

$$\text{RULE } r \subseteq s$$

This is a shorthand for

$$\text{RULE } r \cap s = r$$

We have enhanced the type function $\mathfrak{T}$ and the interpretation function $\mathfrak{I}$ to cover rules as well. If $\mathfrak{T}(r) = \mathfrak{T}(s)$ and $\mathfrak{T}(s) = \langle A, B\rangle$:

$$\mathfrak{T}(\text{RULE } r = s) = \langle A,B\rangle \tag{17}$$
$$\mathfrak{T}(\text{RULE } r \subseteq s) = \langle A,B\rangle \tag{18}$$
$$\mathfrak{I}(\text{RULE } r = s) = \mathfrak{I}(\mathbb{V}_{[A\times B]} - ((s - r) \cup (r - s))) \tag{19}$$
$$\mathfrak{I}(\text{RULE } r \subseteq s) = \mathfrak{I}(\mathbb{V}_{[A\times B]} - (r - s)) \tag{20}$$

We call rule r *satisfied* when $\mathfrak{I}(\text{RULE } r = s) = \mathfrak{I}(\mathbb{V}_{[A \times B]})$. As the population of relations used in r changes with time, the satisfaction of the rule changes accordingly. A software developer, who conceives these rules, must consider how to keep each one of them satisfied. We call a rule *violated* if it is not satisfied. The set $\mathfrak{I}((s - r) \cup (r - s))$ is called the *violation set* of RULE $r = s$. To *resolve* violations means to change the contents of relations such that the rule is satisfied[3]. Each pair in the violation set of a rule is called a violation of that rule.

The software developer must define how to resolve violations when they occur. She does so by inserting and/or deleting pairs in appropriately chosen relations. Whatever choice she makes, she must ensure that her code yields data that satisfies the rules. When we say: "rule r specifies this action" we mean that satisfaction of rule r is the goal of any action specified by rule r.

3 Conceptual Analysis

As a case study, an argument assistance system, MirrorMe, was built. We have chosen to implement the ideas of Toulmin [29], because his book "The uses of Arguments" is still one of the most influential works in the area of legal argument management[4]. Toulmin is regarded as the first scholar in modern history to come up with a usable theory of argumentation. Recent work typically draws on Toulmin, so his doctrine offers a good starting point. Toulmin's ideas have been implemented before, for example in a tool called ArguMed [30].

Software systems that support legal arguments have been around for many decades. Verheij [31] distinguishes between argument assistance systems and automated reasoning systems. Automated reasoning has never gained wide acceptance for making legal decisions, because lawyers and judges alike feel that human judgment should be at the core of any legal decision. Attempts to apply mathematical logic to legal judgments have had limited impact for similar reasons [23]. Legal reasoning differs from logic reasoning because of the human judgments that are involved. The literature on legal reasoning [17] makes it quite clear why mathematical logic alone does not suffice. Argument assistance systems [32] have been more successful, because they respect the professional freedom of legal professionals to construct their own line of argumentation. Such systems offer help in many different ways. They can help by looking up legal references, jurisdiction from the past, scholarly works etc. They can also help to construct and validate arguments by keeping arguments and evidence organized. They can store, disclose and share legal evidence.

In our case study we use logic to reason about the correctness of a program. The argumentation principles of Toulmin are *implemented in* logic rather than *replaced by* logic. The structure of MirrorMe was designed by conceptually analysing ideas of Toulmin, such as claim, warrant, argument, and rebuttal. They appear in MirrorMe as relations. The Ampersand compiler generates a

[3] To *restore invariance* is sometimes used as a synonym to resolving violations. Consequently, a rule is sometimes called *invariant*.

[4] Thanks to Elleke Baecke for pointing us towards this source.

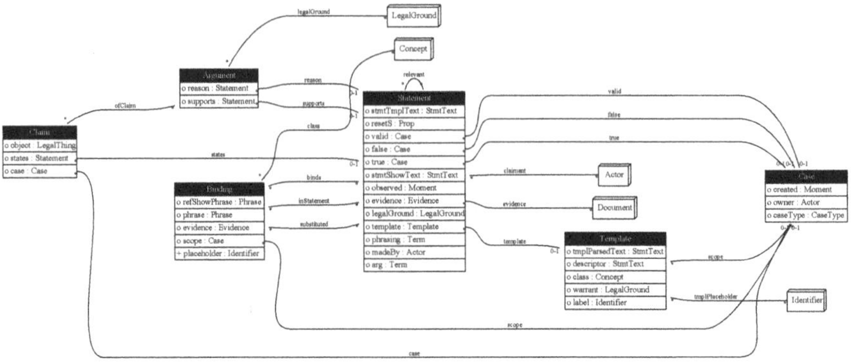

Fig. 1. Conceptual data model

conceptual data model to help the software developer to oversee all relations. Even though our case study yields a model that is a bit too large for this paper, Fig. 1 gives a good impression of what it looks like.

To convey the flavour of software development in relation algebra, it suffices to discuss a tiny part of the whole program. Therefore, we shall discuss the part of the conceptual model that is used in the sequel. The context in which a user creates arguments and reasons about them is a legal case. For this reason, validity, falsehood and truth of statements are related to the case. A *statement* is a phrase in natural language that can be either true or false. An example is "`The employee, John Brown, is entitled to 50 Euros`"[5]. In MirrorMe, a user can define a template, such as "`The employee, [emp], is entitled to [increase]`". The strings "`[emp]`" and "`[increase]`" are called placeholders. The reason for using placeholders is that legal rules are stated in general terms, e.g. "`Every employee is entitled to an increase in salary`". The user of MirrorMe will pick a legal text (from a source he trusts) and substitute parts of that text by placeholders. When the facts are known, the argument can be completed by substituting placeholders by actual phrases.

It is precisely this substitution process that we have chosen to describe in this paper as a case study in programming with relation algebra.

4 Programming in Relation Algebra

At this point we have reached the core of this case study. We focus on the substitution of placeholders when their values change. By focusing on this tiny detail, we can discuss the mechanics "under the hood" of the application generated by Ampersand.

[5] This type of statement is typical for cases. It is valid only in the case where this particular John Brown is known. In other cases, where John Brown is unknown, this statement is meaningless.

First we show how the computer solves the issue by looking at an excerpt of a log file (Fig. 2). It shows an alternating sequence of the computer (ExecEngine) mentioning a rule, followed by insert or delete actions to satisfy that rule. Then we zoom in further, one rule at a time, to explain precisely what each rule looks like and how programming is done. We then discuss the same flow of events by means of a graph (Fig. 3) in which these rules and actions are nodes. This graph serves as an event flow diagram to illustrate the process behind Fig. 2.

```
ExecEngine run started
ExecEngine satisfying rule 'signal phrase update'
InsPair(resetS,Statement,Stat623,Statement,Stat623)
ExecEngine satisfying rule 'flush substitutions'
DelPair(substituted,Binding,Bind625,Statement,Stat623)
ExecEngine satisfying rule 'reset statement text'
InsPair(stmtShowText,Statement,Stat623,StmtText,
        The employee, [emp], is entitled to [increase].)
ExecEngine satisfying rule 'done initializing'
DelPair(resetS,Statement,Stat623,Statement,Stat623)
ExecEngine satisfying rule 'substitute'
InsPair(stmtShowText,Statement,Stat623,StmtText,
        The employee, James, is entitled to [increase].)
InsPair(substituted,Binding,Bind625,Statement,Stat623)
ExecEngine satisfying rule 'fill shownPhrase'
InsPair(shownPhrase,Binding,Bind625,Phrase,James)
ExecEngine run completed
```

Fig. 2. Log file of a substitution

The log file of Fig. 2 has been taken from a computer that carries out the procedure to satisfy all rules. The machine will only act on rules that are violated. The first rule to be violated is "signal phrase update", in which the computer signals that something or someone has made a change in relation *phrase*. The procedure ends by doing the necessary substitutions, ensuring that all statements have the actual phrase of a placeholder in their text.

Let us now study the actual rules to see how these rules cause the right actions to take place in the correct order. We will follow the log file from Fig. 2 in reverse direction, reasoning backwards from the result.

Whether a placeholder has been edited can be observed by comparing its new phrase to the shown phrase[6]. The phrase of a placeholder is kept in the relation *phrase*. The shown phrase is kept in the relation *shownPhrase*. The sole purpose for having the relation *shownPhrase* is to detect a change in *phrase*. Let us introduce *differB* to represent the bindings with an updated phrase:

[6] Note that we use the notion "the phrase of a placeholder" to indicate a pair from $phrase^{\smile}; placeholder$.

$$differB = \mathbb{I}_{Binding} \cap shownPhrase; \overline{\mathbb{I}_{Phrase}}; phrase^{\smile}$$

When the phrase of a placeholder changes, that phrase must be updated in every statement in which the placeholder was used. That update action is specified in Sect. 4.2. Section 4.1 specifies how *shownPhrase* is made equal to *phrase*, after all necessary substitutions are done.

4.1 Rule: Fill shownPhrase

$$\text{RULE } (\mathbb{I}_{Binding} \cap substituted / inStatement); phrase \subseteq shownPhrase$$

This rule says that for each binding that has been substituted in every statement it is used in, the *phrase* must be equal to the *shownPhrase*. When violated, it is satisfied by inserting all violations into *shownPhrase*.

4.2 Rule: Substitute

Let us now look into the process of substituting placeholders by phrases. The relation *tmplParsedText* contains the original text, provided by a user. Placeholders are specified by enclosing them in brackets, e.g. "`The employee, [emp], is entitled to [increase]`". The text in which a placeholder has been substituted by a phrase, e.g. "`The employee, John Brown, is entitled to 50 Euros`", is kept in relation *stmtShowText*. Each substitution that has been done in a statement corresponds to a binding-statement pair in the relation *substituted*. This relation keeps track of all substitutions. After a placeholder has been substituted, it no longer occurs in *stmtShowText*. This poses a problem if we want to substitute the new phrase in *stmtShowText*. For the placeholder that defines the place in the text where to substitute, is no longer in that text. Therefore, substitutions must be done in the original text of the statement. All placeholders in that text must then be substituted again. So, the text in *stmtShowText* must first be reset to the original text from *tmplParsedText*. To keep track of substitutions correctly, all corresponding binding-statement pairs must be removed from the relation *substituted*. Only after resetting is done, the substitutions can be put back in place with the new phrases filled in.

We define a relation *resetS* to register the statements that are being reset. In statements that are not being reset, $\mathbb{I}_{Statement} - resetS$, substitutions can take place. All placeholders that have a binding with a phrase can be substituted. The following rule specifies the action of substituting placeholders.

$$\begin{array}{l}\text{RULE}\\ \quad (\mathbb{I}_{Binding} \cap phrase; phrase^{\smile}); inStatement; (\mathbb{I}_{Statement} - resetS)\\ \subseteq\\ \quad substituted\end{array}$$

Violations of this rule are binding-statement pairs, of which the binding has a phrase and the statement is not being reset. Hence, this rule can be satisfied by inserting every violation into the relation substituted.

4.3 Rule: Done Initializing

Resetting a statement is done when two conditions are met. First, every statement that is (still) being reset may have no bindings in the relation *substituted*. Second, the text in *stmtShowText* corresponds to the text in *tmplParsedText*. So the rule that specifies the action is:

$$\begin{array}{l}\text{RULE}\\ \quad (\mathit{resetS} - \mathit{inStatement}^{\smile}; \mathit{substituted}) \cap \\ \quad \mathit{template}; \mathit{tmplParsedText}; \mathit{stmtShowText}^{\smile} \\ \subseteq \\ \quad \overline{\mathit{resetS}}\end{array}$$

Violations of this rule are statements that are no longer being reset, but are still in the relation *resetS*. The appropriate action is to remove them from *resetS*.

4.4 Rule: Reset Statement Text

To satisfy one condition from Sect. 4.3, the text in *stmtShowText* must be made equal to the original text in the template. The action is specified by the following rule:

$$\text{RULE } \mathit{resetS}; \mathit{template}; \mathit{descriptor} \subseteq \mathit{stmtShowText}$$

Violations of this rule are descriptors of templates that belong to statements that are being reset. These violations can be resolved by inserting them in *stmtShowText*.

4.5 Rule: Flush Substitutions

To satisfy the other condition from Sect. 4.3, the following rule specifies the action to be taken:

$$\text{RULE } \mathbb{V}_{[Binding \times Binding]}; \mathit{inStatement}; \mathit{resetS} \subseteq \overline{\mathit{substituted}}$$

Every binding in a statement that is being reset needs to be removed from the relation *substituted*. The software developer can implement this by deleting all violations of this rule from the relation *substituted*.

4.6 Rule: Signal Phrase Update

When the phrase in a binding is edited and the new phrase differs from the shown phrase, this signals that substitutions must be flushed (see Sect. 4.5), that the statement text must be reset to the original text (see Sect. 4.4), that the reset-state must be revoked (see Sect. 4.3), that substitution must take place (see Sect. 4.2), and finally that the phrase detection is switched off again (see Sect. 4.1). The initial condition occurs if a binding has been used (substituted) in a statement, and the binding satisfies *differB*. The following rule specifies the action that resetting a statement can start:

$$\text{RULE } \mathbb{I}_{Statement} \cap inStatement^{\smile}; differB; substituted \subseteq resetS$$

Violations of this rule are statements of which a substitution must be re-done. The software developer can have these violations added to *resetS* to satisfy this rule. In doing so, the chain of events is triggered that ends when all rules are satisfied.

5 Programming in the Small

Let us take a closer look at the programming process. Working in relation algebra, a software developer must think about satisfying constraints. She considers how violations arise from changing content of relations. And she thinks about insert and delete actions to restore these violations. In our case study, we have reasoned with the event types: Ins ⟨*relation*⟩ and Del ⟨*relation*⟩[7]. Table 1 shows which event types may violate which rules. Recall Sect. 4, where the process of substitution started by updating the value of a placeholder. This meant doing an insert after a delete on the relation *phrase*, causing the updated phrase to appear in *differB*. That was signaled by the rule "signal phrase update". In general, the software developer must decide how to resolve violations that occur as a result of some event. This can be done by choosing among the duals of the event types from Table 1. By systematically swapping Ins for Del and vice-versa, the table shows by which type of events violations can be restored. To ensure progress, the software developer will pick a different relation for restoring than the relation that causes the violation. The software developer can draw a graph that contains all information from Table 1 and the dual information. Figure 3 shows part of that graph. A circle represents a rule, a rectangle represents an event type, and the arrows connect them. The software developer can work her way through the

Table 1. By which type of events can rules be violated?

Rule	Event types
signal phrase update	Ins *inStatement*, Ins *differB*, Ins *substituted*, Del *resetS*
flush substitutions	Ins *inStatement*, Ins *resetS*, Del *substituted*
reset statement text	Ins *resetS*, Ins *template*, Ins *descriptor*, Del *stmtShowText*
done initializing	Ins *resetS*, Del *inStatement*, Del *substituted*, Ins *template*, Ins *tmplParsedText*, Ins *stmtShowText*
substitute	Ins *phrase*, Ins *inStatement*, Del *resetS*, Del *substituted*
fill shownPhrase	Ins *substituted*, Del *inStatement*, Del *shownPhrase*, Ins *phrase*

[7] Ins and Del are called each others duals.

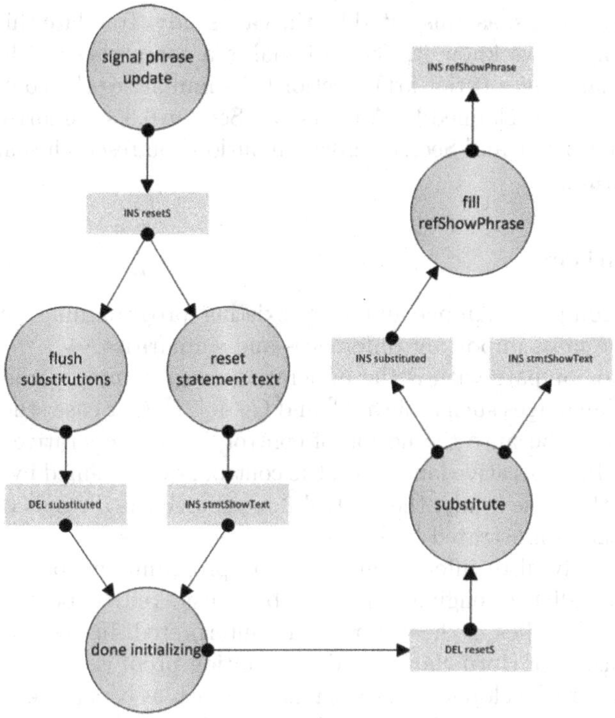

Fig. 3. Event flow

graph, to write code to restore invariants for every event type that might cause violations. Figure 3 shows just that part of the graph that corresponds with the case study in this article.

6 Reflection

Ampersand is built on the belief that software development should be automated. It uses rules to represent requirements, in the firm belief that consistent requirements are essential to the software development process [5]. Being fully aware of the fate of formal methods in computer science, Ampersand is founded on the belief that software development must be done more formally. So Ampersand is building further on the foundations laid by formal specification methods such as Z [26] and Alloy [14]. In contrast however with such methods, Ampersand is equipped with a software generator that generates an information system. Thus, specifying in Ampersand is developing software at the same time. Ampersand complements the RelView approach [4], which also generates software but is stronger in specifying complex computational problems. In retrospect we see that programming in relation algebra yields an unconventional programming experience. This experience consists of inventing rules and choosing event types for resolving violations, as illustrated by the case study. To relate this experience to programming as we know it, Sect. 6.1 makes a comparison with established programming languages (prior art). Section 6.2 summarizes the contributions to software development claimed by Ampersand. Section 6.3 summarizes the use of Ampersand in practice and Sect. 6.4 gives an outlook on research that is required in the near future.

6.1 Comparison

This section compares Ampersand with existing programming paradigms by mentioning the most important differences and similarities.

The first to compare with is the imperative programming paradigm, known from popular languages such as Java [8] and C++ [27]. Our case study cannot be called imperative, because the notion of control flow in imperative languages is very different. In imperative languages, the control flow is defined by the software developer. In this case study, the control flow emerges as a result of changes in relation content as illustrated by Fig. 3.

The case study also differs from the logic programming paradigm of Prolog [18] and of all rule engines that can be considered to be an offspring of Prolog. A difference lies in the way rules are interpreted. In logic programming, a program consists of Horn-clauses and a resolution-proof is constructed on runtime. The software developer works with notions such as backtracking (backward chaining) and unification, which are absent in Ampersand. Ampersand is not restricted to Horn-clauses; any relation-algebraic equation over relations can be used as a rule.

The case study also differs from functional programming, of which Haskell and Scala are prominent representants. A core idea in functional programming is to evaluate a program as a function from input to output [2]. A functional program consists of function definitions, that are evaluated by a term- or graph-rewriter, using various strategies such as lazy or eager evaluation. In contrast, one might interpret our case study as a relaxation of the constraint that everything is a function. In relation algebra, everything is a relation and a function is a restricted form of a relation. A similarity to functional programming is the declarative style, because substitution of equal terms without changing the semantics is a property we see in both worlds.

A difference with database programming is found in the type of algebra that is used. Relational databases are founded on relational algebra [6]. They are typically programmed in SQL. In contrast with database programming, Ampersand implements heterogeneous relation algebra [25]. A software developer working with relational algebra sees n-ary tables, while Ampersand is restricted to binary relations. The comparison between relation*al* algebra and relation algebra might relate to comparing relational databases with graph databases [33], although no literature was found to corroborate this.

In the tradition of formal specification, there are many relational approaches, such as Z [26], CSP [24], LOTOS [10], VDM [12]. Where formal specification techniques typically analyse and diagnose specifications, Ampersand actually synthesizes (generates) information systems.

If Ampersand represents a programming style at all, we might call it "a relation-algebraic style of programming". That style would be characterized by a programmer who specifies constraints and a computer trying to satisfy these constraints by resolving violations.

6.2 Contribution

Contributions of Ampersand to the software development process are:

- Ampersand has the usual benefits of a declarative language: This means that terms can be manipulated by Tarski's axioms without changing their semantics [34]. It also means that the order in which rules are written has no consequence for their semantics.
- Heterogeneous relation algebra has a straightforward interpretation in natural language [13]. We have used that to formalize business requirements without exposing business stakeholders to any formal notation.
- Heterogeneous relation algebra in Ampersand is statically typed [16]. There is much evidence for significantly lower software maintenance cost due to static typing as opposed to dynamic typing [7,22].
- Heterogeneous relation algebra is well studied. As a consequence, many tools that are readily available in the public domain can be put to good use. For executives of large organizations it can be reassuring that the formalism is free of childhood diseases.
- Relation algebra facilitates composing software from reusable components, because a program consists of rules. Since the union of sets of rules is a set of rules, compositionality comes from the union operator. In practice, when components are brought together in larger assemblies, hardly any adjustments have to be made[8].

Ampersand also has disadvantages. It appears to be difficult to learn for large groups of software professionals. Research [20] shows that this is largely due to deficits in prerequisite knowledge, especially skills in discrete mathematics. Also, programming appears to be difficult in practice.

6.3 Ampersand in Practice

Ampersand has been used in practice both in education (Open University of the Netherlands) and in industry (Ordina and TNO-ICT). For example, Ordina designed a proof-of-concept in 2007 of the INDIGO-system. This design was based on Ampersand, to obtain correct, detailed results in the least amount of time. Today INDIGO is in use as the core information system of the Dutch immigration authority, IND. More recently, Ampersand was used to design an

[8] This is (unsubstantiated) experience collected from projects we have done with Ampersand.

information system called DTV for the Dutch food authority, NVWA. A prototype of DTV was built in Ampersand and was used as a model to build the actual system. TNO-ICT, a major Dutch industrial research laboratory, is using Ampersand for research purposes. For example, TNO-ICT did a study of international standardizations efforts such as RBAC (Role Based Access Control) in 2003 and architecture (IEEE 1471-2000) [9] in 2004. Several inconsistencies were found in the last (draft) RBAC standard [1]. TNO-ICT has also used the technique in conceiving several patents[9]. At the Open University of the Netherlands, Ampersand is being taught in a course called Rule Based Design [13]. In this course, students use a platform called RAP, which has been built in Ampersand [20]. RAP has been the first Ampersand-application that has run in production.

6.4 Further Research

Further research on this topic is required to bring relation algebra still closer to the community of practitioners. Further use of relation algebra can be made by incorporating a model checker, such as the Alloy analyser [14], to detect inconsistent rules. An exciting new development is Amperspiegel [15], which brings notational flexibility at the fingertips of the user. Developments in the Ampersand-compiler are going towards a rule-repository (written in Ampersand itself). This will make collaborative information systems development in Ampersand easier, because the repository can assist in automating the software development process further. Other research is needed towards a comprehensive theory of information systems. Currently, there is no theory (in the mathematical meaning of the word) for information systems. In the Ampersand project, a sub-project called "Formal Ampersand" is being conducted to achieve this goal.

References

1. ANSI, INCITS 359: Information Technology: Role Based Access Control Document Number: ANSI/INCITS 359–2004. InterNational Committee for Information Technology Standards (formerly NCITS) (2004)
2. Backus, J.: Can programming be liberated from the von Neumann style?: A functional style and its algebra of programs. Commun. ACM **21**(8), 613–641 (1978). doi:10.1145/359576.359579
3. Berghammer, R., Ehler, H., Zierer, H.: Towards an algebraic specification of code generation. Sci. Comput. Program. **11**(1), 45–63 (1988). doi:10.1016/0167-6423(88)90064-0
4. Berghammer, R., Neumann, F.: RelView – an OBDD-based computer algebra system for relations. In: Ganzha, V.G., Mayr, E.W., Vorozhtsov, E.V. (eds.) CASC 2005. LNCS, vol. 3718, pp. 40–51. Springer, Heidelberg (2005). doi:10.1007/11555964_4

[9] e.g. patents DE60218042D, WO2006126875, EP1727327, WO2004046848, EP1563361, NL1023394C, EP1420323, WO03007571, and NL1013450C.

5. Boehm, B.W.: Software Engineering Economics. Advances in Computing Science and Technology. Prentice Hall PTR, Upper Saddle River (1981)
6. Codd, E.F.: A relational model of data for large shared data banks. Commun. ACM **13**(6), 377–387 (1970). http://doi.acm.org/10.1145/362384.362685
7. Hanenberg, S., Kleinschmager, S., Robbes, R., Tanter, É., Stefik, A.: An empirical study on the impact of static typing on software maintainability. Empirical Softw. Eng. **19**(5), 1335–1382 (2014). doi:10.1007/s10664-013-9289-1
8. Harms, D., Fiske, B.C., Rice, J.C.: Web Site Programming With Java. McGraw-Hill, New York City (1996). http://www.incunabula.com/websitejava/index.html
9. IEEE: Architecture Working Group of the Software Engineering Committee: Standard 1471–2000: Recommended Practice for Architectural Description of Software Intensive Systems. IEEE Standards Department (2000)
10. ISO: ISO 8807: Information processing systems - open systems interconnection - LOTOS - a formal description technique based on the temporal ordering of observational behaviour. Standard, International Standards Organization, Geneva, Switzerland. 1st edn. (1987)
11. Jackson, D.: A comparison of object modelling notations: Alloy, UML and Z. Technical report (1999). http://sdg.lcs.mit.edu/publications.html
12. Jones, C.B.: Systematic Software Development Using VDM. Prentice Hall International (UK) Ltd., Hertfordshire (1986)
13. Joosten, S., Wedemeijer, L., Michels, G.: Rule Based Design. Open Universiteit, Heerlen (2013)
14. Jackson, D.: Software Abstractions: Logic, Language, and Analysis. The MIT Press, Cambridge (2006)
15. Joosten, S.J.C.: Parsing and printing of and with triples. In: Pous, D., Struth, G., Höfner, P. (eds.) RAMiCS 2017. LNCS, vol. 10226, pp. 159–176. Springer International Publishing, Berlin (2017)
16. Joosten, S.M.M., Joosten, S.J.C.: Type checking by domain analysis in ampersand. In: Kahl, W., Winter, M., Oliveira, J.N. (eds.) RAMICS 2015. LNCS, vol. 9348, pp. 225–240. Springer, Cham (2015). doi:10.1007/978-3-319-24704-5_14
17. Lind, D.: Logic and Legal Reasoning. The National Judicial College Press, Beijing (2007)
18. Lloyd, J.W.: Foundations of Logic Programming. Springer, New York (1984)
19. Maddux, R.: Relation Algebras. Elsevier Science, Studies in Logic and the Foundations of Mathematics (2006)
20. Michels, G.: Development environment for rule-based prototyping. Ph.D. thesis, Open University of the Netherlands (2015)
21. Michels, G., Joosten, S., van der Woude, J., Joosten, S.: Ampersand. In: Swart, H. (ed.) RAMICS 2011. LNCS, vol. 6663, pp. 280–293. Springer, Heidelberg (2011). doi:10.1007/978-3-642-21070-9_21
22. Petersen, P., Hanenberg, S., Robbes, R.: An empirical comparison of static and dynamic type systems on API usage in the presence of an IDE: Java vs. groovy with eclipse. In: Proceedings of the 22nd International Conference on Program Comprehension, ICPC 2014, pp. 212–222. ACM, New York (2014). doi:10.1145/2597008.2597152
23. Prakken, H.: Ai & law, logic and argument schemes. Argumentation **19**(3), 303–320 (2005). doi:10.1007/s10503-005-4418-7
24. Roscoe, A.W., Hoare, C.A.R., Bird, R.: The Theory and Practice of Concurrency. Prentice Hall PTR, Upper Saddle River (1997)

25. Schmidt, G., Hattensperger, C., Winter, M.: Heterogeneous relation algebra. In: Brink, C., Kahl, W., Schmidt, G. (eds.) Relational Methods in Computer Science, pp. 39–53. Springer, New York (1997). ISBN 3-211-82971-7
26. Spivey, J.: The Z Notation: A Reference Manual. International Series in Computer Science, 2nd edn. Prentice Hall, New York (1992)
27. Stroustrup, B.: The C++ Programming Language, 3rd edn. Addison-Wesley Professional, Boston (1997)
28. Swart, H., Berghammer, R., Rusinowska, A.: Computational social choice using relation algebra and relview. In: Berghammer, R., Jaoua, A.M., Möller, B. (eds.) RelMiCS 2009. LNCS, vol. 5827, pp. 13–28. Springer, Heidelberg (2009). doi:10.1007/978-3-642-04639-1_2
29. Toulmin, S.E.: The Uses of Argument. Cambridge University Press, Cambridge (1958). http://www.amazon.com/exec/obidos/redirect?tag=citeulike-20&path=ASIN/0521534836
30. Verheij, B.: Automated argument assistance for lawyers. In: Proceedings of the 7th International Conference on Artificial Intelligence and Law, ICAIL 1999, pp. 43–52. ACM, New York (1999). doi:10.1145/323706.323714
31. Verheij, B.: Artificial argument assistants for defeasible argumentation. Artif. Intell. **150**(1), 291–324 (2003). doi:10.1016/S0004-3702(03)00107-3. http://www.sciencedirect.com/science/article/pii/S0004370203001073
32. Verheij, B.: Virtual Arguments. On the Design of Argument Assistants for Lawyers and Other Arguers. T.M.C. Asser Press, The Hague (2005)
33. Vicknair, C., Macias, M., Zhao, Z., Nan, X., Chen, Y., Wilkins, D.: A comparison of a graph database and a relational database: a data provenance perspective. In: Proceedings of the 48th Annual Southeast Regional Conference, ACM SE 2010, pp. 42:1–42:6. ACM, New York (2010). doi:10.1145/1900008.1900067
34. van der Woude, J., Joosten, S.: Relational heterogeneity relaxed by subtyping. In: Swart, H. (ed.) RAMICS 2011. LNCS, vol. 6663, pp. 347–361. Springer, Heidelberg (2011). doi:10.1007/978-3-642-21070-9_25

Allegories and Collagories for Transformation of Graph Structures Considered as Coalgebras

Wolfram Kahl(✉)

McMaster University, Hamilton, ON, Canada
kahl@mcmaster.ca

Abstract. Although coalgebras are widely used to model dynamic systems with infinite behaviours, they are actually also a more natural tool than algebras to model the static systems that are the main subject of the "algebraic approach" to graph transformation and model transformation: many variants of graph structures and object webs are more easily modelled as coalgebras than as algebras. By characterising the kinds of coalgebras that give rise to different kinds of allegories, we make the tools of the relation-algebraic approach and also of the category-theoretic "adhesive" "algebraic approach" available to the transformations of coalgebras.

Keywords: Relation-algebraic approach to graph transformation · Allegories of coalgebras · Meet-preserving relators · Adhesive categories of coalgebras

1 Introduction

The "algebraic approach to graph transformation" (Ehrig et al. 2006) models graph structures as unary algebras, which have nice category-theoretic properties that algebras with non-unary function symbols do not share. As a way to make the results of the algebraic approach available to more varied structures, the relevant category-theoretic properties have been abstracted into "adhesive categories" (Lack and Sobocinski 2004, 2005; Ehrig et al. 2006).

It turns out that many graph-like structures that require more-or-less adhoc treatments in the context of the adhesive approach can be modelled as coalgebras. For example, the following is a signature for directed hypergraphs where each hyperedge has a sequence (implemented using the type constructor List) of source nodes and a sequence of target nodes, and each node is labelled with an element of the constant set L:

$$\begin{aligned} \mathsf{sigDHG} := \langle \textbf{sorts: } & \mathsf{N}, \mathsf{E} \\ \textbf{ops: } & \mathsf{src} : \mathsf{E} \rightarrow \mathsf{List\ N} \\ & \mathsf{trg} : \mathsf{E} \rightarrow \mathsf{List\ N} \\ & \mathsf{nlab} : \mathsf{N} \rightarrow L \quad \rangle \end{aligned}$$

This signature is coalgebraic: The argument type of each operation is exactly one sort. It gives rise to a coalgebra in the product category $Set \times Set$ (details in

P. Höfner et al. (Eds.): RAMiCS 2017, LNCS 10226, pp. 193–208, 2017.
DOI: 10.1007/978-3-319-57418-9_12

Sect. 6). While constant sets like L are perfectly standard as results in coalgebras, modelling labelled graphs as algebras always has to employ the trick of declaring the label sets as additional sorts, and then considering the subcategory that has algebras with a fixed choice for these label sorts, and morphisms that map them only with the identity. Similarly, list-valued source and target functions are frequently considered for algebraic graph transformation, but with ad-hoc definitions for morphisms and custom proofs of their properties.

There is significant attention to attributed graphs in the graph transformation literature, including symbolically attributed graphs, where node and/or edge attributes are taken from the term algebra over some term signature Σ. The problem the algebraic approach faces here is that usually, an individual graph forms a complete (normally finite) algebra, but here, this algebra ends up including the infinite term algebra. This has led to ad-hoc solutions including partitioning of the sorts and adding attribute-carrier sorts (Löwe et al. 1993), or partitioning the function symbols (Heckel et al. 2002). The following is a natural signature for edge-labelled (with label set ELab) and node-attributed graphs, with symbolic attributes taken from the set of Σ terms over variables from the carrier set for sort V:

$$\begin{aligned} \mathsf{sigSNAG}_\Sigma := \langle & \textbf{sorts: } \mathsf{N}, \mathsf{E}, \mathsf{V} \\ & \textbf{ops: } \mathsf{src} : \mathsf{E} \rightarrow \mathsf{N} \\ & \qquad\ \mathsf{trg} : \mathsf{E} \rightarrow \mathsf{N} \\ & \qquad\ \mathsf{lab} : \mathsf{E} \rightarrow \mathsf{ELab} \\ & \qquad\ \mathsf{attr} : \mathsf{N} \rightarrow \mathcal{T}_\Sigma\ \mathsf{V} \quad \rangle \end{aligned}$$

Recognising this signature as coalgebraic solves the separation between graph structure and attribute algebra in a much more natural way, and without additional overhead.

A first exploration of this coalgebraic approach to graph structure modelling started in (Kahl 2014, 2015), modifying the category of coalgebras by moving into the Kleisli category of a monad, and using this monad component for modelling symbolically attributed graphs with morphisms that can substitute for variables occurring in attribute terms.

In the current paper we address a different angle (and do not consider monadic coalgebra homomorphisms): we investigate "relational homomorphisms" between coalgebras. In the case of algebras, the restriction to unary signatures is required to move from allegories to distributive allegories; it turns out that for coalgebras, no corresponding restriction is necessary, producing Kleene collagories and distributive allegories for a large class of coalgebra signatures.

After providing necessary category-theoretic notations and background on "relational categories and their functors", namely different flavours of allegories, and relators, in Sect. 2, we quickly present basic concepts of algebraic graph transformation in Sect. 3, and background on relation-algebraic "treatment" of pushouts and pullbacks (Sect. 4) and products and sums (Sect. 5). Then we define coalgebras and their "relational homomorphisms" in Sect. 6, and obtain allegories of coalgebras in Sect. 7. We proceed to add join (union) and zero morphisms in

Sect. 8, and show creation of tabulations and cotabulations in Sect. 9 making a wide range of relation-algebraic reasoning tools available for coalgebras over a large class of relators. In Sect. 11 we provide a more practical and accessible way to define classes of coalgebras, using more conventional signatures, and show how these give rise to the well-behaved relators of the previous sections. In Sect. 10 we show that direct products can only be calculated componentwise in extremely restricted circumstances.

At the time of writing, all theorems in Sects. 6–10 have mechanically-checked calculational proofs, quite similar in style and readability to those found in (Bird and de Moor 1997), written in the dependently-typed language Agda (Norell 2007); these are available via http://RelMiCS.McMaster.ca/RATH-Agda/.

2 Notation and Background: Categories, Allegories, Collagories

We assume familiarity with the basics of category theory; for notation, we write "$f : A \to B$" to declare that morphism f goes from object A to object B, and use ";" as the associative binary *forward composition* operator that maps two morphisms $f : A \to B$ and $g : B \to C$ to $(f; g) : A \to C$. The identity morphism for object A is written $\mathbb{I}_A$, or frequently just $\mathbb{I}$ where the object can be inferred from the context.

We assign ";" higher priority than other binary operators, and assign unary operators higher priority than all binary operators.

The category of sets and functions is denoted by *Set*.

A *functor* $\mathcal{F}$ from one category to another maps objects to objects and morphisms to morphisms respecting the structure generated by $\to$, $\mathbb{I}$, and composition; we denote functor application by juxtaposition both for objects, $\mathcal{F}\ A$, and for morphisms, $\mathcal{F}\ f$.

A *bifunctor* is a functor where the source is a product category. An important example is the coproduct bifunctor $+ : \mathbf{C} \times \mathbf{C} \to \mathbf{C}$ for a category $\mathbf{C}$ with a choice of coproducts. Functors with more than two arguments can be handled similarly, but will be considered as taking arguments from right-nested binary products.

An *OCC*, short for "ordered category with converse" (Kahl 2004), is a category where each homset is partially ordered via morphism *inclusion* $\sqsubseteq$, and that has an involutory *converse* operator $\breve{}$ that induces a contravariant identity-on-objects endofunctor, that is, $(R\breve{})\breve{} = R$ and $(R; S)\breve{} = S\breve{}; R\breve{}$.

In an OCC, a morphism $F : A \to B$ is called a *mapping* iff it is

- *univalent*, that is, $F\breve{}; F \sqsubseteq \mathbb{I}$, and
- *total*, that is, $\mathbb{I} \sqsubseteq F; F\breve{}$.

Dually, F is called *bijective* iff it is

- *injective*, that is, $F; F\breve{} \sqsubseteq \mathbb{I}$, and
- *surjective*, that is, $\mathbb{I} \sqsubseteq F\breve{}; F$.

An *allegory* (Freyd and Scedrov 1990) is an OCC where each homset has binary *meets* $\sqcap$ with respect to inclusion $\sqsubseteq$, and where the Dedekind rule holds:

$$Q \sqcap R ; S \sqsubseteq (R \sqcap Q ; S^{\smile}) ; (S \sqcap R^{\smile} ; Q)$$

The allegory (or OCC, or collagory...) of sets and relations is denoted by *Rel*, and has *Set* as its subcategory of mappings.

The language of allegories where each homset is a distributive lattice, with binary join operator $\sqcup$, and where composition distributes from both sides over binary joins has just enough expressivity to define the basic gluing constructions used for algebraic graph transformation; we therefore refer to such allegory as a *collagory*, from the Greek $\kappa\acute{o}\lambda\lambda\alpha$ for glue (Kahl 2011).

With respect to the hierarchy introduced by Freyd and Scedrov (1990), collagories are an intermediate structure between allegories and *distributive allegories*, which can now be seen as collagories with zero morphisms, that is, where each homset has a least morphism $\bot$ that also acts as left- and right-zero for composition.

A *Kleene collagory* (Kahl 2011) is a collagory that has a Kleene star (axiomatised following Kozen (1994)) on each homset of endomorphisms. (A Kleene collagory is therefore not necessarily a Kleene category, or typed Kleene algebra following Kozen (1998), since it may not have zero morphisms.)

A *division allegory* (Freyd and Scedrov 1990) is a distributive allegory with left- and right residuals of composition. The two residuals are dual to each other; we show only the definition of right residual $(Q \setminus S) : B \to C$ of two morphisms $Q : A \to B$ and $S : A \to C$:

$$X \sqsubseteq Q \setminus S \qquad \text{iff} \qquad Q ; X \sqsubseteq S$$

A *relator*[1] is a functor between OCCs that preserves inclusion and commutes with converse. (Bird and de Moor (1997, Theorem 5.1) prove that, for tabular allegories, these two conditions are equivalent, but we prefer to avoid the additional assumptions in favour of added generality.)

Lemma 2.1. Each relator $\mathcal{R}$ satisfies the following properties:

1. If the morphism F is univalent, respectively total, injective, or surjective, then $\mathcal{R}\,F$ is univalent, respectively total, injective, or surjective, too. ✓[2]
2. Relators on allegories sub-distribute over meets:

$$\mathcal{R}(R \sqcap_1 S) \sqsubseteq_2 (\mathcal{R}\ R) \sqcap_2 (\mathcal{R}\ S) \checkmark$$

If also $\mathcal{R}(R \sqcap_1 S) \sqsupseteq_2 (\mathcal{R}\ R) \sqcap_2 (\mathcal{R}\ S)$, then we say that $\mathcal{R}$ *preserves meets.*

[1] Originally introduced as "I-functor" by Kawahara (1973).

[2] These "check marks" indicate that the associated fact has a proof in the Agda development, with RATH-Agda theory location possibly visible in PDF readers supporting "tool tips".

3. Between collagories, relators super-distribute over joins:

$$\mathcal{R}(R \sqcup_1 S) \sqsupseteq_2 (\mathcal{R}\ R) \sqcup_2 (\mathcal{R}\ S) \checkmark$$

If also $\mathcal{R}(R \sqcup_1 S) \sqsubseteq_2 (\mathcal{R}\ R) \sqcup_2 (\mathcal{R}\ S)$, then we say that $\mathcal{R}$ *preserves joins.*

4. Between Kleene collagories, star sub-distributes over relators:

$$(\mathcal{R}\ R)^{*_2} \sqsubseteq_2 \mathcal{R}(R^{*_1}) \checkmark$$

5. Between division allegories, relators sub-distribute over division:

$$\mathcal{R}(Q \setminus_1 S) \sqsubseteq_2 (\mathcal{R}\ Q) \setminus_2 (\mathcal{R}\ S) \checkmark$$

If also $\mathcal{R}(Q \setminus_1 S) \sqsupseteq_2 (\mathcal{R}\ Q) \setminus_2 (\mathcal{R}\ S)$, then we say that $\mathcal{R}$ *preserves right-residuals.* □

3 Background: Algebraic Graph Transformation

Mainly for the purpose of motivating our attention to pushouts and pullbacks, we now quickly highlight the central concepts of the algebraic approach to graph transformation. (However, the details of this section are only relevant for the statement of Theorem 11.5 and the material thereafter.)

The double-pushout (DPO) approach to high-level rewriting (Ehrig et al. 2006) uses transformation rules that are spans $L \xleftarrow{l} G \xrightarrow{r} R$ in an appropriate category between the left-hand side L, gluing object G, and right-hand side R. A direct transformation step starts from a match m to an application object A, identifies (if possible) a host object H as *pushout complement* for $G \xrightarrow{l} L \xrightarrow{m} A$, and then concludes by constructing the pushout for the span $H \xleftarrow{h} G \xrightarrow{r} R$, producing the result object B.

$$\begin{array}{ccccc} L & \xleftarrow{l} & G & \xrightarrow{r} & R \\ {\scriptstyle m}\downarrow & & {\scriptstyle h}\downarrow & & \downarrow{\scriptstyle n} \\ A & \xleftarrow{a} & H & \xrightarrow{b} & B \end{array}$$

Abstracting this approach from concrete categories of graphs to general "high level replacement (HLR) systems" started notably with Ehrig et al. (1991); recently, Lack and Sobocinski (2004, 2005) introduced adhesive categories as a useful abstraction for frequently studied HLR properties. The following two definitions are taken from there:

Definition 3.1. A *van Kampen square* (i) is a pushout which satisfies the following condition: given a commutative cube (ii) of which (i) forms the bottom face and the back faces are pullbacks (where C is considered to be in the back), the front faces are pullbacks if and only if the top face is a pushout.

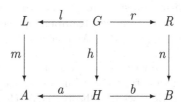

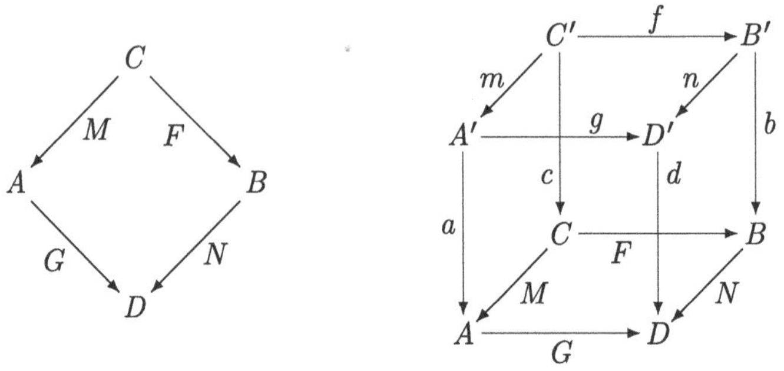

(i) (ii)

□

Definition 3.2. A category **C** is said to be *adhesive* if

- **C** has pushouts along monomorphisms;
- **C** has pullbacks;
- pushouts along monomorphisms are van Kampen squares. □

Two weaker variants are given as Definitions 4.9 and 4.13 by Ehrig et al. (2006):

Definition 3.3. A category **C** with a morphism class $\mathcal{M}$ is called a *(weak) adhesive HLR category* if

- $\mathcal{M}$ is a class of monomorphisms closed under isomorphisms, composition, and decomposition (i.e., if $f;g \in \mathcal{M}$ and $g \in \mathcal{M}$, then $f \in \mathcal{M}$);
- **C** has pushouts and pullbacks along $\mathcal{M}$-morphisms, and $\mathcal{M}$-morphisms are closed under pushouts and pullbacks;
- **C** has pullbacks;
- pushouts in **C** along $\mathcal{M}$-morphisms are (weak) van Kampen squares, where in a weak van Kampen square, the van Kampen square property holds for all commutative cubes with $M, F \in \mathcal{M}$ or $M, a, b, d \in \mathcal{M}$. □

The categories of sets with functions *Set* and of standard graphs with graph homomorphisms are adhesive categories, while other categories of interest are only adhesive HLR or even only weak adhesive HLR (Ehrig et al. 2006, Theorem 4.6 & Sect. 4.2).

For DPO rewriting in a (weak) adhesive HLR categories $(\mathbf{C}, \mathcal{M})$, both rule morphisms l and r are restricted to belong to $\mathcal{M}$.

4 Tabulations and Cotabulations

Tabulations and cotabulations are the "relation-algebraic essence" of pullbacks and pushouts, and the relation-algebraic approach to pullbacks and pushouts is practically useful since it not only produces the theoretical results of the category-theoretic universal characterisations, but also produces construction mechanisms that are useful for implementations.

The term "tabulation" appears to have been coined by Freyd and Scedrov (1990); we provide an equivalent characterisation here:

Definition 4.1. In an allegory **A**, the span $B \xleftarrow{P} A \xrightarrow{Q} C$ is called a *tabulation* of $V : B \to C$ if and only if the following equations hold:

$$P^{\smile};Q = V \qquad \begin{array}{c} P^{\smile};P = \mathbb{I} \sqcap V;V^{\smile} \\ Q^{\smile};Q = \mathbb{I} \sqcap V^{\smile};V \end{array} \qquad P;P^{\smile} \sqcap Q;Q^{\smile} = \mathbb{I}_A. \quad \square$$

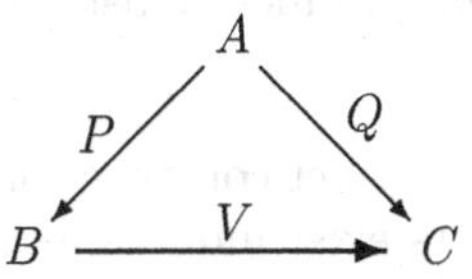

If a co-span $B \xrightarrow{R} D \xleftarrow{S} C$ of mappings is given, then, if a tabulation of $R;S^{\smile}$ exists, it is a *pullback* in the category of mappings in **A** (Freyd and Scedrov 1990, 2.147).

Since the properties connecting joins with converse in collagories or distributive allegories are not exactly dual to those connecting meet with converse in allegories, the original characterisation of Freyd and Scedrov (1990) does not directly dualise to a useful characterisation of collagories, while the characterisation above does dualise:

Definition 4.2. In a collagory **C**, the co-span $B \xrightarrow{R} D \xleftarrow{S} C$ is called a *cotabulation* of $W : B \to C$ iff the following equations hold:

$$R;S^{\smile} = W \qquad \begin{array}{c} R;R^{\smile} = \mathbb{I} \sqcup W;W^{\smile} \\ S;S^{\smile} = \mathbb{I} \sqcup W^{\smile};W \end{array} \qquad R^{\smile};R \sqcup S^{\smile};S = \mathbb{I}_D. \quad \square$$

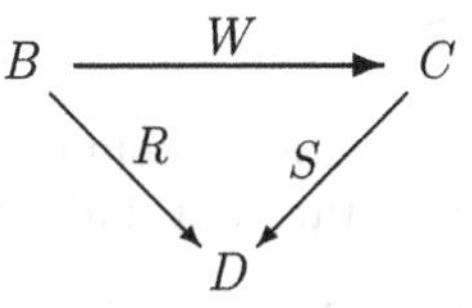

If a span $B \xleftarrow{P} A \xrightarrow{Q} C$ of mappings is given, then, if a cotabulation of the difunctional closure of $P^{\smile};Q$ exists, it is a *pushout* in the category of mappings in **C**. The difunctional closure is necessary due to commutativity: The other three conditions imply that R and S are mappings; this, together with $W = R;S^{\smile}$, implies that W is difunctional, whereas $P^{\smile};Q$ is not necessarily difunctional.

For more information about cotabulations, see (Kahl 2011).

5 Direct Products and Sums

Direct products and sums, following the nomenclature of Schmidt and Ströhlein (1993), are the relation-algebraic characterisations of Cartesian products respectively disjoint unions, and therefore correspond to category-theoretic products and coproducts, which are (co-)limits of discrete diagrams. In relation-algebraic settings, these discrete diagrams translate into least (respectively greatest) morphisms that can serve as starting points for (co-)tabultations:

We define a *direct product* of objects A and B in an allegory to be a tabulation $A \xleftarrow{\pi} P \xrightarrow{\rho} B$ of a $\mathbb{T}_{A,B}$, provided that this greatest morphism $\mathbb{T}_{A,B}$ exists. In an allegory where all direct products exist, these are the products in the subcategory of mappings.

Proposition 5.1. The direct product construction in an allegory with top morphisms where all direct products exist induces a relator ✓ and this relator preserves meets. ✓ □

Dually, in a collagory, we define a *direct sum* of objects A and B to be a co-tabulation of $\perp\!\!\!\perp_{A,B}$, if that least morphism exists. In a distributive allegory where all direct sums exist, these are the coproducts in the subcategory of mappings.

Proposition 5.2. In a distributive allegory where all direct sums exist, the direct sum construction induces a relator ✓, and this relator preserves meets. ✓ □

Join preservation for the direct sum relator would mean that

$$(\mathsf{R}_1 \sqcup \mathsf{S}_1) \otimes (\mathsf{R}_2 \sqcup \mathsf{S}_2) \sqsubseteq (\mathsf{R}_1 \otimes \mathsf{R}_2) \sqcup (\mathsf{S}_1 \otimes \mathsf{S}_2),$$

which is in general obviously not valid, and analogously with $\otimes$ for the direct product relator. Fortunately, none of the constructions we consider requires join-preserving relators, and we list the join preservation facts only for completeness.

6 Coalgebras in OCCs

For defining coalgebras and their relational homomorphisms, the setting of OCCs is sufficient. Now let $\mathcal{R}$ be an arbitrary but fixed endo-relator on an OCC.

Definition 6.1. An *$\mathcal{R}$-coalgebra* (T, op) consists of

- a *carrier* object T, and
- an *operation* mapping $op : T \to \mathcal{R}\,T$. □

Definition 6.2. A *<u>relational</u> coalgebra homomorphism* from $\mathsf{A} = (T_\mathsf{A}, op_\mathsf{A})$ to $(T_\mathsf{B}, op_\mathsf{B})$ is a morphism $\varphi : T_\mathsf{A} \to T_\mathsf{B}$ such that the L-simulation condition of de Roever and Engelhardt (1998) holds:

$$op_\mathsf{A}^{\smile} ; \varphi \sqsubseteq \mathcal{R}\,\varphi ; op_\mathsf{B}^{\smile}$$

Since op_{A} and op_{B} are mappings, the L-simulation condition is equivalent to

$$\varphi ; op_{\mathsf{B}} \sqsubseteq op_{\mathsf{A}} ; \mathcal{R}_{\varphi},$$

and our Agda formalisation of relational coalgebra homomorphisms actually uses the latter. (If φ is restricted to be a mapping, then this is further equivalent to the equality $\varphi ; op_{\mathsf{B}} = op_{\mathsf{A}} ; \mathcal{R}_{\varphi}$, which is the standard definition of ("non-relational") coalgebra homomorphisms.)

For example, the hypergraph signature sigDHG mentioned in the introduction induces (in a way to be made explicit in Sect. 11) the endo-relator $\mathcal{R}_{\mathsf{sigDHG}}$ on the product allegory *Rel* × *Rel* with (let List denote the standard list relator on *Rel*):

$$\mathcal{R}_{\mathsf{sigDHG}}\ (N\ ,\ E)\ =\ (L\ ,\ ((\mathsf{List}\ N) \times (\mathsf{List}\ N)))$$

A *directed hypergraph* (DHG), defined as a $\mathcal{R}_{\mathsf{sigDHG}}$-coalgebra, therefore consists of

- a carrier object, that is, pair of sets, (N, E), and
- an operation mapping $op : (N, E) \rightarrow (L\ ,\ ((\mathsf{List}\ N) \times (\mathsf{List}\ N)))$,

which can be reorganised into

- two *carrier sets* N and E, and
- three operation mappings: $\mathsf{nlab} : N \rightarrow L$
  $\mathsf{src} : E \rightarrow \mathsf{List}\ N$
  $\mathsf{trg} : E \rightarrow \mathsf{List}\ N$

A *relational DHG homomorphism* from $\mathsf{A} = (N_{\mathsf{A}}, E_{\mathsf{A}}, \mathsf{nlab}_{\mathsf{A}}, \mathsf{src}_{\mathsf{A}}, \mathsf{trg}_{\mathsf{A}})$ to $\mathsf{B} = (N_{\mathsf{B}}, E_{\mathsf{B}}, \mathsf{nlab}_{\mathsf{B}}, \mathsf{src}_{\mathsf{B}}, \mathsf{trg}_{\mathsf{B}})$ then is a product allegory morphism $\varphi : (N_{\mathsf{A}}, E_{\mathsf{A}}) \rightarrow (N_{\mathsf{B}}, E_{\mathsf{B}})$, that is, a pair of relations (*Rel*-morphisms) $\varphi_{\mathsf{N}} : N_{\mathsf{A}} \leftrightarrow N_{\mathsf{B}}$ and $\varphi_{\mathsf{E}} : E_{\mathsf{A}} \leftrightarrow E_{\mathsf{B}}$ satisfying:

- $\varphi_{\mathsf{N}} ; \mathsf{nlab}_{\mathsf{B}} \sqsubseteq \mathsf{nlab}_{\mathsf{A}}$
- $\varphi_{\mathsf{E}} ; \mathsf{src}_{\mathsf{B}} \sqsubseteq \mathsf{src}_{\mathsf{A}} ; \mathsf{List}\ \varphi_{\mathsf{N}}$
- $\varphi_{\mathsf{E}} ; \mathsf{src}_{\mathsf{B}} \sqsubseteq \mathsf{src}_{\mathsf{A}} ; \mathsf{List}\ \varphi_{\mathsf{N}}$

It is easy to see that $\mathcal{R}$-coalgebras form an OCC ✓.

Definition 6.3. We let $U_{\mathcal{R}}$ denote the forgetful relator ✓ from the OCC of $\mathcal{R}$-coalgebras and relational coalgebra homomorphisms to the underlying OCC. □

Then inclusion and converse of coalgebra homomorphisms are in fact *created* by $U_{\mathcal{R}}$, that is, they are defined by inclusion and converse of the underlying allegory.

In the following, we will state many additional properties of $\mathcal{R}$-coalgebra OCCs as "$U_{\mathcal{R}}$ creates ..."; this category-theoretic phrase pattern essentially means that if a "..." situation in the underlying OCC is mapped to by $U_{\mathcal{R}}$, then the preimage constitutes a "..." situation in the OCC of $\mathcal{R}$-coalgebras.

7 Coalgebras in Allegories

For obtaining more structure than only OCCs, a precondition on the relator $\mathcal{R}$ is required, and moving into the context of an underlying allegory:

Theorem 7.1. If $\mathcal{R}$ preserves meets, then $U_{\mathcal{R}}$ creates meets. ✓
Therefore, if $\mathcal{R}$ preserves meets, then $\mathcal{R}$-coalgebras form an allegory. ✓ □

Proposition 7.2. Constant relators and the identity relator preserve meets. ✓
For two meet-preserving relators $\mathcal{R}_1 : \mathcal{A}_0 \to \mathcal{A}_1$ and $\mathcal{R}_2 : \mathcal{A}_1 \to \mathcal{A}_2$, the composed relator $\mathcal{R}_1 ; \mathcal{R}_2 : \mathcal{A}_0 \to \mathcal{A}_2$ is meet-preserving, too. ✓
For two allegories $\mathcal{A}_1$ and $\mathcal{A}_2$, the projection relators from the product allegory $\mathcal{A}_1 \times \mathcal{A}_2$ to $\mathcal{A}_1$ respectively $\mathcal{A}_2$ are meet-preserving. ✓
For two meet-preserving relators $\mathcal{R}_1 : \mathcal{A}_0 \to \mathcal{A}_1$ and $\mathcal{R}_2 : \mathcal{A}_0 \to \mathcal{A}_2$, the "fork" relator $\mathcal{R}_1 \nabla \mathcal{R}_2 : \mathcal{A}_0 \to \mathcal{A}_1 \times \mathcal{A}_2$ to the product allegory $\mathcal{A}_1 \times \mathcal{A}_2$ is meet-preserving, too. ✓ □

Type and Cotype Relators

Now consider a bi-relator $\mathcal{R}$ on an allegory $\mathbf{A}$, that is, a relator from the product allegory $\mathbf{A} \times \mathbf{A}$ to $\mathbf{A}$. If $\mathcal{R}$ has initial algebras $(T_{\mathcal{R}}, \alpha_{\mathcal{R}})$, that is, for every object X an algebra $(T_{\mathcal{R}}\, X, \alpha_{\mathcal{R},X} : \mathcal{R}(X, T_{\mathcal{R}}\, X) \to T_{\mathcal{R}}\, X)$ that is the initial object in the category of $\mathcal{R}(X, \_)$-algebras as witnessed by *catamorphisms* (also called "fold" functions) (Meijer et al. 1991), then Bird and de Moor (1997, 5.5) show that $T_{\mathcal{R}}$ extends to a relator by invoking the fact that in tabular allegories, converse-preserving functors are also monotone (inclusion-preserving). They mention that for meet-preserving $\mathcal{R}$, the type relator $T_{\mathcal{R}}$ is meet-preserving again; we use this fact to prove monotony of the relator:

Theorem 7.3. If a bi-relator $\mathcal{R}$ on an allegory preserves meets, then the type relator $T_{\mathcal{R}}$ is a relator and preserves meets. ✓ □

Changing from initial algebras to final coalgebras $(C_{\mathcal{R}}\, X, \gamma_{\mathcal{R},X} : C_{\mathcal{R}}\, X \to \mathcal{R}(X, C_{\mathcal{R}}\, X))$ actually is cleanly dual for this purpose, so we also have:

Theorem 7.4. If a bi-relator $\mathcal{R}$ on an allegory preserves meets, then the co-type relator $C_{\mathcal{R}}$ is a relator and preserves meets. ✓ □

8 Coalgebras in Collagories

Moving into an underlying collagory, we observe that Lemma 2.1.(3), the generally true part of the join preservation property, is the opposite inclusion as in the case of meets, and this is sufficient for creation of joins of $\mathcal{R}$-coalgebra homomorphisms without precondition on $\mathcal{R}$:

Theorem 8.1. If $\mathcal{R}$ is an endo-relator on a collagory, then $U_{\mathcal{R}}$ creates joins. ✓
Therefore, if $\mathcal{R}$ on a collagory preserves meets, then $\mathcal{R}$-coalgebras form a collagory. ✓ □

Proof. For showing creation of joins, for relational coalgebra homomorphisms F and G from $\mathsf{A} = (T_{\mathsf{A}}, op_{\mathsf{A}})$ to $(T_{\mathsf{B}}, op_{\mathsf{B}})$, it suffices to show:

$$\begin{array}{ll} & (F \sqcup G); op_{\mathsf{B}} \\ \sqsubseteq & \{\text{Distributivity of; over } \sqcup\text{; rel. hom. property of } F \text{ and } G\} \\ & op_{\mathsf{A}}; \mathcal{R}F \ \sqcup \ op_{\mathsf{A}} \ ; \mathcal{R}\,G \\ = & \{\text{Distributivity of ; over } \sqcup\} \\ & op_{\mathsf{A}}; (\mathcal{R}\,F \ \sqcup \mathcal{R}\,G) \\ \sqsubseteq & \{\text{Lemma 2.1.(3)}\} \\ & op_{\mathsf{A}}; \mathcal{R}\,(F \sqcup G) \end{array}$$

□

Proposition 8.2. Constant relators and the identity relator on collagories preserve joins. ✓

The projection relators for product collagories preserve joins. ✓

Relator composition ; (Proposition 7.2) preserves join preservation. ✓

The "fork" relator combinator ∇ (Proposition 7.2) preserves join preservation. ✓ □

Theorem 8.3. If $\mathcal{R}$ is an endo-relator on a distributive allegory, then $U_{\mathcal{R}}$ creates zero morphisms. ✓

Therefore, if $\mathcal{R}$ on a distributive allegory preserves meets, then $\mathcal{R}$-coalgebras form a distributive allegory. ✓ □

As we will see below, all polynomial functors preserve meets, so we have collagories and distributive allegories of coalgebras for all polynomial functors. Contrast this with the situation for algebras, where more-than-unary function symbols break distributivity over joins (Kahl 2001, Example 4.2.3), and where zero-ary function symbols break the zero-laws, so that we obtain distributive allegories only for signatures where all function symbols are unary; the resulting algebras have been coined "graph structures" by Löwe (1993).

Kleene star preserves the coalgebra homomorphism property: ✓

$$\begin{array}{ll} & F^* \,; op \\ \subseteq & \{\text{Kleene star commutation with Definition 6.2}\} \\ & op\,; (\mathcal{R}\,F)^* \\ \subseteq & \{\text{Lemma 2.1.(4)}\} \\ & op\,; \mathcal{R}\,(F^*) \end{array}$$

Therefore we finally have:

Theorem 8.4. If $\mathcal{R}$ is an endo-relator on a Kleene collagory, then $U_{\mathcal{R}}$ creates Kleene stars. ✓

Therefore, if $\mathcal{R}$ on a Kleene collagory preserves meets, then $\mathcal{R}$-coalgebras form a Kleene collagory. ✓ □

In the statement of this theorem we did not refer to distributive allegories since the presence of zero morphisms, or even of least morphisms, is not relevant to the creation of Kleene stars.

9 Tabulations and Cotabulations of Relational Collagory Homomorphisms

Theorem 9.1. If $\mathcal{R}$ is a meet-preserving endo-relator on an allegory $\mathcal{A}$, then $U_{\mathcal{R}}$ creates tabulations. ✓

Therefore, if $\mathcal{A}$ has tabulations, then the allegory of $\mathcal{R}$-coalgebras has tabulations, too. ✓ □

Theorem 9.2. If $\mathcal{R}$ is a meet-preserving endo-relator on a collagory $\mathcal{A}$, then $U_{\mathcal{R}}$ creates cotabulations. ✓

Therefore, if $\mathcal{A}$ has cotabulations, then the collagory of $\mathcal{R}$-coalgebras has cotabulations, too. ✓ □

In summary:

Corollary 9.3. If $\mathcal{A}$ is a bitabular collagory, then the collagory of $\mathcal{R}$-coalgebras is bitabular, too. □

10 Creation of Top Morphisms and of Direct Products

Given an endo-relator $\mathcal{R}$ on an allegory $\mathbf{A}$ where each homset has a top morphisms $\top$, then $U_{\mathcal{R}}$ *creates top morphisms* if for all coalgebras A and B, $\top_{\mathsf{A},\mathsf{B}}$ extends to a coalgebra homomorphism, that is, $\top ; op_{\mathsf{B}} \sqsubseteq op_{\mathsf{A}} ; \mathcal{R}\,\top$.

For the identity relator, U_{Identity} creates top morphisms. However, for constant relators, the relator $U_{\mathsf{Const}\ X}$ in general does not.

Creation of top morphisms is sufficient for creation of direct products:

Theorem 10.1. Let a meet-preserving endo-relator $\mathcal{R}$ on an allegory $\mathbf{A}$ with top morphisms $\top$ and direct products be given. If $U_{\mathcal{R}}$ creates top morphisms, then $U_{\mathcal{R}}$ also creates direct products: for any two coalgrebras A and B, the coalgebra $(T_{\mathsf{A}} \times T_{\mathsf{B}}, \pi ; op_{\mathsf{A}} ; \mathcal{R}\pi^{\smile} \sqcap \rho ; op_{\mathsf{B}} ; \mathcal{R}\rho^{\smile})$ is well-defined, and a direct product for the projections π and ρ, which extend to coalgebra homomorphism. ✓ □

Creation of top morphisms is needed first for establishing that the operation of the product object is total, and second for the homomorphism property of the projections.

Since $\mathsf{Const}\ X\ \top = \mathbb{I}_X$, construction of counterexamples essentially only requires an object X corresponding to a set with at least two elements.

11 Coalgebraic Graph Structure Transformation

After collecting the basic ingredients for data structure definition, we now provide the linguistic means to build graph structure collagories from descriptions like sigDHG from the introduction. We first present the "internal syntax" side:

A *type signature* consists of

- a countable set of *object constants*, and
- for each positive arity $n : \mathbb{N} - \{0\}$ a countable set of *n-ary functor symbols*.

The set of *types* over a countable set $\mathcal{X}$ of *type variables* is defined inductively as follows:

- Each type variable from $\mathcal{X}$ is a type over $\mathcal{X}$.
- Each object constant is a type over $\mathcal{X}$.
- For each n-ary functor symbol f and each sequence $t_1, \ldots, t_n$ of n types over $\mathcal{X}$, the *application* $f(t_1, \ldots, t_n)$ is a type over $\mathcal{X}$.
- For each type variable x from $\mathcal{X}$ and each type t over $\mathcal{X}$, the constructs $\mu\, x \,.\, t$ (for least fixed-point) and $\nu\, x \,.\, t$ (for greatest fixed-point) are types over $\mathcal{X} - \{x\}$.

A *type interpretation* consists of

- an allegory $\mathbf{A}$
- a *object constant interpretation* $[\![\_]\!]_O$ that maps each object constant to an object of $\mathbf{A}$.
- a *functor symbol interpretation* $[\![\_]\!]_F$ that maps each n-ary functor symbol f to a meet-preserving n-ary endo-relator on $\mathbf{A}$, that is, to a meet-preserving relator $[\![f]\!]_F : \mathbf{A}^n \to \mathbf{A}$ from the n-ary product allegory of $\mathbf{A}$ to $\mathbf{A}$, where $\mathbf{A}^n$ is constructed from right-nested binary product allegory constructions.

The *type semantics* $[\![\_]\!]$ then maps each type t over an m-element (finite) sequence of distinct type variables $\langle x_1, \ldots x_m \rangle$ to an m-ary endo-relator $[\![t]\!] : \mathbf{A}^m \to \mathbf{A}$, and each sequence $t_1, \ldots, t_n$ of n types over an m-element sequence $\mathcal{X}$ of distinct type variables to a functor $[\![t_1, \ldots, t_n]\!] : \mathbf{A}^m \to \mathbf{A}^n$ as follows:

- For each type variable x_i from $\mathcal{X}$, $[\![x_i]\!] = \mathsf{Proj}_i$, where the i-th projection functor $\mathsf{Proj}_i : \mathbf{A}^m \to \mathbf{A}$ can be constructed from Proj_1 and Proj_2 according to the nesting of $\mathbf{A}^m$.
- For each object constant c, we have $[\![c]\!] = [\![c]\!]_O$.
- For each n-ary functor symbol f and each sequence $\langle t_1, \ldots, t_n \rangle$ of n types over $\mathcal{X}$:

$$[\![f(t_1, \ldots, t_n)]\!] = [\![f]\!]_F [\![\langle t_1, \ldots, t_n \rangle]\!]$$

- For each type variable x_i from $\mathcal{X}$ and each type t over $\mathcal{X}$:

$$[\![\mu\; x_i . \, t]\!] = T_{\mathsf{swap}_i ;[\![t]\!]} \qquad\qquad [\![\nu\; x_i \,.\, t]\!] = C_{\mathsf{swap}_i ;[\![t]\!]}$$

where swap_i is defined from projection functors and ∇ to implement the tuple permutation

$$\langle x_i, \quad x_1, \ldots, x_{i-1}, \quad x_{i+1}, \ldots, x_n \rangle \quad \mapsto \quad \langle x_1, \ldots, x_{i-1}, x_i, x_{i+1}, \ldots, x_n \rangle .$$

From Proposition 7.2, Theorems 7.3, and 7.4, it is then quite straightworward to obtain:

Theorem 11.1. For each type sequence $t_1, \ldots, t_n$ of n types over an m-element sequence of distinct type variables, the relator $[\![t_1, \ldots, t_n]\!] : \mathbf{A}^m \to \mathbf{A}^n$ is meet-preserving. □

Definition 11.2. A *raw coalgebra signature* $\Sigma = (\langle s_1, \ldots, s_n \rangle, \langle t_1, \ldots, t_n \rangle)$ over a type signature Θ is a sequence $\langle s_1, \ldots, s_n \rangle$ of *sorts* together with an equally-long sequence $\langle t_1, \ldots, t_n \rangle$ of types over $\langle s_1, \ldots, s_n \rangle$.

We define $\mathcal{R}_\Sigma := [\![\langle t_1, \ldots, t_n \rangle]\!]$ to be the *signature relator* for Σ. □

The signature relator $\mathcal{R}_\Sigma$ is by definition a meet-preserving endo-relator on $\mathbf{A}^n$.

We also enable a more readable presentation of coalgebras as follows:

Definition 11.3. A *coalgebra signature* $\Sigma = (\mathcal{S}, \mathcal{F})$ over a type signature Θ consists of

- a sequence $\mathcal{S} = \langle\ s_1, \ldots, s_n\ \rangle$ of *sorts*, and
- a sequence $\mathcal{F}$ of *function symbol signatures*, where each function symbol signature is a triple (f_j, i_j, t_j) written $f_j : s_{i_j} \to t_j$ and consisting of a *function symbol* f_j, a source sort index $i_j \in \{1, \ldots, n\}$, and a type t_j over $\mathcal{S}$. □

For example, consider again the signature for directed hypergraphs from Sect. 1:

$$\begin{aligned} \mathsf{sigDHG} := \langle \textbf{sorts: } \mathsf{N}, \mathsf{E}\ ; \quad \textbf{ops: } & \mathsf{src} : \mathsf{E} \to \mathsf{List}\ \mathsf{N} \\ & \mathsf{trg} : \mathsf{E} \to \mathsf{List}\ \mathsf{N}\ ; \quad \mathsf{nlab} : \mathsf{N} \to L \rangle \end{aligned}$$

The coalgebra functor corresponding to sigDHG is a functor between product categories, because of the two sorts:

$$F_{\mathsf{sigDHG}}\ (N\ ,\ E) \;=\; (L\ ,\ ((\mathsf{List}\ N) \times (\mathsf{List}\ N)))$$

For edges, the underlying translation into a raw coalgebra signature had to "collect" the two function symbols src and trg into the single type $\mathsf{List}\ N \times \mathsf{List}\ N$, where $\mathsf{List}\ N$ is really the initial algebra functor $\mu\ x\ .\ \mathbb{1} + N \times X$.

Working out the details of this translation from coalgebra signatures to raw coalgebra signature is straightforward, and we obtain:

Theorem 11.4. Let $\mathbf{A}$ be a distributive allegory, respectively a Kleene collagory. For every coalgebra signature Σ, the coalgebras over the signature relator $\mathcal{R}_\Sigma$ form a distributive allegory, respectively a Kleene collagory.

Proof. After translating Σ into a raw coalgebra signature Σ', the signature relator $\mathcal{R}_{\Sigma'}$ according to Definition 11.2 is meet-preserving according to Theorem 11.1. The statement then follows by Theorem 8.3, respectively by Theorem 8.4, □

Since Kleene collagories are sufficient for the formalisation of the relation-algebraic approach to graph transformation (Kahl 2001, 2010), this is immediately applicable to a wide range of coalgebras. Furthermore, we have:

Theorem 11.5. (Kahl 2011, Corollary 4.2.5) For a bi-tabular collagory **A** where all monos in the subcategory **M** of mappings are injective in **A**, the mapping category **M** is adhesive. □

Together with Theorem 11.4, this yields:

Corollary 11.6. Let **A** be a bi-tabular Kleene collagory where all monos in the subcategory of mappings are injective in **A**. For every coalgebra signature Σ, category of coalgebras over the signature relator $\mathcal{R}_\Sigma$ (obtained as the subcategory of mappings in the Kleene collagory of Theorem 11.4) is adhesive.

For example, the directed hypergraph category DHG constructed as the mapping category of sigDHG-coalgebras over the bitabular Kleene collagory *Rel* is exactly a node-labelled variant of the **HyperGraphs** category of (Ehrig et al. 2006, Fact 4.17) (where only adhesive HLR is argued, not adhesive). Whereas there, in the setting of algebras, the src and trg lists required special ad-hoc treatment, in our coalgebra setting we obtain full adhesiveness, and we obtain it as a simple instance of a general theorem. The resulting double-pushout rewriting concept is therefore exactly what would normally be expected for DPO rewriting of directed hypergraphs.

12 Conclusion and Outlook

We showed that "relational homomorphisms" between coalgebras over all polynomial functors, and additionally over functors constructed also using initial algebra and final coalgebra constructors, give rise to distributive allegories and Kleene collagories, and therewith make many relation-algebraic reasoning and specification tools available for coalgebras. Indirectly, this also makes the theorems of the adhesive approach to graph transformation available—this latter part still awaits formalisation in Agda.

However, even the theories already developed can, instantiated with appropriate base allegories, directly form the foundation and central parts of the implementation of an executable coalgebra-based graph transformation system, and of its correctness proofs.

References

Bird, R.S., de Moor, O.: Algebra of Programming. International Series in Computer Science, vol. 100. Prentice Hall, Upper Saddle River (1997)

de Roever, W.-P., Engelhardt, K.: Data Refinement: Model-Oriented Proof Methods and Their Comparison. Cambridge University Press, Cambridge (1998)

Ehrig, H., Habel, A., Kreowski, H.J., Parisi-Presicce, F.: From graph grammars to high level replacement systems. In: Ehrig, H., Kreowski, H.-J., Rozenberg, G. (eds.) Graph Grammars 1990. LNCS, vol. 532, pp. 269–287. Springer, Heidelberg (1991)

Ehrig, H., Ehrig, K., Prange, U., Taentzer, G.: Fundamentals of Algebraic Graph Transformation. Springer, Heidelberg (2006). doi:10.1007/3-540-31188-2

Freyd, P.J., Scedrov, A.: Categories, Allegories, vol. 39. North-Holland Mathematical Library, Amsterdam (1990)

Heckel, R., Küster, J.M., Taentzer, G.: Confluence of typed attributed graph transformation systems. In: Corradini, A., Ehrig, H., Kreowski, H.-J., Rozenberg, G. (eds.) ICGT 2002. LNCS, vol. 2505, pp. 161–176. Springer, Heidelberg (2002). doi:10.1007/3-540-45832-8_14

Kahl, W.: A relation-algebraic approach to graph structure transformation, 2001. Habil thesis, Fakultät für Informatik, Univ. der Bundeswehr München, Technical report 2002–03. http://relmics.mcmaster.ca/~kahl/Publications/RelRew/

Kahl, W.: Refactoring heterogeneous relation algebras around ordered categories and converse. J. Relational Methods Comput. Sci. **1**, 277–313 (2004). http://www.jormics.org/

Kahl, W.: Amalgamating pushout and pullback graph transformation in collagories. In: Ehrig, H., Rensink, A., Rozenberg, G., Schürr, A. (eds.) ICGT 2010. LNCS, vol. 6372, pp. 362–378. Springer, Heidelberg (2010). doi:10.1007/978-3-642-15928-2_24

Kahl, W.: Collagories: relation-algebraic reasoning for gluing constructions. J. Logic Algebraic Programming **80**(6), 297–338 (2011). doi:10.1016/j.jlap.2011.04.006

Kahl, W.: Categories of coalgebras with monadic homomorphisms. In: Bonsangue, M.M. (ed.) CMCS 2014. LNCS, vol. 8446, pp. 151–167. Springer, Heidelberg (2014). doi:10.1007/978-3-662-44124-4_9. Agda theories at http://RelMiCS.McMaster.ca/RATH-Agda/

Kahl, W.: Graph transformation with symbolic attributes via monadic coalgebra homomorphisms. ECEASST **71**, 5.1–5.17 (2015). doi:10.14279/tuj.eceasst.71.999

Kawahara, Y.: Notes on the universality of relational functors. Mem. Fac. Sci. Kyushu Univ. Ser. A **27**(2), 275–289 (1973)

Kozen, D.: A completeness theorem for Kleene algebras and the algebra of regular events. Inf. Comput. **110**(2), 366–390 (1994)

Kozen, D.: Typed Kleene algebra. Technical report 98–1669, Computer Science Department, Cornell University (1998)

Lack, S., Sobociński, P.: Adhesive categories. In: Walukiewicz, I. (ed.) FoSSaCS 2004. LNCS, vol. 2987, pp. 273–288. Springer, Heidelberg (2004). doi:10.1007/978-3-540-24727-2_20

Lack, S., Sobociński, P.: Adhesive and quasiadhesive categories. RAIRO Inform. Théor. Appl. **39**(3), 511–545 (2005). doi:10.1051/ita:2005028

Löwe, M.: Algebraic approach to single-pushout graph transformation. Theoret. Comput. Sci. **109**(1–2), 181–224 (1993). doi:10.1016/0304-3975(93)90068-5

Löwe, M., Korff, M., Wagner, A.: An algebraic framework for the transformation of attributed graphs. In: Sleep, M., Plasmeijer, M., van Eekelen, M. (eds.) Term Graph Rewriting: Theory and Practice, pp. 185–199. Wiley, Hoboken (1993)

Meijer, E., Fokkinga, M., Paterson, R.: Functional programming with bananas, lenses, envelopes and barbed wire. In: Hughes, J. (ed.) FPCA 1991. LNCS, vol. 523, pp. 124–144. Springer, Heidelberg (1991). doi:10.1007/3540543961_7

Norell, U.: Towards a practical programming language based on dependent type theory. Ph.D. thesis, Department of Computer Science and Engineering, Chalmers University of Technology (2007). See also http://wiki.portal.chalmers.se/agda/pmwiki.php

Schmidt, G., Ströhlein, T.: Relations and Graphs, Discrete Mathematics for Computer Scientists. EATCS-Monographs on Theoretical Computer Science. Springer, Heidelberg (1993)

Aggregation of Votes with Multiple Positions on Each Issue

Lefteris Kirousis[1](✉), Phokion G. Kolaitis[2], and John Livieratos[1]

[1] Department of Mathematics, National and Kapodistrian University of Athens, Athens, Greece
{lkirousis,jlivier89}@math.uoa.gr

[2] Computer Science Department, UC Santa Cruz and IBM Research - Almaden, Santa Cruz, CA, USA
kolaitis@cs.ucsc.edu

Abstract. We consider the problem of aggregating votes cast by a society on a fixed set of issues, where each member of the society may vote for one of several positions on each issue, but the combination of votes on the various issues is restricted to a set of feasible voting patterns. We require the aggregation to be supportive, i.e., for every issue, the corresponding component of every aggregator, when applied to a tuple of votes, must take as value one of the votes in that tuple. We prove that, in such a set-up, non-dictatorial aggregation of votes in a society of an arbitrary size is possible if and only if a non-dictatorial binary aggregator exists or a non-dictatorial ternary aggregator exists such that, for each issue, the corresponding component of the aggregator, when restricted to two-element sets of votes, is a majority operation or a minority operation. We then introduce a notion of a uniform non-dictatorial aggregator, which is an aggregator such that on every issue, and when restricted to arbitrary two-element subsets of the votes for that issue, differs from all projection functions. We first give a characterization of sets of feasible voting patterns that admit a uniform non-dictatorial aggregator. After this and by making use of Bulatov's dichotomy theorem for conservative constraint satisfaction problems, we connect social choice theory with the computational complexity of constraint satisfaction by proving that if a set of feasible voting patterns has a uniform non-dictatorial aggregator of some arity, then the multi-sorted conservative constraint satisfaction problem on that set (with each issue representing a different sort) is solvable in polynomial time; otherwise, it is NP-complete.

1 Introduction

Kenneth Arrow initiated the theory of aggregation by establishing his celebrated General Possibility Theorem (also known as Arrow's Impossibility Theorem) [1], which asserts that it is impossible, even under mild conditions, to aggregate in a non-dictatorial way the preferences of a society. Wilson [16] introduced aggregation on general attributes, rather than just preferences, and proved Arrow's result in this context. Later on, Dokow and Holzman [7] adopted a framework

P. Höfner et al. (Eds.): RAMiCS 2017, LNCS 10226, pp. 209–225, 2017.
DOI: 10.1007/978-3-319-57418-9_13

similar to Wilson's in which the voters have a binary position on a number of issues, and an individual voter's feasible position patterns are restricted to lie in a domain X. Dokow and Holzman discovered a necessary and sufficient condition for X to have a non-dictatorial aggregator that involves a property called *total blockedness*, which was originally introduced in [10]. Roughly speaking, a domain X is totally blocked if "any position on any issue can be deduced from any position on any issue" (the precise definition is given in Sect. 3). In other words, total blockedness is a property that refers to the propagation of individuals' positions from one issue to another.

After this, Dokow and Holzman [8] extended their earlier work by allowing the positions to be non-Boolean (non-binary). By generalizing the notion of a domain being totally blocked to the non-Boolean framework, they gave a sufficient (but not necessary) condition for non-dictatorial aggregation, namely, they showed that if a domain is not totally blocked, then it is a possibility domain. Recently, Szegedy and Xu [14] discovered necessary and sufficient conditions for non-dictatorial aggregation. Quite remarkably, their approach relates aggregation theory with universal algebra, specifically with the structure of the space of *polymorphisms*, that is, functions under which a relation is closed. It should be noted that properties of polymorphisms have been successfully used towards the delineation of the boundary between tractability and intractability for the Constraint Satisfaction Problem (for an overview, see, e.g., [6]).

Szegedy and Xu [14] distinguished the *supportive* (also known as *conservative*) case, where the social position must be equal to the position of at least one individual, from the *idempotent* (also known as *Paretian*) case, where the social position need not agree with any individual position, unless the votes are unanimous. In the idempotent case, they gave a necessary and sufficient condition for possibility of non-dictatorial aggregation that involves no propagation criterion (such as the domain being totally blocked), but only refers to the possibility of non-dictatorial aggregation for societies of a fixed cardinality (as large as the space of positions). In the supportive case, however, their necessary and sufficient conditions still involve the notion of the domain being totally blocked.

Here, we follow Szegedy and Xu's idea of deploying the algebraic "toolkit" [14] and we prove that, in the supportive case, non-dictatorial aggregation is possible for all societies of some cardinality if and only if a non-dictatorial binary aggregator exists or a non-dictatorial ternary aggregator exists such that on every issue j, the corresponding component f_j is a majority operation, i.e., for all x and y, it satisfies the equations

$$f_j(x, x, y) = f_j(x, y, x) = f_j(y, x, x) = x$$

or f_j is a minority operation, i.e., for all x and y, it satisfies the equations

$$f_j(x, x, y) = f_j(x, y, x) = f_j(y, x, x) = y.$$

(For additional information about the notions of majority and minority operations, see Szendrei [15, p. 24].)

We also show that a domain is totally blocked if and only if it admits no non-dictatorial binary aggregator; thus, the notion of a domain being totally blocked is, in a precise sense, a weak form of an impossibility domain.

After this, we introduce the notion of *uniform* non-dictatorial aggregator, which is an aggregator that on every issue, and when restricted to an arbitrary two-element subset of the votes for that issue, differs from all projection functions. We first give a characterization of sets of feasible voting patterns that admit uniform non-dictatorial aggregators. Then, making use of Bulatov's dichotomy theorem for conservative constraint satisfaction problems (see [2–4]), we connect social choice theory with the computational complexity of constraint satisfaction by proving that if a set of feasible voting patterns X has a uniform non-dictatorial aggregator of some arity, then the multi-sorted conservative constraint satisfaction problem on X, in the sense introduced by Bulatov and Jeavons [5], with each issue representing a sort, is tractable; otherwise it is NP-complete.

Due to space limitations, the proofs of almost all of our results were omitted. They can be found in [9].

2 Basic Concepts and Earlier Work

2.1 Basic Concepts

In all that follows, we have a fixed set $I = \{1, \ldots, m\}$ of issues. Let $\mathcal{A} = \{A_1, \ldots, A_m\}$ be a family of finite sets, each of cardinality at least 2, representing the possible positions (voting options) on the issues $1, \ldots, m$, respectively. If every A_j has cardinality exactly 2 (i.e., if for every issue only a "yes" or "no" vote is allowed), we say that we are in the *binary* or the *Boolean framework*; otherwise, we say that we are in the *non-binary* or the *non-Boolean* framework.

Let X be a non-empty subset of $\prod_{j=1}^{m} A_j$ that represents the feasible voting patterns. We write X_j, $j = 1 \ldots, m$, to denote the j-th projection of X. From now on, we assume that each X_j has cardinality at least 2 (this is a *non-degeneracy* condition). Throughout the rest of the paper, unless otherwise declared, X will denote a set of feasible voting patterns on m issues, as we just described.

Let $n \geq 2$ be an integer representing the number of voters. The elements of X^n can be viewed as $n \times m$ matrices, whose rows correspond to voters and whose columns correspond to issues. We write x_j^i to denote the entry of the matrix in row i and column j; clearly, it stands for the vote of voter i on issue j. The row vectors of such matrices will be denoted as $x^1, \ldots, x^n$, and the column vectors as $x_1, \ldots, x_m$.

Let now $\bar{f} = (f_1, \ldots, f_m)$ be an m-tuple of n-ary functions $f_j : A_j^n \mapsto A_j$.

An m-tuple of functions $\bar{f} = (f_1, \ldots, f_m)$ as above is called *supportive (conservative)* if for all $j = 1 \ldots m$, we have that:

$$\text{if } x_j = (x_j^1, \ldots, x_j^n) \in A_j^n, \text{ then } f_j(x_j) = f_j(x_j^1, \ldots, x_j^n) \in \{x_j^1, \ldots, x_j^n\}.$$

An m-tuple $\bar{f} = (f_1, \ldots, f_m)$ of (n-ary) functions as above is called an (n-ary) *aggregator for* X if it is supportive and, for all $j = 1, \ldots, m$ and for all $x_j \in A_j^n, j = 1, \ldots, m$, we have that:

$$\text{if } (x^1, \ldots, x^n) \in X^n, \text{ then } (f_1(x_1), \ldots, f_m(x_m)) \in X.$$

Note that $(x^1, \ldots, x^n)$ is an $n \times m$ matrix with rows $x^1, \ldots, x^n$ and columns $x_1, \ldots, x_m$, whereas $(f_1(x_1), \ldots, f_m(x_m))$ is a row vector required to be in X. The fact that aggregators are defined as m-tuples of functions $A_j^n \mapsto A_j$, rather than a single function $X^n \mapsto X$, reflects the fact that the social vote is assumed to be extracted issue-by-issue, i.e., the aggregate vote on each issue does not depend on voting data on other issues.

An aggregator $\bar{f} = (f_1, \ldots, f_m)$ is called *dictatorial on* X if there is a number $d \in \{1, \ldots, n\}$ such that $(f_1, \ldots f_m) \upharpoonright X = (\text{pr}_d^n, \ldots, \text{pr}_d^n) \upharpoonright X$, i.e., $(f_1, \ldots f_m)$ restricted to X is equal to $(\text{pr}_d^n, \ldots, \text{pr}_d^n)$ restricted to X, where pr_d^n is the n-ary projection on the d-th coordinate; otherwise, $\bar{f}$ is called *non-dictatorial on* X. We say that X *has a non-dictatorial aggregator* if, for some $n \geq 2$, there is a non-dictatorial n-ary aggregator on X.

A set X of feasible voting patterns is called a *possibility domain* if it has a non-dictatorial aggregator. Otherwise, it is called an *impossibility domain.* A possibility domain is, by definition, one where aggregation is possible for societies of some cardinality, namely, the arity of the non-dictatorial aggregator.

Aggregators do what their name indicates, that is, they aggregate positions on m issues, $j = 1, \ldots, m$, from data representing the voting patterns of n individuals on all issues. The fact that aggregators are assumed to be supportive (conservative) reflects the restriction of our model that the social vote for every issue should be equal to the vote cast on this issue by at least one individual. Finally, the requirement of non-dictatorialness for aggregators reflects the fact that the aggregate vote should not be extracted by adopting the vote of a single individual designated as a "dictator".

Example 1. Suppose that X is a cartesian product $X = Y \times Z$, where $Y \subseteq \prod_{j=1}^{l} A_j$ and $Z \subseteq \prod_{j=l+1}^{m} A_j$, with $1 \leq l < m$. It is easy to see that X is a possibility domain.

Indeed, for every $n \geq 2$, the set X has non-dictatorial n-ary aggregators of the form $(f_1, \ldots, f_l, f_{l+1}, \ldots, f_m)$, where for some d and d' with $d \neq d'$, we have $f_j = \text{pr}_d^n$, for $j = 1, \ldots, l$, and also $f_j = \text{pr}_{d'}^n$, for $j = l+1, \ldots, m$. Thus, every cartesian product of two sets of feasible patterns is a possibility domain. □

Now, following Szendrei [15, p. 24], we define the notions of a majority operation and of a minority operation.

Definition 1. *A ternary operation* $f : A^3 \mapsto A$ *on an arbitrary set* A *is a* majority *operation if for all* x *and* y *in* A,

$$f(x, x, y) = f(x, y, x) = f(y, x, x) = x,$$

and it is a minority *operation if for all* x *and* y *in* A,

$$f(x, x, y) = f(x, y, x) = f(y, x, x) = y.$$

We also define what it means for a set to admit a majority operation and a minority operation. (Since the arity of an aggregator is the arity of its component functions, a ternary aggregator is an aggregator with components of arity three.)

Definition 2. *Let X be a set of feasible voting patterns.*

- *X admits a* majority aggregator *if it admits a ternary aggregator $\bar{f} = (f_1, \ldots, f_m)$ such that f_j is a majority operation on X_j, for all $j = 1, \ldots, m$.*
- *X admits a* minority aggregator *if it admits a ternary aggregator $\bar{f} = (f_1, \ldots, f_m)$ such that f_j is a minority operation on X_j, for all $j = 1, \ldots, m$.*

Clearly, X admits a majority aggregator if and only if there is a ternary aggregator $\bar{f} = (f_1, \ldots, f_m)$ for X such that, for all $j = 1, \ldots, m$ and for all two-element subsets $B_j \subseteq X_j$, we have that $f_j \restriction B_j = \mathrm{maj}$, where

$$\mathrm{maj}(x, y, z) = \begin{cases} x & \text{if } x = y \text{ or } x = z, \\ y & \text{if } y = z. \end{cases}$$

Also, X admits a minority aggregator if and only if there is a ternary aggregator $\bar{f} = (f_1, \ldots, f_m)$ for X such that, for all $j = 1, \ldots, m$ and for all two-element subsets $B_j \subseteq X_j$, we have that $f_j \restriction B_j = \oplus$, where

$$\oplus(x, y, z) = \begin{cases} z & \text{if } x = y, \\ x & \text{if } y = z, \\ y & \text{if } x = z. \end{cases}$$

It is known that in the Boolean framework (in which for all issues only "yes" or "no" votes are allowed), a set X admits a majority aggregator if and only if X is a bijunctive logical relation, i.e., a subset of $\{0, 1\}^m$ that is the set of satisfying assignments of a 2CNF-formula. Moreover, X admits a minority aggregator if and only if X is an affine logical relation, i.e., a subset of $\{0, 1\}^m$ that is the set of solutions of linear equations over the two-element field (see Schaefer [13]).

Example 2. $X = \{(a, a, a), (b, b, b), (c, c, c), (a, b, b), (b, a, a), (a, a, c), (c, c, a)\}$ admits a majority aggregator.

To see this, let $\bar{f} = (f, f, f)$, where $f : \{a, b, c\} \rightarrow \{a, b, c\}$ is as follows:

$$f(u, v, w) = \begin{cases} a & \text{if } u,\ v, \text{ and } w \text{ are pairwise different;} \\ \mathrm{maj}(u, v, w) & \text{otherwise.} \end{cases}$$

Clearly, if B is a two-element subset of $\{a, b, c\}$, then $f \restriction B = \mathrm{maj}$. So, to show that X admits a majority aggregator, it remains to show that $\bar{f} = (f, f, f)$ is an aggregator for X. In turn, this amounts to showing that $\bar{f}$ is supportive and that X is closed under f. It is easy to check that $\bar{f}$ is supportive. To show that X is closed under f, let $x = (x_1, x_2, x_3), y = (y_1, y_2, y_3), z = (z_1, z_2, z_3)$ be three elements of X. We have to show that $(f(x_1, y_1, z_1), f(x_2, y_2, z_2), f(x_3, y_3, z_3))$ is also in X. The only case that needs to be considered is when x, y, and z are pairwise distinct. Several subcases need to be considered. For instance, if $x = (a, b, b)$, $y = (a, a, c)$, $z = (c, c, a)$, then $\bar{f}(x, y, z) = (f(a, a, c), f(b, a, c), f(b, c, a)) = (a, a, a) \in X$; the remaining combinations are left to the reader. □

Example 3. $X = \{(a, b, c), (b, a, a), (c, a, a)\}$ admits a minority aggregator.

To see this, let $\bar{f} = (f, f, f)$, where $f : \{a, b, c\} \to \{a, b, c\}$ is as follows:

$$f(u, v, w) = \begin{cases} a & \text{if } u,\ v, \text{ and } w \text{ are pairwise different;} \\ \oplus(u, v, w) & \text{otherwise.} \end{cases}$$

Clearly, if B is a two-element subset of $\{a, b, c\}$, then $f \restriction B = \oplus$. So, to show that X admits a minority aggregator, it remains to show that $\bar{f} = (f, f, f)$ is an aggregator for X. In turn, this amounts to showing that $\bar{f}$ is supportive and that X is closed under f. It is easy to check that $\bar{f}$ is supportive. To show that X is closed under f, let $x = (x_1, x_2, x_3), y = (y_1, y_2, y_3), z = (z_1, z_2, z_3)$ be three elements of X. We have to show that $(f(x_1, y_1, z_1), f(x_2, y_2, z_2), f(x_3, y_3, z_3))$ is also in X. The only case that needs to be considered is when x, y, and z are distinct, say, $x = (a, b, c)$, $y = (b, a, a)$, $z = (c, a, a)$. In this case, we have that $(f(a, b, c), f(b, a, a), f(c, a, a)) = (a, b, c) \in X$; Since f is not affected by permutations of the input, the proof is complete. □

So far, we have given examples of possibility domains only. Next, we give an example of an impossibility domain in the Boolean framework.

Example 4. Let $W = \{(1, 0, 0), (0, 1, 0), (0, 0, 1)\}$ be the 1-in-3 relation, i.e., the set of all Boolean tuples of length 3 in which exactly one 1 occurs.

We claim that W is an impossibility domain. It is not hard to show that W is not affine and that it does not admit a non-dictatorial binary aggregator. Theorem 2 in the next section implies that W is an impossibility domain. □

Every logical relation $X \subseteq \{0, 1\}^m$ gives rise to a generalized satisfiability problem in the context studied by Scheafer [13]. We point out that the property of X being a possibility domain in the Boolean framework is not related to the tractability of the associated generalized satisfiability problem. For example, the set W is an impossibility domain and its associated generalized satisfiability problem is the NP-complete problem POSITIVE 1-IN-3-SAT. As discussed earlier, the cartesian product $W \times W$ is a possibility domain. Using the results in [13], however, it can be verified that the generalized satisfiability problem arising from $W \times W$ is NP-complete. At the same time, the set $\{0, 1\}^m$ is trivially a possibility domain and gives rise to a trivially tractable satisfiability problem. Thus, the property of X being a possibility domain is not related to the tractability of the generalized satisfiability problem arising from X.

Nonetheless, in Sect. 3 we establish the equivalence between the stronger notion of X being a *uniform possibility domain* and the weaker notion of the tractability of the *multi-sorted* generalized satisfiability problem arising from X, where each issue is taken as a different sort. Actually, we establish this equivalence not only for satisfiability problems but also for constraint satisfaction problems whose variables range over arbitrary finite sets.

2.2 Earlier Work

There has been a significant body of earlier work on possibility domains. Here, we summarize some of the results that relate the notion of a possibility domain to the notion of a set being *totally blocked*, a notion originally introduced in the context of the Boolean framework by Nehring and Puppe [10]. As stated earlier, a set X of possible voting patterns is totally blocked if, intuitively, "any position on any issue can be deduced from any position on any issue"; this intuition is formalized by asserting that a certain directed graph G_X associated with X is strongly connected. The precise definition of this notion is given in Sect. 3.

In the case of the Boolean framework, Dokow and Holzman [7] obtained the following necessary and sufficient condition for a set to be a possibility domain.

Theorem A (Dokow and Holzman [7, Theorem 2.2]). *Let $X \subseteq \{0,1\}^m$ be a set of feasible voting patterns. The following statements are equivalent.*

- *X is a possibility domain.*
- *X is affine or X is not totally blocked.*

For the non-Boolean framework, Dokow and Holzman [8] found the following connection between the notions of totally blocked and possibility domain.

Theorem B (Dokow and Holzman [8, Theorem 2]). *Let X be a set of feasible voting patterns. If X is not totally blocked, then X is a possibility domain; in fact, there is a non-dictatorial n-ary aggregator, for every $n \geq 2$.*

Note that, in the case of the Boolean framework, Theorem B was stated and proved as Claim 3.6 in [7].

For the non-Boolean framework, Szegedy and Xu [14] obtained a sufficient and necessary condition for a totally blocked set X to be a possibility domain.

Theorem C (Szegedy and Xu [14, Theorem 8]). *Let X be a set of feasible voting patterns that is totally blocked. The following statements are equivalent.*

- *X is a possibility domain.*
- *X admits a binary non-dictatorial aggregator or a ternary non-dictatorial aggregator.*

Note that, in the case of the Boolean framework, Theorem C follows from the preceding Theorem A (Theorem 2.2 in [7]).

A binary non-dictatorial aggregator can also be viewed as a ternary one, where one of the arguments is ignored. By considering whether or not X is totally blocked, Theorems B and C imply the following corollary, which characterizes possibility domains without involving the notion of total blockedness; to the best of our knowledge, this result has not been explicitly stated previously.

Corollary 1. *Let X be a set of feasible voting patterns. The following statements are equivalent.*

1. *X is a possibility domain.*
2. *X has a non-dictatorial binary aggregator or a non-dictatorial ternary aggregator.*
3. *X has a non-dictatorial ternary aggregator.*

3 Results

3.1 Possibility Domains

Our first result is a necessary and sufficient condition for a set of feasible voting patterns to be a possibility domain (for the proof see [9]).

Theorem 1. *Let X be a set of feasible voting patterns. The following statements are equivalent.*

1. *X is a possibility domain.*
2. *X admits a majority aggregator or it admits a minority aggregator or it has a non-dictatorial binary aggregator.*

Theorem 1 is stronger than the preceding Corollary 1 because, unlike Corollary 1, it gives explicit information about the nature of the components f_j of non-dictatorial ternary aggregators $\bar{f} = (f_1, \ldots, f_m)$, when the components are restricted to a two-element subset $B_j \subseteq X_j$ of the set of positions on issue j, information that is necessary to relate results in aggregation theory with complexity theoretic results (besides the three projections, there are 61 supportive ternary functions on a two element set). Observe also that if $\bar{f} = (f_1, \ldots, f_m)$ is a binary aggregator, then every component f_j is necessarily a projection function or the function $\wedge$ or the function $\vee$, when restricted to a two-element subset $B_j \subseteq X_j$ (identified with the set $\{0, 1\}$). So, for binary aggregators, the information about the nature of their components is given *gratis*.

Only the direction $1 \Longrightarrow 2$ of Theorem 1 requires proof. Towards this goal, we introduce a new notion, state three lemmas whose proofs can be found in [9], and then use them to prove Theorem 1.

Let X be a set of feasible voting patterns and let $\bar{f} = (f_1, \ldots, f_m)$ be an n-ary aggregator for X.

Definition 3. *We say that $\bar{f}$ is* locally monomorphic *if for all indices i and j with $1 \leq i, j \leq m$, for all two-element subsets $B_i \subseteq X_i$ and $B_j \subseteq X_j$, for every bijection $g : B_i \mapsto B_j$, and for all column vectors $x_i = (x_i^1, \ldots, x_i^n) \in B_i^n$, we have that*

$$f_j(g(x_i^1), \ldots, g(x_i^n)) = g(f_i(x_i^1, \ldots, x_i^n)).$$

Intuitively, the above definition says that, no matter how we identify the two elements of B_i and B_j with 0 and 1, the restrictions $f_i \restriction B_i$ and $f_j \restriction B_j$ are equal as functions.

The first lemma gives a sufficient condition for all aggregators of all arities to be locally monomorphic.

Lemma 1. *Let X be a set of feasible voting patterns. If every binary aggregator for X is dictatorial on X, then, for every $n \geq 2$, every n-ary aggregator for X is locally monomorphic.*

Next, we state a technical lemma whose proof was inspired by a proof in Dokow and Holzman [8, Proposition 5].

Lemma 2. *Assume that for all integers $n \geq 2$ and for every n-ary aggregator $\bar{f} = (f_1, \ldots, f_m)$, there is an integer $d \leq n$ such that for every integer $j \leq m$ and every two-element subset $B_j \subseteq X_j$, the restriction $f_j \restriction B_j$ is equal to pr_d^n, the n-ary projection on the d-th coordinate. Then for all integers $n \geq 2$ and for every n-ary aggregator $\bar{f} = (f_1, \ldots, f_m)$ and for all $s \geq 2$, there is an integer $d \leq n$ such that for every integer $j \leq m$ and every subset $B_j \subseteq X_j$ of cardinality at most s, the restriction $f_j \restriction B_j$ is equal to pr_d^n.*

Next, we bring into the picture some basic concepts and results from universal algebra; we refer the reader to Szendrei's monograph [15] for additional information and background. A *clone* on a finite set A is a set $\mathcal{C}$ of finitary operations on A (i.e., functions from a power of A to A) such that $\mathcal{C}$ contains all projection functions and is closed under arbitrary compositions (superpositions). The proof of the next lemma is straightforward.

Lemma 3. *Let X be a set of feasible voting patterns. For every j with $1 \leq j \leq m$ and every subset $B_j \subseteq X_j$, the set $\mathcal{C}_{B_j}$ of the restrictions $f_j \restriction B_j$ of the j-th components of aggregators $\bar{f} = (f_1, \ldots, f_m)$ for X is a clone on B_j.*

Post [12] classified all clones on a two-element set (for more recent expositions of Post's pioneering results, see, e.g., [15] or [11]). One of Post's main findings is that if $\mathcal{C}$ is a clone of conservative functions on a two-element set, then either $\mathcal{C}$ contains only projection functions or $\mathcal{C}$ contains one of the following operations: the binary operation $\wedge$, the binary operation $\vee$, the ternary operation $\oplus$, the ternary operation maj. We use this result below.

Proof (Proof of Theorem 1). As stated earlier, only the direction $1 \Longrightarrow 2$ requires proof. In the contrapositive, we will prove that if X does not admit a majority or a minority aggregator, and it does not admit a non-dictatorial binary aggregator, then X does not have an n-ary non-dictatorial aggregator, for any n. Towards this goal, and assuming that X is as stated, we will first show that the hypothesis of Lemma 2 holds. Once this is established, the conclusion will follow from Lemma 2 by taking $s = \max\{|X_j| : 1 \leq j \leq m\}$.

Given $j \leq m$ and a two-element subset $B_j \subseteq X_j$, consider the clone $\mathcal{C}_{B_j}$. If $\mathcal{C}_{B_j}$ contained one of the binary operations $\wedge$ or $\vee$ then X would have a binary non-dictatorial aggregator, a contradiction. If, on the other hand, $\mathcal{C}_{B_j}$ contained the ternary operation $\oplus$ or the ternary operation maj, then, by Lemma 1, X would admit a minority or a majority aggregator, a contradiction as well. So, by the aforementioned Post's result, all elements of $\mathcal{C}_{B_j}$, no matter what their arity is, are projection functions. By Lemma 1 again, since X has no binary non-dictatorial aggregator, we have that for every n and for every n-ary aggregator

$\bar{f} = (f_1, \ldots, f_m)$, there exists an integer $d \leq n$ such that for every $j \leq m$ and every two-element set $B_j \subseteq X_j$, the restriction $f_j \restriction B_j$ is equal to pr_d^n, the n-ary projection on the d-th coordinate. This concludes the proof of Theorem 1. □

In the case of the Boolean framework, Theorem 1 takes the stronger form of Theorem 2 below. Although this result for the Boolean framework is implicit in Dokow and Holzman [7], we give an independent proof in [9].

Theorem 2 (Dokow and Holzman). *Let $X \subseteq \{0,1\}^m$ be a set of feasible voting patterns. The following statements are equivalent.*

1. *X is a possibility domain.*
2. *X is affine (i.e., X admits a minority aggregator) or X has a non-dictatorial binary aggregator.*

As discussed in the preceding section, much of the earlier work on possibility domains used the notion of a set being totally blocked. Our next result characterizes this notion in terms of binary aggregators and, in many respects, "explains" the role of this notion in the earlier results about possibility domains.

We begin by giving the precise definition of what it means for a set X of feasible voting patterns to be totally blocked. We will follow closely the notation and terminology used by Dokow and Holzman [8].

Let X be a set of feasible voting patterns.

- Given subsets $B_j \subseteq X_j, j = 1, \ldots, m$, the product $B = \prod_{j=1}^m B_j$ is called a *sub-box*. It is called a *2-sub-box* if $|B_j| = 2$, for all j.
  Elements of a box B that belong also to X will be called *feasible evaluations within* B (in the sense that each issue $j = 1, \ldots, m$ is "evaluated" within B).
- Let K be a subset of $\{1, \ldots, m\}$ and let x be a tuple in $\prod_{j \in K} B_j$.
  We say that x is a *feasible partial evaluation within* B if there exists a feasible $y \in B$ that extends x, i.e. $x_j = y_j$, for all $j \in K$; otherwise, we say that x is an *infeasible partial evaluation within* B.
  We say that x is a *B-Minimal Infeasible Partial Evaluation* (B-MIPE) if x is an infeasible partial evaluation within B and if for every $j \in K$, there is a $b_j \in B_j$ such that changing the j-th coordinate of x to b_j results into a feasible partial evaluation within B.
- We define a directed graph G_X as follows.
  The vertices of G_X are the pairs of *distinct* elements u, u' in X_j, for all $j = 1, \ldots m$. Each such vertex is denoted by uu'_j.
  Two vertices uu'_k, vv'_l with $k \neq l$ are connected by a directed edge from uu'_k to vv'_l if there exists a 2-sub-box $B = \prod_{j=1}^m B_j$, a $K \subseteq \{1, \ldots, m\}$ and a B-MIPE $x = (x_j)_{j \in K}$ such that $k, l \in K$ and $B_k = \{u, u'\}$ and $B_l = \{v, v'\}$ and $x_k = u$ and $x_l = v'$. Each such directed edge is denoted by $uu'_k \underset{B,x,K}{\longrightarrow} vv'_l$ (or just $uu'_k \rightarrow vv'_l$, in case B, x, K are understood from the context).
- We say that X is *totally blocked* if the graph G_X is strongly connected, i.e., every two distinct vertices uu'_k, vv'_l are connected by a directed path (this must hold even if $k = l$). This notion, defined in Dokow and Holzman [8],

is a generalization to the case where the A_j's are allowed to have arbitrary cardinalities of a corresponding notion for the Boolean framework (every A_j has cardinality 2), originally given in [10].

We are now ready to state the following result (for the proof see [9]).

Theorem 3. *Let X be a set of feasible voting patterns. The following statements are equivalent.*

1. *X is totally blocked.*
2. *X has no non-dictatorial binary aggregator.*

Observe that Theorem 2 is also an immediate consequence of Theorems A and 3. In view of Theorem B by Dokow and Holzman [8], only the direction $1 \Longrightarrow 2$ of Theorem 3 requires proof. In [9], we prove both directions of Theorem 3 for completeness.

Before proceeding further, we point out that the three types of non-dictatorial aggregators in Theorem 1 are, in a precise sense, independent of each other.

Example 5. Consider the set $X = \{0,1\}^3 \setminus \{(1,1,0)\}$ of satisfying assignments of the Horn clause $(\neg x \vee \neg y \vee z)$.

It is easy to see that X is closed under the binary operation $\wedge$, but it is not closed under the ternary majority operation maj or the ternary minority operation $\oplus$.

Thus, X is a possibility domain admitting a non-dictatorial binary aggregator, but not a majority aggregator or a minority aggregator. □

Example 6. Consider the set $X = \{(0,0,1),(0,1,0),(1,0,0),(1,1,1)\}$ of solutions of the equation $x + y + z = 1$ over the two-element field.

It is easy to see that X is closed under the ternary minority operation $\oplus$, but it is not closed under the ternary majority operation maj. Moreover, Dokow and Holzman [7, Example 3] pointed out that X is totally blocked, hence Theorem 3 implies that X does not admit a non-dictatorial binary aggregator.

Thus, X is a possibility domain admitting a minority aggregator, but not a majority aggregator or a non-dictatorial binary aggregator. □

Example 7. Consider the set $X = \{(0,1,2),(1,2,0),(2,0,1),(0,0,0)\}$.

This set was studied in [8, Example 4]. It can be shown that X admits a majority aggregator. To see this, consider the ternary operator $\overline{f} = (f_1, f_2, f_3)$ such that $f_j(x,y,z)$ is the majority of x, y, z, if at least two of the three values are equal, or it is 0 otherwise. Notice that in the latter case the value 0 must be one of the x, y, z, so this operator is indeed supportive. It is easy to verify that X is closed under (f_1, f_2, f_3). Moreover, if one of the f_j's is restricted to a two-element domain (i.e., to one of $\{0,1\}$, $\{(1,2)\}$, $\{0,2\}$), then it must be the majority function by its definition, so $\overline{f}$ is indeed a majority aggregator on X.

Dokow and Holzman argued that X is totally blocked, hence Theorem 3 implies that X does not admit a non-dictatorial binary aggregator.

Next, we claim that X does not admit a minority aggregator. Towards a contradiction, assume it admits the minority aggregator $\overline{g} = (g_1, g_2, g_3)$. By applying $\overline{g}$ to the triples $(0,1,2)$, $(1,2,0)$, $(0,0,0)$ in X, we infer that the triple $(g_1(0,1,0)$, $g_2(1,2,0)$, $g_3(2,0,0))$ must be in X. By the assumption that this aggregator is the minority operator on two-element domains, we have that $g_1(0,1,0) = 1$ and $g_3(2,0,0) = 2$, so X contains a triple of the form $(1, g_2(1,2,0), 2)$; however, X contains no triple whose first coordinate is 1 and its third coordinate is 2, so we have arrived at a contradiction.

Thus, X is a possibility domain admitting a majority aggregator, but not a minority aggregator or a non-dictatorial binary aggregator. □

Observe that the possibility domains in Examples 5 and 6 are in the Boolean framework, while the possibility domain in Example 7 is not. This is no accident, because it turns out that, in the Boolean framework, if a set admits a majority aggregator, then it also admits a non-dictatorial binary aggregator. This property is shown as a Claim in the proof of Theorem 2 in [9]. Note also that this explains why admitting a majority aggregator is not part of the characterization of possibility domains in the Boolean framework in Theorem 2.

3.2 Uniform Possibility Domains

In this Section, we connect aggregation theory with multi-sorted constraint satisfaction problems. Towards this goal, we introduce the following stronger notion of a non-dictatorial aggregator.

Definition 4. *Let X be a set of feasible voting patterns.*

- *We say that an aggregator $\overline{f} = (f_1, \ldots, f_m)$ for X is* uniform non-dictatorial *if for every $j = 1, \ldots, m$ and every two-element subset $B_j \subseteq X_j$, we have that $f_j \restriction B_j$ is not a projection function.*
- *We say that X is a* uniform possibility domain *if X admits a uniform non-dictatorial aggregator of some arity.*

The next example shows that the notion of a uniform possibility domain is stricter than the notion of a possibility domain.

Example 8. Let $W = \{(1,0,0), (0,1,0), (0,0,1)\}$ be the 1-in-3 relation, considered in Example 4. As seen earlier, the cartesian product $W \times W$ is a possibility domain. We claim that $W \times W$ is not a uniform possibility domain in the sense of Definition 4. Indeed, since W is an impossibility domain, it follows easily that for every n, all n-ary aggregators of $W \times W$ are of the form

$$(pr^n_d, pr^n_d, pr^n_d, pr^n_{d'}, pr^n_{d'}, pr^n_{d'}), \text{ for } d, d' \in \{1, \ldots, n\}. \qquad \square$$

It is obvious that every set X that admits a majority aggregator or a minority aggregator is a uniform possibility domain. The next example states that uniform possibility domains are closed under cartesian products.

Example 9. If X and Y are uniform possibility domains, then so is their cartesian product $X \times Y$.

Assume that $X \subseteq \prod_{j=1}^{l} A_j$ and $Z \subseteq \prod_{j=l+1}^{m} A_j$, where $1 \leq l < m$. Let $(f_1, \ldots, f_l)$ be a uniform non-dictatorial aggregator for X and let $(f_{l+1} \ldots, f_m)$ be a uniform non-dictatorial aggregator for X. Then

$$(f_1, \ldots, f_l, f_{l+1}, \ldots, f_m)$$

is a uniform non-dictatorial aggregator for $X \times Y$. □

Let B be an arbitrary two-element set, viewed as the set $\{0, 1\}$, and consider the binary logical operations $\wedge$ and $\vee$ on B (since we will always deal with both these logical operations concurrently, it does not matter which element of B we take as 0 and which as 1). For notational convenience. we define two ternary operations on B as follows:

$$\wedge^{(3)}(x, y, z) = x \wedge y \wedge z \quad \text{and} \quad \vee^{(3)}(x, y, z) = x \vee y \vee z.$$

We now state the following result (for the proof see [9]).

Theorem 4. *Let X be a set of feasible voting patterns. The following statements are equivalent.*

1. *X is a uniform possibility domain.*
2. *For every $j = 1, \ldots, m$ and for every two-element subset $B_j \subseteq X_j$, there is an aggregator $\bar{f} = (f_1, \ldots, f_m)$ (that depends on j and B_j) of some arity such that $f_j \restriction B_j$ is not a projection function.*
3. *There is a ternary aggregator $\bar{f} = (f_1, \ldots, f_m)$ such that for all $j = 1, \ldots, m$ and all two-element subsets $B_j \subseteq X_j$, we have that $f_j \restriction B_j$ is one of the ternary operations $\wedge^{(3)}$, $\vee^{(3)}$, maj, $\oplus$ (to which of these four ternary operations the restriction $f_j \restriction B_j$ is equal to depends on j and B_j).*
4. *There is a ternary aggregator $\bar{f} = (f_1, \ldots, f_m)$ such that for all $j = 1, \ldots, m$ and all $x, y \in X_j$, we have that $f_j(x, y, y) = f_j(y, x, y) = f_j(y, y, x)$.*

See also the related result by Bulatov on "three basic operations" [3, Proposition 3.1], [4, Proposition 2.2] (that result however considers only operations of arity two or three). Some of the techniques employed in the proof of Theorem 4 (see [9]) had been used in the aforementioned works by Bulatov.[1]

To state our result that connects the property of X being a uniform possibility domain with the property of tractability of a multi-sorted constraint satisfaction problems, we first introduce some notions following closely [3,5].

As before, we consider a fixed set $I = \{1, , \ldots, m\}$, but this time I represents *sorts*. We also consider a family $\mathcal{A} = \{A_1, \ldots, A_m\}$ of finite sets, each of cardinality at least 2, representing the values the corresponding sorts can take.

[1] This came to the attention of the authors only after the work reported here had been essentially completed.

- Let $(i_1, \ldots, i_k)$ be a list of (not necessarily distinct) indices from I. A *multi-sorted relation* over $\mathcal{A}$ with arity k and signature $(i_1, \ldots, i_k)$ is a subset R of $A_{i_1} \times \cdots \times A_{i_k}$, together with the list $(i_1, \ldots, i_k)$. The signature of such a multi-sorted language R will be denoted $\sigma(R)$.
- A *multi-sorted constraint language* Γ over $\mathcal{A}$ is a set of multi-sorted relations over $\mathcal{A}$.

Definition 5 (Multi-sorted CSP). *Let Γ be a multi-sorted constraint language over a family $\mathcal{A} = \{A_1, \ldots, A_m\}$ of finite sets. The multi-sorted constraint satisfaction problem* MCSP(Γ) *is the following decision problem.*

An instance of MCSP(Γ) *is a quadruple $(V, \mathcal{A}, \delta, \mathcal{C})$, where V is a finite set of variables; δ is a mapping from V to I, called the sort-assignment function (v belongs to the sort $\delta(v)$); $\mathcal{C}$ is a set of constraints where each constraint $C \in \mathcal{C}$ is a pair (s, R), such that $s = (v_1, \ldots, v_k)$ is a tuple of variables of length k, called the constraint scope; R is a k-ary multi-sorted relation over $\mathcal{A}$ with signature $(\delta(v_1), \ldots, \delta(v_k))$, called the constraint relation.*

The question is whether a value-assignment exists, i.e., a mapping $\phi : V \mapsto \bigcup_{i=1}^{m} A_i$, such that, for each variable $v \in V$, we have that $\phi(v) \in A_{\delta(v)}$, and for each constraint $(s, R) \in \mathcal{C}$, with $s = (v_1, \ldots, v_k)$, we have that the tuple $(\phi(v_1), \ldots, \phi(v_k))$ belongs to R.

A multi-sorted constraint language Γ over $\mathcal{A}$ is called *conservative* if for all sets $A_j \in \mathcal{A}$ and all subsets $B \subseteq A_j$, we have that $B \in \Gamma$ (as a relation over A_j).

If $X \subseteq \prod_{j=1}^{m} A_j$ is a set of feasible voting patterns, then X can be considered as multi-sorted relation with signature $(1, \ldots, m)$ (one sort for each issue). We write Γ_X^{cons} to denote the multi-sorted conservative constraint language consisting of X and all subsets of every $A_j, j = 1, \ldots, m$, the latter considered as relations over A_j.

If the sets A_j are equal to each other and $|I| = 1$, i.e., if there is no differentiation between sorts, then MCSP(Γ) is denoted the constraint satisfaction problem CSP(Γ). If the sets of votes for all issues are equal, then it is possible to consider a feasible set of votes X as a one-sorted relation (all issues are of the same sort). In this framework, and in case all A_j's are equal to $\{0, 1\}$, we have that CSP(Γ_X^{cons}) coincides with the problem introduced by Schaefer [13], which he called the "generalized satisfiability problem with constants" and denoted by $\text{SAT}_\text{C}(\{X\})$. Note that the presence of the sets $\{0\}$ and $\{1\}$ in the constraint language amounts to allowing constants, besides variables, in the constraints.

Schaefer [13] proved a prototypical dichotomy theorem for the complexity of the generalized satisfiability problem with constants. Bulatov [3, Theorem 2.16] proved a dichotomy theorem for conservative multi-sorted constraint languages We now state the following dichotomy theorem.

Theorem 5. *If X is a uniform possibility domain, then* MCSP(Γ_X^{cons}) *is solvable in polynomial time; otherwise it is* NP*-complete.*

Theorem 5 is obtained by combining Theorem 4 with the aforementioned Bulatov's dichotomy theorem for conservative multi-sorted constraint languages

[3, Theorem 2.16] (an exact statement of Bulatov's dichotomy theorem tailored to our needs as well as a complete proof of Theorem 5 are given in [9]).

Example 10. Let $Y = \{0,1\}^3 \setminus \{(1,1,0)\}$ be the set of satisfying assignments of the clause $(\neg x \vee \neg y \vee z)$ and let $Z = \{(1,1,0),(0,1,1),(1,0,1),(0,0,0)\}$ be the set of solutions of the equation $x+y+z=0$ over the two-element field.

We claim that Y and Z are uniform possibility domains, hence, by Example 9, the cartesian product $X = Y \times Z$ is also a uniform possibility domain. From Theorem 5, it follows that $\text{MCSP}(\Gamma_X^{\text{cons}})$ is solvable in polynomial time. However, the generalized satisfiability problem with constants $\text{SAT}_\text{C}(\{X\})$ (equivalently $\text{CSP}(\Gamma_X^{\text{cons}})$) is NP-complete.

Indeed, in Schaefer's [13] terminology, the set Y is Horn (equivalently, it is coordinate-wise closed under $\wedge$); however, it is not dual Horn (equivalently, it is not coordinate-wise closed under $\vee$), nor affine (equivalently, it does not admit a minority aggregator) nor bijunctive (equivalently, it does not admit a majority aggregator). Therefore, by coordinate-wise closure under $\wedge$, we have that Y is a uniform possibility domain. Also, Z is affine, but not Horn, nor dual Horn neither bijunctive. So, being affine, Z is a uniform possibility domain. The NP-completeness of $\text{SAT}_\text{C}(\{X\})$ (equivalently, the NP-completeness of $\text{CSP}(\Gamma_X^{\text{cons}})$) follows from Schaefer's dichotomy theorem [13], because X is not Horn, dual Horn, affine, nor bijunctive. □

4 Concluding Remarks

In this paper, we used algebraic tools to investigate the structural properties of possibility domains, that is, domains that admit non-dictatorial aggregators. We also established a connection between the stronger notion of a uniform possibility domain and multi-sorted constraint satisfaction. We conclude by discussing two algorithmic problems that underlie the notions of a possibility domain and a uniform possibility domain.

Given a family $\mathcal{A} = \{A_1, \ldots, A_m\}$ and a subset $X \subseteq \prod_{j=1}^m A_j$ as input, adopting a terminology used in computational complexity theory, we call *meta-problems* the following two questions:

(i) Is X a possibility domain?
(ii) Is X a uniform possibility domain?

Theorem 1 (in fact, even Corollary 1) and, respectively, Theorem 4, easily imply that the meta-problem (i) and, respectively, the meta-problem (ii), is in NP. Indeed, we only have to guess suitable ternary or binary operations and check for closure. However, even if the sizes of all A_j's are bounded by a constant (but the number m of issues/sorts, is unbounded), it is conceivable that the problems are not in polynomial time, as there are exponentially many ternary or binary aggregators. The question of pinpointing the exact complexity of these two meta-problems is the object of ongoing research. Of course, if, besides the cardinality of all sets A_j, their number m is also bounded, then Theorem 1 (in fact, even

Corollary 1) and respectively, Theorem 4 imply that the meta-problem (i) and, respectively, the meta-problem (ii)) is solvable in polynomial time (for the first meta-problem, this was essentially observed by Szegedy and Xu [14]). Note that, in the preceding considerations, it is assumed that X is given by listing explicitly its elements. If X is given implicitly in a succinct way (e.g., as the set of satisfying assignments of a given Boolean formula), then the upper bound for the meta-problems is higher. The exact complexity of the aforementioned meta-problems with X represented succinctly remains to be investigated.

Acknowledgments. We are grateful to Mario Szegedy for sharing with us an early draft of his work on impossibility theorems and the algebraic toolkit. We are also grateful to Andrei Bulatov for bringing to our attention his "three basic operations" proposition [3, Proposition 3.1], [4, Proposition 2.2]. We sincerely thank the anonymous reviewers of RAMiCS 2017 for their very helpful comments.

Part of this research was carried out while Lefteris Kirousis was visiting the Computer Science Department of UC Santa Cruz during his sabbatical leave from the National and Kapodistrian University of Athens in 2015. Part of the research and the writing of this paper was done while Phokion G. Kolaitis was visiting the Simons Institute of Theory of Computing in the fall of 2016. Lefteris Kirousis' participation to RAMICS 2017 was funded by the Special Account for Research Grants of the National and Kapodistrian University of Athens.

References

1. Arrow, K.J.: Social Choice and Individual Values. Wiley, New York (1951)
2. Barto, L.: The dichotomy for conservative constraint satisfaction problems revisited. In: Proceedings of the 26th Annual IEEE Symposium on Logic in Computer Science (LICS), pp. 301–310. IEEE (2011)
3. Bulatov, A.A.: Complexity of conservative constraint satisfaction problems. ACM Trans. Comput. Logic (TOCL) **12**(4) (2011). Article No. 24
4. Bulatov, A.A.: Conservative constraint satisfaction re-revisited. J. Comput. Syst. Sci. **82**(2), 347–356 (2016)
5. Bulatov, A.A., Jeavons, P.: An algebraic approach to multi-sorted constraints. In: Rossi, F. (ed.) CP 2003. LNCS, vol. 2833, pp. 183–198. Springer, Heidelberg (2003). doi:10.1007/978-3-540-45193-8_13
6. Bulatov, A.A., Valeriote, M.A.: Recent results on the algebraic approach to the CSP. In: Creignou, N., Kolaitis, P.G., Vollmer, H. (eds.) Complexity of Constraints: An Overview of Current Research Themes. LNCS, vol. 5250, pp. 68–92. Springer, Heidelberg (2008). doi:10.1007/978-3-540-92800-3_4
7. Dokow, E., Holzman, R.: Aggregation of binary evaluations. J. Econ. Theory **145**(2), 495–511 (2010)
8. Dokow, E., Holzman, R.: Aggregation of non-binary evaluations. Adv. Appl. Math. **45**(4), 487–504 (2010)
9. Kirousis, L.M., Kolaitis, P.G., Livieratos, J.: Aggregation of votes with multiple positions on each issue. CoRR, abs/1505.07737v2 (2016)
10. Nehring, K., Puppe, C.: Strategy-proof social choice on single-peaked domains: possibility, impossibility and the space between (2002). University of California at Davis. http://vwl1.ets.kit.edu/puppe.php

11. Pelletier, F.J., Martin, N.M.: Post's functional completeness theorem. Notre Dame J. Formal Logic **31**(2), 462–475 (1990)
12. Post, E.L.: The Two-Valued Iterative Systems of Mathematical Logic. Annals of Mathematics Studies, vol. 5. Princeton University Press, Princeton (1941)
13. Schaefer, T.J.: The complexity of satisfiability problems. In: Proceedings of the Tenth Annual ACM Symposium on Theory of Computing, pp. 216–226. ACM (1978)
14. Szegedy, M., Xu, Y.: Impossibility theorems and the universal algebraic toolkit. CoRR, abs/1506.01315 (2015)
15. Szendrei, Á.: Clones in Universal Algebra, vol. 99. Presses de l'Université de Montréal, Montréal (1986)
16. Wilson, R.: On the theory of aggregation. J. Econ. Theory **10**(1), 89–99 (1975)

Complete Solution of an Optimization Problem in Tropical Semifield

Nikolai Krivulin(✉)

Saint Petersburg State University, Saint Petersburg 199034, Russia
nkk@math.spbu.ru

Abstract. We consider a multidimensional optimization problem that is formulated in the framework of tropical mathematics to minimize a function defined on vectors over a tropical semifield (a semiring with idempotent addition and invertible multiplication). The function, given by a matrix and calculated through a multiplicative conjugate transposition, is nonlinear in the tropical mathematics sense. We show that all solutions of the problem satisfy a vector inequality, and then use this inequality to establish characteristic properties of the solution set. We examine the problem when the matrix is irreducible. We derive the minimum value in the problem, and find a set of solutions. The results are then extended to the case of arbitrary matrices. Furthermore, we represent all solutions of the problem as a family of subsets, each defined by a matrix that is obtained by using a matrix sparsification technique. We describe a backtracking procedure that offers an economical way to obtain all subsets in the family. Finally, the characteristic properties of the solution set are used to provide a complete solution in a closed form.

Keywords: Tropical semifield · Tropical optimization · Matrix sparsification · Complete solution · Backtracking

1 Introduction

Tropical (idempotent) mathematics, which deals with the theory and applications of semirings with idempotent addition [2,6–8,10,11,23,24], offers a useful analytical framework to solve many actual problems in operations research, computer science and other fields. These problems can be formulated and solved as optimization problems in the tropical mathematics setting, referred to as the tropical optimization problems. Examples of the application areas of tropical optimization include project scheduling [1,17,18,22,25,27], location analysis [9,14,15,26], and decision making [3–5,19,21].

Many tropical optimization problems are formulated to minimize or maximize functions defined on vectors over idempotent semifields (semirings with multiplicative inverses). These problems may have functions to optimize (objective functions), which can be linear or non-linear in the tropical mathematics sense, and constraints, which can take the form of vector inequalities and equalities. Some problems have direct, explicit solutions obtained under general assumptions. For other problems, only algorithmic solutions under restrictive conditions

P. Höfner et al. (Eds.): RAMiCS 2017, LNCS 10226, pp. 226–241, 2017.
DOI: 10.1007/978-3-319-57418-9_14

are known, which apply iterative numerical procedures to find a solution if it exists, or to indicate infeasibility of the problem otherwise. A short overview of tropical optimization problems and their solutions can be found in [16].

In this paper, we consider the tropical optimization problem as to

$$\text{minimize} \quad (\boldsymbol{A}\boldsymbol{x})^{-}\boldsymbol{x},$$

where $\boldsymbol{A}$ is a given square matrix, $\boldsymbol{x}$ is an unknown vector, and the minus sign in the superscript serves to specify conjugate transposition of vectors.

A partial solution of the problem was obtained in [12]. The main purpose of this paper is to continue the investigation of the problem to derive a complete solution. We follow an approach developed in [20] and based on a characterization of the solution set. We show that all solutions of the problem satisfy a vector inequality, and then use this inequality to establish characteristic properties of the solution set. The solutions are represented as a family of solution subsets, each defined by a matrix that is obtained by using a matrix sparsification technique. We describe a backtracking procedure that offers an economical way to obtain all subsets in the family. Finally, the characteristic properties of the solution set are applied to provide a complete solution in a closed form. The results obtained are illustrated with illuminating numerical examples.

The rest of the paper is organized as follows. In Sect. 2, we give a brief overview of basic definitions and preliminary results of tropical algebra. Section 3 formulates the tropical optimization problem under study, and performs a preliminary analysis of the problem. The analysis includes the evaluation of the minimum of the objective function, the derivation of a partial solution, and the investigation of the characteristic properties of the solutions. In Sect. 4, a complete solution to the problem is given as a family of subsets, and then represented in a compact closed vector form. Finally, Sect. 5 offers concluding remarks and suggestions for further research.

2 Preliminary Definitions and Results

We start with a brief overview of the preliminary definitions and results of tropical algebra to provide an appropriate formal background for the development of solutions for the tropical optimization problems in the subsequent sections. The overview is mainly based on the results in [12,13,17,18,22], which offer a useful framework to obtain solutions in a compact vector form, ready for further analysis and practical implementation. Additional details on tropical mathematics at both introductory and advanced levels can be found in many recent publications, including [2,6–8,10,11,23,24].

2.1 Idempotent Semifield

An idempotent semifield is a system $(\mathbb{X}, \mathbb{0}, \mathbb{1}, \oplus, \otimes)$, where $\mathbb{X}$ is a nonempty set endowed with associative and commutative operations, addition $\oplus$ and multiplication $\otimes$, which have as neutral elements the zero $\mathbb{0}$ and the one $\mathbb{1}$. Addition is idempotent, which implies $x \oplus x = x$ for all $x \in \mathbb{X}$. Multiplication distributes over addition, has $\mathbb{0}$ as absorbing element, and is invertible, which gives any nonzero x its inverse x^{-1} such that $x \otimes x^{-1} = \mathbb{1}$.

Idempotent addition induces on $\mathbb{X}$ a partial order such that $x \leq y$ if and only if $x \oplus y = y$. With respect to this order, both addition and multiplication are monotone, which means that, for all $x, y, z \in \mathbb{X}$, the inequality $x \leq y$ entails that $x \oplus z \leq y \oplus z$ and $x \otimes z \leq y \otimes z$. Furthermore, inversion is antitone to take the inequality $x \leq y$ into $x^{-1} \geq y^{-1}$ for all nonzero x and y. Finally, the inequality $x \oplus y \leq z$ is equivalent to the pair of inequalities $x \leq z$ and $y \leq z$. The partial order is assumed to extend to a total order on the semifield.

The power notation with integer exponents is routinely defined to represent iterated products for all $x \neq \mathbb{0}$ and integer $p \geq 1$ in the form $x^0 = \mathbb{1}$, $x^p = x \otimes x^{p-1}$, $x^{-p} = (x^{-1})^p$, and $\mathbb{0}^p = \mathbb{0}$. Moreover, the equation $x^p = a$ is assumed to be solvable for any a, which extends the notation to rational exponents. In what follows, the multiplication sign $\otimes$ is, as usual, dropped to save writing.

A typical example of the semifield is the system $(\mathbb{R} \cup \{-\infty\}, -\infty, 0, \max, +)$, which is usually referred to as the max-plus algebra. In this semifield, the addition $\oplus$ is defined as max, and the multiplication $\otimes$ is as arithmetic addition. The number $-\infty$ is taken as the zero $\mathbb{0}$, and 0 is as the one $\mathbb{1}$. For each $x \in \mathbb{R}$, the inverse x^{-1} coincides with the conventional opposite number $-x$. For any $x, y \in \mathbb{R}$, the power x^y corresponds to the arithmetic product xy. The order induced by idempotent addition complies with the natural linear order on $\mathbb{R}$.

2.2 Matrix and Vector Algebra

The set of matrices over $\mathbb{X}$ with m rows and n columns is denoted by $\mathbb{X}^{m \times n}$. A matrix with all entries equal to $\mathbb{0}$ is the zero matrix denoted by $\mathbf{0}$. A matrix without zero rows (columns) is called row- (column-) regular.

For any matrices $\boldsymbol{A}, \boldsymbol{B} \in \mathbb{X}^{m \times n}$ and $\boldsymbol{C} \in \mathbb{X}^{n \times l}$, and scalar $x \in \mathbb{X}$, matrix addition, matrix multiplication and scalar multiplication are routinely defined by the entry-wise formulas

$$\{\boldsymbol{A} \oplus \boldsymbol{B}\}_{ij} = \{\boldsymbol{A}\}_{ij} \oplus \{\boldsymbol{B}\}_{ij}, \quad \{\boldsymbol{A}\boldsymbol{C}\}_{ij} = \bigoplus_{k=1}^{n} \{\boldsymbol{A}\}_{ik}\{\boldsymbol{C}\}_{kj}, \quad \{x\boldsymbol{A}\}_{ij} = x\{\boldsymbol{A}\}_{ij}.$$

For any nonzero matrix $\boldsymbol{A} = (a_{ij}) \in \mathbb{X}^{m \times n}$, the conjugate transpose is the matrix $\boldsymbol{A}^- = (a_{ij}^-) \in \mathbb{X}^{n \times m}$, where $a_{ij}^- = a_{ji}^{-1}$ if $a_{ji} \neq \mathbb{0}$, and $a_{ij}^- = \mathbb{0}$ otherwise.

The properties of the scalar addition, multiplication and inversion with respect to the order relations are extended entry-wise to the matrix operations.

Consider square matrices in the set $\mathbb{X}^{n \times n}$. A matrix is diagonal, if its off-diagonal entries are all equal to $\mathbb{0}$. A diagonal matrix with all diagonal entries equal to $\mathbb{1}$ is the identity matrix denoted by $\boldsymbol{I}$. The power notation with non-negative integer exponents serves to represent repeated multiplication as $\boldsymbol{A}^0 = \boldsymbol{I}$, $\boldsymbol{A}^p = \boldsymbol{A}\boldsymbol{A}^{p-1}$ and $\mathbf{0}^p = \mathbf{0}$ for any non-zero matrix $\boldsymbol{A}$ and integer $p \geq 1$.

If a row-regular matrix $\boldsymbol{A}$ has exactly one non-zero entry in each row, then the inequalities $\boldsymbol{A}^-\boldsymbol{A} \leq \boldsymbol{I}$ and $\boldsymbol{A}\boldsymbol{A}^- \geq \boldsymbol{I}$ hold (corresponding, in the context of relational algebra, to the univalent and total properties of a relation $\boldsymbol{A}$).

The trace of any matrix $\boldsymbol{A} = (a_{ij})$ is routinely defined as

$$\operatorname{tr} \boldsymbol{A} = \bigoplus_{i=1}^{n} a_{ii},$$

and retains the standard properties of traces with respect to matrix addition and to matrix and scalar multiplications.

To represent solutions proposed in the subsequent sections, we exploit the function, which takes any matrix $\boldsymbol{A} \in \mathbb{X}^{n\times n}$ to the scalar

$$\operatorname{Tr}(\boldsymbol{A}) = \bigoplus_{m=1}^{n} \operatorname{tr} \boldsymbol{A}^{m}.$$

Provided that the condition $\operatorname{Tr}(\boldsymbol{A}) \leq \mathbb{1}$ holds, the asterisk operator (also known as the Kleene star) maps $\boldsymbol{A}$ to the matrix

$$\boldsymbol{A}^{*} = \bigoplus_{m=0}^{n-1} \boldsymbol{A}^{m}.$$

If $\operatorname{Tr}(\boldsymbol{A}) \leq \mathbb{1}$, then the inequality $\boldsymbol{A}^{k} \leq \boldsymbol{A}^{*}$ holds for all integer $k \geq 0$.

The description of the solutions also involves the matrix $\boldsymbol{A}^{+}$ which is obtained from $\boldsymbol{A}$ as follows. First, we assume that $\operatorname{Tr}(\boldsymbol{A}) \leq \mathbb{1}$, and calculate the matrices $\boldsymbol{A}^{*}$ and $\boldsymbol{A}\boldsymbol{A}^{*} = \boldsymbol{A} \oplus \cdots \oplus \boldsymbol{A}^{n}$. Then, the matrix $\boldsymbol{A}^{+}$ is constructed by taking those columns in the matrix $\boldsymbol{A}\boldsymbol{A}^{*}$ which have the diagonal entries equal to $\mathbb{1}$.

A scalar $\lambda \in \mathbb{X}$ is an eigenvalue and a non-zero vector $\boldsymbol{x} \in \mathbb{X}^{n}$ is a corresponding eigenvector of a square matrix $\boldsymbol{A} \in \mathbb{X}^{n\times n}$ if they satisfy the equality

$$\boldsymbol{A}\boldsymbol{x} = \lambda \boldsymbol{x}.$$

Any matrix that consists of one row (column) is considered a row (column) vector. All vectors are assumed to be column vectors, unless otherwise specified. The set of column vectors of order n is denoted $\mathbb{X}^{n}$. A vector with all zero elements is the zero vector $\boldsymbol{0}$. A vector is regular if it has no zero elements.

For any non-zero vector $\boldsymbol{x} = (x_j) \in \mathbb{X}^{n}$, the conjugate transpose is the row vector $\boldsymbol{x}^{-} = (x_j^{-})$, where $x_j^{-} = x_j^{-1}$ if $x_j \neq \mathbb{0}$, and $x_j^{-} = \mathbb{0}$ otherwise.

For any non-zero vector $\boldsymbol{x}$, the equality $\boldsymbol{x}^{-}\boldsymbol{x} = \mathbb{1}$ is obviously valid.

For any regular vectors $\boldsymbol{x}, \boldsymbol{y} \in \mathbb{X}^{n}$, the matrix inequality $\boldsymbol{x}\boldsymbol{y}^{-} \geq (\boldsymbol{x}^{-}\boldsymbol{y})^{-1}\boldsymbol{I}$ holds and becomes $\boldsymbol{x}\boldsymbol{x}^{-} \geq \boldsymbol{I}$ when $\boldsymbol{y} = \boldsymbol{x}$.

A vector $\boldsymbol{b}$ is said to be linearly dependent on vectors $\boldsymbol{a}_1, \ldots, \boldsymbol{a}_n$ if the equality $\boldsymbol{b} = x_1\boldsymbol{a}_1 \oplus \cdots \oplus x_n\boldsymbol{a}_n$ holds for some scalars $x_1, \ldots, x_n \in \mathbb{X}$. The vector $\boldsymbol{b}$ is linearly dependent on $\boldsymbol{a}_1, \ldots, \boldsymbol{a}_n$ if an only if the condition $(\boldsymbol{A}(\boldsymbol{b}^{-}\boldsymbol{A})^{-})^{-}\boldsymbol{b} = \mathbb{1}$ is valid, where $\boldsymbol{A}$ is the matrix with the vectors $\boldsymbol{a}_1, \ldots, \boldsymbol{a}_n$ as its columns.

A system of vectors $\boldsymbol{a}_1, \ldots, \boldsymbol{a}_n$ is linearly dependent if at least one vector is linearly dependent on others, and linearly independent otherwise.

Suppose that the system $\boldsymbol{a}_1, \ldots, \boldsymbol{a}_n$ is linearly dependent. To construct a maximal linearly independent system, we use a procedure that sequentially reduces the system until it becomes linearly independent. The procedure applies the above condition to examine the vectors one by one to remove a vector if it is linearly dependent on others, or to leave the vector in the system otherwise.

2.3 Reducible and Irreducible Matrices

A matrix $\boldsymbol{A} \in \mathbb{X}^{n\times n}$ is reducible if simultaneous permutations of its rows and columns can transform it into a block-triangular normal form, and irreducible otherwise. The lower block-triangular normal form of the matrix $\boldsymbol{A}$ is given by

$$\boldsymbol{A} = \begin{pmatrix} \boldsymbol{A}_{11} & \boldsymbol{0} & \dots & \boldsymbol{0} \\ \boldsymbol{A}_{21} & \boldsymbol{A}_{22} & & \boldsymbol{0} \\ \vdots & \vdots & \ddots & \\ \boldsymbol{A}_{s1} & \boldsymbol{A}_{s2} & \dots & \boldsymbol{A}_{ss} \end{pmatrix}, \tag{1}$$

where, in each block row $i = 1, \dots, s$, the diagonal block $\boldsymbol{A}_{ii}$ is either irreducible or the zero matrix of order n_i, the off-diagonal blocks $\boldsymbol{A}_{ij}$ are arbitrary matrices of size $n_i \times n_j$ for all $j < i$, and $n_1 + \cdots + n_s = n$.

Any irreducible matrix $\boldsymbol{A}$ has only one eigenvalue, which is calculated as

$$\lambda = \bigoplus_{m=1}^{n} \mathrm{tr}^{1/m}(\boldsymbol{A}^m). \tag{2}$$

From (2) it follows, in particular, that $\mathrm{tr}(\boldsymbol{A}^m) \leq \lambda^m$ for all $m = 1, \dots, n$.

All eigenvectors of the irreducible matrix $\boldsymbol{A}$ are regular, and given by

$$\boldsymbol{x} = (\lambda^{-1}\boldsymbol{A})^{+}\boldsymbol{u},$$

where $\boldsymbol{u}$ is any regular vector of appropriate size.

Note that every irreducible matrix is both row- and column-regular.

Let $\boldsymbol{A}$ be a matrix represented in the form (1). Denote by λ_i the eigenvalue of the diagonal block $\boldsymbol{A}_{ii}$ for $i = 1, \dots, s$. Then, the scalar $\lambda = \lambda_1 \oplus \cdots \oplus \lambda_s$ is the maximum eigenvalue of the matrix $\boldsymbol{A}$, which is referred to as the spectral radius of $\boldsymbol{A}$ and calculated as (2). For any irreducible matrix, the spectral radius coincides with the unique eigenvalue of the matrix.

Without loss of generality, the normal form (1) can be assumed to order all block rows, which have non-zero blocks on the diagonal and zero blocks elsewhere, before the block rows with non-zero off-diagonal blocks. Moreover, the rows, which have non-zero blocks only on the diagonal, can be arranged in increasing order of the eigenvalues of diagonal blocks. Then, the normal form is refined as

$$\boldsymbol{A} = \begin{pmatrix} \boldsymbol{A}_{11} & & \boldsymbol{0} & \boldsymbol{0} & \dots & \boldsymbol{0} \\ & \ddots & & \vdots & & \vdots \\ \boldsymbol{0} & & \boldsymbol{A}_{rr} & \boldsymbol{0} & \dots & \boldsymbol{0} \\ \boldsymbol{A}_{r+1,1} & \dots & \boldsymbol{A}_{r+1,r} & \boldsymbol{A}_{r+1,r+1} & & \boldsymbol{0} \\ \vdots & & \vdots & \vdots & \ddots & \\ \boldsymbol{A}_{s1} & \dots & \boldsymbol{A}_{sr} & \boldsymbol{A}_{s,r+1} & \dots & \boldsymbol{A}_{ss} \end{pmatrix}, \tag{3}$$

where the eigenvalues of $\boldsymbol{A}_{11}, \dots, \boldsymbol{A}_{rr}$ satisfy the condition $\lambda_1 \leq \cdots \leq \lambda_r$, and each row $i = r+1, \dots, s$ has a block $\boldsymbol{A}_{ij} \neq \boldsymbol{0}$ for some $j < i$.

2.4 Vector Inequalities and Equations

In this subsection, we present solutions to vector inequalities, which appear below in the analysis of the optimization problem under study.

Suppose that, given a matrix $\boldsymbol{A} \in \mathbb{X}^{m\times n}$ and a vector $\boldsymbol{d} \in \mathbb{X}^{m}$, we need to find vectors $\boldsymbol{x} \in \mathbb{X}^{n}$ to satisfy the inequality

$$\boldsymbol{A}\boldsymbol{x} \leq \boldsymbol{d}. \tag{4}$$

A direct solution proposed in [17] can be obtained as follows.

Lemma 1. *For any column-regular matrix* $\boldsymbol{A}$ *and regular vector* $\boldsymbol{d}$, *all solutions to inequality* (4) *are given by the inequality* $\boldsymbol{x} \leq (\boldsymbol{d}^{-}\boldsymbol{A})^{-}$.

Next, we consider the following problem: given a matrix $\boldsymbol{A} \in \mathbb{X}^{n\times n}$, find regular vectors $\boldsymbol{x} \in \mathbb{X}^{n}$ to satisfy the inequality

$$\boldsymbol{A}\boldsymbol{x} \leq \boldsymbol{x}. \tag{5}$$

The following result [13,18] provides a direct solution to inequality (5).

Theorem 1. *For any matrix* $\boldsymbol{A}$, *the following statements hold:*

1. *If* $\mathrm{Tr}(\boldsymbol{A}) \leq \mathbb{1}$, *then all regular solutions to inequality* (5) *are given by* $\boldsymbol{x} = \boldsymbol{A}^{*}\boldsymbol{u}$, *where* $\boldsymbol{u}$ *is any regular vector.*
2. *If* $\mathrm{Tr}(\boldsymbol{A}) > \mathbb{1}$, *then there is no regular solution.*

We conclude this subsection with a solution to a vector equation. Given a matrix $\boldsymbol{A} \in \mathbb{X}^{n\times n}$ and a vector $\boldsymbol{b} \in \mathbb{X}^{n}$, the problem is to find regular vectors $\boldsymbol{x} \in \mathbb{X}^{n}$ that solve the equation

$$\boldsymbol{A}\boldsymbol{x} \oplus \boldsymbol{b} = \boldsymbol{x}. \tag{6}$$

The next statement [13] offers a solution when the matrix $\boldsymbol{A}$ is irreducible.

Theorem 2. *For any irreducible matrix* $\boldsymbol{A}$ *and non-zero vector* $\boldsymbol{b}$, *the following statements hold:*

1. *If* $\mathrm{Tr}(\boldsymbol{A}) < \mathbb{1}$, *then equation* (6) *has the unique regular solution* $\boldsymbol{x} = \boldsymbol{A}^{*}\boldsymbol{b}$.
2. *If* $\mathrm{Tr}(\boldsymbol{A}) = \mathbb{1}$, *then all regular solutions to* (6) *are given by* $\boldsymbol{x} = \boldsymbol{A}^{+}\boldsymbol{u} \oplus \boldsymbol{A}^{*}\boldsymbol{b}$, *where* $\boldsymbol{u}$ *is any regular vector of appropriate size.*
3. *If* $\mathrm{Tr}(\boldsymbol{A}) > \mathbb{1}$, *then there is no regular solution.*

3 Tropical Optimization Problem

We are now in a position to describe the optimization problem under study, and to provide some preliminary solution to the problem. The problem is formulated to minimize a function defined on vectors over a general idempotent semifield. Given a matrix $\boldsymbol{A} \in \mathbb{X}^{n\times n}$, we need to find regular vectors $\boldsymbol{x} \in \mathbb{X}^{n}$ that

$$\text{minimize} \quad (\boldsymbol{A}\boldsymbol{x})^{-}\boldsymbol{x}. \tag{7}$$

A partial solution to the problem for both irreducible and reducible matrices $\boldsymbol{A}$ was given in [12]. The solutions offered below improve the previous results by adding new characterization properties of the solution set and by extending the partial solution given in the case of reducible matrices. The determination of the minimum value of the objective function, the derivation of the partial solution for irreducible matrices and the evaluation of the lower bound on the function for reducible matrices are taken from the previous proof, and presented here for the sake of completeness.

We start with a solution of problem (7) for an irreducible matrix.

Lemma 2. *Let $\boldsymbol{A}$ be an irreducible matrix with spectral radius λ. Then, the minimum value in problem (7) is equal to λ^{-1}, and all regular solutions are given by the inequality*

$$\boldsymbol{x} \leq \lambda^{-1}\boldsymbol{A}\boldsymbol{x}. \tag{8}$$

Specifically, any eigenvector of the matrix $\boldsymbol{A}$, given by $\boldsymbol{x} = (\lambda^{-1}\boldsymbol{A})^{+}\boldsymbol{u}$, where $\boldsymbol{u}$ is any regular vector of appropriate size, is a solution of the problem.

Proof. Let $\boldsymbol{x}_0$ be an eigenvector of the matrix $\boldsymbol{A}$. Since $\boldsymbol{A}$ is irreducible, the vector $\boldsymbol{x}_0$ is regular, and thus $\boldsymbol{x}_0\boldsymbol{x}_0^{-} \geq \boldsymbol{I}$. For any regular $\boldsymbol{x}$, we obtain

$$(\boldsymbol{A}\boldsymbol{x})^{-}\boldsymbol{x} \geq (\boldsymbol{A}\boldsymbol{x}_0\boldsymbol{x}_0^{-}\boldsymbol{x})^{-}\boldsymbol{x} = (\boldsymbol{x}_0^{-}\boldsymbol{x})^{-1}(\boldsymbol{A}\boldsymbol{x}_0)^{-}\boldsymbol{x} = \lambda^{-1}(\boldsymbol{x}_0^{-}\boldsymbol{x})^{-1}\boldsymbol{x}_0^{-}\boldsymbol{x} = \lambda^{-1},$$

which means that λ^{-1} is a lower bound for the objective function.

As the substitution $\boldsymbol{x} = \boldsymbol{x}_0$ yields $(\boldsymbol{A}\boldsymbol{x})^{-}\boldsymbol{x} = (\boldsymbol{A}\boldsymbol{x}_0)^{-}\boldsymbol{x}_0 = \lambda^{-1}\boldsymbol{x}_0^{-}\boldsymbol{x}_0 = \lambda^{-1}$, the lower bound λ^{-1} is strict, and thus presents the minimum in problem (7).

All regular vectors $\boldsymbol{x}$ that solve the problem are determined by the equation $(\boldsymbol{A}\boldsymbol{x})^{-}\boldsymbol{x} = \lambda^{-1}$. Since λ^{-1} is the minimum, we can replace this equation by the inequality $(\boldsymbol{A}\boldsymbol{x})^{-}\boldsymbol{x} \leq \lambda^{-1}$, where the matrix $\boldsymbol{A}$ is irreducible and thus row-regular. Considering that $(\boldsymbol{A}\boldsymbol{x})^{-}$ is then a column-regular matrix, we solve the last inequality by applying Lemma 1 in the form of (8). □

Example 1. Let us examine problem (7), given in terms of the semifield $\mathbb{R}_{\max,+}$ by the irreducible matrix

$$\boldsymbol{A} = \begin{pmatrix} 1 & -1 \\ 3 & -2 \end{pmatrix}.$$

First, we evaluate the minimum in the problem. We successively calculate

$$\operatorname{tr}\boldsymbol{A} = 1, \qquad \boldsymbol{A}^2 = \begin{pmatrix} 2 & 0 \\ 4 & 2 \end{pmatrix}, \qquad \operatorname{tr}\boldsymbol{A}^2 = 2.$$

Since $\lambda = \operatorname{tr}\boldsymbol{A} \oplus \operatorname{tr}^{1/2}(\boldsymbol{A}^2) = 1$, we have the minimum $\lambda^{-1} = -1$.

Furthermore, we obtain the eigenvectors of the matrix $\boldsymbol{A}$, which present a solution to the problem. We define $\boldsymbol{B} = \lambda^{-1}\boldsymbol{A}$ and calculate matrices

$$\boldsymbol{B} = \begin{pmatrix} 0 & -2 \\ 2 & -3 \end{pmatrix}, \qquad \boldsymbol{B}^{*} = \boldsymbol{B}\boldsymbol{B}^{*} = \boldsymbol{B}^{+} = \begin{pmatrix} 0 & -2 \\ 2 & 0 \end{pmatrix}.$$

Considering that both columns in the matrix $\boldsymbol{B}$ are collinear, we can take one of them to represent a solution to the problem in the form

$$\boldsymbol{x} = \begin{pmatrix} -2 \\ 0 \end{pmatrix} u, \qquad u \in \mathbb{R}.$$

We now extend this result to arbitrary matrices, and then obtain two useful consequences. For simplicity, we concentrate on the matrices in the block-triangular form (3), which have no zero rows. The case of matrices with zero rows follows the same arguments with minor technical modifications.

Theorem 3. *Let $\boldsymbol{A}$ be a matrix in the refined block-triangular normal form (3), where the diagonal blocks $\boldsymbol{A}_{ii}$ for all $i = 1, \ldots, r$ have eigenvalues $\lambda_i > \mathbb{0}$.*

Then, the minimum value in problem (7) is equal to λ_1^{-1}, and all regular solutions are characterized by the inequality

$$\boldsymbol{x} \leq \lambda_1^{-1} \boldsymbol{A} \boldsymbol{x}. \tag{9}$$

Specifically, any block vector $\boldsymbol{x} = (\boldsymbol{x}_1^T, \ldots, \boldsymbol{x}_s^T)^T$ with the blocks $\boldsymbol{x}_i$ defined successively for each $i = 1, \ldots, s$ by the conditions

$$\boldsymbol{x}_i = \begin{cases} (\lambda_i^{-1} \boldsymbol{A}_{ii})^{+} \boldsymbol{u}_i, & \text{if } \lambda_i \geq \lambda_1; \\ \lambda_1^{-1} (\lambda_1^{-1} \boldsymbol{A}_{ii})^{*} \bigoplus_{j=1}^{i-1} \boldsymbol{A}_{ij} \boldsymbol{x}_j, & \text{if } \lambda_i < \lambda_1; \end{cases}$$

where $\boldsymbol{u}_i$ are regular vectors of appropriate size, is a solution of the problem.

Proof. Considering the refined block-triangular form of the matrix $\boldsymbol{A}$ with s rows, where all blocks above the diagonal are zero matrices, we write

$$(\boldsymbol{A}\boldsymbol{x})^{-} \boldsymbol{x} = \bigoplus_{i=1}^{s} \left(\bigoplus_{j=1}^{i} \boldsymbol{A}_{ij} \boldsymbol{x}_j \right)^{-} \boldsymbol{x}_i = \bigoplus_{i=1}^{r} (\boldsymbol{A}_{ii} \boldsymbol{x}_i)^{-} \boldsymbol{x}_i \oplus \bigoplus_{i=r+1}^{s} \left(\bigoplus_{j=1}^{i} \boldsymbol{A}_{ij} \boldsymbol{x}_j \right)^{-} \boldsymbol{x}_i.$$

An application of Lemma 2 and the condition that $\lambda_i^{-1} \leq \lambda_1^{-1}$ for $i \leq r$ yield

$$(\boldsymbol{A}\boldsymbol{x})^{-} \boldsymbol{x} \geq \bigoplus_{i=1}^{r} (\boldsymbol{A}_{ii} \boldsymbol{x}_i)^{-} \boldsymbol{x}_i \geq \bigoplus_{i=1}^{r} \lambda_i^{-1} = \lambda_1^{-1},$$

which means that λ_1^{-1} is a lower bound for the objective function in the problem.

To verify that the bound λ_1^{-1} is strict, and hence is the minimum value of the objective function, we need to present a vector $\boldsymbol{x}$ that produces this bound.

We have to solve the inequality $(\boldsymbol{A}\boldsymbol{x})^{-} \boldsymbol{x} \leq \lambda_1^{-1}$, which, due to the block-triangular form of $\boldsymbol{A}$, is equivalent to the system of inequalities

$$\left(\bigoplus_{j=1}^{i} \boldsymbol{A}_{ij} \boldsymbol{x}_j \right)^{-} \boldsymbol{x}_i \leq \lambda_1^{-1}, \qquad i = 1, \ldots, s.$$

We successively define a sequence of vectors $\boldsymbol{x}_i$ for $i = 1, \ldots, s$. If $\lambda_i \geq \lambda_1$ we take $\boldsymbol{x}_i$ to be an eigenvector of $\boldsymbol{A}_{ii}$, given by the equation $\lambda_i^{-1}\boldsymbol{A}_{ii}\boldsymbol{x}_i = \boldsymbol{x}_i$, which is solved as $\boldsymbol{x}_i = (\lambda_i^{-1}\boldsymbol{A}_{ii})^{+}\boldsymbol{u}_i$, where $\boldsymbol{u}_i$ is a regular vector of appropriate size.

Note that the condition $\lambda_i \geq \lambda_1$ is fulfilled if $i \leq r$. With this condition, we have

$$\bigoplus_{j=1}^{i} \boldsymbol{A}_{ij}\boldsymbol{x}_j = \bigoplus_{j=1}^{i-1} \boldsymbol{A}_{ij}\boldsymbol{x}_j \oplus \boldsymbol{A}_{ii}\boldsymbol{x}_i \geq \boldsymbol{A}_{ii}\boldsymbol{x}_i = \lambda_i\boldsymbol{x}_i \geq \lambda_1\boldsymbol{x}_i$$

and therefore,

$$\left(\bigoplus_{j=1}^{i} \boldsymbol{A}_{ij}\boldsymbol{x}_j\right)^{-} \boldsymbol{x}_i \leq \lambda_1^{-1}.$$

If $\lambda_i < \lambda_1$ we define $\boldsymbol{x}_i$ as the solution to the equation

$$\lambda_1^{-1} \bigoplus_{j=1}^{i-1} \boldsymbol{A}_{ij}\boldsymbol{x}_j \oplus \lambda_1^{-1}\boldsymbol{A}_{ii}\boldsymbol{x}_i = \boldsymbol{x}_i.$$

Since, in this case, $\mathrm{Tr}(\lambda_1^{-1}\boldsymbol{A}_{ii}) = \lambda_1^{-1}\operatorname{tr}\boldsymbol{A}_{ii} \oplus \cdots \oplus \lambda_1^{-n}\operatorname{tr}(\boldsymbol{A}_{ii}^{n}) < \mathbb{1}$, the equation is solved by Theorem 2 in the form

$$\boldsymbol{x}_i = \lambda_1^{-1}(\lambda_1^{-1}\boldsymbol{A}_{ii})^{*} \bigoplus_{j=1}^{i-1} \boldsymbol{A}_{ij}\boldsymbol{x}_j.$$

With the solution vector $\boldsymbol{x}_i$, we write

$$\lambda_1\boldsymbol{x}_i = \bigoplus_{j=1}^{i-1} \boldsymbol{A}_{ij}\boldsymbol{x}_j \oplus \boldsymbol{A}_{ii}\boldsymbol{x}_i = \bigoplus_{j=1}^{i} \boldsymbol{A}_{ij}\boldsymbol{x}_j,$$

and then have

$$\left(\bigoplus_{j=1}^{i} \boldsymbol{A}_{ij}\boldsymbol{x}_j\right)^{-} \boldsymbol{x}_i = \lambda_1^{-1}.$$

Combining all obtained vectors $\boldsymbol{x}_i$ together yields

$$(\boldsymbol{A}\boldsymbol{x})^{-}\boldsymbol{x} = \bigoplus_{i=1}^{s} \left(\bigoplus_{j=1}^{i} \boldsymbol{A}_{ij}\boldsymbol{x}_j\right)^{-} \boldsymbol{x}_i \leq \lambda_1^{-1},$$

which shows that λ_1^{-1} is the minimum value of the problem.

Finally, the application of Lemma 1 to solve the inequality $(\boldsymbol{A}\boldsymbol{x})^{-}\boldsymbol{x} \leq \lambda_1^{-1}$ with respect to $\boldsymbol{x}$ leads to inequality (9). □

Example 2. Consider problem (7) defined in terms of $\mathbb{R}_{\max,+}$ with the matrix

$$\boldsymbol{A} = \begin{pmatrix} 1 & \mathbb{0} \\ 3 & -2 \end{pmatrix}.$$

Note that the matrix $\boldsymbol{A}$ is reducible, and has the block-triangular form (3) with the diagonal blocks given by (1×1)-matrices. The eigenvalues of the diagonal blocks are easily found to be $\lambda_1 = 1$ and $\lambda_2 = -2$.

By Theorem 3, the minimum in the problem is equal to $\lambda^{-1} = -1$. The solution offered by the theorem is given by the vector $\boldsymbol{x} = (x_1, x_2)^T$, where $x_1 = u$ for all $u \in \mathbb{R}$. The element x_2 is defined by the equation $x_2 = 2x_1 \oplus (-3)x_2$, which reduces to the equality $x_2 = 2x_1$. In vector form, the solution becomes

$$\boldsymbol{x} = \begin{pmatrix} 0 \\ 2 \end{pmatrix} u, \qquad u \in \mathbb{R}.$$

We now consider a special case of the problem, where the partial solution given by the previous theorem takes a more compact form.

Corollary 1. *Under the conditions of Theorem 3, if $\lambda_1 \leq \lambda_i$ for all $i = 1, \ldots, s$, then the vector*

$$\boldsymbol{x} = \boldsymbol{D}\boldsymbol{u}, \qquad \boldsymbol{D} = \begin{pmatrix} (\lambda_1^{-1}\boldsymbol{A}_{11})^+ & & \boldsymbol{0} \\ & \ddots & \\ \boldsymbol{0} & & (\lambda_s^{-1}\boldsymbol{A}_{ss})^+ \end{pmatrix},$$

where $\boldsymbol{u}$ is any regular vector of appropriate size, is a solution of the problem.

Proof. It follows from Theorem 3 that the vector $\boldsymbol{x} = (\boldsymbol{x}_1^T, \ldots, \boldsymbol{x}_s^T)^T$, which have, for all $i = 1, \ldots, s$, the blocks $\boldsymbol{x}_i = (\lambda_i^{-1}\boldsymbol{A}_{ii})^+\boldsymbol{u}_i$, where $\boldsymbol{u}_i$ are regular vectors of appropriate size, is a solution of the problem.

It remains to introduce the block vector $\boldsymbol{u} = (\boldsymbol{u}_1^T, \ldots, \boldsymbol{u}_s^T)^T$ and the block-diagonal matrix $\boldsymbol{D} = \mathrm{diag}((\lambda_1^{-1}\boldsymbol{A}_{11})^+, \ldots, (\lambda_s^{-1}\boldsymbol{A}_{ss})^+)$ to finish the proof. □

The next result shows a useful property of the solutions of problem (7).

Corollary 2. *Under the conditions of Theorem 3, the set of solution vectors of problem* (7) *is closed under vector addition and scalar multiplication.*

Proof. Suppose that vectors $\boldsymbol{x}$ and $\boldsymbol{y}$ are solutions of the problem, which implies, by Theorem 3, that $\boldsymbol{x} \leq \lambda_1^{-1}\boldsymbol{A}\boldsymbol{x}$ and $\boldsymbol{y} \leq \lambda_1^{-1}\boldsymbol{A}\boldsymbol{y}$. We take arbitrary scalars α and β, and consider the vector $\boldsymbol{z} = \alpha\boldsymbol{x} \oplus \beta\boldsymbol{y}$. Since

$$\boldsymbol{z} = \alpha\boldsymbol{x} \oplus \beta\boldsymbol{y} \leq \alpha\lambda_1^{-1}\boldsymbol{A}\boldsymbol{x} \oplus \beta\lambda_1^{-1}\boldsymbol{A}\boldsymbol{y} = \lambda_1^{-1}\boldsymbol{A}(\alpha\boldsymbol{x} \oplus \beta\boldsymbol{y}) = \lambda_1^{-1}\boldsymbol{A}\boldsymbol{z},$$

the vector $\boldsymbol{z}$ is a solution of the problem, which proves the statement. □

4 Derivation of Complete Solution

It follows from the results from the previous section that, under the assumptions of Theorem 3, all solutions of problem (7) are given by inequality (9). Below, we derive all solutions of the inequality, which we represent, without loss of generality, in a form without a scalar factor on the right-hand side.

Given a matrix $\boldsymbol{A} \in \mathbb{X}^{n \times n}$, we consider the inequality

$$\boldsymbol{x} \leq \boldsymbol{A}\boldsymbol{x}. \tag{10}$$

4.1 Solution via Matrix Sparsification

We start with the description of all solutions in the form of a family of solution sets, each defined by means of sparsification of the matrix $\boldsymbol{A}$.

Theorem 4. *Let $\boldsymbol{A}$ be a matrix in the refined block-triangular normal form* (3), *where the diagonal block $\boldsymbol{A}_{11}$ has eigenvalue $\lambda_1 > \mathbb{1}$.*

Denote by $\mathcal{A}$ the set of matrices $\boldsymbol{A}_1$ that are obtained from $\boldsymbol{A}$ by fixing one non-zero entry in each row and by setting the others to $\mathbb{0}$, and that satisfy the condition $\mathrm{Tr}(\boldsymbol{A}_1^{-}(\boldsymbol{A} \oplus \boldsymbol{I})) \leq \mathbb{1}$.

Then, all regular solutions of inequality (10) *are given by the conditions*

$$\boldsymbol{x} = (\boldsymbol{A}_1^{-}(\boldsymbol{A} \oplus \boldsymbol{I}))^{*}\boldsymbol{u}, \qquad \boldsymbol{u} > \boldsymbol{0}, \qquad \boldsymbol{A}_1 \in \mathcal{A}. \tag{11}$$

Proof. First we note that, under the conditions of the theorem, regular solutions to inequality (10) exist. Indeed, using similar arguments as in Theorem 3, one can see that the block vector $\boldsymbol{x} = (\boldsymbol{x}_1^T, \ldots, \boldsymbol{x}_s^T)^T$, where, for all $i = 1, \ldots, s$, we take $\boldsymbol{x}_i$ to be an eigenvector of the matrix $\boldsymbol{A}_{ii}$ if $\lambda_i \geq \mathbb{1}$, or to be a solution of the equation $\boldsymbol{A}_{i1}\boldsymbol{x}_1 \oplus \cdots \oplus \boldsymbol{A}_{ii}\boldsymbol{x}_i = \boldsymbol{x}_i$ otherwise, satisfies the inequality.

To prove the theorem, we show that any regular solution of inequality (10) can be represented as (11), and vice versa. Assume $\boldsymbol{x} = (x_j)$ to be a regular solution of (10) with a matrix $\boldsymbol{A} = (a_{ij})$, and consider the scalar inequality

$$x_k \leq a_{k1}x_1 \oplus \cdots \oplus a_{kn}x_n, \tag{12}$$

which corresponds to row k in the matrix $\boldsymbol{A}$.

If this inequality holds for some $x_1, \ldots, x_n$, then, as the order defined by the relation $\leq$ is assumed linear, there is a term in the sum on the right-hand side that provides the maximum of the sum. Suppose that the maximum is attained at the pth term $a_{kp}x_p$, and hence $a_{kp} > \mathbb{0}$. Under this condition, we can replace the above inequality by the two inequalities $a_{kp}x_p \geq a_{k1}x_1 \oplus \cdots \oplus a_{kn}x_n$ and $a_{kp}x_p \geq x_k$, or, equivalently, by one inequality

$$a_{kp}x_p \geq a_{k1}x_1 \oplus \cdots \oplus (a_{kk} \oplus \mathbb{1})x_k \oplus \cdots \oplus a_{kn}x_n. \tag{13}$$

Now assume that we determine maximum terms in all scalar inequalities in (10). Similarly as above, we replace each inequality by an inequality with the maximum term isolated on the left side.

To represent the new inequalities in a vector form, we introduce a matrix $\boldsymbol{A}_1$ that is obtained from $\boldsymbol{A}$ by fixing one entry, which corresponds to the maximum term, in each row, and by setting the other entries to $\mathbb{0}$. With the matrix $\boldsymbol{A}_1$, the scalar inequalities are written in vector form as $\boldsymbol{A}_1\boldsymbol{x} \geq (\boldsymbol{A} \oplus \boldsymbol{I})\boldsymbol{x}$.

Let us verify that this inequality is equivalent to the inequality

$$\boldsymbol{x} \geq \boldsymbol{A}_1^{-}(\boldsymbol{A} \oplus \boldsymbol{I})\boldsymbol{x}.$$

We multiply the former inequality by $\boldsymbol{A}_1^{-}$ on the left. Since $\boldsymbol{A}_1^{-}\boldsymbol{A}_1 \leq \boldsymbol{I}$, we have $\boldsymbol{x} \geq \boldsymbol{A}_1^{-}\boldsymbol{A}_1\boldsymbol{x} \geq \boldsymbol{A}_1^{-}(\boldsymbol{A} \oplus \boldsymbol{I})\boldsymbol{x}$, which gives the latter one. At the same time, the multiplication of the latter inequality by $\boldsymbol{A}_1$ on the left, and the condition $\boldsymbol{A}_1\boldsymbol{A}_1^- \geq \boldsymbol{I}$ result in the former inequality as $\boldsymbol{A}_1\boldsymbol{x} \geq \boldsymbol{A}_1\boldsymbol{A}_1^-(\boldsymbol{A}\oplus\boldsymbol{I})\boldsymbol{x} \geq (\boldsymbol{A}\oplus\boldsymbol{I})\boldsymbol{x}$.

By Theorem 1, the last inequality has regular solutions if and only if the condition $\mathrm{Tr}(\boldsymbol{A}_1^-(\boldsymbol{A}\oplus\boldsymbol{I})) \leq \mathbb{1}$ holds. All solutions are given by

$$\boldsymbol{x} = (\boldsymbol{A}_1^-(\boldsymbol{A}\oplus\boldsymbol{I}))^*\boldsymbol{u}, \qquad \boldsymbol{u} > \boldsymbol{0},$$

which means that the vector $\boldsymbol{x}$ is represented in the form of (11).

Now suppose that a vector $\boldsymbol{x}$ is defined by the conditions at (11). To verify that $\boldsymbol{x}$ satisfies (10), we first use the condition $\mathrm{Tr}(\boldsymbol{A}_1^-(\boldsymbol{A}\oplus\boldsymbol{I})) \leq \mathbb{1}$ to see that $(\boldsymbol{A}_1^-(\boldsymbol{A}\oplus\boldsymbol{I}))^* \geq (\boldsymbol{A}_1^-(\boldsymbol{A}\oplus\boldsymbol{I}))^n$. Then, we write

$$\boldsymbol{A}(\boldsymbol{A}_1^-(\boldsymbol{A}\oplus\boldsymbol{I}))^* = \boldsymbol{A}\bigoplus_{m=0}^{n-1}(\boldsymbol{A}_1^-(\boldsymbol{A}\oplus\boldsymbol{I}))^m = \bigoplus_{m=0}^{n}\boldsymbol{A}(\boldsymbol{A}_1^-(\boldsymbol{A}\oplus\boldsymbol{I}))^m.$$

Considering that $\boldsymbol{A} \geq \boldsymbol{A}_1$, we have $\boldsymbol{A}\boldsymbol{A}_1^- \geq \boldsymbol{A}_1\boldsymbol{A}_1^- \geq \boldsymbol{I}$. For each $m \geq 1$, we obtain $\boldsymbol{A}(\boldsymbol{A}_1^-(\boldsymbol{A}\oplus\boldsymbol{I}))^m \geq (\boldsymbol{A}\oplus\boldsymbol{I})(\boldsymbol{A}_1^-(\boldsymbol{A}\oplus\boldsymbol{I}))^{m-1} \geq (\boldsymbol{A}_1^-(\boldsymbol{A}\oplus\boldsymbol{I}))^{m-1}$, from which it follows that

$$\begin{aligned}\boldsymbol{A}(\boldsymbol{A}_1^-(\boldsymbol{A}\oplus\boldsymbol{I}))^* &= \boldsymbol{A}\oplus\bigoplus_{m=1}^{n}\boldsymbol{A}(\boldsymbol{A}_1^-(\boldsymbol{A}\oplus\boldsymbol{I}))^m \\ &\geq \bigoplus_{m=1}^{n}(\boldsymbol{A}_1^-(\boldsymbol{A}\oplus\boldsymbol{I}))^{m-1} = \bigoplus_{m=0}^{n-1}(\boldsymbol{A}_1^-(\boldsymbol{A}\oplus\boldsymbol{I}))^m = (\boldsymbol{A}_1^-(\boldsymbol{A}\oplus\boldsymbol{I}))^*.\end{aligned}$$

Since, in this case, $\boldsymbol{A}\boldsymbol{x} = \boldsymbol{A}(\boldsymbol{A}_1^-(\boldsymbol{A}\oplus\boldsymbol{I}))^*\boldsymbol{u} \geq (\boldsymbol{A}_1^-(\boldsymbol{A}\oplus\boldsymbol{I}))^*\boldsymbol{u} = \boldsymbol{x}$, we conclude that $\boldsymbol{x}$ satisfies inequality (10). □

4.2 Backtracking Procedure of Generating Solution Sets

Note that, although the generation of the sparsified matrices according to the solution described above is a quite simple task, the number of the matrices in practical problems may be excessively large. Below, we propose a backtracking procedure that allows to reduce the number of matrices under examination.

The procedure successively checks rows $i = 1, \ldots, n$ of the matrix $\boldsymbol{A}$ to find and fix one non-zero entry a_{ij} for $j = 1, \ldots, n$, and to set the other entries to zero. On selection of an entry in a row, we examine the remaining rows to modify their non-zero entries by setting to $\mathbb{0}$, provided that these entries do not affect the current solution. One step of the procedure is completed when a non-zero entry is fixed in the last row, and hence a sparsified matrix is fully defined.

To prepare the next step, we take the next non-zero entry in the row, provided that such an entry exists. If there is no non-zero entries left in the row, the procedure has to go back to the previous row. It cancels the last selection of non-zero entry, and rolls back the modifications made to the matrix in accordance with the selection. Then, the procedure fixes the next non-zero entry in this row if it exists, or continues back to the previous rows until a new unexplored non-zero entry is found, otherwise. If the new entry is fixed in a row, the procedure continues forward to fix non-zero entries in the next rows, and to modify the remaining rows. The procedure is completed when no more non-zero entries can be selected in the first row.

To describe the modification routine implemented in the procedure, assume that there are non-zero entries fixed in rows $i = 1, \ldots, k-1$, and we now select the entry a_{kp} in row k. Since this selection implies that $a_{kp}x_p$ is considered the maximum term in the right-hand side of inequality (12), it follows from (13) that $x_p \geq a_{kp}^{-1}a_{kj}x_j$ for all $j \neq k$, and $x_p \geq a_{kp}^{-1}(a_{kk} \oplus \mathbb{1})x_k$ for $j = k$.

Let us examine the inequality $x_i \leq a_{i1}x_1 \oplus \cdots \oplus a_{in}x_n$ for $i = k+1, \ldots, n$. If the condition $a_{ip}a_{kp}^{-1}a_{ki} \geq \mathbb{1}$ holds, then the inequality is fulfilled at the expense of its pth term alone, because $a_{ip}x_p \geq a_{ip}a_{kp}^{-1}a_{ki}x_i \geq x_i$. Since, in this case, the contribution of the other terms is of no concern, we can set the entries a_{ij} for all $j \neq p$ to $\mathbb{0}$ without changing the solution set under construction.

Suppose that the above condition is not satisfied. Then, we can verify the conditions $a_{ip}a_{kp}^{-1}a_{kj} \geq a_{ij}$ for all $j \neq p, k$, and $a_{ip}a_{kp}^{-1}(a_{kk} \oplus \mathbb{1}) \geq a_{ik}$ for $j = k$. If these conditions are satisfied for some $j \neq k$ or $j = k$, then we have $a_{ip}x_p \geq a_{ip}a_{kp}^{-1}a_{kj}x_j \geq a_{ij}x_j$ or $a_{ip}x_p \geq a_{ip}a_{kp}^{-1}(a_{kk} \oplus \mathbb{1})x_k \geq a_{ik}x_k$. This means that term p dominates over term j. As before, considering that the last term does not affect the right-hand side of the inequality, we put $a_{ij} = \mathbb{0}$.

4.3 Closed-Form Representation of Complete Solution

We conclude with the representation of the solution to the entire optimization problem under investigation, in a compact closed form.

Theorem 5. *Let $\boldsymbol{A}$ be a matrix in the refined block-triangular normal form (3), where the diagonal blocks $\boldsymbol{A}_{ii}$ for all $i = 1, \ldots, r$ have eigenvalues $\lambda_i > \mathbb{0}$.*

Define $\boldsymbol{B} = \lambda_1^{-1}\boldsymbol{A}$, and denote by $\mathcal{B}$ the set of matrices $\boldsymbol{B}_1$ that are obtained from $\boldsymbol{B}$ by fixing one non-zero entry in each row and by setting the other entries to $\mathbb{0}$, and that satisfy the condition $\mathrm{Tr}(\boldsymbol{B}_1^{-}(\boldsymbol{B} \oplus \boldsymbol{I})) \leq \mathbb{1}$.

Let $\boldsymbol{S}$ be the matrix, which is constituted by the maximal linear independent system of columns in the matrices $\boldsymbol{S}_1 = (\boldsymbol{B}_1^{-}(\boldsymbol{B} \oplus \boldsymbol{I}))^*$ for all $\boldsymbol{B}_1 \in \mathcal{B}$.

Then, the minimum value in problem (7) is equal to λ_1^{-1}, and all regular solutions are given by

$$\boldsymbol{x} = \boldsymbol{S}\boldsymbol{v}, \qquad \boldsymbol{v} > \boldsymbol{0}.$$

Proof. By Theorem 3, we find the minimum in the problem to be λ_1^{-1}, and characterize all solutions by the inequality $\boldsymbol{x} \leq \boldsymbol{B}\boldsymbol{x}$ with the matrix $\boldsymbol{B} = \lambda_1^{-1}\boldsymbol{A}$.

Application of Theorem 4 involves defining a set $\mathcal{B}$ of matrices $\boldsymbol{B}_1$ that are obtained from $\boldsymbol{B}$ by fixing one non-zero entry in each row together with setting the others to $\mathbb{0}$, and such that $\mathrm{Tr}(\boldsymbol{B}_1^{-}(\boldsymbol{B} \oplus \boldsymbol{I})) \leq \mathbb{1}$. The theorem yields a family of solutions $\boldsymbol{x} = \boldsymbol{S}_1\boldsymbol{u}$ with $\boldsymbol{S}_1 = (\boldsymbol{B}_1^{-}(\boldsymbol{B} \oplus \boldsymbol{I}))^*$ for all $\boldsymbol{u} > \boldsymbol{0}$ and $\boldsymbol{B}_1 \in \mathcal{B}$.

Considering that each solution $\boldsymbol{x} = \boldsymbol{S}_1\boldsymbol{u}$ defines a subset of vectors generated by the columns of the matrix $\boldsymbol{S}_1$, we apply Corollary 2 to represent all solutions as the linear span of the columns in the matrices $\boldsymbol{S}_1$, corresponding to all $\boldsymbol{B}_1 \in \mathcal{B}$.

Finally, we reduce the set of all columns by eliminating those, which are linearly dependent on others. We take the remaining columns to form a matrix $\boldsymbol{S}$, and then write the solution as $\boldsymbol{x} = \boldsymbol{S}\boldsymbol{v}$, where $\boldsymbol{v}$ is any regular vector. □

Example 3. We now apply the results offered by Theorem 4 to derive all solutions of the problem considered in Example 1. We take the matrix $\boldsymbol{B}$ and replace one element in each row of $\boldsymbol{B}$ by $\mathbb{0} = -\infty$ to produce the sparsified matrices

$$\boldsymbol{B}_1 = \begin{pmatrix} 0 & \mathbb{0} \\ 2 & \mathbb{0} \end{pmatrix}, \quad \boldsymbol{B}_2 = \begin{pmatrix} \mathbb{0} & -2 \\ 2 & \mathbb{0} \end{pmatrix}, \quad \boldsymbol{B}_3 = \begin{pmatrix} 0 & \mathbb{0} \\ \mathbb{0} & -3 \end{pmatrix}, \quad \boldsymbol{B}_4 = \begin{pmatrix} \mathbb{0} & -2 \\ \mathbb{0} & -3 \end{pmatrix}.$$

To find those sparsified matrices, which satisfy the conditions of the theorem, we need to calculate the matrices

$$\boldsymbol{B}_1^-(\boldsymbol{B} \oplus \boldsymbol{I}) = \begin{pmatrix} 0 & -2 \\ \mathbb{0} & \mathbb{0} \end{pmatrix}, \qquad \boldsymbol{B}_2^-(\boldsymbol{B} \oplus \boldsymbol{I}) = \begin{pmatrix} 0 & -2 \\ 2 & 0 \end{pmatrix},$$

$$\boldsymbol{B}_3^-(\boldsymbol{B} \oplus \boldsymbol{I}) = \begin{pmatrix} 0 & -2 \\ 5 & 3 \end{pmatrix}, \qquad \boldsymbol{B}_4^-(\boldsymbol{B} \oplus \boldsymbol{I}) = \begin{pmatrix} \mathbb{0} & \mathbb{0} \\ 5 & 3 \end{pmatrix}.$$

Furthermore, we obtain $\mathrm{Tr}(\boldsymbol{B}_1^-(\boldsymbol{B}\oplus\boldsymbol{I})) = \mathrm{Tr}(\boldsymbol{B}_2^-(\boldsymbol{B}\oplus\boldsymbol{I})) = 0 = \mathbb{1}$. Since the matrices $\boldsymbol{B}_1$ and $\boldsymbol{B}_2$ satisfy the conditions, they are accepted. Considering that $\mathrm{Tr}(\boldsymbol{B}_3^-(\boldsymbol{B} \oplus \boldsymbol{I})) = \mathrm{Tr}(\boldsymbol{B}_4^-(\boldsymbol{B} \oplus \boldsymbol{I})) = 3 > \mathbb{1}$, the last two matrices are rejected.

To represent all solutions of the problem, we calculate the matrices

$$\boldsymbol{S}_1 = \boldsymbol{B}_1^-(\boldsymbol{B} \oplus \boldsymbol{I})^* = \begin{pmatrix} 0 & -2 \\ \mathbb{0} & 0 \end{pmatrix}, \qquad \boldsymbol{S}_2 = \boldsymbol{B}_2^-(\boldsymbol{B} \oplus \boldsymbol{I})^* = \begin{pmatrix} 0 & -2 \\ 2 & 0 \end{pmatrix},$$

and then combine their columns to form a matrix that generates the solutions.

Taking into account that both columns of the matrix $\boldsymbol{S}_2$ are collinear to the second column of $\boldsymbol{S}_1$, we drop these columns, and define $\boldsymbol{S} = \boldsymbol{S}_1$. As a result, we represent the solution of the problem as

$$\boldsymbol{x} = \boldsymbol{S}\boldsymbol{u}, \qquad \boldsymbol{S} = \begin{pmatrix} 0 & -2 \\ \mathbb{0} & 0 \end{pmatrix}, \qquad \boldsymbol{u} \in \mathbb{R}^2.$$

Example 4. Finally, we apply Theorem 4 to the problem in Example 2. We first calculate the matrix

$$\boldsymbol{B} = \lambda_1^{-1}\boldsymbol{A} = \begin{pmatrix} 0 & \mathbb{0} \\ 2 & -3 \end{pmatrix}.$$

We can derive two sparsified matrices $\boldsymbol{B}_1$ and $\boldsymbol{B}_2$ from $\boldsymbol{B}$, and write

$$\boldsymbol{B}_1 = \begin{pmatrix} 0 & \mathbb{0} \\ 2 & \mathbb{0} \end{pmatrix}, \qquad \boldsymbol{B}_1^-(\boldsymbol{B} \oplus \boldsymbol{I}) = \begin{pmatrix} 0 & -2 \\ \mathbb{0} & \mathbb{0} \end{pmatrix}, \qquad \mathrm{Tr}(\boldsymbol{B}_1^-(\boldsymbol{B} \oplus \boldsymbol{I})) = 0;$$

$$\boldsymbol{B}_2 = \begin{pmatrix} 0 & \mathbb{0} \\ \mathbb{0} & -3 \end{pmatrix}, \qquad \boldsymbol{B}_2^-(\boldsymbol{B} \oplus \boldsymbol{I}) = \begin{pmatrix} \mathbb{0} & \mathbb{0} \\ 5 & 3 \end{pmatrix}, \qquad \mathrm{Tr}(\boldsymbol{B}_2^-(\boldsymbol{B} \oplus \boldsymbol{I})) = 3.$$

The matrix $\boldsymbol{B}_2$ does not satisfy the condition of the theorem, and thus is rejected. Note that the matrix $\boldsymbol{B}_1$ coincides with the corresponding matrix in Example 3. Since the matrix $\boldsymbol{B}_1$ completely determines the computations, the solution to the problem has the same form as that obtained in this example.

5 Conclusions

The paper focused on the development of methods and techniques for the complete solution of an optimization problem, formulated in the framework of tropical mathematics to minimize a nonlinear function defined by a matrix on vectors over idempotent semifield. As the starting point, we have taken our previous result, which offers a partial solution to the problem with both irreducible and reducible matrices. To further extend this result, we have first obtained a new partial solution of the problem in the case of a reducible matrix, and derived a characterization of the solutions in the form of a vector inequality. We have developed an approach to describe all solutions of the problem as a family of solution subsets by using a matrix sparsification technique. To generate all members of the family in a reasonable way, we have proposed a backtracking procedure. Finally, we have offered a representation for the complete solution of the problem in a compact vector form, ready for further analysis and calculation.

The results obtained were illustrated with illuminating numerical examples.

The directions of future research will include the development of real-world applications of the solutions proposed. A detailed analysis of the computational complexity of the backtracking procedure is of particular interest. Various extensions of the solution to handle other classes of optimization problems with different objective functions and constraints, and in different algebraic settings are also considered promising lines of future investigation.

Acknowledgments. This work was supported in part by the Russian Foundation for Humanities (grant number 16-02-00059). The author is very grateful to three referees for their extremely valuable comments and suggestions, which have been incorporated into the revised version of the manuscript.

References

1. Aminu, A., Butkovič, P.: Non-linear programs with max-linear constraints: a heuristic approach. IMA J. Manag. Math. **23**(1), 41–66 (2012)
2. Butkovič, P.: Max-linear Systems, Springer Monographs in Mathematics. Springer, London (2010)
3. Elsner, L., van den Driessche, P.: Max-algebra and pairwise comparison matrices. Linear Algebra Appl. **385**(1), 47–62 (2004)
4. Elsner, L., van den Driessche, P.: Max-algebra and pairwise comparison matrices, II. Linear Algebra Appl. **432**(4), 927–935 (2010)
5. Gavalec, M., Ramík, J., Zimmermann, K.: Decision Making and Optimization. Lecture Notes in Economics and Mathematical Systems. Springer, Cham (2015)
6. Golan, J.S.: Semirings and Affine Equations Over Them, Mathematics and Its Applications, vol. 556. Kluwer Academic Publishers, Dordrecht (2003)
7. Gondran, M., Minoux, M.: Graphs, Dioids and Semirings. Operations Research/ Computer Science Interfaces. Springer, New York (2008)
8. Heidergott, B., Olsder, G.J., van der Woude, J.: Max Plus at Work. Princeton Series in Applied Mathematics. Princeton University Press, Princeton (2006)

9. Hudec, O., Zimmermann, K.: Biobjective center - balance graph location model. Optimization **45**(1–4), 107–115 (1999)
10. Itenberg, I., Mikhalkin, G., Shustin, E.: Tropical Algebraic Geometry. Oberwolfach Seminars, vol. 35. Birkhäuser, Basel (2007)
11. Kolokoltsov, V.N., Maslov, V.P.: Idempotent Analysis and Its Applications, Mathematics and Its Applications, vol. 401. Kluwer Academic Publishers, Dordrecht (1997)
12. Krivulin, N.K.: Eigenvalues and eigenvectors of matrices in idempotent algebra. Vestnik St. Petersburg Univ. Math. **39**(2), 72–83 (2006)
13. Krivulin, N.K.: Solution of generalized linear vector equations in idempotent algebra. Vestnik St. Petersburg Univ. Math. **39**(1), 16–26 (2006)
14. Krivulin, N.K., Plotnikov, P.V.: On an algebraic solution of the Rawls location problem in the plane with rectilinear metric. Vestnik St. Petersburg Univ. Math. **48**(2), 75–81 (2015)
15. Krivulin, N.: Complete solution of a constrained tropical optimization problem with application to location analysis. In: Höfner, P., Jipsen, P., Kahl, W., Müller, M.E. (eds.) RAMICS 2014. LNCS, vol. 8428, pp. 362–378. Springer, Cham (2014). doi:10.1007/978-3-319-06251-8_22
16. Krivulin, N.: Tropical optimization problems. In: Petrosyan, L.A., Romanovsky, J.V., Yeung, D.W.K. (eds.) Advances in Economics and Optimization, pp. 195–214. Nova Science Publishers, New York (2014). Economic Issues, Problems and Perspectives
17. Krivulin, N.: Extremal properties of tropical eigenvalues and solutions to tropical optimization problems. Linear Algebra Appl. **468**, 211–232 (2015)
18. Krivulin, N.: A multidimensional tropical optimization problem with nonlinear objective function and linear constraints. Optimization **64**(5), 1107–1129 (2015)
19. Krivulin, N.: Rating alternatives from pairwise comparisons by solving tropical optimization problems. In: Tang, Z., Du, J., Yin, S., He, L., Li, R. (eds.) 2015 12th International Conference on Fuzzy Systems and Knowledge Discovery (FSKD), pp. 162–167. IEEE (2015)
20. Krivulin, N.: Solving a tropical optimization problem via matrix sparsification. In: Kahl, W., Winter, M., Oliveira, J.N. (eds.) RAMICS 2015. LNCS, vol. 9348, pp. 326–343. Springer, Cham (2015). doi:10.1007/978-3-319-24704-5_20
21. Krivulin, N.: Using tropical optimization techniques to evaluate alternatives via pairwise comparisons. In: Gebremedhin, A.H., Boman, E.G., Ucar, B. (eds.) 2016 Proceedings of 7th SIAM Workshop on Combinatorial Scientific Computing, pp. 62–72. SIAM, Philadelphia (2016)
22. Krivulin, N.: Direct solution to constrained tropical optimization problems with application to project scheduling. Comput. Manag. Sci. **14**(1), 91–113 (2017)
23. Maclagan, D., Sturmfels, B.: Introduction to Tropical Geometry. Graduate Studies in Mathematics, vol. 161. AMS, Providence (2015)
24. McEneaney, W.M.: Max-Plus Methods for Nonlinear Control and Estimation. Systems and Control, Foundations and Applications. Birkhäuser, Boston (2006)
25. Tam, K.P.: Optimizing and Approximating Eigenvectors in Max-Algebra. Ph.D. thesis. The University of Birmingham, Birmingham (2010)
26. Tharwat, A., Zimmermann, K.: One class of separable optimization problems: solution method, application. Optimization **59**(5), 619–625 (2010)
27. Zimmermann, K.: Disjunctive optimization, max-separable problems and extremal algebras. Theor. Comput. Sci. **293**(1), 45–54 (2003)

Concurrency-Preserving Minimal Process Representation

Adrián Puerto(✉)

Dipartimento di Informatica Sistemistica e Comunicazione, Università degli Studi di Milano-Bicocca, Milan, Italy
adrian.puertoaubel@disco.unimib.it

Abstract. We propose a method for reducing a partially ordered set, in such a way that the lattice derived from a closure operator based on concurrency is changed as little as possible. In fact, we characterize in which cases it remains unchanged, and prove minimality of the resulting reduced poset. In these cases, we can complete this poset so as to obtain a causal net on which the closure operator will lead to the same lattice.

Keywords: Partial order · Concurrency · Closure operator · Lattice · Atom · Causal net

1 Introduction

In this work, we explore canonical representability of processes based on their concurrency features. We study the closure operator based on concurrency developed in [1–3]. These works focus on a paradigm that considers a set of events, and the set of conditions they relate to as a bipartite directed graph called a Petri net. Acyclic Petri nets are a suitable representation of asynchronous processes, in that their elements are partially ordered [4]. Indeed, partially ordered sets are a common characteristic in different models of true concurrent processes, such as event structures [11, 14] or Mazurkiewicz traces [9]. All these models share the fact that maximal totally ordered subsets (or chains) represent sequential subprocesses, whereas maximal subsets of pairwise unordered elements (or antichains) represent global states. Several subset operators have been defined on these structures. In particular, *Nielsen et al.* have presented downwards closed subsets of event structures, or configurations, as forming interesting spaces when ordered by inclusion: domains [11, 14]. Intuitively, these configurations gather information on the history of the subprocess leading to a particular set of events. In this sense they can be understood as partial states, uniquely determined by the past of a given subset, thus relying solely on causal dependence relations. As a matter of fact, inclusion in a domain represents a chronological ordering of possible observations. The closure operator studied in this paper however, evolves around concurrency relations and independence of subprocesses. Like in domains, it provides a family of subsets, and endows it with a structure. Nevertheless, unlike

P. Höfner et al. (Eds.): RAMiCS 2017, LNCS 10226, pp. 242–257, 2017.
DOI: 10.1007/978-3-319-57418-9_15

configurations, it presents a notion of complementation determined by concurrency. Any element is by definition concurrent to all the elements in its complement, and so it could not distinguish one of them from another without a global clock. From its local point of view its whole complement subset behaves as a single local state. In the structure obtained from this closure operator, inclusion is not to be interpreted as a chronology, but rather conveys the idea of a coarser point of view. Informally, if we consider that sequential subprocesses can only share information by synchronising or splitting, then this operator determines which information about the process is available locally.

Even though this closure operator was originally defined excluding conflict situations [2], the authors then extended their results to include these [1]. In the same line of ideas, the results presented in this paper do not consider conflict, as we intend to consider them further on. Also, this operator is described in the frame of acyclic Petri nets, but the original authors generalise the result to partial orders. One of their core contributions is the identification of a local property of these partial orders, N-density (see Remark 2), with a property of the space of subsets obtained from the closure operator, orthomodularity (see Definition 4). Analogously, the results of this paper are framed in terms of Petri nets, but presented for more general partially ordered sets, so that the results could be applied to other models of concurrent processes, such as the above mentioned. In fact, instead of N-density and orthomodularity, we here consider weaker notions, so that our results are presented in a slightly more general form.

In this work, we propose to use this closure operator for reducing a partial order. We characterize the cases in which this reduction preserves the structure of closed sets, and we prove minimality of the obtained partial order. Intuitively, the obtained partial order is an abstraction of the process, carrying only the information which is available locally. In this sense, it is the skeleton of the interactions between sequential subprocesses. Indeed, all the elements of the reduced partial order are involved either in a synchronisation, or a branching, or both.

In Sect. 2, the theoretical background is presented. Partial orders, together with their in-line and concurrency relations are formally defined, so as to introduce the notions of complement, and closure operator. Well known results relating this type of operators with lattice theory are presented, and finally, the central feature of our novel construction is defined: atoms, or minimal closed subsets. Section 3 presents the reduction: we try to obtain a normalised version of the partial order in which the previously defined atoms become single elements. The first propositions show some properties of atoms, that will allow us to order them partially in a way consistent with the original poset. The obtained partial order is a minimal representation of causal dependencies such that the overall concurrency structure is respected as much as possible. Section 4 is devoted to show under which assumptions, and to which extent this structure is fully preserved. To this aim, we characterise the concurrency relation on the reduced poset in terms of closures and complements of the original one. We then construct the corresponding lattice of closed sets, and explicitly map it to the closed set-lattice of the original poset with a homomorphism. We then show that this homomorphism is an embedding, and under which conditions it is an actual isomorphism. Finally, minimality of the reduced partial order is proven. In Sect. 5 we provide an example on how to use this reduction. Starting from a rather general (although N-dense) partial order, we apply the reduction technique. We then perform a completion of the reduced partial order so as to obtain a Petri net. In this new structure, we see that the elements of the reduced poset correspond to the conditions of the Petri net. This fact should clarify the idea that the considered closed sets abstract sequential subprocesses as local states.

The proofs are omitted in this version, for space reasons. The extended version containing the full proofs can be found here:
(www.mc3.disco.unimib.it/pub/P17.pdf)

2 Posets, Closure Operators and Lattices

A natural way to represent a process is by ordering a set of events, in some cases together with local states, or conditions. We do so by means of an order relation that should be interpreted as a causal dependence between these.

Definition 1 ***Partially Ordered Set.*** *A* Partially Ordered Set, *or* poset *is a set P equipped with a* reflexive, antisymmetric, *and* transitive *relation $\leq_P \subseteq P^2$ that we call* order. *Whenever two elements $x, y \in P$ are ordered either $x \leq_P y$ or $y \leq_P x$, we say they are in* line: $\mathbf{li}_P := \leq_P \cup \leq_P^{-1}$. *Conversely, if a pair of elements is not in the order relation we say they are* concurrent: $\mathbf{co}_P := P^2 \setminus \mathbf{li}_P$.

We may define the covering relation *associated to $\leq_P$, $\prec_P := \leq_P \setminus \leq_P^2$. We say $(P, \leq_P)$ is* combinatorial *whenever the order is the transitive and reflexive closure of the covering relation: $\leq_P = (\prec_P)^\star$.*

Example 1. Figure 1 shows the following poset: $P_1 = \{x, y, u, v\}$ with $\leq_{P_1} = \{(x, y), (x, u), (v, y)\}$. We can clearly see that x $\mathbf{co}_P$ v, u $\mathbf{co}_P$ y and u $\mathbf{co}_P$ v.

We pay special attention to the **co** relation. Since $\leq$ is reflexive, **co** must be *irreflexive*, and **li** being *symmetric*, so must also be **co**.

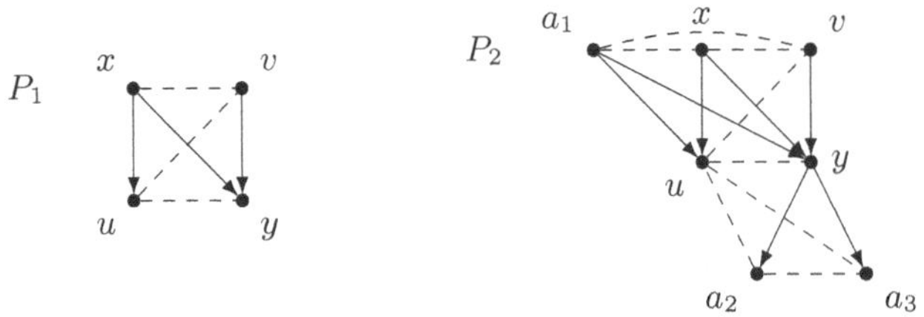

Fig. 1. Two *combinatorial posets* P_1 and P_2. Arrows represent $\prec P_i$, and dashed lines represent $\mathbf{co}_{P_i}$, $i \in \{1, 2\}$. On the left, $P_1 = \{x, y, u, v\}$ with $\leq_{P_1} = \{(x, y), (x, u), (v, y)\}$. We can clearly see that (x, v), (u, y), $(u, v) \in \mathbf{co}_{P_1}$, as represented by the dashed lines.

We would like to identify which elements of the partial order are concurrent to the same elements. To this aim, we may extend the concurrency relation to the power set $\mathcal{P}(P) = \{S \mid S \subseteq P\}$ of P as follows.

Definition 2. *For any subset $S \subseteq P$, we define the* polarity *induced by* $\mathbf{co}_P$ *as*

$$(\cdot)' : \mathcal{P}(P) \longrightarrow \mathcal{P}(P)$$
$$S \longmapsto S' := \{y \in P \mid \forall x \in S : x \,\mathbf{co}_P\, y\}$$

We will henceforth refer to this operator simply as a polarity, *and refer to S' as the* polar *of S.*

It is a well known result that applying such a polarity two times yields a closure operator on $\mathcal{P}(P)$. In fact, $((\cdot)', (\cdot)')$ is a Galois connection [5, Chap. V Sect. 7].

Definition 3. *For any subset $S \subseteq P$, we define the* closure *induced by* $\mathbf{co}_P$ *as*

$$(\cdot)'' : \mathcal{P}(P) \longrightarrow \mathcal{P}(P)$$
$$S \longmapsto S'' := (S')'$$

This operator is indeed:

- extensive $\forall S \in \mathcal{P}(P) : S \subseteq S''$
- monotone $\forall S_1, S_2 \in \mathcal{P}(P) : (S_1 \subseteq S_2 \Rightarrow S_1'' \subseteq S_2'')$, *and*
- idempotent $\forall S \in \mathcal{P}(P) : (S'')'' = S''$

We will from this point on refer to S'' as the closure *of S, and consider the space $L(P) = \{S'' \mid S \in \mathcal{P}(P)\}$ of closed subsets of P.*

The empty set $\emptyset$ and the full poset P are trivially closed, and polar to each other. As for any structure defined on a power set, there is a natural ordering of its elements induced by inclusion. In this way, a set precedes another if it is contained in it. When ordered in such a way, $L(P)$ forms again a poset, and as such it is common practice to represent it as a Hasse diagram.

Example 2. In Fig. 2 we can see the closed sets of P_1 from Fig. 1. Since $u \,\mathbf{co}_{P_1}\, v$ and $u \,\mathbf{co}_{P_1}\, y$, we have that $\{u\}' = \{v, y\}$, and no other element is concurrent to both v and y, so $\{u\}'' = \{u\}$. If we consider $\{x\}' = \{v\}$, since $u \,\mathbf{co}_{P_1}\, v$ we have that $u \in \{x\}''$. In fact $\{x\}'' = \{x, u\}$. We may represent $L(P_1)$ as a Hasse diagram. We note that P_1 is common in the literature and often referred to as the *N Poset*. It is, in particular, the paradigmatic poset that is not series-parallel.

It is a well known result [5, Chap. V Sect. 7] that, when considering the closure operator associated to a symmetric relation, the resulting collection of closed sets forms a complete lattice. Furthermore, since the **co** relation we handle is also irreflexive, $L(P)$ is actually an *orthocomplemented lattice* or shortly, *ortholattice*. As such, $L(P)$ is endowed with a set of operations related to the order relation induced by inclusion. Since it is not our purpose here to discuss general aspects of lattice theory, we will consider only the case of interest, and refer the inquisitive reader to [6].

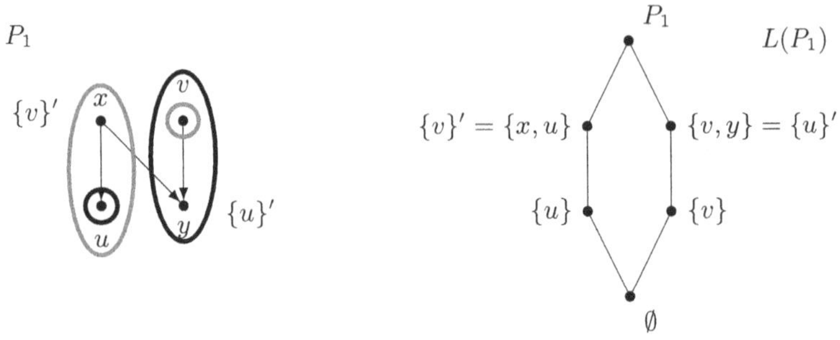

Fig. 2. Closure operator induced by concurrency relation on P_1 generates $L(P_1)$. On the left: closed sets as subsets of P_1. On the right: Hasse diagram of $L(P_1)$.

Remark 1. $L(P)$ with the usual set operations and $(\cdot)'$ forms an ortholattice: $\langle L(P), \leq_{L(P)}, \wedge^{L(P)}, \vee^{L(P)}, (\cdot)', \emptyset, P\rangle$

$$\text{order: } \leq_{L(P)} := \{(s_1, s_2) \in L(P)^2 \mid s_1 \subseteq s_2\}$$
$$\text{meet: } \forall S \subseteq L(P) : \bigwedge_{s \in S}{}^{L(P)} s = \bigcap_{s \in S} s$$
$$\text{join: } \forall S \subseteq L(P) : \bigvee_{s \in S}{}^{L(P)} s = (\bigcup_{s \in S} s)''$$
$$\text{orthocomplement: } \forall s \in L(P) : s' \in L(P)$$

We note that $\emptyset$, and P behave respectively as least and largest elements, often called *bottom* and *top* in lattice theory. These elements are often represented by $\bot$ and $\top$ in the literature. We however prefer to use 0_L and 1_L, to avoid confusion, so as to use the $\bot$ symbol consistently with [8], for the following relation. Orthocomplementation induces a binary relation called *orthogonality*:

$$\bot_P := \{(s_1, s_2) \in L(P)^2 \mid s_1 \leq_{L(P)} s_2'\}$$

Orthocomplements naturally behave such that $s_1 \leq_{L(P)} s_2' \Leftrightarrow s_2 \leq_{L(P)} s_1'$, so that orthogonality is symmetric and irreflexive. Intuitively, two sets are orthogonal, whenever they represent concurrent subprocesses. In fact it is apparent that $s_1 \bot_P s_2 \Leftrightarrow s_1 \times s_2 \subseteq \mathbf{co}_P$, which allows us to extend this relation to the whole of $\mathcal{P}(P)$.

We may now consider a particular type of closed sets, namely the smallest possible non-empty ones.

Definition 4. *An* atom *is a closed set which contains no proper non-empty closed subset. Given* $L(P)$*, we may consider its set of atoms:*

$$\mathcal{A}(P, \leq) = \mathcal{A}_P := \{A \subseteq P \mid A \neq \emptyset \wedge \ \forall S \subseteq A : (S \neq \emptyset \Rightarrow S'' = A)\}$$

In terms of lattices, an atom is an element such that each other element less or equal than it is either the bottom element, or the atom itself:

$$\forall a \in L(P) : (a \in \mathcal{A}_P \Leftrightarrow a \neq \emptyset \wedge (\forall s \in L(P) : (s \leq_{L(P)} a \Rightarrow (s = \emptyset \vee s = a))))$$

We say a lattice is atomic *if every element is greater or equal to some atom:*

$$L(P) \text{ is atomic} \Leftrightarrow \forall s \in L(P) : (\exists a \in \mathcal{A}_P : a \leq_{L(P)} s)$$

In an atomic lattice, we may consider the set of atoms under any given element:

$$\forall s \in L(P) : \mathcal{A}_s := \{a \in \mathcal{A}_P \mid a \leq_{L(P)} s\}$$

An atomic lattice is said to be atomistic *if every element can be expressed as the join of the atoms under it:*

$$L(P) \text{ is atomistic} \Leftrightarrow \forall s \in L(P) : s = \bigvee_{a \in \mathcal{A}_s}{}^{L(P)} a$$

Finally, an ortholattice is said to be orthomodular, *whenever* $\forall x, y \in L(P)$:

$$x \leq_{L(P)} y \Rightarrow y = x \vee^{L(P)} (x' \wedge^{L(P)} y)$$

Non-atomic lattices are those in which at least one closed set contains an infinite sequence of closed sets such that each is properly contained in the previous one. An atomistic lattice is one in which each closed set can be uniquely characterized by the atoms contained in it. In an orthomodular lattice, if a set properly contains another, then the former must intersect the orthocomplement of the latter. An orthomodular lattice is atomistic, and every atomistic lattice must be atomic [8, Chap. 3 Sect. 10]. The following examples should help clarify these notions.

Example 3. Consider the poset P_1 as previously defined. In Fig. 2 we can clearly identify the atoms: $\mathcal{A}_{P_1} = \{\{u\}, \{v\}\}$. Obviously, $L(P_1)$ is atomic. However, we cannot make the difference between $\{x, u\}$ and $\{u\}$ in terms of atoms, so it is not atomistic. Indeed $\mathcal{A}_{\{x,u\}} = \mathcal{A}_{\{u\}} = \{\{u\}\}$, so $\bigvee_{a \in \mathcal{A}_{\{u\}}}^{L(P)} a = \{u\} = \bigvee_{a \in \mathcal{A}_{\{x,u\}}}^{L(P)} a \neq \{x, u\}$. Since it is not atomistic, neither is it orthomodular, and we see that $\{u\} \subseteq \{x, u\}$ but $\{u\}' \cap \{x, u\} = \{v, y\} \cap \{x, u\} = \emptyset$. On Fig. 3 we see P_2 on which atoms are depicted, together with a couple of their complements. Clearly, $L(P_2)$ is atomic. On the Hasse diagram of $L(P_2)$ we can see that all elements are above different sets of atoms, hence it is atomistic. However, it is not orthomodular. Indeed $\{u\}$ is a closed set contained in $\{v\}'$, but $\{v\}' \cap \{u\}' = \emptyset$, so $\{u\} \vee^{L(P)} (\{u\}' \wedge^{L(P)} \{v\}') = \{u\} \vee^{L(P)} \emptyset = \{u\} \neq \{v\}'$.

Remark 2. It is worth noting that posets that lead to an orthomodular lattice have been characterized [2], assuming that the poset is combinatorial [4, Chap. 2 Sect. 2]. Under this condition, whenever a poset P has a subposet (i.e. a subset where order is preserved) isomorphic to P_1 from our example, then $L(P)$ will only be orthomodular if there is an element $a \in P$ such that $x \leq_P a$, $a \leq_P y$, and u $\mathbf{co}_P$ a $\mathbf{co}_P$ v. Such a property is called *N-density* [4, Chap. 2 Sect. 3]. Therefore, N-density is also a sufficient condition on a poset for its closed-set lattice to be atomistic. The poset P_3 in Fig. 4 is *N-dense*. Also, note that N-dense posets may contain the N poset (P_1) as a subposet. Hence, N-density is a weaker notion than N-freeness, and the class of N-dense posets is wider than that of series-parallel posets.

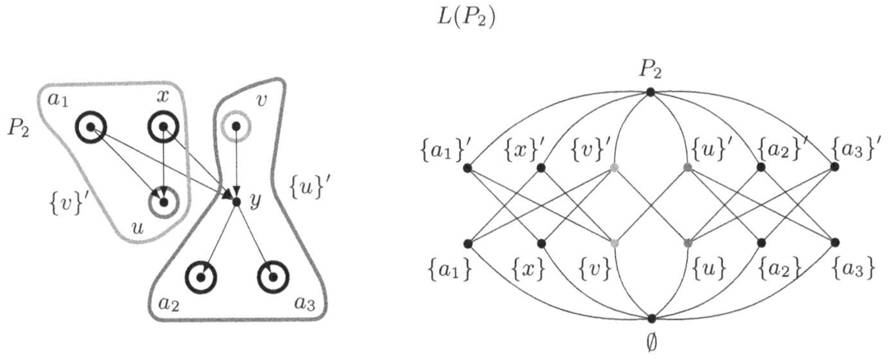

Fig. 3. On the left: some closed sets of P_2. On the right: Hasse diagram of $L(P_2)$

Example 4. In Fig. 4 we see an N-dense poset P_3 that contains P_1 as a subposet. On the figure, one of the ways we find P_1 as subposet of P_3 has been represented by labeling the corresponding points according to Fig. 1. Clearly, the element a is in the configuration described in the previous remark. The corresponding lattice $L(P_3)$ is orthomodular, and so atomistic.

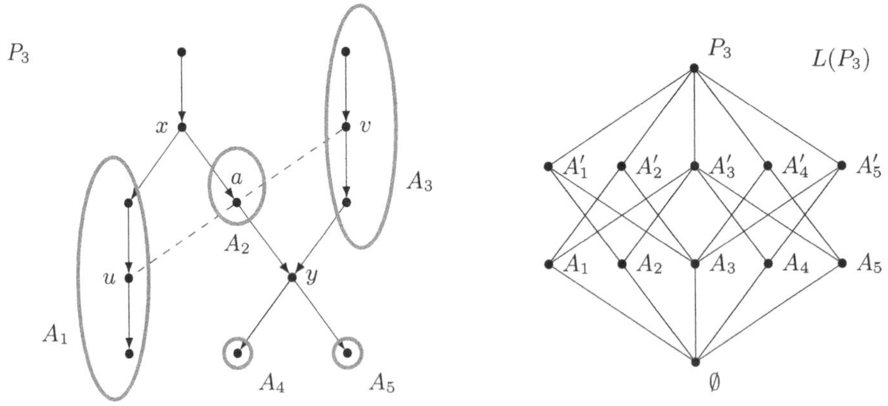

Fig. 4. On the left: N-dense poset P_3 and its atoms. On the right: $L(P_3)$ is orthomodular.

3 Poset of Atoms

In this section, we will start presenting the contribution of this paper, by paying special attention to atoms. We will consider them as subsets of a partial order, and we will see how this order can be extended to them, so that the set of atoms forms a subposet of the former.

As a first remark, we may notice that they are all pairwise disjoint.

Proposition 1. $\forall A_1, A_2 \in \mathcal{A}_P : (A_1 \cap A_2 \neq \emptyset \Rightarrow A_1 = A_2)$.

Another interesting property of atoms is that all the elements inside one of them relate identically to the elements outside of it, either by the order relation, or the concurrency one. Such a property of subsets is rather common in the literature, and referred to diversely according to the subject (See *D-autonomous* sets defined on pre-orders in [10] for additional references). We prove this property in Propositions 2 through 5. This will subsequently allow us to define a consistent order on the set of atoms.

In the case of concurrency, the result is quite straightforward:

Proposition 2. *Let* $A \in \mathcal{A}_P, x \in A, y \in P \backslash A : y\ \mathbf{co}_P\ x$. *Then* $\forall z \in A : y\ \mathbf{co}_P\ z$.

We will use the following result to prove the counterpart of Proposition 2 for the ordering relation.

Proposition 3. *Closed sets are convex. Formally:*

$$\forall S \subseteq P,\ \forall x, y \in S'' :\ (x \leq z \leq y) \Rightarrow (z \in S'')$$

We are now able to see that if an element of an atom is ordered before an element outside of it, then all the elements of this atom must be ordered accordingly.

Proposition 4. *Let* $A \in \mathcal{A}_P, x \in A, y \in P \setminus A$. *If* $x \leq_P y$, *then for each* $z \in A : z \leq_P y$.

An analogous proof leads to the following result.

Proposition 5. *Let* $A \in \mathcal{A}_P, x \in A, y \in P \setminus A$. *If* $y \leq_P x$, *then for each* $z \in A : y \leq_P z$.

For the sake of clarity, we summarize these results as follows. Let $A \in \mathcal{A}_P$, $y \in P \setminus A$:

- $(\exists x \in A : x\ \mathbf{co}_P\ y) \Rightarrow (\forall z \in A : z\ \mathbf{co}_P\ y)$
- $(\exists x \in A : x \leq_P y) \Rightarrow (\forall z \in A : z \leq_P y)$
- $(\exists x \in A : y \leq_P x) \Rightarrow (\forall z \in A : y \leq_P z)$

Under these conditions, we may define $\leq_{\mathcal{A}_P}$ as follows: Let $A_1, A_2 \in \mathcal{A}_P$. Then $A_1 \leq_{\mathcal{A}_P} A_2 :\Leftrightarrow: \exists x \in A_1, \exists y \in A_2 : x \leq_P y$. Then clearly $A_1 \leq_{\mathcal{A}_P} A_2 \Leftrightarrow \forall x \in A_1, \forall y \in A_2 : x \leq_P y$. Furthermore, $A_1\ \mathbf{co}_{\mathcal{A}_P}\ A_2 \Leftrightarrow A_1 \perp_P A_2$.

$\leq_{\mathcal{A}_P}$ is rather trivially an order on $\mathcal{A}_P$. It inherits reflexivity and transitivity from $\leq_P$, and if it were not antisymmetric, neither would $\leq_P$. As a matter of fact, any choice function f, defined as follows, is an order embedding:

$$\begin{aligned} f : \mathcal{A}_P &\longrightarrow P \\ A &\longmapsto f(A) := x \in A \end{aligned} \tag{1}$$

And so, $(\mathcal{A}_P, \leq_{\mathcal{A}_P})$ can be embedded into $(P, \leq_P)$. Indeed, provided f exists, its injectivity comes as a consequence of atoms being pairwise disjoint (see Proposition 1). This justifies the idea that $(\mathcal{A}_P, \leq_{\mathcal{A}_P})$ is a reduced version of $(P, \leq_P)$. Intuitively, the reduction can be seen as a collapsing of the atoms. As we will show in the next section, atoms are sufficient to recover the lattice of closed sets. On the other hand, Propositions 2 to 5 imply not only that atoms are totally ordered subsets, but also that no branchings occur at any of their elements. Indeed, no element of an atom has more than one predecessor and one successor. This should clarify that the inner structures of atoms provide no information on the interactions between the different sequential subprocesses represented in the poset. Instead, all this information is condensed in their outer structure: the reduced poset $(\mathcal{A}_P, \leq_{\mathcal{A}_P})$. For instance, a totally ordered set, representing a single sequential process will consist of one single atom, and its reduced version will then be a single isolated element. A set of n non interacting sequential processes would consist of the corresponding n atoms, leading to a reduced version of n pairwise concurrent elements. Naturally, the structure of the reduced poset would grow more complex as sequential processes interact, stepping away from these trivial examples.

In the following section, we will see that the concurrency structure of $(P, \leq_P)$ is preserved in $(\mathcal{A}_P, \leq_{\mathcal{A}_P})$ by means of orthogonality, and under which conditions this statement fully holds.

4 Atoms as Lattice Generators

The results presented in this section rely heavily on the fact that an atomistic lattice is uniquely determined by its set of atoms, and their orthogonality relation. We study how this implies that $L(\mathcal{A}_P)$ and $L(P)$ are isomorphic. We will see that $L(\mathcal{A}_P)$ is always atomistic, and after inspecting the orthogonality relation in both lattices, build the actual isomorphism.

The idea behind the following result arises from the observation that $\forall A_1, A_2 \in \mathcal{A}_P : \{A_1\} \perp_{\mathcal{A}_P} \{A_2\} \Leftrightarrow A_1 \ \mathbf{co}_{\mathcal{A}} \ A_2 \Leftrightarrow A_1 \perp_P A_2$, which can in fact be extended to arbitrary subsets of $\mathcal{A}_P$.

Proposition 6. *Let $B_1, B_2 \subseteq \mathcal{A}_P$. Then $B_1 \perp_{\mathcal{A}_P} B_2 \Leftrightarrow (\bigcup_{A \in B_1} A) \perp_P (\bigcup_{A \in B_2} A)$.*

We can now define $L(\mathcal{A}_P)$ in an analogous manner to $L(P)$, to the point of showing that $L(\mathcal{A})$ can be embedded into $L(P)$. We shall later see under which conditions, this embedding is actually an isomorphism.

To this aim, the following two propositions show that $L(\mathcal{A}_P)$ is an atomistic lattice.

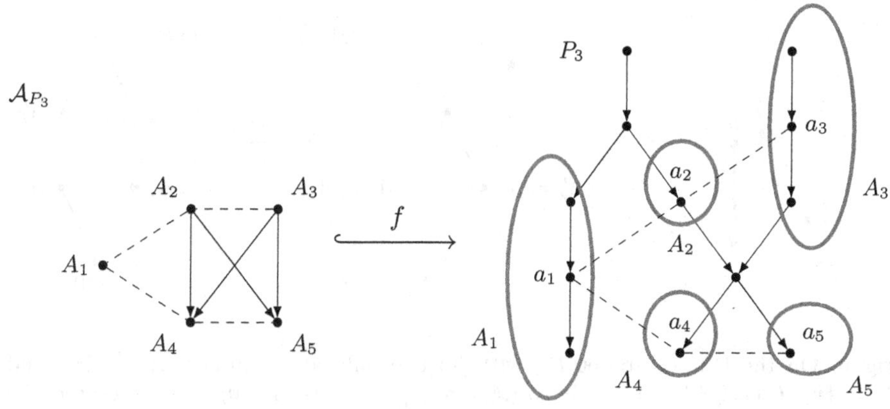

Fig. 5. On the left, $\mathcal{A}_{P_3} = \mathcal{A}(P_3, \leq_{P_3})$, obtained from P_3 on the right. Consider $f : \mathcal{A}_{P_3} \longrightarrow P_3$ such that $f(A_i) = a_i \forall i \in 1,..,5$. The elements in its image have been labeled showing the actual embedding of $\mathcal{A}_{P_3}$ into P_3. Clearly, f both preserves and reflects concurrency. Note that since $L(P_3)$ is atomistic (see Fig. 4), it is isomorphic to $L(\mathcal{A}_{P_3})$.

Proposition 7. $\forall A \in \mathcal{A}_P : \{A\} \in L(\mathcal{A}_P)$.

This trivially implies that $L(\mathcal{A}_P)$ is atomistic.

Proposition 8. $L(\mathcal{A}_P)$ *is* atomistic, *formally:*

$$\forall B \in L(\mathcal{A}_P) : B = \bigvee_{A \in B} \{A\} = (\bigcup_{A \in B} \{A\})''$$

Or equivalently:
$\forall B_1, B_2 \in L(\mathcal{A}_P) : (\{A \in \mathcal{A}(\mathcal{A}_P, \leq_{\mathcal{A}_P}) \mid \{A\} \subseteq B_1\} = \{A \in \mathcal{A}(\mathcal{A}_P, \leq_{\mathcal{A}_P}) \mid \{A\} \subseteq B_2\}) \Rightarrow B_1 = B_2$

At this point, we show that $L(\mathcal{A}_P)$ can be embedded into $L(P)$. We do so by defining a map between them, and showing that it is an order homomorphism in Proposition 9, and that it is injective in Proposition 10.

Definition 5. *We define the morphism ϕ that will turn out to be an injective order homomorphism (i.e. an embedding)*

$$\begin{aligned} \phi : L(\mathcal{A}_P) &\longrightarrow L(P) \\ B &\longmapsto \phi(B) := \bigvee_{A \in B} A = (\bigcup_{A \in B} A)'' \end{aligned}$$

We start by proving that ϕ preserves and reflects the order of the corresponding lattices.

Proposition 9. $\forall B_1, B_2 \in L(\mathcal{A}_P) : B_1 \leq_{L(\mathcal{A}_P)} B_2 \Leftrightarrow \phi(B_1) \leq_{L(P)} \phi(B_2)$.

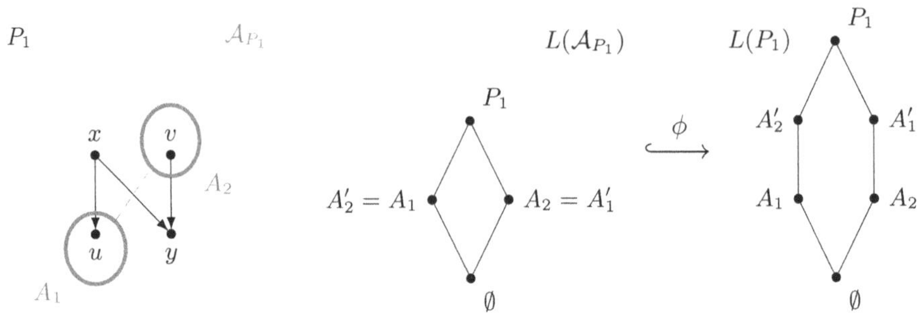

Fig. 6. On the left, the poset P_1, with its two only atoms drawn: $A_1 = \{u\}$ and $A_2 = \{v\}$. Clearly $A_1 \ \mathbf{co}_{\mathcal{A}_{P_1}} \ A_2$ so $(\mathcal{A}_{P_1}, \leq_{\mathcal{A}_{P_1}}) = (\{A_1, A_2\}, \emptyset)$. On the center, the corresponding lattice $L(\mathcal{A}_{P_1})$, which is embeddable into $L(P_1)$, on the right. Note that $A_2 \leq_{L(P_1)} A'_1 \Rightarrow A_2 \perp_{L(P_1)} A_1$, so the embedding ϕ preserves orthogonality.

This implies that ϕ is an order homomorphism. So if it were injective it would be an order embedding. Let us confirm this by proving the injectivity of ϕ.

Proposition 10. *Let* $B_1, B_2 \in L(\mathcal{A}_P)$ *such that* $\phi(B_1) = \phi(B_2)$, *then* $B_1 = B_2$.

Thus, ϕ^{-1} is a well defined function on the codomain of ϕ. As a matter of fact, provided a closed set of the codomain, it simply returns the set formed by the atoms under it : $\phi^{-1}(S) = \{A \in \mathcal{A}_P \mid A \leq_{L(P)} S\}$.

We can now positively state that $L(\mathcal{A}_P)$ can be embedded into $L(P)$:

$$\phi : L(\mathcal{A}_P) \hookrightarrow L(P)$$

Our purpose is however to go further and have $L(\mathcal{A}_P)$ and $L(P)$ isomorphic, for which we require ϕ to be surjective. However, in general this is not the case, as we may see in Fig. 6. As we can see, the fact preventing ϕ from being surjective, seems to be that A'_2 (respectively A'_1) cannot be differentiated from A_1 (A_2) solely in terms of atoms. This observation leads naturally to the following result:

Proposition 11. ϕ *is surjective iff* $L(P)$ *is atomistic.*

We naturally would like to characterize the posets $(P, \leq)$ for which $L(P)$ is atomistic, in this sense it seems clear that a necessary and sufficient condition will be:

$$\forall S \subseteq P : (\exists A_1 \subseteq P : A''_1 \subsetneq S'') \Rightarrow (\exists A_2 \subseteq P : A''_2 \subsetneq S'' \text{ and } A''_1 \neq A''_2) \quad (2)$$

We note that this condition is strictly weaker than N-density, which in turn is weaker than N-freeness. For instance, poset P_2 of Figs. 1 and 3 verifies it, although it is not N-dense, (and therefore neither N-free). When $L(P)$ is atomistic, we are able to assert that ϕ is an order isomorphism. Furthermore, when this is the case, Proposition 13 will prove that ϕ is even an orthocomplemented lattice isomorphism, preserving not only order, but orthogonality as well. Propositions 14 and 15 will then show that ϕ preserves lattice operations.

In order to do this, we first require the following technical result.

Proposition 12. *If* $L(P)$ *is atomistic, then*

$$\forall S \in L(P): \; S' = (\bigcup_{A \in \mathcal{A}_P : A \perp_P S} A)''$$

We shall now prove that ϕ preserves orthocomplementation.

Proposition 13. *If* $L(P)$ *is atomistic then* $\forall B \in L(\mathcal{A}_P): \; \phi(B') = \phi(B)'$.

Therefore, ϕ is an ortholattice isomorphism, hence $L(P) \simeq L(\mathcal{A}_P)$. We may confirm this by giving the two following results.

Proposition 14. $\forall B_1, B_2 \in L(\mathcal{A}_P) : \phi(B_1 \vee^{L(\mathcal{A}_P)} B_2) = \phi(B_1) \vee^{L(P)} \phi(B_2)$.

The following result is however not as trivial, and requires $L(P)$ to be atomistic as well.

Proposition 15. $\forall B_1, B_2 \in L(\mathcal{A}_P) : \phi(B_1 \wedge^{L(\mathcal{A}_P)} B_2) = \phi(B_1) \wedge^{L(P)} \phi(B_2)$.

At this point, we are willing to state that $(\mathcal{A}_P, \leq_{\mathcal{A}_P})$ is the smallest poset embeddable into $(P, \leq_P)$ such that $L(P) \simeq L(\mathcal{A}_P)$, provided $L(P)$ is atomistic.

We formalise this notion of minimality, as follows: Let $(P', \leq_{P'})$ be an arbitrary poset. If $(P', \leq_{P'})$ can be embedded into $(P, \leq_P)$, and $L(P) \simeq L(P')$, then $(\mathcal{A}_P, \leq_{\mathcal{A}_P})$ can be embedded into $(P', \leq_{P'})$.

Proposition 16. *Let* $(P, \leq_P)$ *and* $(P', \leq_{P'})$ *be two posets such that* $L(P)$ *and* $L(P')$ *are atomistic and isomorphic, and there exists an order embedding* $g : (P', \leq_{P'}) \hookrightarrow (P, \leq_P)$. *Let* $\mathcal{A}_P = \mathcal{A}(P, \leq_P)$, *then* $(\mathcal{A}_P, \leq_{\mathcal{A}_P})$ *can be embedded into* $(P', \leq_{P'})$.

5 Completion and Causal Nets

We have provided a way to reduce a poset to a minimal form that preserves the lattice obtained from the closure induced by $\mathbf{co}_P$. This minimality has some implications in terms of local structure, which we shall exploit to offer a possible application of the presented reduction.

Occurrence nets are a prevalent paradigm for modelling distributed processes. In this case, we leave conflict situations for further study, and consider only causal nets: the class of conflict-free occurrence nets.

Definition 6. *A* Petri net $N = (B, E, \mathcal{F})$ *is a bipartite directed graph with the set of vertices partitioned in into the disjoint sets* B *and* E, *and* $\mathcal{F} \subseteq (B \times E) \cup (E \times B)$ *as the set of edges. Elements in* B *are called* places, *those in* E transitions. *Edges can only link places with transitions.*

Whenever it is acyclic, one may consider the reflexive and transitive closure $(\mathcal{F})^*$ *of* $\mathcal{F}$*, so that* $(B \cup E, (\mathcal{F})^*)$ *is a partial order.*

A causal net *is an acyclic net which presents no conflict. This is the case whenever the following property holds:* $\forall b \in B : |\{e \in E \mid (e,b) \in \mathcal{F}\}| \leq 1$ *and* $|\{e \in E \mid (b,e) \in \mathcal{F}\}| \leq 1$*. We furthermore require that* $\forall e \in E : \{b \in B \mid (b,e) \in \mathcal{F}\} \neq \emptyset$ *and* $\{b \in B \mid (e,b) \in \mathcal{F}\} \neq \emptyset$*.*

Typically, in graphical representations, places are displayed as circles, and transitions as rectangles, as in Fig. 7. When considering causal nets as partial orders, two properties are most relevant [4]. First, the order relation is *combinatorial*, since it is the transitive and reflexive closure of the covering relation $\mathcal{F}$. Second, all causal nets are *N-dense*, a notion mentioned in Remark 2.

Definition 7. *A combinatorial poset* $(P, \leq_P)$ *is* N-dense *iff* $\forall x, y, u, v \in P$ *such that* $x \leq_P y, x \leq_P u, v \leq_P y, x \ \mathbf{co}_P\ v, u \ \mathbf{co}_P\ v$*, and* $u \ \mathbf{co}_P\ y$*, there is an element* $a \in P$ *verifying* $a \ \mathbf{co}_P\ u, a \ \mathbf{co}_P\ v$ *and* $x \leq_P a \leq_P y$*.*

The poset P_1 in Fig. 1 is not N-dense, whereas P_3 in Fig. 4 is.

Combinatorialness, and N-density are however not sufficient for a partial order to be a causal net. Under these conditions one could still be unable to find a suitable bipartition of the elements.

We show, however, that the atoms of a poset, as defined in this paper, structurally behave like places in the following sense. Given a combinatorial and N-dense poset $(P, \leq_P)$, the poset $(\mathcal{A}_P, \leq_{\mathcal{A}_P})$ resulting from reduction can be completed with transitions, so as to obtain a causal net.

We will first note that whenever $\leq_P$ is combinatorial, so must be $\leq_{\mathcal{A}_P}$. On the other hand, reduction also preserves N-density.

Proposition 17. *Let* $(P, \leq_P)$ *be combinatorial, and N-dense. Then* $(\mathcal{A}_P, \leq_{\mathcal{A}_P})$ *is N-dense.*

We will henceforth assume that both $(P, \leq_p)$, and $(\mathcal{A}_P, \leq_{\mathcal{A}_P})$ are combinatorial and N-dense.

Our purpose is now to perform a Dedekind-MacNeille completion on $(\mathcal{A}_P, \leq_{\mathcal{A}_P})$ (see for example [6, Chap. 7 Sect. 36]). To this aim, we introduce some notation:

Definition 8. *Let* $(P, \leq_P)$ *be a poset, and define, for each* $S \subseteq P$*:*
The upset of S*:* $\uparrow S := \{x \in P \mid \forall s \in S : s \leq_P x\}$
The downset of S*:* $\downarrow S := \{x \in P \mid \forall s \in S : x \leq_P s\}$
The Dedekind Mac-Neille completion of $(P, \leq_P)$ *is* $\mathbf{DM}(P) := \{S \subseteq P \mid \downarrow(\uparrow S) = S\}$ *with the order induced by inclusion.*

The following statements are known results, we refer to [6, Chap. 7 Sect. 36–44] for the full proofs. We may first note that $(\uparrow, \downarrow)$ is a Galois connection, hence $\downarrow(\uparrow \cdot)$ is a closure operator. $\mathbf{DM}(P)$ is a complete lattice, thus justifying the name. It contains the intersection of any of its elements. The empty set and P are trivially in $\mathbf{DM}(P)$, and it is common practice not to include them in $\mathbf{DM}(P)$, which we shall do in this paper. On the other hand, $\downarrow(\uparrow \cdot)$ constitutes a closure operator, so that $\forall S \subseteq P : S \subseteq \downarrow(\uparrow S)$. On top of that, it holds that $\forall x \in P : \downarrow(\uparrow(\downarrow\{x\})) = \downarrow\{x\}$ hence $\forall x \in P : \exists! \downarrow\{x\} \in \mathbf{DM}(P)$. This way, $(P, \leq)$ can be embedded into $(DM(P), \subseteq)$ so that order is both preserved and reflected. Naturally, $\mathbf{co}_{DM(P)} := \{(s_1, s_2) \in \mathbf{DM}(P) \mid s_1 \not\subseteq s_2 \text{ and } s_2 \not\subseteq s_1\}$.

It is worth noting that, in the non-trivial case, $(\mathcal{A}_P, \leq_{\mathcal{A}_P})$ has no maximal, (nor minimal) element. Indeed, if there is an $x \in \mathcal{A}_P$ such that $\forall a \in \mathcal{A}_P : x \leq_{\mathcal{A}_P} a$ $(a \leq_{\mathcal{A}_P} x)$ then clearly $x' = \emptyset$, so $x = \mathcal{A}_P$, and $(\mathcal{A}_P, \leq_{\mathcal{A}_P}) = (\{\mathcal{A}_P\}, \emptyset)$.

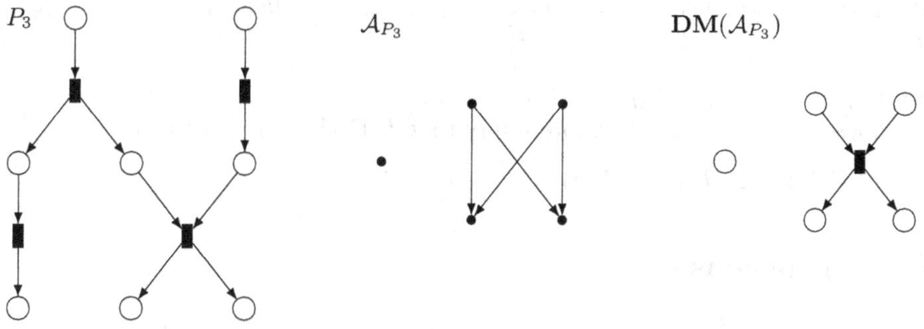

Fig. 7. From left to right: poset P_3 of previous examples seen as a causal net, its atomic reduction, and its completion.

In the following we will show that $\mathbf{DM}(\mathcal{A}_P)$ is a causal net. To this aim we introduce some notation. Let:

- $B_N = \{\downarrow\{a\} \mid a \in \mathcal{A}_P\}$ be the set of principal ideals, those elements of $\mathbf{DM}(\mathcal{A}_P)$ that can be identified with the ones of $\mathcal{A}_P$;
- $E_N = \mathbf{DM}(\mathcal{A}_P) \setminus \{\downarrow\{a\} \mid a \in \mathcal{A}_P\} = \mathbf{DM}(\mathcal{A}_P) \setminus B_N$ be the elements introduced by the completion; and
- $N = (B_N, E_N, \prec_{DM(\mathcal{A}_P)})$

The following two propositions prove, on one hand that the order on $\mathbf{DM}(\mathcal{A}_P)$ is nowhere dense (i.e. $\leq_{DM(\mathcal{A}_P)} = (\prec_{DM(\mathcal{A}_P)})^\star$); and on the other hand, that it is bipartite, hence $(\mathbf{DM}(\mathcal{A}_P), \prec_{DM(\mathcal{A}_P)}) \simeq N = (B_N, E_N, \prec_{DM(\mathcal{A}_P)})$ is an ocurrence net.

Proposition 18. *Let* $a_1, a_2 \in \mathcal{A}_P : a_1 \prec_{\mathcal{A}_P} a_2$.
Then $\exists! s \in \mathbf{DM}(\mathcal{A}_P) : \downarrow\{a_1\} \subsetneq s \subsetneq \downarrow\{a_2\}$.

Proposition 19. *Let* $S \in E_N$, *and* $a_i, a_f \in \mathcal{A}_P$ *be such that* $\downarrow\{a_i\} \subseteq S \subseteq \downarrow\{a_f\}$. *Then* $\exists a, a' \in \mathcal{A}_P$ *such that* $a \prec_{\mathcal{A}_P} a'$, *and* $\downarrow\{a_i\} \subseteq \downarrow\{a\} \subseteq S \subseteq \downarrow\{a'\} \subseteq \downarrow\{a_f\}$.

These results imply that $\mathbf{DM}(\mathcal{A}_P)$ is combinatorial, and that $\forall a_1, a_2 \in \mathcal{A}_P : a_1 \not\prec_{DM(\mathcal{A}_P)} a_2$. Furthermore, suppose $s_1, s_2 \in E_N : s_1 \prec_{DM(\mathcal{A}_P)} s_2$. Then $\exists a_1 \in s_2 \setminus s_1 : s_1 \subsetneq \downarrow\{a_1\} \subsetneq s_2$ which is absurd. And so $N_{\mathcal{A}_P} = (B_N, E_N, \prec_{DM(\mathcal{A}_P)})$ is a Petri net, which is certainly acyclic. Furthermore, we have that for $s \in E_N$, either $s = \emptyset$, $s = \mathcal{A}_P$, or $\exists a_1, a_2 \in \mathcal{A}_P$ such that $\downarrow\{a_1\} \prec_{DM(\mathcal{A}_P)} S \prec_{DM(\mathcal{A}_P)} \downarrow\{a_2\}$.

We shall now see that N is not only an occurrence nets, but actually a causal net. This is achieved by showing that it is conflict-free, in other words, that all forks and joins happen at elements of E_N. Proposition 20 proves that no forks can happen at B_N, whereas Proposition 21 shows the respective result for joins.

Proposition 20. *Let* $a \in \mathcal{A}_P$, *and* $s_1, s_2 \in E_N$ *such that* $\downarrow\{a\} \prec_{DM(\mathcal{A}_P)} s_1$, *and* $\downarrow\{a\} \prec_{DM(\mathcal{A}_P)} s_2$. *Then* $s_1 = s_2$.

Proposition 21. *Let* $a \in \mathcal{A}_P$, *and* $s_1, s_2 \in E_N$ *such that* $s_1 \prec_{DM(\mathcal{A}_P)} \downarrow\{a\}$, *and* $s_2 \prec_{DM(\mathcal{A}_P)} \downarrow\{a\}$. *Then* $s_1 = s_2$.

So $N_{\mathcal{A}_P} = (B_N, E_N, \prec_{DM(\mathcal{A}_P)})$ is a causal net.

These last results furthermore imply that $L(\mathbf{DM}(\mathcal{A}_P)) \simeq L(\mathcal{A}_P)$.

Proposition 22. $L(\mathbf{DM}(\mathcal{A}_P)) \simeq L(\mathcal{A}_P)$.

6 Conclusions

We have provided a way to reduce a poset to a minimal form which preserves the lattice of closed sets. The obtained poset is of particular interest, because it depicts only the interactions between sequential subprocesses modeled in the original poset. In this sense, operations which are internal to a given sequential subprocess are considered superfluous, and therefore not expressed in the reduced version. We have furthermore shown that, under certain assumptions, the elements of the reduced poset behave structurally as the places of a causal net. As a matter of fact, it can be shown to be *jump-free*, and consists of *gaps* (see [4, Chap. 2 Sect. 4]). Intuitively, the term gap makes reference to a missing element [13, Chap. 8 Sect. 4]. There are two kinds of gaps, those isomorphic to P_1 (see Fig. 1), and those isomorphic to $\{A_2, A_3, A_4, A_5\}$ in $\mathcal{A}_{P_3}$ as in Fig. 5. According to Smith [13], the first corresponds to a missing local state, whereas the latter to a missing event. Clearly, by requiring N-density, we leave room only for the latter. We note that jump-freeness implies that the causal net presented in the last section is DC-continuous [12].

The extension of the presented procedure to posets which present conflict is ongoing research. Further research on this topic will involve the following conjectures. We claim that there is a bijection between the maximal chains of the poset and those of the reduced one. This would imply that the two posets have the same *dimension* (as defined, for example, in [7]). We furthermore believe that poset dimension is an invariant on the classes of posets characterised by their closed set lattices. We also claim that the maximal anti-chains, or *cuts*, in the reduced poset are equivalence classes on the set of cuts of the original poset, so that the reduction defines a partition on the cuts of the poset.

Acknowledgments. Work partially supported by MIUR.

References

1. Bernardinello, L., Ferigato, C., Haar, S., Pomello, L.: Closed sets in occurrence nets with conflicts. Fund. Inform. **133**(4), 323–344 (2014)
2. Bernardinello, L., Pomello, L., Rombolà, S.: Closure operators and lattices derived from concurrency in posets and occurrence nets. Fund. Inform. **105**, 211–235 (2010)
3. Bernardinello, L., Pomello, L., Rombolà, S.: Orthomodular algebraic lattices related to combinatorial posets. In: Proceedings of the 15th Italian Conference on Theoretical Computer Science, Perugia, Italy, 17–19 September 2014, pp. 241–245 (2014)
4. Best, E., Fernandez, C.: Nonsequential Processes-A Petri Net View. Monographs in Theoretical Computer Science. An EATCS Series, vol. 13. Springer, Heidelberg (1988)
5. Birkhoff, G.: Lattice Theory, 3rd edn. American Mathematical Society, Providence (1979)
6. Davey, B.A., Priestley, H.A.: Introduction to Lattices and Order. Cambridge University Press, Cambridge (1990)
7. Dushnik, B., Miller, E.W.: Partially ordered sets. Am. J. Math. **63**(3), 600–610 (1941)
8. Kalmbach, G.: Orthomodular Lattices. Academic Press, New York (1983)
9. Mazurkiewicz, A.: Trace theory. In: Brauer, W., Reisig, W., Rozenberg, G. (eds.) ACPN 1986. LNCS, vol. 255, pp. 278–324. Springer, Heidelberg (1987). doi:10.1007/3-540-17906-2_30
10. Möhring, R.H.: Algorithmic aspects of comparability graphs, interval graphs. In: Rival, I. (ed.) Graphs, Order: The Role of Graphs in the Theory of Ordered Sets and Its Applications, pp. 41–101. Springer Netherlands, Dordrecht (1985)
11. Nielsen, M., Plotkin, G.D., Winskel, G.: Petri nets, event structures and domains, part I. Theor. Comput. Sci. **13**, 85–108 (1981)
12. Petri, C.A., Smith, E.: Concurrency and continuity. In: Rozenberg, G. (ed.) APN 1986. LNCS, vol. 266, pp. 273–292. Springer, Heidelberg (1987). doi:10.1007/3-540-18086-9_30
13. Smith, E.: Carl Adam Petri: Life and Science. Springer, Heidelberg (2015)
14. Winskel, G.: Event structures. In: Brauer, W., Reisig, W., Rozenberg, G. (eds.) ACPN 1986. LNCS, vol. 255, pp. 325–392. Springer, Heidelberg (1987). doi:10.1007/3-540-17906-2_31

Embeddability into Relational Lattices Is Undecidable

Luigi Santocanale(✉)

LIF, CNRS UMR 7279, Aix-Marseille Université, Marseille, France
luigi.santocanale@lif.univ-mrs.fr

Abstract. The natural join and the inner union operations combine relations of a database. Tropashko and Spight realized that these two operations are the meet and join operations in a class of lattices, known by now as the relational lattices. They proposed then lattice theory as an algebraic approach to the theory of databases alternative to the relational algebra. Litak et al. proposed an axiomatization of relational lattices over the signature that extends the pure lattice signature with a constant and argued that the quasiequational theory of relational lattices over this extended signature is undecidable.

We prove in this paper that embeddability is undecidable for relational lattices. More precisely, it is undecidable whether a finite subdirectly-irreducible lattice can be embedded into a relational lattice. Our proof is a reduction from the coverability problem of a multimodal frame by a universal product frame and, indirectly, from the representability problem for relation algebras.

As corollaries we obtain the following results: the quasiequational theory of relational lattices over the pure lattice signature is undecidable and has no finite base; there is a quasiequation over the pure lattice signature which holds in all the finite relational lattices but fails in an infinite relational lattice.

1 Introduction

The natural join and the inner union operations combine relations (i.e. tables) of a database. Most of today's web programs query their databases making repeated use of the natural join and of the union, of which the inner union is a mathematically well behaved variant. Tropashko and Spight realized [22,23] that these two operations are the meet and join operations in a class of lattices, known by now as the class of relational lattices. They proposed then lattice theory as an algebraic approach, alternative to Codd's relational algebra [3], to the theory of databases.

An important first attempt to axiomatize these lattices is due to Litak, Mikulás, and Hidders [13]. These authors propose an axiomatization, comprising equations and quasiequations, in a signature that extends the pure lattice signature with a constant, the header constant. A main result of that paper is that

Extended abstract, see [21] for a full version of this paper.

P. Höfner et al. (Eds.): RAMiCS 2017, LNCS 10226, pp. 258–273, 2017.
DOI: 10.1007/978-3-319-57418-9_16

the quasiequational theory of relational lattices is undecidable in this extended signature. Their proof mimics Maddux's proof that the equational theory of cylindric algebras of dimension $n \geq 3$ is undecidable [14].

We have investigated in [20] equational axiomatizations for relational lattices using as tool the duality theory for finite lattices developed in [19]. A conceptual contribution from [20] is to make explicit the similarity between the developing theory of relational lattices and the well established theory of combination of modal logics, see e.g. [11]. This was achieved on the syntactic side, but also on the semantic side, by identifying some key properties of the structures dual to the finite atomistic lattices in the variety generated by the relational lattices, see [20, Theorem 7]. These properties make the dual structures into frames for commutator multimodal logics in a natural way.

In this paper we exploit this similarity to transfer results from the theory of multidimensional modal logics to lattice theory. Our main result is that *it is undecidable whether a finite subdirectly irreducible lattice can be embedded into a relational lattice.* We prove this statement by reducing to it the coverability problem of a frame by a universal $S5^3$-product frame, a problem shown to be undecidable in [10]. As stated there, the coverability problem is—in light of standard duality theory—a direct reformulation of the representability problem of finite simple relation algebras, problem shown to be undecidable by Hirsch and Hodkinson [9].

Our main result and its proof allow us to derive further consequences. Firstly, we refine the undecidability theorem of [13] and prove that *the quasiequational theory of relational lattices in the pure lattice signature is undecidable* as well and *has no finite base.* Then we argue that *there is a quasiequation that holds in all the finite relational lattices, but fails in an infinite one.* For the latter result, we rely on the work by Hirsch et al. [10] who constructed a finite 3-multimodal frame which has no finite p-morphism from a finite universal $S5^3$-product frame, but has a p-morphism from an infinite one. On the methodological side, we wish to point out our use of generalized ultrametric spaces to tackle these problems. A key idea in the proof of the main result is the characterization of universal $S5^A$-product frames as pairwise complete generalized ultrametric spaces with distance valued in the Boolean algebra $P(A)$, a characterization that holds when A is finite.

The paper is structured as follows. We recall in Sect. 2 some definitions and facts on frames and lattices. Relational lattices are introduced in Sect. 3. In Sect. 4 we outline the proof of our main result—embeddability of a finite subdirectly-irreducible lattice into a relational lattice is undecidable—and derive then the other results. In Sect. 5 we show how to construct a lattice from a frame and use functoriality of this construction to argue that such lattice embeds into a relational lattice whenever the frame is a p-morphic image of a universal product frame. The proof of the converse statement is carried out in Sect. 7. Among the technical tools needed to prove the converse, the theory of generalized ultrametric spaces over a powerset Boolean algebra and the aforementioned characterization of universal $S5^A$-product frames as pairwise complete spaces over $P(A)$ are developed in Sect. 6.

Due to the lack of space, we omit most of the technical proofs on lattices and ultrametric spaces; these proofs are accessible via the preprint [21].

2 Frames and Lattices

Frames. Let A be a set of actions. An A *-multimodal frame* (briefly, an A*-frame* or a *frame*) is a structure $\mathfrak{F} = \langle X_{\mathfrak{F}}, \{R_a \mid a \in A\}\rangle$ where, for each $a \in A$, R_a is a binary relation on $X_{\mathfrak{F}}$. We say that an A-frame is S4 if each R_a is reflexive and transitive. If $\mathfrak{F}_0$ and $\mathfrak{F}_1$ are two A-frames, then a p-morphism from $\mathfrak{F}_0$ to $\mathfrak{F}_1$ is a function $\psi : X_{\mathfrak{F}_0} \longrightarrow X_{\mathfrak{F}_1}$ such that, for each $a \in A$,

- if $xR_a y$, then $\psi(x)R_a\psi(y)$,
- if $\psi(x)R_a z$, then $xR_a y$ for some y with $\psi(y) = z$.

Let us mention that A-multimodal frames and p-morphisms form a category.

A frame $\mathfrak{F}$ is said to be rooted (or *initial*, see [18]) if there is $f_0 \in X_{\mathfrak{F}}$ such that every other $f \in X_{\mathfrak{F}}$ is reachable from f_0. We say that an A-frame $\mathfrak{F}$ is *full* if, for each $a \in A$, there exists $f, g \in X_{\mathfrak{F}}$ such that $f \neq g$ and $fR_a g$. If $G = (V, D)$ is a directed graph, then we shall say that G is rooted if it is rooted as a unimodal frame.

A particular class of frames we shall deal with are the *universal* $\mathrm{S5}^A$ *-product frames*. These are the frames $\mathfrak{U}$ with $X_{\mathfrak{U}} = \prod_{a \in A} X_a$ and $xR_a y$ if and only if $x_i = y_i$ for each $i \neq a$, where $x :=\langle x_i \mid i \in A\rangle$ and $y :=\langle y_i \mid i \in A\rangle$.

Orders and Lattices. We assume some basic knowledge of order and lattice theory as presented in standard monographs [4,7]. Most of the tools we use in this paper originate from the monograph [6] and have been further developed in [19].

A *lattice* is a poset L such that every finite non-empty subset $X \subseteq L$ admits a smallest upper bound $\bigvee X$ and a greatest lower bound $\bigwedge X$. A lattice can also be understood as a structure $\mathfrak{A}$ for the functional signature $(\vee, \wedge)$, such that the interpretations of these two binary function symbols both give $\mathfrak{A}$ the structure of an idempotent commutative semigroup, the two semigroup structures being connected by the absorption laws $x \wedge (y \vee x) = x$ and $x \vee (y \wedge x) = x$. Once a lattice is presented as such structure, the order is recovered by stating that $x \leq y$ holds if and only if $x \wedge y = x$.

A lattice L is *complete* if any subset $X \subseteq L$ admits a smallest upper bound $\bigvee X$. It can be shown that this condition implies that any subset $X \subseteq L$ admits a greatest lower bound $\bigwedge X$. A lattice is *bounded* if it has a least element $\bot$ and a greatest element $\top$. A complete lattice (in particular, a finite lattice) is bounded, since $\bigvee \emptyset$ and $\bigwedge \emptyset$ are, respectively, the least and greatest elements of the lattice.

If P and Q are partially ordered sets, then a function $f : P \longrightarrow Q$ is *order-preserving* (or *monotone*) if $p \leq p'$ implies $f(p) \leq f(p')$. If L and M are lattices, then a function $f : L \longrightarrow M$ is a *lattice morphism* if it preserves the lattice operations $\vee$ and $\wedge$. A lattice morphism is always order-preserving. A lattice morphism $f : L \longrightarrow M$ between bounded lattices L and M is *bound-preserving* if $f(\bot) = \bot$ and $f(\top) = \top$. A function $g : Q \longrightarrow P$ is said to be *left adjoint* to an order-preserving $f : P \longrightarrow Q$ if $g(q) \leq p$ holds if and only if $q \leq f(p)$ holds; such a left adjoint, when it exists, is unique. If L is finite, M is bounded, and $f : L \longrightarrow M$ is a bound-preserving lattice morphism, then a left adjoint to f always exists and preserves the constant $\bot$ and the operation $\vee$.

A *Moore family on a set* U is a collection $\mathcal{F}$ of subsets of U which is closed under arbitrary intersections. Given a Moore family $\mathcal{F}$ on U, the correspondence sending $Z \subseteq U$ to $\overline{Z} := \bigcap\{Y \in \mathcal{F} \mid Z \subseteq Y\}$ is a closure operator on U, that is, an order-preserving inflationary and idempotent endofunction of $P(U)$. The subsets in $\mathcal{F}$, called the *closed sets*, are exactly the fixpoints of this closure operator. We can give to a Moore family $\mathcal{F}$ a lattice structure by defining

$$\bigwedge X := \bigcap X, \qquad\qquad \bigvee X := \overline{\bigcup X}. \tag{1}$$

Let L be a complete lattice. An element $j \in L$ is *completely join-irreducible* if $j = \bigvee X$ implies $j \in X$, for each $X \subseteq L$; the set of completely join-irreducible elements of L is denoted here $\mathcal{J}(L)$. A complete lattice is *spatial* if every element is the join of the completely join-irreducible elements below it. An element $j \in \mathcal{J}(L)$ is said to be *join-prime* if $j \leq \bigvee X$ implies $j \leq x$ for some $x \in X$, for each finite subset X of L. If x is not join-prime, then we say that x is *non-join-prime*. An *atom* of a lattice L is an element of L such that $\bot$ is the only element strictly below it. A spatial lattice is *atomistic* if every element of $\mathcal{J}(L)$ is an atom.

For $j \in \mathcal{J}(L)$, a *join-cover* of j is a subset $X \subseteq L$ such that $j \leq \bigvee X$. For $X, Y \subseteq L$, we say that X *refines* Y, and write $X \ll Y$, if for all $x \in X$ there exists $y \in Y$ such that $x \leq y$. A join-cover X of j is said to be *minimal* if $j \leq \bigvee Y$ and $Y \ll X$ implies $X \subseteq Y$; we write $j \lhd_{\mathtt{m}} X$ if X is a minimal join-cover of j. In a spatial lattice, if $j \lhd_{\mathtt{m}} X$, then $X \subseteq \mathcal{J}(L)$. If $j \lhd_{\mathtt{m}} X$, then we say that X is a *non-trivial* minimal join-cover of j if $X \neq \{j\}$. Some authors use the word *perfect* for a lattice which is both spatial and dually spatial. We need here something different:

Definition 1. *We say that a complete lattice is* pluperfect *if it is spatial and for each $j \in \mathcal{J}(L)$ and $X \subseteq L$, if $j \leq \bigvee X$, then $Y \ll X$ for some Y such that $j \lhd_{\mathtt{m}} Y$. The* OD-graph *of a pluperfect lattice L is the structure $\langle \mathcal{J}(L), \leq, \lhd_{\mathtt{m}} \rangle$.*

That is, in a pluperfect lattice every cover refines to a minimal one. Notice that every finite lattice is pluperfect. If L is a pluperfect lattice, then we say that $X \subseteq \mathcal{J}(L)$ is *closed* if it is a downset and $j \lhd_{\mathtt{m}} C \subseteq X$ implies $j \in X$. Closed subsets of $\mathcal{J}(L)$ form a Moore family. The interest of considering pluperfect lattices stems from the following representation theorem stated in [16] for finite lattices; its generalization to pluperfect lattices is straightforward.

Theorem 2 Cf. [21, Theorem 2]. *Let L be a pluperfect lattice and let $\mathsf{L}(\mathcal{J}(L), \leq, \lhd_{\mathtt{m}})$ be the lattice of closed subsets of $\mathcal{J}(L)$. The mapping $l \mapsto \{j \in \mathcal{J}(L) \mid j \leq l\}$ is a lattice isomorphism from L to $\mathsf{L}(\mathcal{J}(L), \leq, \lhd_{\mathtt{m}})$.*

3 The Relational Lattices $\mathsf{R}(D, A)$

Throughout this paper we shall use the notation Y^X for the set of functions of domain Y and codomain X, for X and Y any two sets.

Let A be a collection of attributes (or column names) and let D be a set of cell values. A *relation* on A and D is a pair (α, T) where $\alpha \subseteq A$ and $T \subseteq D^{\alpha}$. Elements of the relational lattice $\mathsf{R}(D, A)$[1] are relations on A and D. Informally, a relation (α, T) represents a table of a relational database, with α being the header, i.e. the collection of names of columns, while T is the collection of rows.

Before we define the natural join, the inner union operations, and the order on $\mathsf{R}(D, A)$, let us recall some key operations. If $\alpha \subseteq \beta \subseteq A$ and $f \in D^{\beta}$, then we shall use $f_{\restriction_\alpha} \in D^{\alpha}$ for the restriction of f to α; if $T \subseteq D^{\beta}$, then $T{\upharpoonright}_{\alpha}$ shall denote projection to α, that is, the direct image of T along restriction, $T{\upharpoonright}_{\alpha} := \{f_{\restriction_\alpha} \mid f \in T\}$; if $T \subseteq D^{\alpha}$, then $i_{\beta}(T)$ shall denote cylindrification to β, that is, the inverse image of restriction, $i_{\beta}(T) := \{f \in D^{\beta} \mid f_{\restriction\alpha} \in T\}$. Recall that i_{β} is right adjoint to ${\upharpoonright}_{\alpha}$. With this in mind, the natural join and the inner union of relations are respectively described by the following formulas:

$$
\begin{aligned}
(\alpha_1, T_1) \wedge (\alpha_2, T_2) &:= (\alpha_1 \cup \alpha_2, T) \\
\text{where } T &= \{f \mid f_{\restriction_{\alpha_i}} \in T_i, i = 1, 2\} \\
&= i_{\alpha_1 \cup \alpha_2}(T_1) \cap i_{\alpha_1 \cup \alpha_2}(T_2), \\
(\alpha_1, T_1) \vee (\alpha_2, T_2) &:= (\alpha_1 \cap \alpha_2, T) \\
\text{where } T &= \{f \mid \exists i \in \{1, 2\}, \exists g \in T_i \text{ s.t. } g_{\restriction_{\alpha_1 \cap \alpha_2}} = f\} \\
&= T_1{\upharpoonright}_{\alpha_1 \cap \alpha_2} \cup T_2{\upharpoonright}_{\alpha_1 \cap \alpha_2}.
\end{aligned}
$$

The order is then given by $(\alpha_1, T_1) \leq (\alpha_2, T_2)$ iff $\alpha_2 \subseteq \alpha_1$ and $T_1 {\upharpoonright}_{\alpha_2} \subseteq T_2$.

A convenient way of describing these lattices was introduced in [13, Lemma 2.1]. The authors argued that the relational lattices $\mathsf{R}(D, A)$ are isomorphic to the lattices of closed subsets of $A \cup D^A$, where $Z \subseteq A \cup D^A$ is said to be closed if it is a fixed-point of the closure operator $\overline{(-)}$ defined as

$$\overline{Z} := Z \cup \{f \in D^A \mid A \setminus Z \subseteq Eq(f, g), \text{ for some } g \in Z\},$$

where in the formula above $Eq(f, g)$ is the equalizer of f and g. Letting $\delta(f, g) := \{x \in A \mid f(x) \neq g(x)\}$, the above definition of the closure operator is obviously equivalent to the following one:

$$\overline{Z} := \alpha \cup \{f \in D^A \mid \delta(f, g) \subseteq \alpha, \text{ for some } g \in Z \cap D^A\}, \text{ with } \alpha = Z \cap A.$$

From now on, we rely on this representation of relational lattices. Relational lattices are atomistic pluperfect lattices. The completely join-irreducible elements of $\mathsf{R}(D, A)$ are the singletons $\{a\}$ and $\{f\}$, for $a \in A$ and $f \in D^A$, see [13].

[1] In [13] such a lattice is called *full* relational lattice. The wording "class of relational lattices" is used there for the class of lattices that have an embedding into some lattice of the form $\mathsf{R}(D, A)$.

By an abuse of notation we shall write x for the singleton $\{x\}$, for $x \in A \cup D^A$. Under this convention, we have therefore $\mathcal{J}(\mathsf{R}(D, A)) = A \cup D^A$. Every $a \in A$ is join-prime, while the minimal join-covers are of the form $f \lhd_{\mathtt{m}} \delta(f, g) \cup \{g\}$, for each $f, g \in D^A$, see [20]. The only non-trivial result from [20] that we use later (for Lemma 24 and Theorem 29) is the following:

Lemma 3. *Let L be a finite atomistic lattice in the variety generated by the class of relational lattices. If $\{j\} \cup X \subseteq \mathcal{J}(L)$, $j \leq \bigvee X$, and all the elements of X are join-prime, then j is join-prime.*

The Lemma—which is an immediate consequence of Theorem 7 in [20]—asserts that a join-cover of an element $j \in \mathcal{J}(L)$ which is not join-prime cannot be made of join-prime elements only.

4 Overview and Statement of the Results

For an arbitrary frame $\mathfrak{F}$, we construct in Sect. 5 a lattice $\mathsf{L}(\mathfrak{F})$; if $\mathfrak{F}$ is rooted and full, then $\mathsf{L}(\mathfrak{F})$ is a subdirectly irreducible lattice, see Proposition 16. The key Theorem leading to the undecidability results is the following one.

Theorem 4. *Let A be a finite set and let $\mathfrak{F}$ be an* S4 *finite rooted full A-frame. There is a surjective p-morphism from a universal* $\mathrm{S5}^A$*-product frame $\mathfrak{U}$ to $\mathfrak{F}$ if and only if $\mathsf{L}(\mathfrak{F})$ embeds into some relational lattice $\mathsf{R}(D, B)$.*

Proof (outline). The construction L defined in Sect. 5 extends to a contravariant functor, so if $\mathfrak{U}$ is a universal $\mathrm{S5}^A$-product frame and $\psi : \mathfrak{U} \longrightarrow \mathfrak{F}$ is a surjective p-morphism, then we have an embedding $\mathsf{L}(\psi)$ of $L(\mathfrak{F})$ into $\mathsf{L}(\mathfrak{U})$. We can assume that all the components of $\mathfrak{U}$ are equal, i.e. that the underlying set of $\mathfrak{U}$ is of the form $\prod_{a \in A} X$; if this is the case, then $\mathsf{L}(\mathfrak{U})$ is isomorphic to the relational lattice $\mathsf{R}(X, A)$.

The converse direction, developed from Sect. 6 up to Sect. 7, is subtler. Considering that $\mathsf{L}(\mathfrak{F})$ is subdirectly-irreducible, we argue that if $\psi : \mathsf{L}(\mathfrak{F}) \longrightarrow \mathsf{R}(D, B)$ is a lattice embedding, then we can suppose it preserves bounds; in this case ψ has a surjective left adjoint $\mu : \mathsf{R}(D, B) \longrightarrow \mathsf{L}(\mathfrak{F})$. Let us notice that there is no general reason for ψ to be the image by L of a p-morphism. Said otherwise, the functor L is not full and, in particular, the image of an atom by μ might not be an atom. The following considerations, mostly developed in Sect. 7, make it possible to extract a p-morphism from the left adjoint μ. Since both $\mathsf{L}(\mathfrak{F})$ and $\mathsf{R}(D, B)$ are generated (under possibly infinite joins) by their atoms, each atom $x \in \mathsf{L}(\mathfrak{F})$ has a preimage $y \in \mathsf{R}(D, B)$ which is an atom. The set F_0 of non-join-prime atoms of $\mathsf{R}(D, B)$ such that $\mu(f)$ is a non-join-prime atom of $\mathsf{L}(\mathfrak{F})$ is endowed with a $P(A)$-valued distance δ. The pair (F_0, δ) is shown to be a pairwise complete ultrametric space over $P(A)$. Section 6 recalls and develops some observations on ultrametric spaces valued on powerset algebras. The key ones are Theorem 18 and Proposition 19, stating that—when A is finite—pairwise complete ultrametric spaces over $P(A)$ and universal $\mathrm{S5}^A$-product frames are essentially the same objects. The restriction of μ to F_0 yields then a surjective p-morphism from F_0, considered as a universal $\mathrm{S5}^A$-product frame, to $\mathfrak{F}$. □

The following problem was shown to be undecidable in [10]: given a finite 3-frame $\mathfrak{F}$, does there exists a surjective p-morphism from a universal $S5^3$-product frame $\mathfrak{U}$ to $\mathfrak{F}$? In the introduction we referred to this problem as the coverability problem of a 3-frame by a universal $S5^3$-product frame. The problem was shown to be undecidable by means of a reduction from the representability problem of finite simple relation algebras, shown to be undecidable in [9]. We need to strengthen the undecidability result of [10] with some additional observations—rootedness and fullness—as stated in the following Proposition.

Proposition 5. *It is undecidable whether, given a finite set A with* $\mathrm{card}A \geq 3$ *and an* S4 finite rooted full *A-frame $\mathfrak{F}$, there is a surjective p-morphism from a universal $S5^A$-product $\mathfrak{U}$ to $\mathfrak{F}$.*

Proof. Throughout this proof we assume a minimum knowledge of the theory of relation algebras, see e.g. [15].

The Proposition actually holds if we restrict to the case when $\mathrm{card}A = 3$. Given a finite simple relation algebra $\mathfrak{A}$, the authors of [10] construct a 3-multimodal frame $\mathfrak{F}_{\mathfrak{A},3}$ such that $\mathfrak{A}$ is representable if and only if $\mathfrak{F}_{\mathfrak{A},3}$ is a p-morphic image of some universal $S5^3$-product frame. The frame $\mathfrak{F}_{\mathfrak{A},3}$ is S4 and rooted [10, Claim 8]. We claim that $\mathfrak{F}_{\mathfrak{A},3}$ is also full, unless $\mathfrak{A}$ is the two elements Boolean algebra. To prove this claim, let us recall first that an element of $\mathfrak{F}_{\mathfrak{A},3}$ is a triple (t_0, t_1, t_2) of atoms of $\mathfrak{A}$ such that $t_2^{\smile} \leq t_0; t_1$; moreover, if t, t' are two such triples and $i \in \{0, 1, 2\}$, then tR_it' if and only if t and t' coincide in the i-th coordinate. If a is an atom of $\mathfrak{A}$, then $a \leq e_l; a$ and $a \leq a; e_r$ for two atoms e_l, e_r below the multiplicative unit of $\mathfrak{A}$. Therefore, the triples $t := (e_l, a, a^{\smile})$ and $t' = (a, e_r, a^{\smile})$ are elements of $\mathfrak{F}_{\mathfrak{A},3}$ and tR_2t'. If, for each atom a, these triples are equal, then every atom of $\mathfrak{A}$ is below the multiplicative unit, which therefore coincides with the top element $\top$; since $\mathfrak{A}$ is simple, then relation $\top = \top; x; \top$ holds for each $x \neq \bot$. It follows that $x = \top; x; \top = \top$, for each $x \neq \bot$, so $\mathfrak{A}$ is the two elements Boolean algebra. Thus, if $\mathfrak{A}$ has more than two elements, then $t \neq t'$ and tR_2t' for some $t, t' \in \mathfrak{F}_{\mathfrak{A},3}$. Using the cycle law of relation algebras, one also gets pairs of distinct elements of $\mathfrak{F}_{\mathfrak{A},3}$, call them u, u' and w, w', such that uR_0u' and wR_1w'.

Therefore, if we could decide whether there is a p-morphism from some universal $S5^3$-frame to a given S4 finite rooted full frame $\mathfrak{F}$, then we could also decide whether a finite simple relation algebra $\mathfrak{A}$ is representable, by answering positively if $\mathfrak{A}$ has exactly two elements and, otherwise, by answering the existence problem of a p-morphism to $\mathfrak{F}_{\mathfrak{A},3}$. □

Combining Theorem 4 with Proposition 5, we derive the following undecidability result.

Theorem 6. *It is not decidable whether a finite subdirectly irreducible atomistic lattice embeds into a relational lattice.*

Let us remark that Theorem 6 partly answers Problem 7.1 in [13].

In [13] the authors proved that the quasiequational theory of relational lattices (i.e. the set of all definite Horn sentences valid in relational lattices) in the signature $(\wedge, \vee, H)$ is undecidable. Here H is the header constant, which is interpreted in a relational lattice $\mathsf{R}(D, A)$ as the closed subset A of $A \cup D^A$. Problem 4.10 in [13] asks whether the quasiequational theory of relational lattices in the restricted signature $(\wedge, \vee)$ of pure lattice theory is undecidable as well. We positively answer this question.

Theorem 7. *The quasiequational theory of relational lattices in the pure lattice signature is undecidable.*

It is a general fact that if the embeddability problem of finite subdirectly-irreducible algebras in a class $\mathcal{K}$ is undecidable, then the quasiequational theory of $\mathcal{K}$ is undecidable as well. We thank a colleague for pointing out to us how this can be derived from Evans' work [5]. We add here the proof of this fact, since we shall need it later in the proof of Theorem 10.

Proof. Given a finite subdirectly-irreducible algebra A with least non trivial congruence $\theta(\hat{a}, \bar{a})$, we construct a quasiequation ϕ_A with the following property: for any other algebra (in the same signature) K, $K \not\models \phi_A$ if and only if A has an embedding into K.

The construction is as follows. Let $X_A = \{x_a \mid a \in A\}$ be a set of variables in bijection with the elements of A. For each function symbol f in the signature Ω, let $T_{A,f}$ be its table, that is the formula

$$T_{A,f} = \bigwedge_{(a_1,\ldots,a_{ar(f)}) \in A^{ar(f)}} f(x_{a_1}, \ldots, x_{ar(f)}) = x_{f(a_1,\ldots,a_{ar(f)})}.$$

We let ϕ_A be the universal closure of $\bigwedge_{f \in \Omega} T_{A,f} \Rightarrow x_{\hat{a}} = x_{\bar{a}}$. We prove next that an algebra K sastifies ϕ_A if and only if there is no embedding of A into K.

If $K \models \phi_A$ and $\psi : A \longrightarrow K$, then $v(x_a) = \psi(a)$ is a valuation such that $K, v \models \bigwedge_{f \in \Omega} T_{A,f}$, so $\psi(\hat{a}) = v(x_{\hat{a}}) = v(x_{\bar{a}}) = \psi(\bar{a})$ and ψ is not injective.

Conversely, suppose $K \not\models \phi_A$ and let v be a valuation such that $K, v \models \bigwedge_{f \in \Omega} T_{A,f}$ and $K, v \not\models x_{\hat{a}} = x_{\bar{a}}$. Define $\psi : A \longrightarrow K$ as $\psi(a) = v(x_a)$, then ψ is a morphism, since $K, v \models T_{A,f}$ for each $f \in \Omega$. Let $Ker_\psi = \{(a, a') \mid \psi(a) = \psi(a')\}$ so, supposing that ψ is not injective, Ker_ψ is a non-trivial congruence. Then $(\hat{a}, \bar{a}) \in \theta(\hat{a}, \bar{a}) \subseteq Ker_\psi$, so $v(x_{\hat{a}}) = \psi(\hat{a}) = \psi(\bar{a}) = v(x_{\bar{a}})$, a contradiction. We have therefore $Ker_\psi = \{(a, a) \mid a \in A\}$, which shows that ψ is injective.

Let now $\mathcal{K}$ be a class of algebras in the same signature. We have then

$$\begin{aligned}\mathcal{K} \not\models \phi_A \text{ iff } & K \not\models \phi_A \text{ for some } K \in \mathcal{K}\\ \text{iff } & \text{there is an embedding of } A \text{ into } K, \text{ for some } K \in \mathcal{K}.\end{aligned}$$

Thus, if the embeddability problem of finite subdirectly-irreducible algebras into some algebra in $\mathcal{K}$ is undecidable, then the quasiequational theory of $\mathcal{K}$ is undecidable as well. □

Following [10], let us add some further observations on the quasiequational theory of relational lattices.

Lemma 8. *The class of lattices that have an embedding into a relational lattice is closed under ultraproducts.*

Proof. Let us say that a sublattice L of a lattice $\mathsf{R}(D, A)$ is H-closed if the subset A belongs to L. Let $\mathcal{R}$ denote the closure under isomorphisms of the class of H-closed sublattices of some $\mathsf{R}(D, A)$. It is proved in [13, Corollary 4.2] that $\mathcal{R}$ is closed under ultraproducts. It immediately follows from this result that the class of lattices that have an embedding into some relational lattice is closed under ultraproducts, as follows. Let $\{L_i \longrightarrow \mathsf{R}(D_i, A_i) \mid i \in I\}$ be a family of lattice embeddings and let $\mathcal{F}$ be an ultrafilter over I. The ultraproduct constructions on $\{L_i \mid i \in I\}$ and $\{\mathsf{R}(D_i, A_i) \mid i \in I\}$ yield a lattice embedding $\prod_{\mathcal{F}} L_i \longrightarrow \prod_{\mathcal{F}} \mathsf{R}(D_i, A_i)$. Clearly, each $\mathsf{R}(D_i, A_i)$ belongs to $\mathcal{R}$, whence the ultraproduct $\prod_{\mathcal{F}} \mathsf{R}(D_i, A_i)$ belongs to $\mathcal{R}$ as well: thus $\prod_{\mathcal{F}} \mathsf{R}(D_i, A_i)$ embeds into some $\mathsf{R}(D, A)$, and so does $\prod_{\mathcal{F}} L_i$. □

Theorem 9. *The quasiequational theory of relational lattices is not finitely axiomatizable.*

Proof. A known result in universal algebra—see e.g. [2, Theorem 2.25]—states that a subdirectly-irreducible algebra satisfies all the quasiequations satisfied by a class of algebras if and only if it embeds in an ultraproduct of algebras in this class. Lemma 8 implies that the class of lattices that have an embedding into an ultraproduct of relational lattices and the class of lattices that have an embedding into some relational lattices are the same. Therefore a subdirectly-irreducible lattice L embeds in a relational lattice if and only if it satisfies all the quasiequations satisfied by the relational lattices. If this collection of quasiequations was a logical consequence of a finite set of quasiequations, then we could decide whether a finite subdirectly-irreducible L satisfies all these quasiequations, by verifying whether L satisfies the finite set of quasiequations. In this way, we could also decide whether such an L embeds into some relational lattice. □

Finally, the following Theorem, showing that the quasiequational theory of the finite relational lattices is stronger than the quasiequational theory of all the relational lattices, partly answers Problem 3.6 in [13].

Theorem 10. *There is a quasiequation which holds in all the finite relational lattices which, however, fails in an infinite relational lattice.*

Proof. In the first appendix of [10] an S4 finite rooted full 3-frame $\mathfrak{F}$ is constructed that has no surjective p-morphism from a finite universal $S5^3$-product frame, but has such a p-morphism from an infinite one.

Since $\mathsf{L}(\mathfrak{F})$ is finite whenever $\mathfrak{F}$ is finite, we obtain by using Theorem 4 a subdirectly-irreducible finite lattice L which embeds into an infinite relational lattice, but has no embedding into a finite one.

Let ϕ_L be the quasiequation as in the proof of Theorem 7. We have therefore that, for any lattice K, $K \models \phi_L$ if and only if L does not embed into K.

Correspondingly, any finite relational lattice satisfies ϕ_L and, on the other hand, $K \not\models \phi_L$ if K is the infinite lattice into which L embeds. □

5 The Lattice of a Multimodal Frame

We assume throughout this Section that A is a finite set of actions.

Let $\alpha \subseteq A$, $\mathfrak{F}$ be an A-frame, $x, y \in X_{\mathfrak{F}}$. We define an *$\alpha$-path* from x to y as a sequence $x = x_0 R_{a_0} x_1 \ldots x_{k-1} R_{a_{k-1}} x_k = y$ with $\{a_0, \ldots, a_{k-1}\} \subseteq \alpha$. We use the notation $x \xrightarrow{\alpha} y$ to mean that there is an α-path from x to y. Notice that if $\mathfrak{F}$ is an S4 A-frame, then $x \xrightarrow{\{a\}} y$ if and only if $x R_a y$. Given an A-frame $\mathfrak{F} = \langle X_{\mathfrak{F}}, \{R_a \mid a \in A\}\rangle$, we construct a lattice as follows. For $\alpha \subseteq A$, we say that $Y \subseteq X_{\mathfrak{F}}$ is *α -closed* if $x \in Y$, whenever there is a α-path from x to some $y \in Y$. We say that a subset $Z \subseteq A \cup X_{\mathfrak{F}}$ is *closed* if $Z \cap X_{\mathfrak{F}}$ is $Z \cap A$-closed. It is straightforward to verify that the collection of closed subsets of $A \cup X_{\mathfrak{F}}$ is a Moore family.

Definition 11. *The lattice $\mathsf{L}(\mathfrak{F})$ is the lattice of closed subsets of $A \cup X_{\mathfrak{F}}$.*

The lattice operations on $\mathsf{L}(\mathfrak{F})$ are defined as in the display (1). Actually, $\mathsf{L}(-)$ is a contravariant functor from the category of frames to the category of lattices. Namely, for a p-morphism $\psi : \mathfrak{F}_0 \longrightarrow \mathfrak{F}_1$ and any $Z \subseteq A \cup X_{\mathfrak{F}_1}$, define $\mathsf{L}(\psi)(Z) := (Z \cap A) \cup \psi^{-1}(Z \cap X_{\mathfrak{F}_1})$.

Proposition 12 Cf. [21, Proposition 17]. *$\mathsf{L}(\psi)$ sends closed subsets of $A \cup X_{\mathfrak{F}_1}$ to closed subsets of $A \cup X_{\mathfrak{F}_0}$. Its restriction to $\mathsf{L}(\mathfrak{F}_1)$ yields a bound-preserving lattice morphism $L(\psi) : \mathsf{L}(\mathfrak{F}_1) \longrightarrow \mathsf{L}(\mathfrak{F}_0)$. Moreover, if $\psi : \mathfrak{F}_0 \longrightarrow \mathfrak{F}_1$ is surjective, then $\mathsf{L}(\psi)$ is injective.*

We state next the main result of this Section.

Theorem 13. *If there exists a surjective p-morphism from a universal $\mathrm{S5}^A$-product frame $\mathfrak{U}$ to an A-frame $\mathfrak{F}$, then $\mathsf{L}(\mathfrak{F})$ embeds into a relational lattice.*

Proof. We say that $\mathfrak{U}$ is uniform on X if all the components of $\mathfrak{U}$ are equal to X. Spelled out, this means that $X_{\mathfrak{U}} = \prod_{a \in A} X$. Let $\psi : \mathfrak{U} \longrightarrow \mathfrak{F}$ be a p-morphism as in the statement of the Theorem. W.l.o.g. we can assume that $\mathfrak{U}$ is uniform on some set X. If this is not the case, then we choose $a_0 \in A$ such that X_{a_0} has maximum cardinality and surjective mappings $p_a : X_{a_0} \longrightarrow X_a$, for each $a \in A$. The product frame $\mathfrak{U}'$ on $\prod_{a \in A} X_{a_0}$ is uniform and $\prod_{a \in A} p_a : \mathfrak{U}' \longrightarrow \mathfrak{U}$ is a surjective p-morphism. By pre-composing ψ with this p-morphism, we obtain a surjective p-morphism from the uniform $\mathfrak{U}'$ to $\mathfrak{F}$. Now, if $\mathfrak{U}$ is uniform on X, then $\mathsf{L}(\mathfrak{U})$ is equal to the relational lattice $\mathsf{R}(X, A)$. Then, by functoriality of L, we have a lattice morphism $\mathsf{L}(\psi) : \mathsf{L}(\mathfrak{F}) \longrightarrow \mathsf{L}(\mathfrak{U}) = \mathsf{R}(X, A)$. By Proposition 12 $\mathsf{L}(\psi)$ is an embedding. □

We review next some properties of the lattices $\mathsf{L}(\mathfrak{F})$.

Proposition 14 Cf. [21, Proposition 20]. *The completely join-irreducible elements of $\mathsf{L}(\mathfrak{F})$ are the singletons, so $\mathsf{L}(\mathfrak{F})$ is an atomistic lattice.*

Identifying singletons of with their elements, the previous proposition states that $\mathcal{J}(\mathsf{L}(\mathfrak{F})) = A \cup X_{\mathfrak{F}}$. To state the next Proposition, let us say that an α-path from $x \in X_{\mathfrak{F}}$ to $y \in X_{\mathfrak{F}}$ is *minimal* if there is no β-path from x to y, for each proper subset β of α.

Proposition 15 Cf. [21, Proposition 21]. *$\mathsf{L}(\mathfrak{F})$ is a pluperfect lattice. Each element of A is join-prime, while the minimal join-covers of $x \in X_{\mathfrak{F}}$ are of the form $x \lhd_{\mathtt{m}} \alpha \cup \{y\}$, for a minimal α-path from x to y.*

Before stating the next Proposition, let us recall from [6, Corollary 2.37] that a finite lattice L is subdirectly-irreducible if and only if the directed graph $(\mathcal{J}(L), D)$ is rooted. Here D is the *join-dependency relation* on the join-irreducible elements of L, which, on atomistic finite lattices, can be defined by saying that jDk holds if $j \neq k$ and $j \leq p \vee k$ for some $p \in L$ with $j \not\leq p$.

Proposition 16. *If a finite A-frame $\mathfrak{F}$ is rooted and full, then $\mathsf{L}(\mathfrak{F})$ is a subdirectly-irreducible lattice.*

Proof. We argue that the digraph $(\mathcal{J}(\mathsf{L}(\mathfrak{F})), D)$ is rooted. Observe that $x \in \overline{\{a, y\}} = a \vee y$ whenever $xR_a y$. This implies that xDy and xDa when $x, y \in X_{\mathfrak{F}}$, $a \in A$, $x \neq y$ and $xR_a y$. The fact that of $(\mathcal{J}(\mathsf{L}(\mathfrak{F})), D)$ is rooted follows now from $\mathfrak{F}$ being rooted and full. □

6 Some Theory of Generalized Ultrametric Spaces

Generalized ultrametric spaces over a Boolean algebra $P(A)$ turn out to be a useful tool for relational lattices [13,20]—as well as, we claim here, for universal product frames from multidimensional modal logic [11]. The use of metrics is well known in graph theory, where universal product frames are known as Hamming graphs, see e.g. [8]. Generalized ultrametric spaces over a Boolean algebra $P(A)$ were introduced in [17] to study equivalence relations. The main results of this Section are Theorem 18 and Proposition 19 which together substantiate the claim that when A is finite, universal $\mathrm{S5}^A$-product frames are pairwise complete ultrametric spaces valued in the Boolean algebra $P(A)$. It is this abstract point of view that shall allow us to construct a universal product frame given a lattice embedding $\mathsf{L}(\mathfrak{F}) \longrightarrow \mathsf{R}(D, A)$. We shall develop some observations that are not strictly necessary to prove the undecidability result, which is the main result of this paper. Nonetheless we include them since they are part of a coherent set of results and, as far as we know, they are original.

Definition 17. *An* ultrametric space over $P(A)$ *(briefly, a* space*) is a pair (X, δ), with $\delta : X \times X \longrightarrow P(A)$ such that, for every $f, g, h \in X$,*

$$\delta(f, f) \subseteq \emptyset, \qquad \delta(f, g) \subseteq \delta(f, h) \cup \delta(h, g).$$

That is, we have defined an ultrametric space over $P(A)$ as a category (with a small set of objects) enriched over $(P(A)^{op}, \emptyset, \cup)$, see [12]. We shall assume in this paper that such a space (X, δ) is also *reduced* and *symmetric*, that is, that the following two properties hold for every $f, g \in X$:

$$\delta(f,g) = \emptyset \text{ implies } f = g, \qquad\qquad \delta(f,g) = \delta(g,f).$$

A *morphism* of spaces[2] $\psi : (X, \delta_X) \longrightarrow (Y, \delta_Y)$ is a function $\psi : X \longrightarrow Y$ such that $\delta_Y(\psi(f), \psi(g)) \leq \delta_X(f,g)$, for each $f, g \in X$. If $\delta_Y(\psi(f), \psi(g)) = \delta_X(f,g)$, for each $f, g \in X$, then ψ is said to be an *isometry*. For (X, δ) a space over $P(A)$, $f \in X$ and $\alpha \subseteq A$, the ball centered in f of radius α is defined as usual: $B(f, \alpha) := \{g \in X \mid \delta(f,g) \subseteq \alpha\}$. In [1] a space (X, δ) is said to be *pairwise complete* if, for each $f, g \in X$ and $\alpha, \beta \subseteq A$, $B(f, \alpha \cup \beta) = B(g, \alpha \cup \beta)$ implies $B(f, \alpha) \cap B(g, \beta) \neq \emptyset$. This property is easily seen to be equivalent to:

$$\delta(f,g) \subseteq \alpha \cup \beta \text{ implies } \delta(f,h) \subseteq \alpha \text{ and } \delta(h,g) \subseteq \beta, \quad \text{for some } h \in X.$$

If (X, δ_X) is a space and $Y \subseteq X$, then the restriction of δ_X to Y induces a space (Y, δ_X); we say then that (Y, δ_X) is a *subspace* of X. Notice that the inclusion of Y into X yields an isometry of spaces.

Our main example of space over $P(A)$ is (D^A, δ), with D^A the set of functions from A to D and the distance defined by

$$\delta(f,g) := \{a \in A \mid f(a) \neq g(a)\}. \tag{2}$$

A second example is a slight generalization of the previous one. Given a surjective function $\pi : E \longrightarrow A$, let $\mathsf{Sec}(\pi)$ denote the set of all sections of π, that is the functions $f : A \longrightarrow E$ such that $\pi \circ f = id_A$; the formula in (2) also defines a distance on $\mathsf{Sec}(\pi)$. By identifying $f \in \mathsf{Sec}(\pi)$ with a vector $\langle f_a \in \pi^{-1}(a) \mid a \in A \rangle$, we see that

$$\mathsf{Sec}(\pi) = \prod_{a \in A} X_a, \quad \text{where } X_a := \pi^{-1}(a). \tag{3}$$

That is, the underlying set of a space $(\mathsf{Sec}(\pi), \delta)$ is that of a universal $\mathrm{S5}^A$-product frame. Our next observations are meant to understand the role of the universal $\mathrm{S5}^A$-product frame among all the spaces.

A space is *spherically complete* if the intersection $\bigcap_{i \in I} B(f_i, \alpha_i)$ of every chain $\{B(f_i, \alpha_i) \mid i \in I\}$ of balls is non-empty, see e.g. [1]. In this work the injective objects in the category of spaces are characterized as the pairwise and spherically complete spaces. The next Theorem shows that such injective objects are, up to isomorphism, the "universal product frames".

Theorem 18 Cf. [21, Proposition 24 and Theorem 25]. *The spaces of the form $(\mathsf{Sec}(\pi), \delta)$ are pairwise and spherically complete. Moreover, every space (X, δ) over $P(A)$ has an isometry into some $(\mathsf{Sec}(\pi), \delta)$ and if (X, δ) is pairwise and spherically complete, then this isometry is an isomorphism.*

---

[2] As $P(A)$ is not totally ordered, we avoid calling a morphism "*non expanding map*" as it is often done in the literature.

We develop next the minimal theory needed to carry out the proof of undecidability. We shall assume in particular that A is a finite set. It was shown in [17] that, when A is finite, every space over $P(A)$ is spherically complete–so, from now on, this property will not be of concern to us.

Observe now that in the display (3), the transition relations of the universal product frame $\prod_a X_a$ and the metric of the space $\mathsf{Sec}(\pi)$ are interdefinable. Indeed, for each $a \in A$, we have $fR_a g$ iff $\delta(f,g) \subseteq \{a\}$. On the other hand, since A is finite, the metric is completely determined from the transition relations of the frame, using the notion of α-path introduced in Sect. 5, as follows: $\delta(f,g) = \bigcap\{\alpha \subseteq A \mid f \xrightarrow{\alpha} g\}$. We cast our observations in a Proposition:

Proposition 19. *If A is finite, then there is a bijective correspondence between spaces over $P(A)$ of the form $(\mathsf{Sec}(\pi), \delta)$ and universal* $\mathrm{S5}^A$*-product frames. Universal* $\mathrm{S5}^A$*-product frames are, up to isomorphism, the pairwise complete spaces over $P(A)$.*

We assume in the rest of this section that (X, δ) is a fixed pairwise complete space. We say that a function $v : X \longrightarrow P(A)$ is a *module* if $v(f) \subseteq \delta(f,g) \cup v(g)$. In enriched category theory "module" is a standard naming for an enriched functor (here, a space morphism) from an enriched category to the base category enriched on itself. Here a module can be seen as a space morphism from (X, δ) to the space $(P(A), \Delta)$, where Δ is the symmetric difference. Given a module v, let us define $\mathsf{S}_v := \{x \in X \mid v(x) = \emptyset\}$.

Lemma 20 Cf. [21, Corollary 28]. *For each module v, S_v is a pairwise complete subspace of (X, δ).*

It is possible to directly define a lattice $\mathsf{L}(X, \delta)$ for each space (X, δ). For simplicity, we shall use $\mathsf{L}(X, \delta)$ here to denote the lattice structure corresponding to $\mathsf{L}(\mathfrak{U})$, where $\mathfrak{U}$ is a universal product frame corresponding to (X, δ).

7 From Lattice Embeddings to Surjective *p*-morphisms

We prove in this Section the converse of Theorem 13:

Theorem 21. *Let A be a finite set, let $\mathfrak{F}$ be a finite rooted full* S4 *A-frame. If $\mathsf{L}(\mathfrak{F})$ embeds into a relational lattice $\mathsf{R}(D, B)$, then there exists a universal* $\mathrm{S5}^A$*-product frame $\mathfrak{U}$ and a surjective p-morphism from $\mathfrak{U}$ to $\mathfrak{F}$.*

To prove the Theorem, we study bound-preserving embeddings of finite atomistic lattices into lattices of the form $\mathsf{R}(D, B)$. Let in the following $i : L \longrightarrow \mathsf{R}(D, B)$ be a fixed bound-preserving lattice embedding, with L a finite atomistic lattice. Since L is finite, i has a left adjoint $\mu : \mathsf{R}(D, B) \longrightarrow L$. By abuse of notation, we shall also use the same letter μ to denote the restriction of this left adjoint to the set of completely join-irreducible elements of $\mathsf{R}(D, B)$ which, we recall, is identified with $B \cup D^B$. It is a general fact—and the main ingredient of Birkhoff's duality for finite distributive lattices—that left adjoints to bound-preserving lattice morphism preserve join-prime elements. Thus we have:

Lemma 22. *If $b \in B$, then $\mu(b)$ is join-prime.*

It is not in general true that left adjoints send join-irreducible elements to join-irreducible elements, and this is a main difficulty towards a proof of Theorem 21. Yet, the following statements hold:

Lemma 23. *For each $x \in \mathcal{J}(L)$ there exists $y \in B \cup D^B$ such that $\mu(y) = x$.*

Lemma 24. *Let $g \in D^B$ such that $\mu(g)$ is join-reducible in L. There exists $h \in D^B$ such that $\mu(h) \in \mathcal{J}(L)$ and $\mu(g) = \bigvee \mu(\delta(g, h)) \vee \mu(h)$; moreover, $\mu(h)$ is non-join-prime whenever L is not a Boolean algebra.*

Let A be the set of atoms of L that are join-prime. While (D^B, δ) is a space over $P(B)$, we need to transform D^B into a space over $P(A)$. To this end, we define a $P(A)$-valued distance δ_A on D^B by $\delta_A(f, g) := \mu(\delta(f, g))$. Because of Lemma 22, we have $\delta_A(f, g) \subseteq A$.

Proposition 25. *(D^B, δ_A) is a pairwise complete ultrametric space over $P(A)$.*

We define next $v : D^B \longrightarrow P(A)$ by letting $v(f) := \{a \in A \mid a \leq \mu(f)\}$.

Lemma 26 Cf. [21, Proposition 47]. *The map $v : D^B \longrightarrow P(A)$ is a module on (D^B, δ_A). Moreover $v(f) = \emptyset$ if and only if $\mu(f) \in \mathcal{J}(L) \setminus A$.*

Using Lemmas 20 and 26, we derive:

Corollary 27. *The subspace of (D^B, δ_A) induced by $F_0 := \{f \in D^B \mid \mu(f) \in \mathcal{J}(L) \setminus A\}$ is pairwise complete.*

The following Proposition, which ends the study of bound-preserving lattice embeddings into relational lattices, shows that modulo the shift of the codomain to the lattice of a universal product frame, such a lattice embedding can always be normalized, meaning that join-irreducible elements are sent to join-irreducible elements by the left adjoint.

Proposition 28 Cf. [21, Proposition 51]. *Let L be a finite atomistic lattice and let A be the set of its join-prime elements. If L is not a Boolean algebra and $i : L \longrightarrow \mathsf{R}(D, B)$ is a bound-preserving lattice embedding, then there exists a bound-preserving lattice embedding $j : L \longrightarrow \mathsf{L}(F_0, \delta_A)$, where (F_0, δ_A) is the pairwise complete ultrametric space defined in Corollary 27. Moreover, the left adjoint ν to j satisfies the following condition: for each $k \in A \cup F_0$, if $k \in A$ then $\nu(k) = k$ and, otherwise, $\nu(k) \in \mathcal{J}(L) \setminus A$.*

The following Theorem asserts that we can assume that a lattice embedding is bound-preserving, when its domain is a finite subdirectly-irreducible lattice. It is needed in Theorem 7 to exclude the constants $\perp$ and $\top$ from the signature of lattice theory.

Theorem 29 Cf. [21, Sect. 7]. *If L is a finite subdirectly-irreducible atomistic lattice which has a lattice embedding into some relational lattice $\mathsf{R}(D, A)$, then there exists a bound-preserving embedding of L into some other relational lattice $\mathsf{R}(D, B)$.*

We conclude next the proof of the main result of this Section, Theorem 21.

Proof (of Theorem 21). Since $\mathfrak{F}$ is rooted and full, $\mathsf{L}(\mathfrak{F})$ is a finite atomistic subdirectly-irreducible lattice by Proposition 16. Therefore, if $i : \mathsf{L}(\mathfrak{F}) \longrightarrow \mathsf{R}(D, B)$ is a lattice embedding, then we can assume, using Theorem 29, that i preserves the bounds. Also, if $\mathsf{L}(\mathfrak{F})$ is a Boolean algebra, then it is the two elements Boolean algebra, since we are assuming that $\mathsf{L}(\mathfrak{F})$ is subdirectly-irreducible. But then, $\mathfrak{F}$ is a singleton, and the statement of the Theorem trivially holds in this case.

We can therefore assume that $\mathsf{L}(\mathfrak{F})$ is not a Boolean algebra. Let us recall that A is the set of join-prime elements of $\mathsf{L}(\mathfrak{F})$, see Proposition 15. Let (F_0, δ_A) be the pairwise complete space over $P(A)$ and let $j : \mathsf{L}(\mathfrak{F}) \longrightarrow \mathsf{L}(F_0, \delta_A)$ be the lattice morphism with the properties stated in Proposition 28; let ν be the left adjoint to j. We can also assume that $\mathsf{L}(F_0, \delta_A) = \mathsf{L}(\mathfrak{U})$ for some universal $\mathrm{S5}^A$-product frame $\mathfrak{U}$.

To avoid confusions, we depart from the convention of identifying singletons with their elements. We define $\psi : X_{\mathfrak{U}} \longrightarrow X_{\mathfrak{F}}$ by saying that $\psi(x) = y$ when $\nu(\{x\}) = \{y\}$. This is well defined since in $\mathsf{L}(\mathfrak{U})$ (respectively $\mathsf{L}(\mathfrak{F})$) the non-join-prime join-irreducible-elements are the singletons $\{x\}$ with $x \in X_{\mathfrak{F}_{\mathfrak{U}}}$ (resp. $x \in X_{\mathfrak{F}}$); moreover, we have $X_{\mathfrak{U}} = F_0$ and each singleton $\{x\}$ with $x \in F_0$ is sent by ν to a singleton $\{y\} \in \mathcal{J}(\mathsf{L}(\mathfrak{F})) \setminus \{\{a\} \mid a \in A\} = \{\{x\} \mid x \in X_{\mathfrak{F}}\}$. The function ψ is surjective since every non-join-prime atom $\{x\}$ in $\mathsf{L}(\mathfrak{F})$ has a preimage by ν an atom $\{y\}$ and such a preimage cannot be join-prime, so $y \in X_{\mathfrak{U}}$.

We are left to argue that ψ is a p-morphism. To this end, let us remark that, for each $a \in A$ and $x, y \in X_{\mathfrak{F}}$ (or $x, y \in X_{\mathfrak{U}}$), the relation $xR_a y$ holds exactly when there is an $\{a\}$-path from x to y, i.e. when $\{x\} \subseteq \overline{\{a, y\}} = \{a\} \vee \{y\}$ (we need here that $\mathfrak{F}$ and $\mathfrak{U}$ are S4 frames).

Thus, let $x, y \in X_{\mathfrak{U}}$ be such that $xR_a y$. Then $\{x\} \subseteq \{a\} \vee \{y\}$ and $\nu(\{x\}) \subseteq \nu(\{a\}) \vee \nu(\{y\}) = \{a\} \vee \nu(\{y\})$. We have therefore $\psi(x) R_a \psi(y)$. Conversely, let $x \in X_{\mathfrak{U}}$ and $z \in X_{\mathfrak{F}}$ be such that $\psi(x) R_a z$. We have therefore $\nu(\{x\}) \subseteq \{a\} \vee \{z\}$, whence, by adjointness,

$$\begin{aligned}\{x\} \subseteq j(\{a\} \vee \{z\}) &= j(\{a\}) \vee j(\{z\}) \\ &= \{a\} \vee \{y \mid \nu(\{y\}) = \{z\}\} \\ &= \overline{\{a\} \cup \{y \mid \nu(\{y\}) = \{z\}\}}.\end{aligned}$$

But this means that there is some $y \in X_{\mathfrak{U}}$ with $\psi(y) = z$ and a $\{a\}$-path from x to y. But then, we also have $xR_a y$. □

References

1. Ackerman, N.: Completeness in generalized ultrametric spaces. P-Adic Numbers Ultrametric Anal. Appl. **5**(2), 89–105 (2013)
2. Burris, S., Sankappanavar, H.: A Course in Universal Algebra. Dover Publications, Incorporated, Mineola (2012)
3. Codd, E.F.: A relational model of data for large shared data banks. Commun. ACM **13**(6), 377–387 (1970)
4. Davey, B.A., Priestley, H.A.: Introduction to Lattices and Order. Cambridge University Press, New York (2002)
5. Evans, T.: Embeddability and the word problem. J. London Math. Soc. **28**, 76–80 (1953)
6. Freese, R., Ježek, J., Nation, J.: Free lattices. American Mathematical Society, Providence (1995)
7. Grätzer, G.: General Lattice Theory. Birkhäuser Verlag, Basel (1998). New appendices by the author with Davey, B.A., Freese, R., Ganter, B., Greferath, M., Jipsen, P., Priestley, H.A., Rose, H., Schmidt, E.T., Schmidt, S.E., Wehrung, F., Wille, R
8. Hammack, R., Imrich, W., Klavzar, S.: Handbook of Product Graphs, 2nd edn. CRC Press Inc, Boca Raton (2011)
9. Hirsch, R., Hodkinson, I.: Representability is not decidable for finite relation algebras. Trans. Am. Math. Soc. **353**, 1403–1425 (2001)
10. Hirsch, R., Hodkinson, I., Kurucz, A.: On modal logics between K × K × K and S5 × S5 × S5. J. Symb. Log. **67**, 221–234 (2002)
11. Kurucz, A.: Combining modal logics. In: Patrick Blackburn, J.V.B., Wolter, F. (eds.) Handbook of Modal Logic Studies in Logic and Practical Reasoning, pp. 869–924. Elsevier, Amsterdam (2007)
12. Lawvere, F.W.: Metric spaces, generalized logic, closed categories. Rendiconti del Seminario Matematico e Fisico di Milano **XLIII**, 135–166 (1973)
13. Litak, T., Mikuls, S., Hidders, J.: Relational lattices: from databases to universal algebra. J. Logical Algebraic Methods Program. **85**(4), 540–573 (2016)
14. Maddux, R.: The equational theory of CA_3 is undecidable. J. Symbolic Logic **45**(2), 311–316 (1980)
15. Maddux, R.: Relation Algebras Studies in Logic and the Foundations of Mathematics. Elsevier, Amsterdam (2006)
16. Nation, J.B.: An approach to lattice varieties of finite height. Algebra Universalis **27**(4), 521–543 (1990)
17. Priess-Crampe, S., Ribemboim, P.: Equivalence relations and spherically complete ultrametric spaces. C. R. Acad. Sci. Paris **320**(1), 1187–1192 (1995)
18. Sambin, G.: Subdirectly irreducible modal algebras and initial frames. Studia Log. **62**, 269–282 (1999)
19. Santocanale, L.: A duality for finite lattices. Preprint, September (2009). http://hal.archives-ouvertes.fr/hal-00432113
20. Santocanale, L.: Relational lattices via duality. In: Hasuo, I. (ed.) CMCS 2016. LNCS, vol. 9608, pp. 195–215. Springer, Cham (2016). doi:10.1007/978-3-319-40370-0_12
21. Santocanale, L.: The quasiequational theory of relational lattices, in the pure lattice signature. Preprint, July (2016). https://hal.archives-ouvertes.fr/hal-01344299
22. Spight, M., Tropashko, V.: Relational lattice axioms. Preprint (2008). http://arxiv.org/abs/0807.3795
23. Tropashko, V.: Relational algebra as non-distributive lattice. Preprint (2006). http://arxiv.org/abs/cs/0501053

Tower Induction and Up-to Techniques for CCS with Fixed Points

Steven Schäfer(✉) and Gert Smolka

Saarland University, Saarbrücken, Germany
{schaefer,smolka}@ps.uni-saarland.de

Abstract. We present a refinement of Pous' companion-based coinductive proof technique and apply it to CCS with general fixed points. We construct companions based on inductive towers and thereby obtain a powerful induction principle. Our induction principle implies a new sufficient condition for soundness of up-to techniques subsuming respectfulness and compatibility. For bisimilarity in CCS, companions yield a notion of relative bisimilarity. We show that relative bisimilarity is a congruence, a basic result implying soundness of bisimulation up to context. The entire development is constructively formalized in Coq.

1 Introduction

Coinductive definitions and their associated reasoning principles are one of the basic tools for studying equivalences in programming languages and process calculi. For process calculi, the idea has been developed by Milner [7] in the form of bisimilarity and the bisimulation proof method. In the context of programming languages, coinductive simulations are an important tool in the field of compiler verification [6].

Coinductive definitions can be realized as greatest fixed points of monotone functions on complete lattices. Tarski's [16] construction of greatest fixed points yields a primitive coinduction principle, which is dual to structural induction. Unfortunately, this coinduction principle can be inconvenient in practice because it requires the construction of an often involved invariant. In the context of bisimilarity, these invariants are known as bisimulations and can become quite complicated. This is especially cumbersome in the context of interactive theorem proving. The construction of an appropriate bisimulation is often the lion's share of a bisimilarity proof.

Fortunately, there are several enhancements of the coinductive proof method, which mitigate these problems.

One useful enhancement consists in changing the function underlying a coinductive definition to simplify proofs by coinduction. There is significant room for improvement here, as many different functions can have the same greatest fixed point. These enhancements of the coinductive proof method are known as up-to techniques [7,13]. As we will see, the gains from using up-to techniques can be dramatic. There are simple examples (Sect. 4) of bisimilarity proofs, where the smallest bisimulation is infinite, yet there is a finite bisimulation up-to.

P. Höfner et al. (Eds.): RAMiCS 2017, LNCS 10226, pp. 274–289, 2017.
DOI: 10.1007/978-3-319-57418-9_17

Recently, Hur et al. [5] have introduced another enhancement of the coinductive proof method in the form of parameterized coinduction. Parameterized coinduction yields both modular and incremental proof principles for coinduction. In addition, Hur et al. show how to combine up-to techniques with parameterized coinduction, which yields an easier way to apply up-to techniques in coinductive proofs.

Pous [12] starts from this combination of parameterized coinduction and up-to techniques to introduce another substantial simplification and extension of the coinductive proof method. Pous shows that for every monotone function there is a canonical best up-to function, its *companion*. Not only does the use of companions lead to simple proof principles for up-to techniques, it also subsumes the work of Hur et al. and allows for a smooth integration of parameterized coinduction. In this context, up-to techniques for the companion are simply functions below the companion.

If we apply the companion construction to bisimilarity, we obtain a notion of *relative bisimilarity*. Intuitively, two processes are bisimilar relative to a relation R if we can show that they are bisimilar under the assumption that R–related processes are bisimilar. Up-to techniques correspond to properties of relative bisimilarity. For example, the statement that relative bisimilarity is a congruence implies the soundness of bisimulation up to context. In fact, the congruence property of relative bisimilarity is a stronger result.

Despite these advances it remains difficult to show the soundness of up-to techniques. Our aim in this paper is to introduce new proof techniques for the companion, which simplify soundness proofs for up-to techniques. We demonstrate our proof techniques with a non-trivial case study.

The key to our results is a novel inductive construction of companions (Sect. 3). Our construction yields the *tower induction* principle for companions (Theorem 6), which implies a complete characterization of companion-based up-to functions (Lemma 8).

We apply our construction to strong bisimilarity for CCS with fixed points (Sect. 4). Our main result is that relative bisimilarity extended to open terms is a congruence. This roughly corresponds to the soundness of bisimulation up-to context. To the best of our knowledge, this result has not appeared in the literature before. Our proofs make extensive use of our characterization of up-to functions for the companion.

Beyond these case studies, we combine parameterized coinduction with our inductive construction of companions. This leads to the *parameterized tower induction* principle (Sect. 5). The accumulation rule of Hur et al. appears as a consequence of parameterized tower induction.

Finally, we report on a Coq formalization of our results (Sect. 6). The main difference to the paper presentation is in our treatment of binders. We use a de Bruijn representation in the form of J_f-relative monads [2] to distinguish open and closed terms. Following [14], we establish all substitution lemmas using the rules of a convergent rewriting system.

Contributions. We consider the tower induction principle for companions and the formal development of open relative bisimilarity for CCS with general recursive definitions the two main contributions of the paper.

2 Lattice Theory Preliminaries

We recall some basic definitions and results about fixed points in complete lattices [4].

A *complete lattice* may be defined as a triple $(A, \leq, \bigsqcup)$, where $\leq$ is a partial order on A, such that every set $M \subseteq A$ has a supremum $\bigsqcup M$. For every $M \subseteq A$ we have:

$$\bigsqcup M \leq y \;\Leftrightarrow\; \forall x \in M.\, x \leq y$$

Every complete lattice has a greatest element $\top$, as well as binary suprema $x \sqcup y$. Arbitrary infima $\bigsqcap M$ can be obtained as suprema of lower bounds. In particular, every complete lattice has a least element $\bot$, along with binary infima $x \sqcap y$.

A function f is *monotone* if $f(x) \leq f(y)$ whenever $x \leq y$. For a monotone function $f : A \to A$ we say that x is a *prefixed point* of f if $f(x) \leq x$ and a *postfixed point* if $x \leq f(x)$. An element x is a *fixed point* of f if $f(x) = x$.

By Tarski's theorem [16], every monotone function on a complete lattice has a complete lattice of fixed points. In particular, every monotone function has a greatest fixed point.

Theorem 1. Let f be a monotone function on a complete lattice. The supremum over all postfixed points $\nu f := \bigsqcup \{\, x \mid x \leq f(x) \,\}$ is the greatest fixed point of f.

3 Towers and Companions

Pous [12] gives a useful characterization of the greatest fixed point as $\nu f = t(\bot)$ using a function t called the companion for f. Parrow and Weber [8] give an ordinal-based construction of the companion in classical set theory. It turns out that the companion can be obtained in constructive type theory with an inductive tower construction [15].

Definition 2. Let f be a function on a complete lattice. The *f-tower* is the inductive predicate T_f defined by the following rules.

$$\frac{x \in T_f}{f(x) \in T_f} \qquad\qquad \frac{M \subseteq T_f}{\bigsqcap M \in T_f}$$

Using T_f we define t_f, the *companion* of f.

$$t_f(x) \;:=\; \bigsqcap \{\, y \in T_f \mid x \leq y \,\}$$

We will omit the index on T_f and t_f when the function f is clear from the context.

Note that $t(x)$ is the least element of T above x, since the tower is closed under infima. The following are further consequences of the closure under infima.

Fact 3. t is a closure operator with image T.

a) t is monotone
b) $x \leq t(x)$
c) $t(t(x)) = t(x)$
d) $x \in T \leftrightarrow t(x) = x$.

Additionally, since the tower is closed under f, we have:

Fact 4. $f(t(x)) = t(f(t(x)))$

As a consequence of our inductive construction of T, we obtain an induction principle for t.

Definition 5. A predicate P is *inf-closed* if $\bigcap M \in P$ whenever $M \subseteq P$.

Theorem 6 (Tower Induction). Let P be an inf-closed predicate such that $P(t(x))$ implies $P(f(t(x)))$ for all x. Then $P(t(x))$ holds for all x.

Proof. Follows from Fact 3, by induction on the derivation of $t_f(x) \in T_f$. ■

Standard inf-closed predicates include $\lambda x.\ y \leq x$ for a fixed y and $\lambda x.\ g(x) \leq x$ for monotone g. Both instantiations yield useful statements about t.

Using the predicate $\lambda x.\ \nu f \leq x$ in Theorem 6, we can reconstruct the greatest fixed point of f in terms of t.

Lemma 7. If f is monotone, then $\nu f = t_f(\bot)$.

Proof. We have $t(\bot) \leq t(f(t(\bot))) = f(t(\bot))$ by monotonicity of t and Fact 4. It follows that $t(\bot) \leq \nu f$.

In the reverse direction we show $\nu f \leq t(x)$ for all x using Theorem 6. It suffices to show that $\nu f \leq x$ implies that $\nu f \leq f(x)$. This follows from $\nu f = f(\nu f)$ and the monotonicity of f.

More generally, we have $t(x) = \nu f$ for all $x \leq \nu f$, since t is monotone and idempotent.

Using the predicate $\lambda x.\ g(x) \leq x$ in Theorem 6, we prove a characterization of the up-to functions for t, i.e., the monotone functions below t.

Lemma 8 (Up-to Lemma). Let g be monotone. Then the following statements are equivalent.

a) $g \leq t$
b) $g \circ t \leq t$
c) $\forall x.\ g(t(x)) \leq t(x) \to g(f(t(x))) \leq f(t(x))$

Proof. The implication from (c) to (b) follows by tower induction. From (b) to (a) we have $g \le g \circ t \le t$, by Fact 3 and the monotonicity of g. The implication from (a) to (c) follows from Fact 4. ∎

In particular, this shows that f is below t.

Lemma 9. Let f be monotone. Then $f(t(x)) \le t(x)$.

Proof. By Lemma 8(b) using the monotonicity of f. ∎

We now relate our companion construction to Pous' construction [12].

Definition 10. A function g is *compatible* for f if it is monotone and $g \circ f \le f \circ g$.

Lemma 11. For monotone f, we have $t_f = \bigcup \{g \mid g \text{ is compatible for} f\}$.

Proof. Let g be compatible for f. We have $g(f(t(x))) \le f(g(t(x)))$ by compatibility, and by Lemma 8 this implies $g \le t$. Additionally, the companion is compatible for f, since it is monotone by Fact 3 and $t \circ f \le t \circ f \circ t = f \circ t$ by Facts 3 and 4. ∎

In light of Lemma 11, we can see most results in this section as rederivations of results from [12]. The exceptions are Theorem 6 and Lemma 8, which are new results for the companion. In the sequel, we will make extensive use of the new results to show soundness of up-to functions.

4 CCS with Recursive Processes

In this section we apply the companion construction to strong bisimilarity in CCS [7] with fixed point expressions. Using the companion we obtain proof principles analogous to bisimulation up-to context for the extension of bisimilarity to open terms. Our proofs are similar to Milner's proof [7] that strong bisimilarity is a congruence for CCS, yet our results are strictly stronger. We illustrate this by giving a straightforward proof of the well-known fact that weakly guarded equations have unique solutions modulo strong bisimilarity.

The syntax of CCS processes and actions is given by the following grammar.

$$P, Q ::= 0 \mid \alpha.P \mid P \parallel Q \mid P + Q \mid (\nu a)P \mid X \mid \mu X.\, P$$
$$\alpha, \beta ::= a \mid \overline{a} \mid \tau$$

For the paper presentation we assume that there is some countably infinite type of variables X, Y, Z. The fixed point expression $\mu X.\, P$ binds the variable X in P. We use the standard notions of free and bound variables. We adopt the Barendregt convention and consider processes up to renaming of bound variables.

A *substitution* σ is a mapping from variables to processes. We can *instantiate* a process P under a substitution σ by replacing all free variables according to σ, while keeping the bound variables fixed. We write $P[\sigma]$ for the process P

$$\frac{}{\alpha.P \overset{\alpha}{\rightsquigarrow} P} \qquad \frac{P \overset{\alpha}{\rightsquigarrow} P'}{P \parallel Q \overset{\alpha}{\rightsquigarrow} P' \parallel Q} \qquad \frac{Q \overset{\alpha}{\rightsquigarrow} Q'}{P \parallel Q \overset{\alpha}{\rightsquigarrow} P \parallel Q'} \qquad \frac{P \overset{a}{\rightsquigarrow} P' \quad Q \overset{\overline{a}}{\rightsquigarrow} Q'}{P \parallel Q \overset{\tau}{\rightsquigarrow} P' \parallel Q'}$$

$$\frac{P \overset{\overline{a}}{\rightsquigarrow} P' \quad Q \overset{a}{\rightsquigarrow} Q'}{P \parallel Q \overset{\tau}{\rightsquigarrow} P' \parallel Q'} \qquad \frac{P \overset{\alpha}{\rightsquigarrow} P'}{P + Q \overset{\alpha}{\rightsquigarrow} P'} \qquad \frac{Q \overset{\alpha}{\rightsquigarrow} Q'}{P + Q \overset{\alpha}{\rightsquigarrow} Q'} \qquad \frac{P \overset{\alpha}{\rightsquigarrow} P' \quad \alpha \neq a, \overline{a}}{(\nu a)P \overset{\alpha}{\rightsquigarrow} (\nu a)P'}$$

$$\frac{P[X \mapsto \mu X.\, P] \overset{\alpha}{\rightsquigarrow} Q}{\mu X.\, P \overset{\alpha}{\rightsquigarrow} Q}$$

Fig. 1. Labeled transition system for CCS.

instantiated under σ. The expression $X \mapsto P$ denotes the substitution that replaces the variable X by P. We combine substitutions by juxtaposition.

A process is *closed* if it does not contain any free variables.

The semantics of CCS is given by a labeled transition system (LTS), i.e., an indexed relation between closed processes $P \overset{\alpha}{\rightsquigarrow} Q$. Intuitively, the relation $P \overset{\alpha}{\rightsquigarrow} Q$ means that process P can reduce to Q and perform the action α in a single step. The labeled transition system for CCS is defined inductively by the rules in Fig. 1.

Fixed point expressions allow us to specify arbitrary recursive processes. For instance, replication can be expressed as $!P := \mu X.\ X \parallel P$, where X is not free in P. Under this encoding, we have $!P \overset{\alpha}{\rightsquigarrow} Q$ whenever $!P \parallel P \overset{\alpha}{\rightsquigarrow} Q$. Note that this yields an infinitely branching LTS for, e.g., $!a.0$.

We define strong bisimilarity as the greatest fixed point of a function b mapping binary relations into binary relations. First, let s be the function expressing one step of simulation.

$$s(R) := \lambda PQ.\ \forall \alpha P'.\ P \overset{\alpha}{\rightsquigarrow} P' \ \rightarrow\ \exists Q'.\, Q \overset{\alpha}{\rightsquigarrow} Q' \wedge R\, P'\, Q'$$

The function b is just simulation in both directions. More precisely, it is the greatest symmetric function below s, where a function between relations is symmetric if it maps symmetric relations to symmetric relations.

$$b(R) := \lambda R.\ s(R) \wedge s(R^\dagger)^\dagger$$
$$R^\dagger := \lambda PQ.\ R(Q, P)$$

Bisimilarity is the greatest fixed point of b. We write t for the companion of b. Bisimilarity between processes P, Q is denoted by $P \sim Q$.

We are trying to develop effective proof techniques for bisimilarity. Before we consider how to show bisimilarity using the companion, let us recall the classical bisimulation proof method. Following the literature, the postfixed points of b are called *bisimulations*. Bisimilarity is the union of all bisimulations by Theorem 1. In order to show that $P \sim Q$ holds, it suffices to find a bisimulation R containing the pair (P, Q). The problem with this proof technique is that R can be arbitrarily complicated and has to be explicitly constructed.

Consider two processes A, B such that

$$A \sim (\overline{a}.B) \parallel (a.!A)$$
$$B \sim (a.!B) \parallel (\overline{a}.A)$$

The processes A and B are obviously bisimilar, as parallel composition is commutative and the only difference beyond this is a simple renaming. Yet the smallest bisimulation containing the pair (A, B) is infinite.

Instead of using the definition of bisimilarity, we can use the companion and tower induction. The companion gives us a notion of *relative bisimilarity*, or R-bisimilarity, which we write as $P \sim_R Q$. Intuitively, processes are bisimilar relative to R, if we can show that they are bisimilar, assuming that all R-related processes are bisimilar. In coinductive proofs, we can frequently assume that some processes are bisimilar after a step of reduction. We can express this in terms of relative bisimilarity, by introducing *guarded assumptions* $\circ R$. Given a relation R, we define $\circ R := b(t(R))$.

$$P \sim Q := (P, Q) \in \nu b \qquad \text{bisimilarity}$$
$$P \sim_R Q := (P, Q) \in t(R) \qquad \text{relative bisimilarity}$$
$$P \sim_{\circ R} Q \;= (P, Q) \in t(b(t(R))) \qquad \text{guarded relative bisimilarity}$$

The different notions of bisimilarity are related by the following laws, which instantiate lemmas from Sect. 3.

Fact 12. $\sim_\bot = \sim \subseteq \sim_{\circ R} = \circ R \subseteq \sim_R \supseteq R$

Tower induction gives us a proof principle for showing bisimilarity in terms of relative bisimilarity.

Lemma 13. If $P \sim_R Q$ implies $P \sim_{\circ R} Q$ for all R, then $P \sim Q$.

Proof. By Fact 12 together with tower induction using the inf-closed predicate $\lambda R.\ (P, Q) \in R$. ■

Lemma 13 corresponds to the statement that bisimulation up-to the companion is sound. To show that two processes are bisimilar, it suffices to show that they are bisimilar relative to an assumption which states that they are bisimilar after unfolding at least one reduction step.

Phrased in our vocabulary, Pous [12] has shown that relative bisimilarity is a congruence for CCS with replication. This simplifies the proof of the bisimilarity $A \sim B$. By Lemma 13, it suffices to show $A \sim_{\circ R} B$, assuming that $A \sim_R B$. We have

$$A \sim (\overline{a}.B) \parallel (a.!A) \sim (a.!A) \parallel (\overline{a}.B)$$
$$B \sim (a.!B) \parallel (\overline{a}.A)$$

Since $\sim \subseteq \sim_{\circ R}$, and $\sim_{\circ R}$ is transitive, it thus suffices to show that

$$(a.!A) \parallel (\overline{a}.B) \sim_{\circ R} (a.!B) \parallel (\overline{a}.A)$$

Since $\sim_{\circ R}$ is a congruence, this follows from $a.!A \sim_{\circ R} a.!B$ and $\overline{a}.B \sim_{\circ R} \overline{a}.A$. Unfolding the definition of b, we have to show $!A \sim_R !B$ and $B \sim_R A$. The former follows by compatibility with replication, the latter follows from the symmetry of $\sim_R$. We conclude that $A \sim B$.

In this case, we can further simplify the proof to avoid unfolding the definition of b. We simply strengthen the compatibility with action prefixes to $P \sim_R Q \to \alpha.P \sim_{\circ R} \alpha.Q$ since action prefixes can perform a step of reduction.

As defined, relative bisimilarity is not a congruence for CCS with recursive processes, since it is only defined for closed terms. In order to proceed, we lift relative bisimilarity to open terms.

Two processes P, Q are in open bisimilarity $P \dot{\sim} Q$ if they are bisimilar under all closing substitutions, i.e., substitutions which replace all free variables by closed processes. We write θ for closing substitutions. As before, we also consider the relative variant of open bisimilarity. To distinguish open bisimilarity from ordinary bisimilarity, we will refer to the latter as closed bisimilarity.

$$P \dot{\sim} Q := \forall \theta.\ P[\theta] \sim Q[\theta] \qquad \text{open bisimilarity}$$
$$P \dot{\sim}_R Q := \forall \theta.\ P[\theta] \sim_R Q[\theta] \qquad \text{open relative bisimilarity}$$

Open and closed (relative) bisimilarity coincide for closed processes.

Even though open bisimilarity is not defined coinductively, we obtain a reasoning principle analogous to Lemma 13 using tower induction.

Lemma 14. If $P \dot{\sim}_R Q$ implies $P \dot{\sim}_{\circ R} Q$ for all R, then $P \dot{\sim} Q$.

Proof. Tower induction with $P(R) = \forall \theta.\ (P[\theta], Q[\theta]) \in R$. ■

In the remainder of this section we show that relative open bisimilarity is a congruence.

Lemma 15. Relative open bisimilarity is an equivalence relation.

Proof. By the definition of relative open bisimilarity, it suffices to show that relative bisimilarity is an equivalence relation. This follows by tower induction. The intersection of a family of equivalence relations is an equivalence relation and it is easy to see that $b(R)$ is an equivalence relation if R is. ■

For the compatibility with the various connectives of CCS, we define local context operators as follows:

$$\begin{aligned}
c^{\cdot}(R) &:= \{\, (\alpha.P, \alpha.Q) \mid R(P,Q) \,\} \\
c^{\parallel}(R) &:= \{\, (P_1 \parallel Q_1, P_2 \parallel Q_2) \mid R(P_1,P_2), R(Q_1,Q_2) \,\} \\
c^{+}(R) &:= \{\, (P_1 + Q_1, P_2 + Q_2) \mid R(P_1,P_2), R(Q_1,Q_2) \,\} \\
c^{\nu}(R) &:= \{\, ((\nu a)P, (\nu a)Q) \mid R(P,Q) \,\}
\end{aligned}$$

Compatibility under all contexts not containing fixed point expressions corresponds to the statements $c(\sim_R) \subseteq \sim_R$ for all local context operators. These results have already been shown in [12] using a second order companion construction. We give alternative proofs using the up-to lemma (Lemma 8).

As in [12] we encapsulate some of the symmetries in the problem using the following lemma.

Fact 16 [12]. Let g be symmetric and $g(b(R)) \leq s(R)$. Then $g(b(R)) \leq b(R)$.

For compatibility with action prefixes we show a slightly stronger statement.

Lemma 17. If $P \mathrel{\dot{\sim}_R} Q$, then $\alpha.P \mathrel{\dot{\sim}_{\circ R}} \alpha.Q$ and $\alpha.P \mathrel{\dot{\sim}_R} \alpha.Q$.

Proof. We have $c^{\cdot} \leq b$, by unfolding the definitions. The statement follows using Fact 12, since $c^{\cdot}(t(R)) \subseteq b(t(R)) \subseteq \sim_R$. ■

All remaining proofs follow the same pattern. After applying Lemma 8 and Fact 16 (which allows us to focus on establishing the simulation condition), our work boils down to a simple case analysis. We illustrate only the case of parallel composition in detail.

Lemma 18. $P_1 \mathrel{\dot{\sim}_R} P_2 \to Q_1 \mathrel{\dot{\sim}_R} Q_2 \to P_1 \parallel Q_1 \mathrel{\dot{\sim}_R} P_2 \parallel Q_2$

Proof. We show $c^{\parallel}(\sim_R) \subseteq \sim_R$ using Lemma 8. Assume that the statement holds for a relation R and all P_1, P_2, Q_1, Q_2. We will refer to this as the *coinductive hypothesis.*

We have to show that $P_1 \sim_{\circ R} P_2$ and $Q_1 \sim_{\circ R} Q_2$ imply $P_1 \parallel Q_1 \sim_{\circ R} P_2 \parallel Q_2$. By Fact 16 it suffices to show this for one step of simulation. Let $P_1 \parallel Q_1 \overset{\alpha}{\leadsto} U$, we have to find a process V such that $P_2 \parallel Q_2 \overset{\alpha}{\leadsto} V$ and $U \sim_R V$.

We proceed by case analysis on $P_1 \parallel Q_1 \overset{\alpha}{\leadsto} U$. Formally, there are four cases to consider of which two follow by symmetry.

- *Communication between* P_1 *and* Q_1. We have $P_1 \overset{a}{\leadsto} P_1'$, $Q_1 \overset{\overline{a}}{\leadsto} Q_1'$, $\alpha = \tau$, and $U = P_1' \parallel Q_1'$. By the assumptions $P_1 \sim_{\circ R} P_2$ and $Q_1 \sim_{\circ R} Q_2$ there are P_2', Q_2' such that $P_2 \overset{a}{\leadsto} P_2'$, $Q_2 \overset{\overline{a}}{\leadsto} Q_2'$, $P_1' \sim_R P_2'$ and $Q_1' \sim_R Q_2'$. We pick $V = P_2' \parallel Q_2'$, as $P_2 \parallel Q_2 \overset{\tau}{\leadsto} P_2' \parallel Q_2$. The statement $P_1' \parallel Q_1' \sim_R P_2' \parallel Q_2'$ follows from the coinductive hypothesis.
- *Reduction in* P_1. We have $P_1 \overset{\alpha}{\leadsto} P_1'$ and $U = P_1' \parallel Q_1$. By assumption, $P_2 \overset{\alpha}{\leadsto} P_2'$ and $P_1' \sim_R P_2'$. We pick $V = P_2' \parallel Q_2$, as $P_2 \parallel Q_2 \overset{\alpha}{\leadsto} P_2' \parallel Q_2$. The statement $P_1' \parallel Q_1 \sim_R P_2' \parallel Q_2$ follows from the coinductive hypothesis if we can show $Q_1 \sim_R Q_2$. This follows from the assumption that $Q_1 \sim_{\circ R} Q_2$ and Fact 12. ■

Finally, we have to show that bisimilarity is compatible with fixed point expressions. Schematically, the proof remains similar to the proof of Lemma 18, except that we replace the case analysis on the reduction relation by a nested induction.

Substitutions add an additional complication, however, since reducing a fixed point expression instantiates a variable. Intuitively, this means that we have to show that open bisimilarity is compatible with fixed points and instantiation at the same time.

First, let us consider the following context operators on closed relative bisimilarity.

$$\begin{aligned} c^{\mu}(R) &:= \{\, (\mu X.\, P,\ \mu X.\, Q) \mid \\ &\qquad \forall S \text{ closed. } R(P[X \mapsto S],\ Q[X \mapsto S]) \,\} \\ c^{[]}(R) &:= \{\, (P[\theta_1], P[\theta_2]) \mid \forall x.\ R(\theta_1(x), \theta_2(x)) \,\} \end{aligned}$$

If $c^{\mu}(R) \subseteq R$, then we can show that two fixed points $\mu X.\ P$, $\mu X.\ Q$ are related if they are related whenever we substitute the same closed process for X in both P and Q. We are implicitly assuming that X is the only free variable in P, Q.

If $c^{[]}(R) \subseteq R$, then R is compatible under related closing substitutions. Specifically, if P is an open process and θ_1, θ_2 are two pointwise related closing substitutions, we can show that $P[\theta_1]$ and $P[\theta_2]$ are related.

One half of the relationship between c^{μ} and $c^{[]}$ is captured by the following lemma.

Lemma 19. If $c^{\mu}(\sim_R) \subseteq \sim_R$, and $\theta_1(x) \sim_R \theta_2(x)$ for all x, then $P[\theta_1] \sim_R P[\theta_2]$ for all P.

Proof. By induction on P. The cases for action prefixes, choice, parallel composition and restriction follow from the compatibility of $\sim_R$ with the structure of P. The case for variables follows from the assumption on θ_1 and θ_2.

Finally, let $P = \mu X.\ Q$. By compatibility with μ it suffices to show that $Q[\theta_1, X \mapsto S] \sim_R Q[\theta_2, X \mapsto S]$. This follows by induction, since the extended substitutions are related by reflexivity of $\sim_R$. ∎

In fact, we do have $c^{\mu}(\sim_R) \subseteq \sim_R$, and the first assumption in the previous lemma is vacuously true.

Lemma 20. If $P[X \mapsto S] \sim_R Q[X \mapsto S]$ holds for all R and closed S, where X is the only free variable in P, Q, then $\mu X.P \sim_R \mu X.Q$.

Proof. We show $c^{\mu}(\sim_R) \subseteq \sim_R$ by Lemma 8. We can assume that the statement holds for a relation R and have to show that it holds for $\circ R$.

It suffices to show that $Q[X \mapsto \mu X.\, P] \sim_{\circ R} Q[X \mapsto \mu X.\, Q]$. The statement then follows from $\mu X.\, P \sim P[X \mapsto \mu X.\, P]$ and transitivity:

$$\mu X.\ P \sim P[X \mapsto \mu X.\ P] \sim_{\circ R} Q[X \mapsto \mu X.\ P] \sim_{\circ R} Q[X \mapsto \mu X.\ Q] \sim \mu X.\ Q$$

where in the second step, we have used the assumption that P and Q are $\circ R$-related under the same closing substitution.

What is left to show is almost compatibility under instantiation with related substitutions. We show that $Q_0[X \mapsto \mu X.\ P] \sim_{\circ R} Q_0[X \mapsto \mu X.\ Q]$ for all Q_0. By Fact 16, it suffices to show this statement for one step of simulation. Let $Q_0[X \mapsto \mu X.\ P] \overset{\alpha}{\rightsquigarrow} Q'$. We have to find Q'' such that $Q_0[X \mapsto \mu X.\ Q] \overset{\alpha}{\rightsquigarrow} Q''$ and $Q' \sim_R Q''$.

We proceed by induction on the derivation of $Q_0[X \mapsto \mu X.\ P] \overset{\alpha}{\rightsquigarrow} Q'$. There are nine cases to consider in total. We illustrate three representative cases.

- $Q_0 = S \parallel T$ and $S[X \mapsto \mu X.\ P] \overset{\alpha}{\rightsquigarrow} S'$. By the inductive hypothesis, there is an S'' such that $S[X \mapsto \mu X.\ Q] \overset{\alpha}{\rightsquigarrow} S''$ and $S' \sim_R S''$. It suffices to show that $S' \parallel T[X \mapsto \mu X.\ P] \sim_R S'' \parallel T[X \mapsto \mu X.\ Q]$.
  This follows from compatibility with parallel composition and instantiation (Lemma 19). We have $S' \sim_R S''$ by assumption and $\mu X. P \sim_R \mu X. Q$ follows from the coinductive hypothesis.
- $Q_0 = \mu Y. S$ and $S[Y \mapsto \mu Y. S][X \mapsto \mu X. P] \overset{\alpha}{\rightsquigarrow} S'$. By the inductive hypothesis, there is an S'' such that $S[Y \mapsto \mu Y. S][X \mapsto \mu X. Q] \overset{\alpha}{\rightsquigarrow} S''$ and $S' \sim_R S''$, which is what we needed to show.
- $Q_0 = X$ and $P[X \mapsto \mu X. P] \overset{\alpha}{\rightsquigarrow} P'$. By the inductive hypothesis, there is a P'' such that $P[X \mapsto \mu X.\ Q] \overset{\alpha}{\rightsquigarrow} P''$ and $P' \sim_R P''$.
  By the assumption on P and Q there is a Q' such that $Q[X \mapsto \mu X.\ Q] \overset{\alpha}{\rightsquigarrow} Q'$ and $P'' \sim_R Q'$. We have $P' \sim_R Q'$ by transitivity of $\sim_R$ and the statement follows. ∎

At this point we have all we need to prove that open relative bisimilarity is a congruence.

Theorem 21. Open relative bisimilarity is a congruence.

Proof. Congruence under action prefixes, choice, parallel composition and restrictions follows from the corresponding statements for relative bisimilarity, and the fact that instantiation is homomorphic in the process structure.

For fixed points, we have to show that $P \dot{\sim}_R Q$ implies $\mu X. P \dot{\sim}_R \mu X. Q$. Unfolding the definitions, we have to show that $(\mu X. P)[\theta] \sim_R (\mu X. Q)[\theta]$. Note that θ leaves X invariant, since X is bound.

By Lemma 20, it suffices to show $P[\theta, X \mapsto S] \sim_R Q[\theta, X \mapsto S]$ for all closed processes S. This follows from $P \dot{\sim}_R Q$, with an extended closing substitution. ∎

Furthermore, we can use Lemma 19 to show that $\dot{\sim}_R$ is compatible with instantiation.

Theorem 22. Let σ_1, σ_2 be substitutions such that $\sigma_1(x) \dot{\sim}_R \sigma_2(x)$ for all x. If $P \dot{\sim}_R Q$, then $P[\sigma_1] \dot{\sim}_R Q[\sigma_2]$.

Proof. By the definition of $P \dot{\sim}_R Q$, we have $P[\sigma_1] \dot{\sim}_R Q[\sigma_1]$.The statement follows from Lemma 19, Lemma 20 and transitivity. ∎

As a small additional application, we use Theorem 21 to show that weakly guarded equations have unique solutions in CCS.

A *context* C is a process with holes.

$$C ::= [\,] \mid P \mid \alpha.C \mid C + C \mid C \parallel C \mid (\nu a)C \mid \mu X.C$$

A context is *weakly guarded* if every hole appears under an action prefix, where the action in question may be τ. A context C can be *filled* with a process P resulting in a process $C[P]$, by replacing every hole in C with P. For example, the context $C = \alpha.[\,] \parallel \tau.[\,]$ is weakly guarded and we have $C[X] = \alpha.X \parallel \tau.X$.

Weakly guarded equations are bisimilarities of the form $P \sim C[P]$, for weakly guarded contexts C. By a result of Milner [7], such equations have unique solutions. Intuitively, this is because reduction must take a step before reaching a hole. We can formalize this intuition in terms of relative bisimilarities, which leads to a simple proof.

Lemma 23. If C is weakly guarded and $P \sim_R Q$, then $C[P] \sim_{\circ R} C[Q]$.

Proof. By induction on C, using Theorem 21 and in particular using Lemma 17 to move from guarded relative bisimilarity to relative bisimilarity. ∎

Lemma 24 (Unique Solutions). If C is weakly guarded, and P, Q are two processes such that $P \sim C[P]$, and $Q \sim C[Q]$ then $P \sim Q$.

Proof. By Lemma 13, it suffices to show $P \sim_{\circ R} Q$, assuming that $P \sim_R Q$. Using Lemma 23, we have $C[P] \sim_{\circ R} C[Q]$ and the statement follows by transitivity, as $P \sim C[P] \sim_{\circ R} C[Q] \sim Q$. ∎

5 Parameterized Tower Induction

We return to the abstract setting and establish an induction principle similar in spirit to Hur et al.'s parameterized coinduction [5].

Lemma 25 (Parameterized Tower Induction). Let u be an element of a complete lattice A, f a monotone endofunction, and P an inf-closed predicate. We have $P(t(u))$ and $P(f(t(u)))$, whenever

$$\forall x.\ u \leq t(x) \rightarrow P(t(x)) \rightarrow P(f(t(x))).$$

Proof. The statement $P(f(t(u)))$ follows from $P(t(u))$ and the assumption together with Fact 3.

To show $P(t(u))$, we generalize the statement to $\forall x.\ Q(t(x))$ for the inf-closed predicate $Q(x) = u \leq x \rightarrow P(x)$. By tower induction, it suffices to show that $P(f(t(x)))$ follows from $u \leq t(x) \rightarrow P(t(x))$ and $u \leq f(t(x))$. From Lemma 9 we know that $u \leq f(t(x)) \leq t(x)$. Thus $P(t(x))$ holds and $P(f(t(x)))$ follows by assumption. ∎

Hur et al. [5] implement parameterized coinduction with an accumulation rule for parameterized fixed points. Pous shows that the same accumulation rule is applicable to the companion. We present a different proof of the accumulation rule by instantiating Lemma 25 with the predicate $\lambda x.\ y \leq x$.

Lemma 26. For monotone f we have $x \leq f(t(x \cup y)) \leftrightarrow x \leq f(t(y))$.

Proof. The right-to-left direction follows from $y \leq x \cup y$ together with the monotonicity of t and f. In the left-to-right direction we use Lemma 25. It suffices to show that

$$\forall z.\ y \leq t(z) \rightarrow x \leq t(z) \rightarrow x \leq f(t(z)).$$

Combining the two assumptions, we have $x \cup y \leq t(z)$. Using Fact 3, this is equivalent to $t(x \cup y) \leq t(z)$. The statement follows from $x \leq f(t(x \cup y))$ and the monotonicity of f. ∎

Together with Lemma 7, Lemma 26 implies a sound and complete coinduction principle.

Fact 27. If f is monotone, then $x \leq f(t(x)) \leftrightarrow x \leq \nu f$.

Proof. We have $\nu f = f(\nu f) = f(t(\bot))$. ∎

Pous observed that every function below the companion is a sound up-to function [13] for f. This is a consequence of Fact 27.

Definition 28. g is a *sound up-to function* for f, if $x \leq \nu f$ whenever $x \leq f(g(x))$.

Lemma 29. If $g \leq t_f$, then g is a sound up-to function for f.

Proof. This follows from Fact 27: $x \leq f(g(x)) \leq f(t(x))$. ∎

6 Coq Formalization

All results in this paper have been formalized in Coq. We make use of the Ssreflect plugin and library, for its improved tactic language and the formalization of finite types (for J_f-relative monads). To avoid working with pre-lattices, we assume propositional and functional extensionality. The development is available at: www.ps.uni-saarland.de/extras/companions.

The main divergence of the formalization from the paper is our treatment of variable binding in CCS with fixed points. We represent variable binding using a de Bruijn representation. Since we often have to distinguish between open and closed terms we index our term language with an upper bound on the number of free variables. This technique was first used by Adams [1], and later thoroughly explained by Alternkirch et al. in the framework of relative monads [2]. Using the terminology of Altenkirch et al., we formalize terms as J_f-relative monads.

In addition to the laws of a J_f-relative monad, we show all equations from [14]. This allows us to show all substitution lemmas by rewriting.

7 Related Work

Coinduction. Hur et al. [5] introduce parameterized coinduction as an incremental proof technique for coinduction. For a monotone function f on a complete lattice, they construct the function $G_f(x) = \nu(\lambda y.\, f(x \cup y))$. They show that G_f can be used for modular and incremental coinductive reasoning and describe several examples and extensions.

One extension of parameterized coinduction incorporates up-to techniques. Specifically, Hur et al. consider respectful up-to functions. Respectfulness is another sufficient criterion for soundness of up-to functions. They use the fact that the set of respectful up-to functions is closed under union to construct the greatest respectful up-to function t. The parameterized fixed point $G_{f \circ t}$ turned out to obey an "unfolding" lemma, which allowed them to freely use any respectful up-to technique in a coinductive proof.

Recently, Pous [12] noticed that the greatest compatible up-to function already admits the parameterized coinduction principle. It turns out that the greatest compatible and the greatest respectful up-to function coincide. Moreover we have $f \circ t = G_{f \circ t}$. This means that the function t is everything we require for incremental and modular coinductive proofs compatible with up-to techniques.

Pous dubbed the greatest compatible up-to function the companion.

At the same time, Parrow and Weber [8] considered the greatest respectful function for strong bisimilarity in the context of classical set theory. Their construction avoids the quantification over respectful functions by using the theory of ordinals in set theory. They use that bisimilarity may be defined by transfinite iteration to construct the companion for bisimilarity.

Formally, the idea is that if κ is an ordinal larger than the cardinality of the underlying lattice, then $f^{\kappa}(\top)$ is the greatest fixed point of f. This can be used to construct the companion as $t_f(x) = \bigcap \{\, f^{\alpha}(\top) \mid x \leq f^{\alpha}(\top),\ \alpha \text{ ordinal} \,\}$.

The tower construction [15] may be seen as the type theoretic analogue of transfinite iteration in set theory. Under this view, we define the set of points reachable from $\top$ by transfinite f-iteration as an inductive predicate $T \approx \{\, f^{\alpha}(\top) \mid \alpha \text{ ordinal} \,\}$.

Up-To Techniques. The study of up-to techniques for bisimilarity originates with Milner [7]. Milner considers bisimulation up-to bisimilarity to keep proofs of bisimilarity manageable. Practical applications usually require combining several different up-to functions. Even our toy example in Sect. 4 requires bisimulation up-to context and bisimilarity to mimic the proof using the companion.

One problem with using only sound up-to functions in the sense of Definition 28 is that sound up-to functions do not compose. This drawback led Sangiorgi [13] to propose the notion of respectful up-to functions. Respectful up-to functions are sound and closed under composition and union.

Sangiorgi [13] studies bisimilarity, but notes that the same definition of respectfulness makes sense in the more general context of greatest fixed points in complete lattices.

Pous [11] extends and simplifies the work of Sangiorgi by abstracting it to the setting of complete lattices and by introducing the notion of compatibility. This abstraction yields concrete gains, as the set of compatible maps forms another complete lattice. In particular, this implies that we can use up-to techniques to establish soundness of up-to techniques. Pous refers to this as "second order techniques".

Recently, Pous [12] adapted this development to the companion. For every companion t, there exists a second-order companion, classifying the compatible up-to functions. Pous uses the second-order companion extensively to show soundness of bisimulation up-to context for CCS with replication and other case studies.

8 Conclusions and Future Work

We have presented a tower based construction of the companion of a monotone function on a complete lattice. The new tower induction principle derived from this construction allows us to show a number of improved results for companions. We instantiate the abstract lattice theoretic development with strong bisimilarity in CCS with general recursive processes. This instantiation yields a particularly simple proof system for bisimilarity and we show the admissibility of reasoning up-to context about bisimilarity. Our results imply the classical soundness result for bisimulation up-to context in CCS with replication.

There are several avenues for future work.

All case studies in this paper consider up-to techniques for strong bisimilarity in CCS. It is well known [10] that the case of weak bisimilarity is much more subtle. If we try to adapt the development in Sect. 4 to weak bisimilarity, we find that relative weak bisimilarity is not transitive and not compatible with choice. This mirrors the failure of soundness of weak bisimulation up-to weak bisimilarity and the fact that weak bisimilarity is not compatible with choice.

Despite these problems, there are useful up-to techniques for weak bisimilarity. Pous [10] developed weak bisimulation up-to elaboration, which combines weak bisimulations with a limited form of unfolding under a termination hypothesis. At this point it is not clear whether these techniques yield corresponding reasoning principles for relative weak bisimilarity.

There are also open questions concerning the companion construction itself.

Assuming the axiom of excluded middle, it can be shown [15] that T is well-ordered. In particular, in classical type theory, we can use this to expand the tower induction principle to all predicates which are closed under infima of well-ordered subsets. However, this principle is not provable in constructive type theory [3].

It might yet be possible to show a slightly weaker statement constructively. Pataraia [9] gives a constructive proof of Tarski's theorem for least-fixed points on directed complete partial orders. We conjecture that a similar construction can be used to extend the tower induction principle to predicates which are closed under infima of lower directed subsets.

References

1. Adams, R.: Formalized metatheory with terms represented by an indexed family of types. In: Filliâtre, J.-C., Paulin-Mohring, C., Werner, B. (eds.) TYPES 2004. LNCS, vol. 3839, pp. 1–16. Springer, Heidelberg (2006). doi:10.1007/11617990_1
2. Altenkirch, T., Chapman, J., Uustalu, T.: Monads need not be endofunctors. In: Ong, L. (ed.) FoSSaCS 2010. LNCS, vol. 6014, pp. 297–311. Springer, Heidelberg (2010). doi:10.1007/978-3-642-12032-9_21
3. Bauer, A., Lumsdaine, P.L.: On the Bourbaki-Witt principle in toposes. In: Mathematical Proceedings of the Cambridge Philosophical Society, vol. 155, pp. 87–99. Cambridge University Press (2013)

4. Davey, B., Priestley, H.: Introduction to Lattices and Order. Cambridge University Press, Cambridge (2002)
5. Hur, C.-K., Neis, G., Dreyer, D., Vafeiadis, V.: The power of parameterization in coinductive proof. In: The 40th Annual ACM SIGPLAN- SIGACT Symposium on Principles of Programming Languages, POPL 2013, Rome, Italy, 23–25 January 2013, pp. 193–206 (2013)
6. Xavier, L.: Formal verification of a realistic compiler. Commun. ACM **52**(7), 107–115 (2009)
7. Milner, R.: Communication and Concurrency, vol. 84. Prentice Hall, Upper Saddle River (1989)
8. Parrow, J., Weber, T.: The largest respectful function. Log. Methods Comput. Sci. **12**(2) (2016)
9. Pataraia, D.: A constructive proof of Tarski's fixed-point theorem for dcpo's. Presented in the 65th Peripatetic Seminar on Sheaves and Logic, Aarhus, Denmark, November 1997
10. Pous, D.: Weak bisimulation up to elaboration. In: Baier, C., Hermanns, H. (eds.) CONCUR 2006. LNCS, vol. 4137, pp. 390–405. Springer, Heidelberg (2006). doi:10.1007/11817949_26
11. Pous, D.: Complete lattices and up-to techniques. In: Shao, Z. (ed.) APLAS 2007. LNCS, vol. 4807, pp. 351–366. Springer, Heidelberg (2007). doi:10.1007/978-3-540-76637-7_24
12. Pous, D.: Coinduction all the way up. In: Proceedings of the 31st Annual ACM/IEEE Symposium on Logic in Computer Science, LICS 2016, pp. 307–316. ACM, New York (2016)
13. Sangiorgi, D.: On the bisimulation proof method. Math. Struct. Comput. Sci. **8**(5), 447–479 (1998)
14. Schäfer, S., Smolka, G., Tebbi, T.: Completeness and decidability of de Bruijn substitution algebra in Coq. In: Proceedings of the Conference on Certified Programs and Proofs, CPP 2015, Mumbai, India, 15–17 January 2015, pp. 67–73. ACM (2015)
15. Smolka, G., Schäfer, S., Doczkal, C.: Transfinite constructions in classical type theory. In: Urban, C., Zhang, X. (eds.) ITP 2015. LNCS, vol. 9236, pp. 391–404. Springer, Cham (2015). doi:10.1007/978-3-319-22102-1_26
16. Tarski, A.: A lattice-theoretical fixpoint theorem and its applications. Pac. J. Math. **5**(2), 285–309 (1955)

Reasoning About Cardinalities of Relations with Applications Supported by Proof Assistants

Insa Stucke(✉)

Institut für Informatik, Christian-Albrechts-Universität zu Kiel, Kiel, Germany
ist@informatik.uni-kiel.de

Abstract. In this paper we prove the correctness of a program for computing vertex colorings in undirected graphs. In particular, we focus on the approximation ratio which is proved by using a cardinality operation for heterogeneous relations based on Y. Kawaharas characterisation.

All proofs are mechanised by using the two proof assistants Coq and Isabelle/HOL. Our Coq formalisation builds on existing libraries providing tools for heterogeneous relation algebras and cardinalities. To formalise the proofs in Isabelle/HOL we have to change over to untyped relations. Thus, we present an axiomatisation of a cardinality operation to reason about cardinalities algebraically also in homogeneous relation algebras and implement this new theoretical framework in Isabelle/HOL. Furthermore, we study the advantages and disadvantages of both systems in our context.

1 Introduction

Relation algebra (as first introduced in [19] and further studied, e.g., in [10,16]) provides an elegant way to reason about many discrete structures. For instance, there is a direct relationship between relations and graphs via adjacency relations. Hence, computational problems on graphs can be expressed and solved by using the relation-algebraic method as shown in [16], for example. The relation-algebraic approach is known for many methodical advantages in contrast to the conventional set-theoretic one. For example, it allows consice problem specifications and hence very formal calculations. Due to this, relation-algebraic reasoning turned out to be well-suited for mechanisation.

Thus, in [2] the authors develop a relational program for computing vertex colorings in undirected (and loop-free) graphs. The correctness proof is given by combining the assertion-based verification method with relation-algebraic calculations. In this context the usability of an automated theorem prover and the proof assistants Coq and Isabelle/HOL is shown and compared. However, the approximation ratio of the underlying Greedy algorithm is not studied at all since there were no obvious tools to tackle proofs involving cardinalities.

In the last years there has been a lot of work concerning the cardinalities of relations, mostly based on a definition of a cardinality operation for heterogeneous relation algebras presented by Kawahara in [9]. For example, in [3] and

P. Höfner et al. (Eds.): RAMiCS 2017, LNCS 10226, pp. 290–306, 2017.
DOI: 10.1007/978-3-319-57418-9_18

[1], the authors present first results about the cardinalities of special relations as points and vectors building the basis for reasoning about approximation ratios algebraically. Furthermore, in [5], a library for Coq providing an implementation of this cardinality operation in heterogeneous relation algebras is developed.

In the present paper the mentioned results about cardinalities of relations are used to prove the approximation ratio of the program presented in [2]. Furthermore, we study the application of the proof assistants Coq and Isabelle/HOL in this context. Therefore, we first use the library developed in [5] for mechanising the correctness proof in Coq. Our implementation in Isabelle/HOL builds on a library for untyped relations (see [17]). Thus, we modify Y. Kawahara's definition of a cardinality operation and present a new theoretical framework for dealing with cardinalities in homogeneous relation algebras. For this framework we develop a library that is eventually applicable for the mechanisation of the programs' correctness proof. As in [2], we compare the advantages and disadvantages of the usability of both tools in this context.

Our Coq proof script and Isabelle/HOL theories are available here [18].

2 Preliminaries

First, we recall the basic principles of relation algebra based on the heterogeneous approach of [6,15,16]. Set-theoretic relations form the standard model of relation algebras. We assume the basic operations on set-theoretic relations, viz. union, intersection, complementation, transposition and composition, in the remainder denoted by $R \cup S$, $R \cap S$, $\overline{R}$, R^{T} and RS for relations R, S of appropriate type. Furthermore, we consider the predicates $R \subseteq S$ (inclusion) and $R = S$ (equality)and the empty, universal and identity relation denoted by O, L and I.

Those operations and constants form a (heterogeneous) *relation algebra* in the sense of [15,16], with typed relations as elements. We write $R : X \leftrightarrow Y$ if R is a relation with source X and target Y and denote the type of R by $X \leftrightarrow Y$. In the case of typed relations we frequently overload the symbols O, L and I, if their type can be inferred from the context. If necessary we use indices as e.g., L_{XY} for L of type $X \leftrightarrow Y$. The axioms of a relation algebra are

(1) the axioms of a Boolean lattice for all same typed relations under the Boolean operations $\cup$, $\cap$ and $\overline{}$, $\subseteq$ and L and O,
(2) the associativity of composition and that identity relations are neutral w.r.t. composition,
(3) the *Schröder rule*, i.e., that for all relations Q, R and S with appropriate types it holds $QR \subseteq S \iff Q^{\mathsf{T}}\overline{S} \subseteq \overline{R} \iff \overline{S}R^{\mathsf{T}} \subseteq \overline{Q}$
(4) the *Tarski rule*, i.e., that for all relations R and all universal relations with appropriate types it holds $R \neq \mathsf{O} \iff \mathsf{L}R\mathsf{L} = \mathsf{L}$.

In the relation-algebraic proofs of this paper we only indicate applications of (3), (4) and consequences of the above axioms that are not obvious. Furthermore, we assume that complementation and transposition bind stronger than composition and composition binds stronger than union and intersection.

In the following we define some specific classes of relations, for more details we refer again to [15,16]. If R is *homogeneous*, i.e., of type $X \leftrightarrow X$, R is called *irreflexive* iff $R \subseteq \bar{\mathsf{I}}$ and *symmetric* iff $R = R^\mathsf{T}$. A homogeneous relation R is *reflexive* iff $\mathsf{I} \subseteq R$, *antisymmetric* iff $R \cap R^\mathsf{T} \subseteq \mathsf{I}$, and *transitive* iff $RR \subseteq R$. A reflexive, antisymmetric and transitive relation R is a *order relation* and if additionally $R \cup R^\mathsf{T} = \mathsf{L}$ holds, i.e., R is *linear*, then R is called a *linear order relation*. A relation R is *univalent* iff $R^\mathsf{T}R \subseteq \mathsf{I}$ and *total* iff $R\mathsf{L} = \mathsf{L}$. A *mapping* is a univalent and total relation.

A *vector* is a relation v with $v = v\mathsf{L}$. For a set-theoretic relation $v : X \leftrightarrow Y$ the equality $v = v\mathsf{L}$ means that v is of the form $v = Z \times Y$ with a subset Z of X. Then we say that v *models the subset* Z of X. Since for this purpose the target of a vector is irrelevant, we use the specific singleton set $\mathbf{1}$ as target. Moreover, a *point* p is a vector that is *injective* and *surjective*, i.e., $pp^\mathsf{T} \subseteq \mathsf{I}$ and $\mathsf{L}p = \mathsf{L}$.

We also assume the following version of the *Point Axiom* of [7] holding for set-theoretic relations, where $\mathcal{P}(v) := \{p \mid p \subseteq v \wedge p \text{ is point}\}$ for all vectors v.

Axiom 2.1. *For all objects X we have $\mathsf{L}_{X\mathbf{1}} = \bigcup_{p \in \mathcal{P}(\mathsf{L}_{X\mathbf{1}})} p$.*

Additionally we have the following lemma which states that this property can be generalised for arbitrary vectors (see [7]).

Lemma 2.1. *If $v : X \leftrightarrow \mathbf{1}$ is a vector, then $v = \bigcup_{p \in \mathcal{P}(v)} p$.* □

In [9], Kawahara investigates the cardinality of set-theoretic relations. The main result is a characterisation of the cardinalities of relations. Considering the properties of this characterisation as axiomatic specification of the cardinality operation $|\cdot|$ leads to the following definition:

Definition 2.1. *For all relations R we denote its cardinality by $|R|$. The following axioms specify the meaning of the cardinality operation, where Q, R and S are arbitrary relations with appropriate types:*

(C1) If R is finite, then $|R| \in \mathbb{N}$ and $|R| = 0$ iff $R = \mathsf{O}$.
(C2) $|R| = |R^\mathsf{T}|$.
(C3) If R and S are finite, then $|R \cup S| = |R| + |S| - |R \cap S|$.
(C4) If Q is univalent, then $|R \cap Q^\mathsf{T}S| \leq |QR \cap S|$ and $|Q \cap SR^\mathsf{T}| \leq |QR \cap S|$.
(C5) $|\mathsf{I}_{\mathbf{11}}| = 1$.

In (C1) and (C3) the occuring relations are assumed to be finite so that the cardinality $|R|$ can be regarded as a natural number, in (C2) and (C4) the notation $|R| = |S|$ (respectively $|R| \leq |S|$) is equivalent to the fact that there exists a bijection between R and S (respectively an injection from R to S) and (C5) says that the identity relation on the set $\mathbf{1}$ contains precisely one pair. In the present paper we assumes in case of an expression $|R|$ the sets of R's type to be finite and thus $|R| \in \mathbb{N}$.

Based on the above axioms in [9] a lot of laws for the cardinality operation are derived in a purely algebraic manner, for instance, the monotonicity of the cardinality operation, i.e., that $R \subseteq S$ implies $|R| \leq |S|$. Futhermore, they imply $|\bigcup_{R \in \mathcal{R}} R| = \sum_{R \in \mathcal{R}} |R|$, for all finite sets $\mathcal{R}$ of pairwise disjoint relations. Other consequences of the axioms we use in the remainder are the following:

Lemma 2.2.

1. *If* R *and* S *are univalent, then* $|RS \cap Q| = |R \cap QS^T|$.
2. *If* R *is univalent and* S *is a mapping, then* $|RS| = |R|$.
3. *If* R *is univalent, then* $|R^\mathsf{T}S| \leq |S|$.

The cardinality of points and vectors of type $X \leftrightarrow \mathbf{1}$ can be studied by using the above results. The next lemma states that a point contains exactly one pair.

Lemma 2.3. *If* $p : X \leftrightarrow \mathbf{1}$ *is a point, then* $|p| = 1$.

Proof. Using cardinality axioms (C2) and (C5) and Lemma 2.2 ($\mathsf{I}_{\mathbf{11}}$ is univalent and $p^\mathsf{T} : \mathbf{1} \leftrightarrow X$ is a mapping), we have the following calculation:

$$|p| = |p^\mathsf{T}| = |\mathsf{I}_{\mathbf{11}} p^\mathsf{T}| = |\mathsf{I}_{\mathbf{11}}| = 1. \qquad \square$$

This lemma allows to show that the cardinality of a vector with target $\mathbf{1}$ is equal to the cardinality of the set of all points it contains.

Lemma 2.4. *For all* $v : X \leftrightarrow \mathbf{1}$ *we have* $|v| = |\mathcal{P}(v)|$.

Note that in the above lemma with $|\mathcal{P}(v)|$ we denote the usual cardinality of the set $\mathcal{P}(v)$. For more details, in particular omitted proofs, and results concerning the cardinality operation as well as applications we refer to [1,3,9].

3 Approximating Minimal Vertex Colorings

In [2] the authors present a relational program for computing vertex colorings in undirected (and loop-free) graphs. The verification tasks arising by applying the assertion-based verification method are supported by the automated theorem prover Prover9 and the proof assistants Coq and Isabelle/HOL. By this example the advantages and disadvantages of these tools are studied and compared.

The presented program is based on the well-known Greedy algorithm that assigns sequentially a proper color to each vertex, i.e., a color that is not already assigned to one of its neighbours. This procedure does not consider the fact that one is usually interested in computing a minimal and not an arbitrary coloring of a graph. Thus, one usually assumes the colors to be ordered so that the algorithm chooses a minimal color for each vertex. By this approach a minimal vertex coloring is approximated with a ratio of $\Delta + 1$, where Δ is the maximum degree of the given graph.

In [2] the approximation ratio is not treated at all. Thus, in the remainder of this section we prove the ratio of the following program with the modified choice of the color q using the results about the cardinality operation presented in Sect. 2:

$$
\begin{array}{l}
C := \mathsf{O};\\
\textbf{while } C\mathsf{L} \neq \mathsf{L} \textbf{ do}\\
\quad \textbf{let } p = point(\overline{C\mathsf{L}});\\
\quad \textbf{let } q = point(\overline{C^\mathsf{T}Ep} \cap \overline{\overline{M}\,\overline{C^\mathsf{T}Ep}});\\
\quad C := C \cup pq^\mathsf{T} \textbf{ od}
\end{array}
$$

The input relations of this program are an *adjacency relation* $E : X \leftrightarrow X$, modelling a given graph G with a set of vertices X, and a linear order relation $M : F \leftrightarrow F$ on a set of colors F. The output relation of the program is $C : X \leftrightarrow F$ representing the vertex coloring, i.e., a mapping so that in addition $CC \subseteq \overline{E}$ holds. The latter condition is called the coloring property. Furthermore, all occuring universal relations in the program have target $\mathbf{1}$. As in [2] we assume the deterministic operation *point* selecting a point to a given nonempty vector v such that $point(v) \subseteq v$. For more details we refer to [2] since the only difference to our program is the choice of the point q. In [2], q is choosen as $point(\overline{C^\mathsf{T}Ep})$, i.e., q is not used for one of p's neighbours. If we choose q as $point(\overline{C^\mathsf{T}Ep} \cap \overline{\overline{M}\,\overline{C^\mathsf{T}Ep}})$ instead we also ensure that q is minimal since $\overline{C^\mathsf{T}Ep} \cap \overline{\overline{M}\,\overline{C^\mathsf{T}Ep}}$ is the vector of all minimal colors w.r.t. the order relation M, see, e.g., [16] for further information.

To formally verify the correctness of the above program we apply the assertion-based verification method. Thus, we first specify the programs' pre- and postcondition. The precondition is the conjunction of the following formulae specifying E as an adjacency relation, i.e., an irreflexive and symmetric relation, and M as a linear, reflexiv and antisymmetric relation (transitivity is not needed here).

$$Pre(E, M) \;:\Longleftrightarrow\; E = E^\mathsf{T} \wedge E \subseteq \overline{\mathsf{I}} \wedge \mathsf{I} \subseteq M \wedge M \cap M^\mathsf{T} \subseteq \mathsf{I} \wedge M \cup M^\mathsf{T} = \mathsf{L}_{FF}$$

In the remainder we furthermore use the abbreviation $\Delta_v := \max\{|Ex| \mid x \in \mathcal{P}(v)\}$ for all vectors v and $\Delta := \Delta_\mathsf{L}$ for the maximum degree of a given graph modelled by E. If we do not specify a universal or empty relation's type in this section we assume its target to be $\mathbf{1}$.

The postcondition is a conjuction of three formulae stating that C is a vertex coloring, i.e., an univalent and total relation fulfilling the coloring property, and a formula saying that the number of used colors is at most $\Delta + 1$:

$$Post(C, E) \;:\Longleftrightarrow\; C^\mathsf{T}C \subseteq \mathsf{I} \wedge C\mathsf{L} = \mathsf{L} \wedge CC^\mathsf{T} \subseteq \overline{E} \wedge |C^\mathsf{T}\mathsf{L}| \leq \Delta + 1$$

The invariant is a conjunction of four formulae, where the first two ensure that C is univalent and fulfills the coloring property and the latter two are essential for proving the desired approximation ratio:

$$Inv(C, E, M) \;:\Longleftrightarrow\; C^\mathsf{T}C \subseteq \mathsf{I} \wedge CC^\mathsf{T} \subseteq \overline{E} \wedge C^\mathsf{T}\mathsf{L} \subseteq \overline{\overline{M}\,\overline{C^\mathsf{T}\mathsf{L}}} \wedge |C^\mathsf{T}\mathsf{L}| \leq \Delta_{C\mathsf{L}} + 1$$

As usual the following proof obligations have to be proved for partial correctness:

(PO1) $Pre(E, M) \Longrightarrow Inv(E, M, \mathsf{O})$
(PO2) $Inv(E, M, C) \wedge C\mathsf{L} = \mathsf{L} \Longrightarrow Post(E, C)$
(PO3) $Pre(E, M) \wedge Inv(E, M, C) \wedge C\mathsf{L} \neq \mathsf{L} \Longrightarrow Inv(E, M, C \cup pq^\mathsf{T})$ (where p and q are defined as in the given program).

Since $Pre(E, M), Post(E, C)$ and $Inv(E, M, \mathsf{O})$ are conjunctions of various formulae the three obligations can be splitted into single statements for each formula. In the remainder we only consider the statements involving cardinalities.

For the omitted proofs we refer to Sects. 4 and 5 and the appendix. Here, we start with proving the first proof obligation (PO1), i.e., the establishment of the last formula of the invariant.

Lemma 3.1. *For all relation E and M it holds* $\mathit{Inv}(E, M, \mathsf{O})$.

Proof. The last formula of the invariant is shown by using cardinality axiom (C1) two times: $|\mathsf{O}^{\mathsf{T}}\mathsf{L}| = 0 \leq 1 \leq \Delta_{\mathsf{OL}} + 1$. □

Next, we prove (PO2), i.e., that the invariant and the negation of the loop-condition imply the postcondition. Again we concentrate on the last formula involving cardinalities.

Lemma 3.2. *Let E, C, M be relations such that E is symmetric and irreflexive, M is reflexive, antisymmetric and linear and $C\mathsf{L} = \mathsf{L}$ and $\mathit{Inv}(E, C, M)$ holds. Then $\mathit{Post}(E, C)$ holds.*

Proof. Using $\mathit{Inv}(E, C, M)$ in the first and $C\mathsf{L} = \mathsf{L}$ in the second step we have the following inequality: $|C^{\mathsf{T}}\mathsf{L}| \leq \Delta_{C\mathsf{L}} + 1 = \Delta_{\mathsf{L}} + 1 = \Delta + 1$. □

For proving (PO3), i.e., the maintenance of the last formula of the invariant, we need the following auxiliary result.

Lemma 3.3. *Let R be a reflexive, anstisymmetric and linear relation. Then $R^{\mathsf{T}} = \mathsf{I} \cup \overline{R}$ holds.*

Proof. Using the antisymmetry of R we have:

$$R \cap R^{\mathsf{T}} \subseteq \mathsf{I} \iff R \cap R^{\mathsf{T}} \cap \overline{\mathsf{I}} \subseteq \mathsf{O} \iff R^{\mathsf{T}} \subseteq \overline{R \cap \overline{\mathsf{I}}} \iff R^{\mathsf{T}} \subseteq \mathsf{I} \cup \overline{R}.$$

By the linearity and reflexivity of R we show:

$$R \cup R^{\mathsf{T}} = \mathsf{L} \iff \overline{R \cup R^{\mathsf{T}}} \subseteq \mathsf{O} \iff \overline{R} \cap \overline{R^{\mathsf{T}}} \subseteq \mathsf{O} \iff \overline{R} \subseteq R^{\mathsf{T}} \implies \mathsf{I} \cup \overline{R} \subseteq R^{\mathsf{T}}.$$ □

Using the latter Lemma we show the maintenance of the invariants' last formula:

Lemma 3.4. *Let E, C and M be relations so that $\mathit{Pre}(E, M)$ and $\mathit{Inv}(E, M, C)$ hold and p, q points with $p \subseteq \overline{C\mathsf{L}}$, $q \subseteq \overline{C^{\mathsf{T}}Ep} \cap \overline{\overline{M}\,\overline{C^{\mathsf{T}}Ep}}$. Then $|(C \cup pq^{\mathsf{T}})^{\mathsf{T}}\mathsf{L}| \leq \Delta_{(C \cup pq^{\mathsf{T}})\mathsf{L}} + 1$ holds.*

Proof. Since p and q are points it holds $qp^{\mathsf{T}}\mathsf{L} = q$ and thus $|(C \cup pq^{\mathsf{T}})^{\mathsf{T}}\mathsf{L}| = |C^{\mathsf{T}}\mathsf{L} \cup q|$. For the same reasons we have $pq^{\mathsf{T}}\mathsf{L} = p$ which implies $\Delta_{(C \cup pq^{\mathsf{T}})\mathsf{L}} = \Delta_{C\mathsf{L} \cup p}$ Hence we have to show $|C^{\mathsf{T}}\mathsf{L} \cup q| \leq \Delta_{C\mathsf{L} \cup p} + 1$.

Using (C3) and Lemma 2.3 we have the following equality:

$$|C^{\mathsf{T}}\mathsf{L} \cup q| = |C^{\mathsf{T}}\mathsf{L}| + |q| - |C^{\mathsf{T}}\mathsf{L} \cap q| = |C^{\mathsf{T}}\mathsf{L}| + 1 - |C^{\mathsf{T}}\mathsf{L} \cap q|.$$

If $q \subseteq C^{\mathsf{T}}\mathsf{L}$ it holds $|C^{\mathsf{T}}\mathsf{L} \cap q| = |q| = 1$. In this case the claim follows immediately with the assumption *Inv*(C, E, M), in particular the last formula of it, and the fact that $\Delta_{C\mathsf{L}} \leq \Delta_{(C \cup pq^{\mathsf{T}})\mathsf{L}}$.
Hence, we consider the case that $q \subseteq \overline{C^{\mathsf{T}}\mathsf{L}}$. Then $C^{\mathsf{T}}\mathsf{L} \cap q = \mathsf{O}$ and it follows$|C^{\mathsf{T}}\mathsf{L} \cup q| = |C^{\mathsf{T}}\mathsf{L}| + 1$. Thus, it is sufficient to show that $|C^{\mathsf{T}}\mathsf{L}| + 1 \leq \Delta_{C\mathsf{L} \cup p} + 1$ holds.So we show that $|C^{\mathsf{T}}\mathsf{L}| \leq \Delta_{C\mathsf{L} \cup p}$, and therefor, $|C^{\mathsf{T}}\mathsf{L}| \leq |Ep|$.

Because of $q \subseteq \overline{C\mathsf{L}}$ and the third formula of *Inv*(E, M, C) we have

$$q \cup \overline{M}q \subseteq \overline{C^{\mathsf{T}}\mathsf{L}} \cup \overline{M}\,\overline{C^{\mathsf{T}}\mathsf{L}} \subseteq \overline{C^{\mathsf{T}}\mathsf{L}}$$

and thus

$$C^{\mathsf{T}}\mathsf{L} \subseteq \overline{q} \cap \overline{\overline{M}q}. \tag{1}$$

Next, we prove

$$\overline{q} \cap \overline{\overline{M}q} \subseteq C^{\mathsf{T}}Ep \tag{2}$$

by the following calculation:

$$\begin{aligned} q \subseteq \overline{\overline{M}\,\overline{C^{\mathsf{T}}Ep}} &\iff \overline{M}\,\overline{C^{\mathsf{T}}Ep} \subseteq \overline{q} \\ &\iff \overline{\overline{M}^{\mathsf{T}}q} \subseteq C^{\mathsf{T}}Ep && \textit{Schröder rule} \\ &\iff \overline{\mathsf{I} \cup \overline{M}q} \subseteq C^{\mathsf{T}}Ep && \text{Lemma 3.3} \\ &\iff \overline{(\mathsf{I} \cup \overline{M})q} \subseteq C^{\mathsf{T}}Ep && q \text{ point} \\ &\iff \overline{q \cup \overline{M}q} \subseteq C^{\mathsf{T}}Ep \\ &\iff \overline{q} \cap \overline{\overline{M}q} \subseteq C^{\mathsf{T}}Ep. \end{aligned}$$

Using (1), (2) and Lemma 2.2.1 (C is univalent because of *Inv*(E, M, C)) as well as the monotonicity of the cardinality operation we obtain the desired inequality:

$$|C^{\mathsf{T}}\mathsf{L}| \leq |\overline{q} \cap \overline{\overline{M}q}| \leq |C^{\mathsf{T}}Ep| \leq |Ep|.$$

□

4 Cardinalities in Coq

In [2] the proofs of the according obligations (PO1)–(PO3) presented in Sect. 3 are mechanised amongst others with the proof assistant Coq using the library *RelationAlgebra* which provides a model for heterogeneous relation algebra and many other related algebraic structures. The library is available via [13], and presented in [14]. For more general information about Coq we refer to [4, 20].

In [5] the authors extend the mentioned library so that a reasoning about cardinalities is possible. *RelationAlgebra* is enriched by the module `relalg` containing the most important definitions of special classes of relations, e.g., those introduced in Sect. 2. For the tools concerning cardinalities a standalone library was developed. To preserve the modularity of *RelationAlgebra* this library provides a separate module for each algebraic structure we defined in Sect. 2. The hierarchy of the modules is illustrated in Fig. 1.

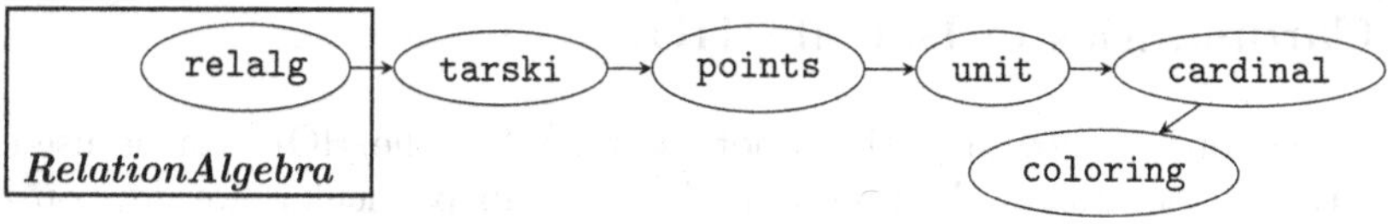

Fig. 1. Hierarchy of the Coq library

To simplify rewriting, the definitions are realized by using classes, for instance, being univalent is defined as follows:

```
Class is_univalent n m (x: X n m) := univalent: xᵀ * x <== 1.
```

Here, the variables n and m specify the type of the relation x and X provides the notions and operations of a relation algebra. The symbols ᵀ,* and <== denote transposition, composition and inclusion. The type of the identity relation denoted by 1 is inferred automatically. In points the *Point Axiom* is assumed and several resulting facts are proved, especially those presented in Sect. 2. The definition of the cardinality operation is given in cardinal and follows the one presented in Sect. 2. A detailed description of each module and the notations can be found in [5].

In cardinal the proofs of all lemmata of Sect. 2 (and many more) are mechanised, for instance, Lemma 2.3 as follows:

```
Lemma card_point X (R: C X unit): is_point R → card R = 1.
Proof. rewrite ←cardcnv, ←dot1x. rewrite card_unimap. apply card1. Qed.
```

Here, is_point specifies the relation R as a point and card denotes the cardinality operation. The Coq proof follows exactly the one of Sect. 2 where cardcnv, card_unimap and card1 correspond to (C2), Lemma 2.2 and (C5).

With the extended library all proofs of Sect. 3 can be done within Coq. In the following we show the formulations of the Lemmata 3.1, 3.4 and 3.2 where inv is the definition of the invariant as given in Sect. 3 (the adjacency and order relation are introduced at the beginning of our Coq module once and for all) and minimal_elements M v the vector of the minimal elements of a vector v w.r.t. a linear order relation M:

```
Lemma PO1: inv (zer n f).
Lemma PO2 (F: X n f) : inv F ∧F*(top' f unit) == top → post (F ∪ p*qᵀ).
Lemma PO3 (F: X n f) (p: X n unit) (q: X f unit):
          is_point p → p <== !(F*top) →
          is_point q → q <== minimal_elements M (!(Fᵀ*E*p)) →
          inv F → inv (F ∪ p*qᵀ).
```

Mainly, the proofs have to be done step by step. At some points we benefit at the one hand from the smart implementation of the specific relations that makes rewriting less difficult and on the other hand from the decision tactics provided by *RelationAlgebra*. A detailed description of those tactics can be found in [14].

5 Cardinalities in Isabelle/HOL

In this section we show how the proof assistant Isabelle/HOL can be used to prove the correctness of the program of Sect. 3. In particular, we develop the required theoretical framework for it.

Compared to the one of Coq the type system of Isabelle/HOL is less powerful. In the end the usage of multi-parameter classes is not possible whereby there is no trivial way to define heterogeneous relation algebras. Thus, our Isabelle/HOL theories built on an existing library, *Relation_Algebra*, for homogeneous relation algebras only, available via the *Archive of Formal Proofs*, see [17]. More general information about Isabelle/HOL can be found, for example, in [8,12].

This limitation makes it impossible to transfer the approach realised in Coq to Isabelle/HOL. Namely, if we consider points of type $X \leftrightarrow \mathbf{1}$, for instance, it was essential for the proofs of Sect. 3 that they have cardinality 1. This fact is mainly based on the cardinality axiom (C5) and the specific type $X \leftrightarrow \mathbf{1}$. When using the *Relation_Algebra* library we are not only restricted to homogeneous relation algebras, but to untyped relations.

Due to this, we have to modify the definition of the cardinality operation of Sect. 2. The first four axioms (C1)–(C4) can be adapted to untyped relations, but (C5) involves the special singleton set $\mathbf{1}$. Thus, we assume the following fifth axiom instead saying that the cardinality of the identity relation equals the number of points (in the relation algebra):

$$\text{(C5')}\ |\mathsf{I}| = |\mathcal{P}(\mathsf{L})|.$$

Note that there are equivalent formulations of (C5'), e.g., $|\mathsf{L}| = |\mathsf{I}|^2$, but for us, the given one is the most intuitive compared to (C5).

In the remainder we also assume a version of the *Point Axiom* for untyped relations. The only difference to Axiom 2.1 is that the occuring universal relation is untyped.

Axiom 5.1. *(Point Axiom). It holds* $\mathsf{L} = \bigcup_{p \in \mathcal{P}(\mathsf{L})} p$.

One can easily check that we get the following corresponding consequences as in Sect. 2.

Lemma 5.1.

1. *For all vectors v we have* $v = \bigcup_{p \in \mathcal{P}(v)} p$.
2. *We have* $\mathsf{I} = \bigcup_{p \in \mathcal{P}(\mathsf{L})} pp^{\mathsf{T}}$.

Furthermore, the Lemma 2.2 also holds in the case of untyped relations. The first important result which is significantly different, due to (C5'), is stated in the following lemma and gives us the cardinality of (untyped) points.

Lemma 5.2. *If p is a point, then* $|p| = |\mathsf{I}|$.

Proof. Using cardinality axioms (C2) and (C5') and Lemma 2.2 (I is univalent and p^{T} is a mapping) we have $|p| = |p^{\mathsf{T}}| = |\mathsf{I}p^{\mathsf{T}}| = |\mathsf{I}|$. □

Obviously, because of the above lemma, points and vectors are no longer suitable for modelling sets if their cardinalities are essential in the context. Thus, in the following and in particular for the formalisation is Isabelle/HOL we use *partial identities*, i.e., relations R with $R \subseteq \mathsf{I}$, instead of vectors to represent sets. In place of points we consider *atoms*, i.e., nonempty relations a with $a\mathsf{L}a^\mathsf{T} \subseteq \mathsf{I}$. We show that the cardinalities of those special relations correspond to the ones of vectors and points. Therefore, we start with a lemma about the cardinality of (untyped) vectors.

Lemma 5.3. *If v is a vector, then $|v| = |\mathcal{P}(v)| \cdot |\mathsf{I}|$.*

Proof. Because of Lemma 5.1, cardinality axioms (C3) and (C1) (the points in $\mathcal{P}(v)$ are pairwise disjoint) and Lemma 5.2 we obtain the claim by

$$|v| = \Big|\bigcup_{p\in\mathcal{P}(v)} p\Big| = \sum_{p\in\mathcal{P}(v)} |p| = \sum_{p\in\mathcal{P}(v)} |\mathsf{I}| = |\mathcal{P}(v)| \cdot |\mathsf{I}|. \qquad \square$$

Note that the above result holds in particular for $v = \mathsf{L}$ since L is a vector. This gives us $|\mathsf{L}| = |\mathsf{I}|^2$ because of (C5').

To prove that every atom has cardinality 1 we need the following technical lemma whose proof we omit due to the lack of space. It states that every atom is the composition of a point and a points' transposed (and vice versa), and that the universal relation can be written as the union of all atoms it contains. Here, we denote the set of all atoms (containted in L) as $\mathcal{A}(\mathsf{L})$.

Lemma 5.4.

1. *It holds* $\mathcal{A}(\mathsf{L}) = \{p;q^T | p,q \in \mathcal{P}(\mathsf{L})\}$.
2. *It holds* $\mathsf{L} = \bigcup_{a\in\mathcal{A}(\mathsf{L})} a$.

From this we get the desired result about the cardinalities of atoms.

Lemma 5.5. *If a is an atom, then $|a| = 1$.*

Proof. For all atoms a it holds $a \neq \mathsf{O}$ and thus $|a| \geq 1$ with cardinality axiom (C1). We prove $|a| = 1$, for all atoms a, by contradiction. Thus, we assume that there exists an atom b with $|b| > 1$. Combining Lemmas 5.3 and 5.4.2 (for $v = \mathsf{L}$) we have $\mathcal{A}(\mathsf{L}) = |\mathsf{I}|^2$. Due to this and again Lemmas 5.3 and 5.4.2 we have

$$\begin{aligned} |\mathsf{I}|^2 = |\mathsf{L}| = |\bigcup_{a\in\mathcal{A}(\mathsf{L})} a| &= \sum_{a\in\mathcal{A}(\mathsf{L})} |a| = |b| + \sum_{a\in\mathcal{A}(\mathsf{L})\setminus\{b\}} |a| \\ &> 1 + \sum_{a\in\mathcal{A}(\mathsf{L})\setminus\{b\}} 1 = 1 + |\mathcal{A}(\mathsf{L}) \setminus \{b\}\,| = 1 + |\mathsf{I}|^2 - 1, \end{aligned}$$

which is a contradiction. $\square$

From this we get that partial points have cardinality 1 which makes them suitable for modelling single elements of sets.

Lemma 5.6. *If p is a partial point, then $|p| = 1$.*

Proof. By definition, p is an atom, thus the claim follows immediately with Lemma 5.5. □

We omit the corresponding proofs of the correctness of the program of Sect. 3, but refer to the Isabelle/HOL formalisation we describe in the remainder of this section and available via the web, see [18].

So far, the library *Relation_Algebra* provides several facts holding in untyped relation algebras as well as theories about functions and vectors with related facts, so that most of the specific relations mentioned in Sect. 2 are already defined. For instance, for vectors this is done in the following way

definition *is_vector* :: "'a ⇒ bool"
  where "*is_vector* $x \equiv x = x; 1$"

In the library the symbols $+$, $\cdot$, $-$, ; and $\smile$ are used for union, intersection, complement, composition and transposition and 1, 0 and $1'$ for the universal, empty and identity relation. For our purpose we import additional theory, e.g., about natural numbers, so that we use $\sqcup$, $\sqcap$, and $'$ for the first three operations and *top*, *bot* and $1'$ for the constants L, O and I. As in the case of Coq neither the *Tarski rule* nor the *Point Axiom* is provided by the library so far. Follow the approach in Coq we develop a separate theory for each structure we define. The dependencies of the main theories are illustrated in Fig. 2.

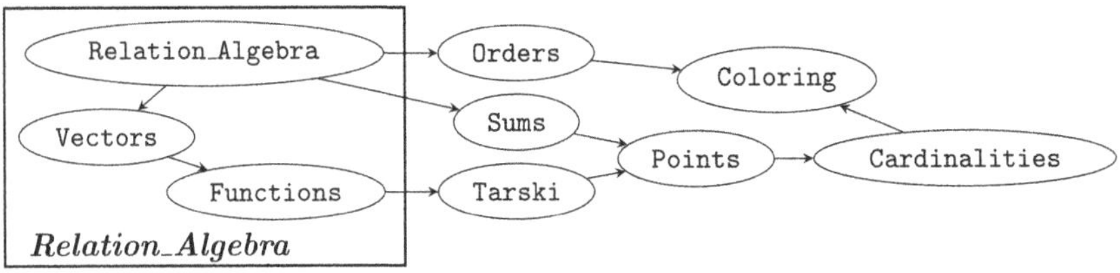

Fig. 2. Hierarchy of the Isabelle/HOL library

First, we extend the class *relation_algebra* by the *Tarski rule* using a class:

class *relation_algebra_tarski* = *relation_algebra* +
  assumes *tarski*: "$x \neq bot \longleftrightarrow top; x; top = top$"

This is done in the theory *Relation_Algebra_Tarski* where we derive some fundametal properties of points, for instance the following one:

lemma *points_surj*: "*is_point* $p \longrightarrow$ *is_sur* p"

The theory *Relation_Algebra_Points* is an extension of the latter providing the Axiom 5.1 and that the number of points is finite:

```
class relation_algebra_fin_points  = relation_algebra_tarski +
  assumes finiteness:           "finite {x. is_point x}"
  and      pointaxiom [simp]: "⨆ {x. is_point x} = top"
```

Here, $\bigsqcup\{x.\ is\_point\ x\}$ is a notation for $\bigcup_{p\in\mathcal{P}(\mathsf{L})} p$ in Isabelle/HOL. In the theory *Relation_Algebra_Sums* we proved a several properties of these finite unions, e.g., monotonicity.

Finally, we have a theory called *Relation_Algebra_Cardinalities* where the cardinality operation is defined in the following way:

```
class cardinal =
  fixes cd ::  "'a ⇒ nat" ("|_|" [30] 999)
class relation_algebra_card  = cardinal + relation_algebra_fin_points  +
  assumes card0       : "|x| = (0 :: nat) ⟷ x = bot"
  and cardcnv [simp] : "|x˘| = |x|"
  and cardcup         : "|x ⊔ y| = |x| + |y| − |x ⊓ y)|"
  and cardded         : "is_p_fun x ⟶ |y ⊓ (x˘; z)| ≤ |(x; y) ⊓ z|"
  and cardded'        : "is_p_fun x ⟶ |x ⊓ (z; y˘)| ≤ |(x; y) ⊓ z|"
  and cardone         : "|1'| = card {x. is_point x}"
```

Here, *card* is the built-in operation for the cardinality of sets. With the given definition we are able to prove the mentioned results about cardinalities, for instance:

```
lemma cardunifun : assumes "is_p_fun x" and "is_fun y" shows "|x; y| = |x|"
lemma cardpoint : assumes "is_point x" shows "|x| = |1'|"
```

The proofs of these lemmata are found automatically by Sledgehammer. From this we get immediately that the cardinality of a point equals the cardinality of the identity relation. In the same way we formalise all lemmata of this section and many more where most of the proofs are heavily supported by Sledgehammer.

For the verification of the program of Sect. 3 we do not only use the above mentioned theories about relation algebra and cardinalities, but also a library for Hoare Logic in Isabelle/HOL, see [11]. This library provides the opportunity to write, for instance, while-programs as theorems as well as tactics generating the proof obligations for partial correctness automatically. Thus we can encode the program as follows.

```
theorem correctness: "VARS e m c p q
 { pre e m}
 c := bot;
 WHILE c • top ≠ top
   INV { inv e m c }
   DO p := point((c • top)ᶜ);
      q := point((c˘ • e • p)ᶜ ⊓ (mᶜ • (c˘ • e • p)ᶜ)ᶜ);
      c := c ⊔ p • q˘
   OD
 { post e c }"
```

Unfortunately, we have to switch to another symbol for composition here since; is already defined in the theory for Hoare logic. Thus, we use • in this context. The pre- and postconditions and the invariant slightly differ from the ones presented in Sect. 3 since we have to use partial points and identities for proving the approximation bound.

definition "*pre e m* $\longleftrightarrow$ *is_irrefl e* $\wedge$ *is_symm e* $\wedge$ *is_lin_order m*"
definition "*post e c* $\longleftrightarrow$ *is_fun c* $\wedge$ *has_color_prop e c* $\wedge$ $|c^{\smile} \bullet top \sqcap 1'| \leq \Delta_{top} + 1$"
definition "*inv e m c* $\longleftrightarrow$ *pre e m* $\wedge$ *is_p_fun c* $\wedge$ *has_color_prop e c*
$\wedge\ c^{\smile} \bullet top \sqsubseteq (m^{c} \bullet (c^{\smile} \bullet top)^{c})^{c} \wedge\ |c^{\smile} \bullet top \sqcap 1'| \leq \Delta_{c \bullet top} + 1$"

One of the big advantages of using the theory for Hoare Logic is that its provides tactis for verification condition generation. In the case of our program or theorem, respectively, we can apply the rule *vcg_simp*. With this rule the three proof obligations w.r.t. the given pre- and postconditions and the loop-invariant are generated as subgoals automatically. In the following we see the resulting subgoals after applying *vcg_simp*.

goal **(3 subgoals)**:
1. $\bigwedge e\ m\ c\ p\ q.\ pre\ e\ m \Longrightarrow inv\ e\ m\ bot$
2. $\bigwedge e\ m\ c\ p\ q.\ inv\ e\ m\ c \wedge c \bullet top \neq top \Longrightarrow$
   $inv\ e\ m\ (c \sqcup point((c \bullet top)^{c}) \bullet point((c^{\smile} \bullet e \bullet p)^{c} \sqcap (m^{c} \bullet (c^{\smile} \bullet e \bullet p)^{c})^{c}))$
3. $\bigwedge e\ m\ c\ p\ q.\ inv\ e\ m\ c \wedge \neg c \bullet top \neq top \Longrightarrow post\ e\ c$

The three statements are shown stepwise by using the theories mentioned in this section. The proofs that are not found by Sledgehammer automatically are given as structured Isar proofs, see [18], so that the reader can follow the basic ideas.

Besides the results presented in this section our library contains over 150 lemmata about finite unions of relations, points and vectors, atoms, and cardinalities of relations. Furthermore, in *Relation_Algebra_Orders* we defined order related relations and proved several facts about them.

6 Comparison of the Implementations

In Sects. 4 and 5 we show how the proof assistants Coq and Isabelle/HOL can be used for formal program verfication and reasoning about relation algebras in general. In this section we want to summarise our experiences with both systems and highlight their advantages and disadvantages from our point of view.

For Coq, we used an existing library that already implements tools for proving results regarding cardinalities. One advantage of the library is that it extends a library including a model for heterogeneous relation algebras and related structures. Here, the implementation of typed relations is possible because of Coq's expressive type system based on the *predicative calculus of inductive constructions*. Such an expressive type system has many common advantages, for instance, it ensures that all expressions and formulae are well-typed. Thus, the Coq proofs mostly correspond to the handwritten ones we gave in this paper.

Coq, and in particular the used library, provides several automated theorem proving tactics and decision procedures, but most of them were not very helpful in our context. Thus, the proofs have mostly to be done step by step using Coqs standard tactics. Unfortunately, a direct link to automated theorem provers is still missing. Furthermore, the formalisation of the proof obligations of the presented program has to be done by hand since there are no tools for an automated generation. For non-experts the Coq code is quite hard to read without using an IDE illustrating proof steps and subgoals.

By contrast, Isabelle/HOL bridges the gap between interactive and automated theorem proving because of its integrated tool Sledgehammer. Due to the limited type system of Isabelle/HOL there is only an existing library for homogeneous relation algebras. For this reason an extension by the cardinality operation, as in the case of Coq, was not possible directly. Thus, we modified the axiomatisation of the operation to make it applicable for homogeneous relations in the first place. We formalised it in Isabelle/HOL and proved the correctness of the relational program heaviliy supported by Sledgehammer. Unlike Coq, Isabelle/HOL provides a library for Hoare Logic including tactics for generating proof obligations automatically. In our context we were able to avoid typed relations by adapting the cardinality operation. In general, reasoning about typed relations can be managed by using, for instance, predicates specifying the source and target of a relation. Such an approach often results in more complicated and longish proofs. In the future, one can benefit from our library containing most of the basic facts that are necessary when dealing with cardinalities. Certainly, invoking Sledgehammer does not always complete proofs successfully. As in Coq, one has to do steps by hand, but Isabelle/HOL supports the proof language Isar. Its intuitive syntax allows to write proofs structured and comprehensible for non-experts.

7 Concluding Remarks

We presented a correctness proof of a relational program for approximating vertex colorings in undirected (and loop-free) graphs. The proof of the approximation ratio we done by using an operation to reason algebraically about cardinalities in heterogeneous relation algebras.

Furthermore, all proofs were mechanised in both proof assistants Coq and Isabelle/HOL and build on existing libraries for relation algebras. In contrast to Coq, there were no tools to tackle cardinalities in Isabelle/HOL so far. To reuse a library for homogeneous relation algebras we presented a new theoretical framework for reasoning about untyped relations. In this context, we not only proved the programs' correctness in Isabelle/HOL, but also developed a library providing over 150 facts about, for instance, points, atoms and cardinalities.

For the future it would be helpful to have a tool for Hoare Logic in Coq so that the generation of a programs' proof obligations has not to be done by hand or external programs. A further investigation of the new axiomatisation of the cardinality operation is also conceivable to see how exhaustive this approach is. In general, it would be interesting to study what is provable without the restriction to finite relations.

Acknowledgement. I thank Walter Guttmann and Damien Pous for their help concerning the use of proof assistants and Rudolf Berghammer for helpful discussions and his support, in general. I thank the unknown referees and Michael Winter for their comments and suggestions which helped to improve the paper.

Appendix

In this appendix we show that the third formula of the invariant $\mathit{Inv}(E, M, C)$ is maintained stated in the following lemma.

Lemma. *Let E, C and M be relations so that $\mathrm{Pre}(E, M)$ and $\mathit{Inv}(E, M, C)$ hold and p, q points with $p \subseteq \overline{C\mathsf{L}}$, $q \subseteq \overline{C^\mathsf{T}Ep} \cap \overline{\overline{M}\,\overline{C^\mathsf{T}Ep}}$. Then $(C \cup pq^\mathsf{T})^\mathsf{T}\mathsf{L} \subseteq \overline{\overline{M}\,\overline{(C \cup pq^\mathsf{T})^\mathsf{T}\mathsf{L}}}$ holds.*

Proof. Since $\mathit{Inv}(E, M, C)$ holds, we have $C^\mathsf{T}\mathsf{L} \subseteq \overline{\overline{M}\,\overline{C^\mathsf{T}\mathsf{L}}}$ and hence

$$\overline{M}\,\overline{C^\mathsf{T}\mathsf{L}} \subseteq \overline{C^\mathsf{T}\mathsf{L}}. \tag{1}$$

The inclusion

$$\overline{M}\,\overline{C^\mathsf{T}\mathsf{L}} \subseteq \overline{q} \tag{2}$$

is shown by the following calculation:

$$\begin{aligned}
C^\mathsf{T}Ep \subseteq C^\mathsf{T}\mathsf{L} &\iff \overline{C^\mathsf{T}\mathsf{L}} \subseteq \overline{C^\mathsf{T}Ep} \\
&\implies \overline{M}\,\overline{C^\mathsf{T}\mathsf{L}} \subseteq \overline{M}\,\overline{C^\mathsf{T}Ep} \\
&\iff \overline{\overline{M}\,\overline{C^\mathsf{T}Ep}} \subseteq \overline{\overline{M}\,\overline{C^\mathsf{T}\mathsf{L}}} \\
&\implies q \subseteq \overline{\overline{M}\,\overline{C^\mathsf{T}\mathsf{L}}} && \text{since } q \subseteq \overline{C^\mathsf{T}Ep} \cap \overline{\overline{M}\,\overline{C^\mathsf{T}Ep}} \\
&\implies \overline{M}\,\overline{C^\mathsf{T}\mathsf{L}} \subseteq \overline{q}
\end{aligned}$$

Furthermore, we have the following:

$$(C \cup pq^\mathsf{T})^\mathsf{T}\mathsf{L} \subseteq \overline{\overline{M}\,\overline{(C \cup pq^\mathsf{T})^\mathsf{T}\mathsf{L}}} \iff \overline{M}\,\overline{(C \cup pq^\mathsf{T})^\mathsf{T}\mathsf{L}} \subseteq \overline{(C \cup pq^\mathsf{T})^\mathsf{T}\mathsf{L}}.$$

We now show that the inclusion above on the right-hand side is true where we use that p, q are points and thus $qp^\mathsf{T}\mathsf{L} = q$ again:

$$\begin{aligned}
\overline{M}\,\overline{(C \cup pq^\mathsf{T})^\mathsf{T}\mathsf{L}} &= \overline{M}\,\overline{C^\mathsf{T}\mathsf{L} \cup q} && qp^\mathsf{T}\mathsf{L} = q \\
&= \overline{M}\,(\overline{C^\mathsf{T}\mathsf{L}} \cap \overline{q})
\end{aligned}$$

$$\begin{aligned}
&\subseteq \overline{M}\,\overline{C^{\mathsf{T}}\mathsf{L}} \cap \overline{M\overline{q}} \\
&\subseteq \overline{M}\,\overline{C^{\mathsf{T}}\mathsf{L}} \\
&\subseteq \overline{C^{\mathsf{T}}\mathsf{L}} \cap \overline{q} && \text{(1) and (2)} \\
&= \overline{C^{\mathsf{T}}\mathsf{L} \cup q} \\
&= \overline{(C \cup pq^{\mathsf{T}})^{\mathsf{T}}\mathsf{L}}. && qp^{\mathsf{T}}\mathsf{L} = q
\end{aligned}$$

□

References

1. Berghammer, R., Danilenko, N., Höfner, P., Stucke, I.: Cardinality of relations with applications. Discret. Math. **339**(12), 3089–3115 (2016)
2. Berghammer, R., Höfner, P., Stucke, I.: Tool-based verification of a relational vertex coloring program. In: Kahl, W., Winter, M., Oliveira, J.N. (eds.) RAMICS 2015. LNCS, vol. 9348, pp. 275–292. Springer, Cham (2015). doi:10.1007/978-3-319-24704-5_17
3. Berghammer, R., Höfner, P., Stucke, I.: Cardinality of relations and relational approximation algorithms. J. Log. Algebraic Methods Program. **85**(2), 269–286 (2016)
4. Bertot, Y., Castéran, P., Huet, G., Paulin-Mohring, C.: Interactive Theorem Proving and Program Development: Coq'Art: The Calculus of iInductive Constructions. Texts in Theoretical Computer Science. Springer, Heidelberg (2004)
5. Brunet, P., Pous, D., Stucke, I.: Cardinalities of finite relations in Coq. In: Blanchette, J.C., Merz, S. (eds.) ITP 2016. LNCS, vol. 9807, pp. 466–474. Springer, Cham (2016). doi:10.1007/978-3-319-43144-4_29
6. Freyd, P., Scedrov, A.: Categories, Allegories. Elsevier Science, Amsterdam (1990). North-Holland Mathematical Library
7. Furusawa, H.: Algebraic formalisations of fuzzy relations and their representation theorems. Ph.D. thesis, Department of Informatics, Kyushu University (1998)
8. Isabelle. https://isabelle.in.tum.de/
9. Kawahara, Y.: On the cardinality of relations. In: Schmidt, R.A. (ed.) RelMiCS 2006. LNCS, vol. 4136, pp. 251–265. Springer, Heidelberg (2006). doi:10.1007/11828563_17
10. Maddux, R.D.: Relation Algebras. Studies in Logic and the Foundations of Mathematics, vol. 150. Elsevier, Amsterdam (2006)
11. Nipkow, T.: Hoare logics in Isabelle/HOL. In: Schwichtenberg, H., Steinbrüggen, R. (eds.) Proof and System-Reliability, pp. 341–367. Kluwer, Dordrecht (2002)
12. Nipkow, T., Wenzel, M., Paulson, L.C.: Isabelle/HOL: A Proof Assistant for Higher-Order Logic. LNCS, vol. 2283. Springer, Heidelberg (2002)
13. Pous, D.: Relation Algebra and KAT in Coq. http://perso.ens-lyon.fr/damien.pous/ra/
14. Pous, D.: Kleene algebra with tests and Coq tools for while programs. In: Blazy, S., Paulin-Mohring, C., Pichardie, D. (eds.) ITP 2013. LNCS, vol. 7998, pp. 180–196. Springer, Heidelberg (2013). doi:10.1007/978-3-642-39634-2_15
15. Schmidt, G.: Relational Mathematics, vol. 132. Cambridge University Press, Cambridge (2011). Encyclopedia of Mathematics and Its Applications

16. Schmidt, G., Ströhlein, T.: Relations and Graphs - Discrete Mathematics for Computer Scientists. EATCS Monographs on Theoretical Computer Science. Springer, Heidelberg (1993)
17. Struth, G., Weber, T.: Relation Algebra. Archive of Formal Proofs (2014). https://www.isa-afp.org/entries/Relation_Algebra.shtml
18. Stucke, I.: Reasoning about Cardinalities Supported by Proof Assistants, Proof Scripts. http://www.rpe.informatik.uni-kiel.de/en/Staff/ist/ramics-2017
19. Tarski, A.: On the calculus of relations. J. Symb. Log. **6**(3), 73–89 (1941)
20. The Coq Proof Assistant. https://coq.inria.fr

Type-n Arrow Categories

Michael Winter(✉)

Department of Computer Science, Brock University,
St. Catharines, ON L2S 3A1, Canada
mwinter@brocku.ca

Abstract. It has been shown that the arrow category of type-2 fuzzy relations with respect to an arrow category $\mathcal{A}$ can be defined as the Kleisli category of $\mathcal{A}$ for a monad based on the concept of the extension of an object. In this paper we want to continue the study of higher-order arrow categories by showing two major results. First, we are going to remove the ad-hoc notion of an extension of an object completely from the construction of higher-order arrow categories. The second result establishes that the newly constructed higher-order arrow category has sufficient structure for constructing further higher-order arrow categories, i.e., that the process of moving from type-n to type-(n+1) arrow categories can always be iterated.

1 Introduction

Allegories and Dedekind categories, in particular, [1,7,8,10,11] provide an adequate categorical and algebraic framework for reasoning about relations. An obvious example for each of these categories is the category **REL** of sets and binary relation. A binary relation can be represented by its characteristic function, i.e., by a function that returns true if the pair is in the relation and false if not. The category **REL** is not the only example of an allegory, of course. Given a complete Heyting algebra L, the category L-**REL** of sets and so called L-relations is also an example. L-relations differ from regular relations by assigning to each pair a degree of membership from L instead of true or false. Certain aspects of L-relations cannot be expressed in allegories or Dedekind categories. For example, consider the special case that an L-relation R returns the smallest or the greatest element of L for each pair. Such relations correspond in an obvious way to regular binary relations, and therefore they are called crisp. Even though several abstract notions of crispness in Dedekind categories have been proposed [2,5,6], it was shown that this property cannot be expressed in the language of allegories or Dedekind categories [14,15]. Therefore, Goguen and arrow categories [14,16] were introduced adding two additional operations to the theory of Dedekind categories covering the notion of crispness.

A higher-order or type-2 L-relation uses membership values from the L^L, i.e., it uses endofunctions on L as the degree of membership for each pair. Since L^L forms a complete Heyting algebra if L does, the category L^L-**REL** also forms an arrow category. To

M. Winter—The author gratefully acknowledges support from the Natural Sciences and Engineering Research Council of Canada.

P. Höfner et al. (Eds.): RAMiCS 2017, LNCS 10226, pp. 307–322, 2017.
DOI: 10.1007/978-3-319-57418-9_19

distinguish between L-**REL** and L^L-**REL** one normally speaks about type-1 L-relations and type-2 L-relations, respectively. By iterating the process we can define type-n L-relations for arbitrary n. In this paper we are interested in the relationship between type-1 and type-2 L-relations and the iteration process leading to type-n L-relations.

In [17, 18] the extension was on object was used to show that the category of type-2 L-relations can be constructed as a Kleisli category of type-1 L relations in the context of arrow categories. Since relations in arrow categories are abstract entities, i.e., morphisms of a category, a simple replacement of L by L^L is not possible. The lattice L is only available implicitly by certain classes of special relations such as scalars. However, moving from type-1 to type-2 relations and back is important in order to model higher-order fuzzy controllers [20]. The extension $A^\sharp$ of a set A is the set $A \times L$ of pairs of A elements and a lattice value. The corresponding isomorphism, i.e., the bijection between $A^\sharp$ and $A \times 1^\sharp$ with unit 1, was shown in the abstract setting. Furthermore, it was shown that the induced product functor together with appropriate natural transformations forms a monad so that the category of type-2 L-relations was obtained as the Kleisli category for this monad. These result were used in [20] to apply the abstract theory of arrow categories in the development of fuzzy controllers.

In this paper we want to show two major results. The notion of an extension of an object is an ad-hoc notion, i.e., has never been used otherwise. Therefore, we first want to remove the extension of an object completely from the construction of higher-order arrow categories. In [17, 18] it was already shown that $A^\sharp$ is isomorphic to $A \times 1^\sharp$, i.e., to a relational product with crisp projections. In this paper we will show that $1^\sharp$ is isomorphic to $\mathcal{P}_L(1)$ where $\mathcal{P}_L(A)$ is the fuzzy power of A. The fuzzy power of A is the abstract version of the L-fuzzy powerset. Please note that this construction is different from relational powers resp. the constructions given in power allegories. As an immediate consequence we obtain the following result in the abstract setting of arrow categories: If an arrow category has a unit, crisp relational products and fuzzy powers, then the arrow category of type-2 L relations can be defined as the Kleisli category induced by the monad above. Our second result shows that this process can be iterated, i.e., it verifies that the arrow category of type-2 L relations again has a unit, crisp relational products and fuzzy powers.

2 Mathematical Preliminaries

2.1 Allegories and Arrow Categories

In this section we want to recall some basic notions from categories, allegories and arrow categories. For further details we refer to [1, 15].

We will write $R : A \to B$ to indicate that a morphism R of a category C has source A and target B. Composition and the identity morphism are denoted by; and $\mathbb{I}_A$ with the convention that composition is from left to right, i.e., $R; S$ means R first, and then S.

Since the category of type-2 fuzzy relations can be obtained as a Kleisli category of the category of type-1 fuzzy relations, we recall the concept of a monad and the Kleisli category derived from a monad. Suppose C is a category. Then a monad on C is a triple (F, η, μ) consisting of a endo-functor $F : C \to C$ and two natural transformations

$\eta : I \to F$, i.e., from the identity functor to F, and $\mu : F^2 \to F$, i.e., from the functor obtained by applying F twice to F, so that the following two diagrams commute:

$$
\begin{array}{ccc}
F(F(F(A))) & \xrightarrow{\mu_{F(A)}} & F(F(A)) \\
\downarrow{\scriptstyle F(\mu_A)} & & \downarrow{\scriptstyle \mu_A} \\
F(F(A)) & \xrightarrow{\mu_A} & F(A)
\end{array}
\qquad
\begin{array}{ccc}
F(A) \xrightarrow{F(\eta_A)} F(F(A)) \xleftarrow{\eta_{F(A)}} F(F(A)) \\
{\scriptstyle I_{F(A)}} \searrow \quad \downarrow{\scriptstyle \mu_A} \quad \swarrow {\scriptstyle I_{F(A)}} \\
F(A)
\end{array}
$$

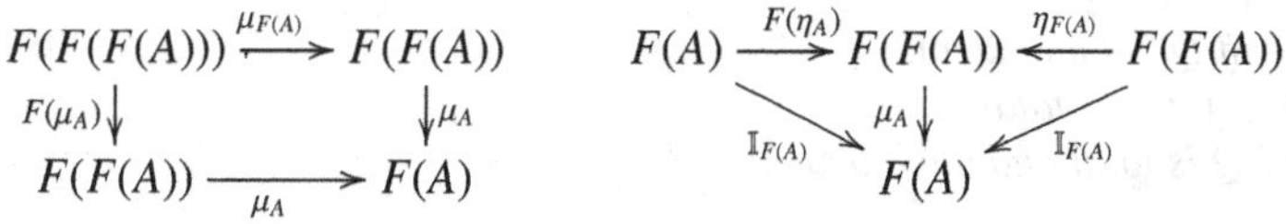

Monads allow one to define new categories based on the additional behavior and.or properties encoded in the functor. The Kleisli category C_F has the same objects as C. A morphism in C_F from A to B is a morphism from A to $F(B)$ in C. η acts as an identity for the Kleisli composition $R \mathbin{\S} S = R; F(S); \mu$.

Now we want to recall some fundamentals on Dedekind categories [7, 8]. Categories of this type are called locally complete division allegories in [1]. In this paper we will call morphisms of a Dedekind category relations.

Definition 1. *A Dedekind category $\mathcal{R}$ is a category satisfying the following:*

1. *For all objects A and B the collection $\mathcal{R}[A, B]$ is a complete Heyting algebra. Meet, join, the induced ordering, the least and the greatest element are denoted by $\sqcap, \sqcup, \sqsubseteq, \amalg_{AB}, \top_{AB}$, respectively.*
2. *There is a monotone operation $\check{\ }$ (called converse) mapping a relation $Q : A \to B$ to $Q^\smile : B \to A$ such that for all relations $Q : A \to B$ and $R : B \to C$ the following holds: $(Q; R)^\smile = R^\smile; Q^\smile$ and $(Q^\smile)^\smile = Q$.*
3. *For all relations $Q : A \to B, R : B \to C$ and $S : A \to C$ the modular law $(Q; R) \sqcap S \sqsubseteq Q; (R \sqcap (Q^\smile; S))$ holds.*
4. *For all relations $R : B \to C$ and $S : A \to C$ there is a relation $S/R : A \to B$ (called the left residual of S and R) such that for all $X : A \to B$ the following holds: $X; R \sqsubseteq S \Leftrightarrow X \sqsubseteq S/R$.*

The residual operation together with converse induce a second residual by defining $Q \backslash R := (R^\smile / Q^\smile)^\smile$. This operation is characterized by the equivalence $Q; X \sqsubseteq R$ iff $X \sqsubseteq Q \backslash R$. Both residuals together allow the definition of the symmetric quotient. This construction is defined by $\mathrm{syQ}(Q, R) := Q \backslash R \sqcap Q^\smile / R^\smile$. Consequently the symmetric quotient is characterized by $Q; X \sqsubseteq R$ and $R; X^\smile \sqsubseteq Q$ iff $X \sqsubseteq \mathrm{syQ}(Q, R)$.

Notice that a complete Heyting algebra has an implication operation $\to$, i.e., we have $X \sqsubseteq Q \to R$ iff $X \sqcap Q \sqsubseteq R$. This implication leads to a notion of equivalence defined by $Q \leftrightarrow R = (Q \to R) \sqcap (R \to Q)$.

Throughout this paper we will use some basic properties of relations such as $\amalg_{AB}^\smile = \amalg_{BA}, \top_{AB}^\smile = \top_{BA}, I_A^\smile = I_A$, the monotonicity of all operations, and the fact that composition distributes over join from both sides without mentioning.

Notice that we have $\top_{AA}; \top_{AB} = \top_{AB}; \top_{BB} = \top_{AB}$, but that the more general equation $\top_{AB}; \top_{BC} = \top_{AC}$ does not necessarily hold [13]. If it does hold for all objects A, B and C, then we call the Dedekind category uniform.

An important class of relations is given by maps.

Definition 2. *Let $\mathcal{R}$ be a Dedekind category. Then a relation $Q : A \to B$ is called*

1. *univalent (or partial function) iff $Q^{\smile};Q \sqsubseteq \mathbb{I}_B$,*
2. *total iff $\mathbb{I}_A \sqsubseteq Q;Q^{\smile}$,*
3. *injective iff $Q^{\smile}$ is univalent,*
4. *surjective iff $Q^{\smile}$ is total,*
5. *a map iff Q is total and univalent.*

It is well-known that Q is total iff $Q;\mathbb{T}_{BC} = \mathbb{T}_{AC}$. We will use this and the corresponding property for surjective relations without mentioning.

In the following lemma we have summarized some important properties of mappings and univalent relations. Again, a proof can be found in [9–13].

Lemma 1. *Let $\mathcal{R}$ be a Dedekind category. Then we have for all $Q, U : A \to B, R : A \to C, S, T : B \to C$ and maps $f : B \to C$*

1. *$Q;f \sqsubseteq R$ iff $Q \sqsubseteq R;f^{\smile}$,*
2. *if Q is univalent, then $Q;(S \sqcap T) = Q;S \sqcap Q;T$,*
3. *if S is univalent, then $(Q \sqcap R;S^{\smile});S = Q;S \sqcap R$.*
4. *If Q is total and U univalent with $Q \sqsubseteq U$, then $Q = U$.*

A unit 1 is an object of a Dedekind category so that $\mathbb{I}_1 = \mathbb{T}_{11}$ and $\mathbb{T}_{A1}$ is total for all objects A. A unit is an abstract version of a singleton set, and, hence, the relational version of a terminal object. In the subcategory of mappings a unit becomes a terminal object. This immediately shows that a unit is unique up to isomorphism.

The abstract version of a cartesian product is given by a relational product. Notice that a relational product is a categorical product in the subcategory of maps but not within the Dedekind category of all relations.

Definition 3. *The relational product of two objects A and B is an object $A \times B$ together with two relations $\pi : A \times B \to A$ and $\rho : A \times B \to B$ so that the following equations hold*

$$\pi^{\smile};\pi \sqsubseteq \mathbb{I}_A, \quad \rho^{\smile};\rho \sqsubseteq \mathbb{I}_B, \quad \pi^{\smile};\rho = \mathbb{T}_{AB}, \quad \pi;\pi^{\smile} \sqcap \rho;\rho^{\smile} = \mathbb{I}_{A\times B}.$$

A Dedekind category has products if the relational product for each pair of objects exists.

We will use the following abbreviations $Q \otimes R := Q;\pi^{\smile} \sqcap R;\rho^{\smile}$, $S \otimes T := \pi;S \sqcap \rho;T$ and $U \otimes V = \pi;U;\pi^{\smile} \sqcap \rho;V;\rho^{\smile}$. We will adopt the convention that composition binds tighter than the operations defined above. Notice that some of the equations below

$$\begin{array}{ll}
(Q \otimes R);\pi = Q \text{ if } R \text{ total} & (Q \otimes R);\rho = R \text{ if } Q \text{ total} \\
\pi^{\smile};(S \otimes T) = S \text{ if } T \text{ surjective} & \rho^{\smile};(S \otimes T) = T \text{ if } S \text{ surjective} \\
(Q \otimes R);(U \otimes V) = Q;U \otimes R;V & (U \otimes V);(S \otimes T) = U;S \otimes V;T \\
(U \otimes V);(W \otimes X) = U;W \otimes V;X & (Q \otimes R);(S \otimes T) = Q;S \sqcap R;T
\end{array}$$

only hold under certain assumptions. This fact is known as the unsharpness problem of relational products. However, in this paper we will assume that all relational products exist implying that all of the equations above are valid.

Using product we can define two shifting operations. If $Q : A \times B \to C$, then we define $\overrightarrow{Q} : A \to C \times B := \pi^{\smile};(Q \otimes \rho)$ and if $R : A \to C \times B$, then we define $\overleftarrow{R} := (R \otimes \rho^{\smile});\pi : A \times B \to C$. These operations satisfy the following shifting conditions:

$$R \sqsubseteq \overrightarrow{Q} \Leftrightarrow \overleftarrow{R} \sqsubseteq Q, \quad \overrightarrow{Q} \sqsubseteq R \Leftrightarrow Q \sqsubseteq \overleftarrow{R}, \quad \overleftarrow{\overrightarrow{Q}} = Q.$$

Given a complete Heyting algebra L an L-relation R between to sets A and B is a function $R : A \times B \to L$. The values in L serve as degree of membership, i.e., they indicate the degree of relationship between two elements from A and B. Notice that regular binary relations between sets are a special case of L-relations where L is the set $\mathbb{B} = \{\text{true}, \text{false}\}$ of truth values. The collection of all L-relations between sets together with the standard definition of the operations forms a Dedekind category, normally denoted by L-Rel. The lattice L itself can even in the abstract case be identified with the collection of scalar relations on an object.

Definition 4. *A relation $\alpha : A \to A$ is called a scalar on A iff $\alpha \sqsubseteq \mathbb{I}_A$ and $\mathbb{T}_{AA};\alpha = \alpha;\mathbb{T}_{AA}$.*

The notion of scalars was introduced by Furusawa and Kawahara [5]. It is equivalent to the notion of ideal elements, i.e., relations $R : A \to B$ that satisfy $\mathbb{T}_{AA};R;\mathbb{T}_{BB} = R$. These relations were introduced by Jónsson and Tarski [4].

A crisp relation is an L-relation that only uses the least element 0 and the greatest element 1 of L as membership values. The subcategory of crisp relations can be identified with the category of regular binary relations, i.e., with the relations using $\mathbb{B}$ as membership values. The language of Dedekind categories is not strong enough to grasp the notion of a crisp relation [14, 15]. As a consequence so-called arrow categories were introduced [15, 16]. These categories add two operations to Dedekind categories. The down-arrow operation maps an L-relation R to the greatest crisp relation included in R and the up-arrow operation maps R to the least crisp relation that includes R.

Definition 5. *An arrow category $\mathcal{A}$ is a Dedekind category with $\mathbb{T}_{AB} \neq \mathbb{L}_{AB}$ for all objects A and B together with two operations $^{\uparrow}$ and $^{\downarrow}$ satisfying the following:*

1. *$R^{\uparrow}, R^{\downarrow} : A \to B$ for all $R : A \to B$.*
2. *$(^{\uparrow}, ^{\downarrow})$ is a Galois correspondence, i.e., $Q^{\uparrow} \sqsubseteq R$ iff $Q \sqsubseteq R^{\downarrow}$ for all $Q, R : A \to B$.*
3. *$(R^{\smile};S^{\downarrow})^{\uparrow} = R^{\uparrow\smile};S^{\downarrow}$ for all $R : B \to A$ and $S : B \to C$.*
4. *If $\alpha \neq \mathbb{L}_{AA}$ is a non-zero scalar then $\alpha^{\uparrow} = \mathbb{I}_A$.*
5. *$(Q \sqcap R^{\downarrow})^{\uparrow} = Q^{\uparrow} \sqcap R^{\downarrow}$ for all $Q, R : A \to B$.*

A relation $R : A \to B$ of an arrow category $\mathcal{A}$ is called crisp iff $R^{\uparrow} = R$ (or equivalently $R^{\downarrow} = R$). The collection of crisp relations is closed under all operations of a Dedekind category, and, hence, forms a sub-Dedekind category of $\mathcal{A}$.

Arrow categories are always uniform [15]. As a consequence all projections are surjective as the computation $\mathbb{T}_{CA} = \mathbb{T}_{CB};\mathbb{T}_{BA} = \mathbb{T}_{CB};\rho^{\smile};\pi \sqsubseteq \mathbb{T}_{CA\times B};\pi$ shows.

In the context of arrow categories we are usually interested in relational products for which the projections are crisp. Note this property does not follow from the definition of a relational product. If it is true, we call the relational product crisp. Working with crisp relational products is not a restriction because there is usually (an isomorphic) crisp version of any product.

The following lemma lists some further properties of relations in arrow categories that we will be using in the remainder of the paper. A proof can be found in [15].

Lemma 2. *Let $\mathcal{A}$ be an arrow category. Then we have for all $Q, Q_i : A \to B$ for $i \in I$ and $R : B \to C$*

1. $(\bigsqcap_{i \in I} Q_i)^{\downarrow} = \bigsqcap_{i \in I} Q_i^{\downarrow}$,
2. $Q^{\smile\downarrow} = Q^{\downarrow\smile}$,
3. *if R is crisp, then $(Q; R)^{\uparrow} = Q^{\uparrow}; R$,*
4. *if $f : C \to A$ is a map, then $(f; Q)^{\downarrow} = f^{\downarrow}; Q^{\downarrow}$.*

2.2 Extension of an Object

The extension $A^{\sharp}$ of an object A was introduced in [17]. This construction is motivated by pairing each element of A with all membership values from L. In addition to representing the membership values by ideals or scalars this construction also allows to obtain those values as crisp points, i.e., as crisp mappings $p : 1 \to 1^{\sharp}$. For further details we refer to [17].

Later we will use the extension of an object to define an arrow category of type-2 fuzziness. This will be done by defining a suitable Kleisli category. The whole approach is based on the following simple idea. A type-2 L-relation, i.e., a L^L-relation, between the sets A and B is a function $R : A \times B \to (L \to L)$. It is well-known that such functions are isomorphic to functions from $A \times (B \times L) \to L$. Notice that the latter are L-relations from A to $B^{\sharp}$.

Definition 6. *Let A be an object of an arrow category. An object $A^{\sharp}$ together with two relations $\eta_A, \nu_A : A \to A^{\sharp}$ is called the extension of A iff*

1. *η_A is crisp,*
2. $\mathbb{T}_{AA}; \nu_A = \nu_A$,
3. $\eta_A; \eta_A^{\smile} = \mathbb{I}_A$,
4. $\nu_A^{\smile\sharp} \sqcap \eta_A^{\smile}; \eta_A = \mathbb{I}_{A^{\sharp}}$,
5. *$Q^{\sharp}; \eta_A^{\smile} = \mathbb{T}_{BA}$ for every relation $Q : B \to A$,*

where $Q^{\sharp} : B \to A^{\sharp}$ is defined by $Q^{\sharp} = ((Q; \eta_A) \leftrightarrow (\mathbb{T}_{BA}; \nu_A))^{\downarrow}$ and $\leftrightarrow$ is the Heyting equivalence.

Some basic properties of extensions are summarized in the following lemma. A proof can be found in [17].

Lemma 3. *Let $Q : A \to B$ be a relation and $f : C \to A$ be a crisp mapping. Then we have*

1. *η_A is total, surjective, and injective,*
2. $(Q; \mathbb{T}_{BC})^{\sharp} = \mathrm{syQ}(\mathbb{T}_{AB}; Q^{\smile}, \mathbb{T}_{AC}; \nu_C)^{\downarrow}$,
3. $f; Q^{\sharp} = (f; Q)^{\sharp}$,
4. *$Q^{\sharp}$ is total,*
5. $\mathrm{syQ}(\nu_1, \nu_1)^{\downarrow} = \mathbb{I}_{1^{\sharp}}$.

The following theorem was shown in [17] and verifies that $A^{\sharp}$ is indeed a relational product.

Theorem 1. *Let $\mathcal{A}$ be an arrow category with extensions and unit* 1. *Then the extension* $A^\sharp$ *of* A *together with the relations* $\pi := \eta_A^\smile$ *and* $\rho := (\nu_A^\smile; \mathbb{T}_{A1})^\sharp$ *is a crisp relational product of* A *and* $1^\sharp$.

For any object L we define an endofunctor $P : \mathcal{R} \to \mathcal{R}$ by $P(X) = X \times L$ and $P(Q) = Q \otimes \mathbb{I}_L$. Furthermore, we define two morphisms $\eta_A : A \to P(A)$ and $\mu_A : P(P(A)) \to P(A)$ by $\eta_A = \pi^\smile$ and $\mu_A = \pi \sqcap \rho;\rho^\smile$. Notice that μ_A is not total, and that η_A is not univalent, i.e., both relations are not mappings.

Theorem 2. *Let* $\mathcal{R}$ *be a Dedekind category and* L *be an object of* $\mathcal{R}$ *so that the product* $A \times L$ *exists for every object* A. *Then* (P, η, μ) *is a monad on* $\mathcal{R}$.

Recall that the composition $⨾$ in the Kleisli category $\mathcal{R}_P$ is defined by $Q ⨾ R = Q; P(R); \mu_C = Q; (R \otimes \rho^\smile)$ with η as identity. This Kleisli category can be made into a Dedekind category by using the meet and join operation from $\mathcal{R}$ and the following definitions of a converse and residual operation: $Q^\cup = \eta_B; \mu_B^\smile; P(Q^\smile) = \overrightarrow{Q^\smile} = \pi^\smile; (Q^\smile \otimes \rho)$ and $S/_P R = S/(P(R); \mu_C)$.

Theorem 3. *Let* $\mathcal{R}$ *be a Dedekind category and* L *be an object of* $\mathcal{R}$ *so that the product* $A \times L$ *exists for every object* A. *Then the Kleisli category* $\mathcal{R}_P$ *together with the operations defined above forms a Dedekind category.*

If $\mathcal{A}$ is an arrow category with a unit, crisp relational products and fuzzy powers, and we choose L to be the unit, then we can define $Q^\Uparrow = (Q; \eta_B^\smile; \eta_B)^\uparrow$ and $Q^\Downarrow = (Q/(\eta_B^\smile; \eta_B))^\downarrow$.

Theorem 4. *Let* $\mathcal{A}$ *be an arrow category and* L *be an object of* $\mathcal{A}$ *so that a crisp product* $A \times L$ *exists for every object* A. *Then the Dedekind category* $\mathcal{A}_P$ *together with the operations defined above forms an arrow category.*

3 Fuzzy Powers

In this section we want to investigate two abstract versions of power sets. The first one is the well-known relational (or direct) power. Furthermore, they are also the relational version of power objects as known in topos theory [1].

Definition 7. *An object* $\mathcal{P}(A)$ *together with a relation* $\varepsilon : A \to \mathcal{P}(A)$ *is called a relational power iff*

$$\mathrm{syQ}(\varepsilon, \varepsilon) = \mathbb{I}_{\mathcal{P}(A)} \quad \textit{and} \quad \mathrm{syQ}(R, \varepsilon) \textit{ is total for every } R : A \to B.$$

The relational power is an abstract version of the power set of a set. In particular, it emphasizes extensionality ($\mathrm{syQ}(\varepsilon, \varepsilon) = \mathbb{I}$) as a basic property of sets.

In a fuzzy context one is usually interested in the L-power set L^A of a set, i.e., the set of (L-characteristic) functions from A to L. This set together with the obvious generalization of the "is element of" relation does not necessarily satisfy the axioms of a relational power. The reason is that ε might not be extensional. For an example, let

$B_4 = \{0, a, b, 1\}$ be the Boolean algebra with four elements and $A = \{x, y\}$ be a set with two elements. Then we have $| B_4^A |= 4^2 = 16$, i.e., 16 L-subsets of A. The relation ε is given by

$$\varepsilon = \begin{bmatrix} 0\,a\,b\,1\,0\,a\,b\,1\,0\,a\,b\,1\,0\,a\,b\,1 \\ 0\,0\,0\,0\,a\,a\,a\,a\,b\,b\,b\,b\,1\,1\,1\,1 \end{bmatrix}.$$

If we compute $\mathrm{syQ}(\varepsilon, \varepsilon)$ we obtain

$$\mathrm{syQ}(\varepsilon, \varepsilon) = \begin{bmatrix}
1 & b & a & 0 & b & b & 0 & 0 & a & 0 & a & 0 & 0 & 0 & 0 & 0 \\
b & 1 & 0 & a & b & b & 0 & 0 & 0 & a & 0 & a & 0 & 0 & 0 & 0 \\
a & 0 & 1 & b & 0 & 0 & b & b & a & 0 & a & 0 & 0 & 0 & 0 & 0 \\
0 & a & b & 1 & 0 & 0 & b & b & 0 & a & 0 & a & 0 & 0 & 0 & 0 \\
b & b & 0 & 0 & 1 & b & a & 0 & 0 & 0 & 0 & 0 & a & 0 & a & 0 \\
b & b & 0 & 0 & b & 1 & 0 & a & 0 & 0 & 0 & 0 & 0 & a & 0 & a \\
0 & 0 & b & b & a & 0 & 1 & b & 0 & 0 & 0 & 0 & a & 0 & a & 0 \\
0 & 0 & b & b & 0 & a & b & 1 & 0 & 0 & 0 & 0 & 0 & a & 0 & a \\
a & 0 & a & 0 & 0 & 0 & 0 & 0 & 1 & b & a & 0 & b & b & 0 & 0 \\
0 & a & 0 & a & 0 & 0 & 0 & 0 & b & 1 & 0 & a & b & b & 0 & 0 \\
a & 0 & a & 0 & 0 & 0 & 0 & 0 & a & 0 & 1 & b & 0 & 0 & b & b \\
0 & a & 0 & a & 0 & 0 & 0 & 0 & 0 & a & b & 1 & 0 & 0 & b & b \\
0 & 0 & 0 & 0 & a & 0 & a & 0 & b & b & 0 & 0 & 1 & b & a & 0 \\
0 & 0 & 0 & 0 & 0 & a & 0 & a & b & b & 0 & 0 & b & 1 & 0 & a \\
0 & 0 & 0 & 0 & a & 0 & a & 0 & 0 & 0 & b & b & a & 0 & 1 & b \\
0 & 0 & 0 & 0 & 0 & a & 0 & a & 0 & 0 & b & b & 0 & a & b & 1
\end{bmatrix}.$$

The relation above is obviously different from the identity. This relation indicates that certain B_4-sets are indeed equal up to a certain degree. For example, the sets $s_1(x) = b$, $s_1(y) = 0$ and $s_2(x) = 1$, $s_2(y) = a$ are equal up to degree b because for every element z we have $s_1(z) \sqcap b = s_2(z) \sqcap b$, i.e., every element is in s_1 with the same degree (up to b) as in s_2.

This leads to an alternative definition of a power capturing the notion of an L-power set.

Definition 8. *An object $\mathcal{P}_L(A)$ together with a relation $\epsilon : A \to \mathcal{P}_L(A)$ is called a fuzzy relational power iff*

$$\mathrm{syQ}(\epsilon, \epsilon)^{\downarrow} = \mathbb{I}_{\mathcal{P}_L(A)} \quad \textit{and} \quad \mathrm{syQ}(R, \epsilon)^{\downarrow} \textit{ is total for every } R : A \to B.$$

First of all, the definition above provides a unique concept.

Lemma 4. *Let $\mathcal{A}$ be an arrow category. Then the fuzzy relational power is unique up to isomorphism.*

Proof. Suppose $\mathcal{P}'_L(A)$ and ϵ' is another fuzzy relational power of A. From

$$\begin{aligned}
\mathrm{syQ}(\epsilon', \epsilon)^{\downarrow}; \mathrm{syQ}(\epsilon, \epsilon')^{\downarrow} &\sqsubseteq (\mathrm{syQ}(\epsilon', \epsilon); \mathrm{syQ}(\epsilon, \epsilon'))^{\downarrow} \\
&= \mathrm{syQ}(\epsilon', \epsilon')^{\downarrow} \\
&= \mathbb{I}_{\mathcal{P}'_L(A)}
\end{aligned}$$

and the fact that $\mathrm{syQ}(\epsilon', \epsilon)^{\downarrow}$ and $\mathrm{syQ}(\epsilon, \epsilon')^{\downarrow}$ are total by definition we conclude = by Lemma 1(4). The second equation $\mathrm{syQ}(\epsilon, \epsilon')^{\downarrow}; \mathrm{syQ}(\epsilon', \epsilon)^{\downarrow} = \mathbb{I}_{\mathcal{P}_L(A)}$ follows analogously. □

As already indicated by the example above, the L-power set is an example of a fuzzy relational power.

Lemma 5. *The L-power set together with the relation* $\epsilon(x, f) = f(x)$ *is a fuzzy relational power.*

Proof. First of all, we have for every $R : A \to B$

$$\begin{aligned}
&\mathrm{syQ}(R, \epsilon)^{\downarrow}(y, g) = 1 \\
&\Leftrightarrow \mathrm{syQ}(R, \epsilon)(y, g) = 1 \\
&\Leftrightarrow (\bigsqcap_{x \in A} R(x, y) \to \epsilon(x, g)) \sqcap (\bigsqcap_{x \in A} \epsilon(x, g) \to R(x, y)) = 1 \\
&\Leftrightarrow \bigsqcap_{x \in A} R(x, y) \to \epsilon(x, g) = 1 \text{ and } \bigsqcap_{x \in A} \epsilon(x, g) \to R(x, y) = 1 \\
&\Leftrightarrow R(x, y) \to \epsilon(x, g) = 1 \text{ and } \epsilon(x, g) \to R(x, y) = 1 && \text{for all } x \in A \\
&\Leftrightarrow R(x, y) \sqsubseteq \epsilon(x, g) \text{ and } \epsilon(x, g) \sqsubseteq R(x, y) && \text{for all } x \in A \\
&\Leftrightarrow R(x, y) = \epsilon(x, g) && \text{for all } x \in A \\
&\Leftrightarrow R(x, y) = g(x) && \text{for all } x \in A.
\end{aligned}$$

This implies

$$\begin{aligned}
\mathrm{syQ}(\epsilon, \epsilon)^{\downarrow}(f, g) = 1 &\Leftrightarrow \epsilon(x, f) = g(x) && \text{for all } x \in A \\
&\Leftrightarrow f(x) = g(x) && \text{for all } x \in A \\
&\Leftrightarrow f = g \\
&\Leftrightarrow \mathbb{I}_{\mathcal{P}(A)}(f, g) = 1.
\end{aligned}$$

Furthermore, it implies that $\mathrm{syQ}(R, \epsilon)^{\downarrow}(y, g) = 1$ iff $g(x) = R(x, y)$ for all x (or $g = R(\_, y)$), i.e., g is the L-set that is related to y by R. Since there is exactly one such set for every y in $\mathcal{P}^{\downarrow}(A)$ we conclude that $\mathrm{syQ}(R, \epsilon)^{\downarrow}$ is a mapping. □

One can obtain a relational power from the set of all L-subsets by using the (L-fuzzy) equivalence classes of the relation $\mathrm{syQ}(\varepsilon, \varepsilon)$. However, notice that this idea uses relations based on different membership values. In fact, each cell in a matrix representation may use a different lattice for the coefficients of the matrix. This corresponds to a variable basis approach. Since arrow categories emphasize a fixed basis approach, i.e., all relations use the same lattice L for membership, we have to state the following theorem in the broader context of Dedekind categories with cutoff operators. For the more general definition and the concept of splittings we refer to [3].

Theorem 5. *Let* $\mathcal{D}$ *be a Dedekind category with cutoff operator and fuzzy relational powers. Then we have the following:*

1. *If* $\mathcal{P}(A)$ *is a relational power, then the crisp function* $\mathrm{syQ}(\varepsilon, \epsilon)^{\downarrow} : \mathcal{P}(A) \to \mathcal{P}_L(A)$ *is injective.*
2. *If* $R : C \to \mathcal{P}_L(A)$ *splits* $\mathrm{syQ}(\epsilon, \epsilon)$*, then* C *together with* $\epsilon; R^{\smile}$ *is a relational power of* A.

Proof. 1. By the definition of a crisp relational power the relation $\mathrm{syQ}(\varepsilon, \epsilon)^{\downarrow}$ is a map. Furthermore, we have $\mathrm{syQ}(\varepsilon, \epsilon)^{\downarrow\smile} \sqsubseteq \mathrm{syQ}(\varepsilon, \epsilon)^{\smile} = \mathrm{syQ}(\epsilon, \varepsilon)$. Since $\mathrm{syQ}(\epsilon, \varepsilon)$ is univalent we obtain that $\mathrm{syQ}(\varepsilon, \epsilon)^{\downarrow}$ is injective.

2. First of all, we have

$$\begin{aligned} \mathrm{syQ}(Q, \epsilon; R^{\smile}) &= \mathrm{syQ}(Q, \epsilon; R^{\smile}); R; R^{\smile} && R \text{ is a splitting} \\ &= \mathrm{syQ}(Q, \epsilon; R^{\smile}; R); R^{\smile} && R^{\smile} \text{ mapping} \\ &= \mathrm{syQ}(Q, \epsilon; \mathrm{syQ}(\epsilon, \epsilon)); R^{\smile} && R \text{ splits } \mathrm{syQ}(\epsilon, \epsilon) \\ &= \mathrm{syQ}(Q, \epsilon); R^{\smile}. \end{aligned}$$

This implies that $\mathrm{syQ}(Q, \epsilon; R^{\smile})$ is total because $\mathrm{syQ}(Q, \epsilon)$ and $R^{\smile}$ are. From the computation

$$\begin{aligned} \mathrm{syQ}(\epsilon; R^{\smile}, \epsilon; R^{\smile}) &= R; R^{\smile}; \mathrm{syQ}(\epsilon; R^{\smile}, \epsilon; R^{\smile}); R; R^{\smile} && R \text{ is a splitting} \\ &= R; \mathrm{syQ}(\epsilon; R^{\smile}; R, \epsilon; R^{\smile}; R); R^{\smile} && R^{\smile} \text{ mapping} \\ &= R; \mathrm{syQ}(\epsilon; \mathrm{syQ}(\epsilon, \epsilon), \epsilon; \mathrm{syQ}(\epsilon, \epsilon)); R^{\smile} && R \text{ splits } \mathrm{syQ}(\epsilon, \epsilon) \\ &= R; \mathrm{syQ}(\epsilon, \epsilon); R^{\smile} \\ &= R; R^{\smile}; R; R^{\smile} && R \text{ splits } \mathrm{syQ}(\epsilon, \epsilon) \\ &= \mathbb{I}_C && R \text{ is a splitting} \end{aligned}$$

we obtain the second property of a relational power. □

4 Replacing the Extension

In this section we want to show that the monad used in the construction of a higher-order arrow category can be defined without the ad-hoc concept of an extension of an object. As already stated as Theorem 1 it was shown in [17] that $A^{\sharp}$ is isomorphic to the crisp relational product $A \times 1^{\sharp}$. We now want to show that $1^{\sharp}$ is actually the fuzzy power of 1.

Lemma 6. *Let* $\mathcal{A}$ *be an arrow category with a unit and extensions. Then* $1^{\sharp}$ *together with the relation* $\nu_1 : 1 \to 1^{\sharp}$ *is a fuzzy power of* 1.

Proof. Suppose $Q : 1 \to A$. Then we have

$$\begin{aligned} \mathrm{syQ}(Q, \nu_1)^{\downarrow} &= \mathrm{syQ}(\pi_{11}; Q, \pi_{11}; \nu_1)^{\downarrow} && \mathbb{I}_1 = \pi_{11} \\ &= (Q^{\smile}; \pi_{11})^{\sharp} && \text{Lemma 3(2)} \\ &= Q^{\smile\sharp}. && \mathbb{I}_1 = \pi_{11} \end{aligned}$$

By Lemma 3(4) the relation $Q^{\smile\sharp}$ is total, and for $Q = \nu_1$ we obtain $\mathrm{syQ}(\nu_1, \nu_1)^{\downarrow} = \nu_1^{\smile\sharp} = \mathbb{I}_{1^{\sharp}}$ using Lemma 3(5). □

Note that the previous lemma together with Theorem 1 shows that $A^\sharp$ is isomorphic to the crisp relational product $A \times \mathcal{P}_L(1)$. More precisely, it shows that if $A^\sharp$ exists, then the constructions on the right-hand side exist and the two objects are isomorphic. We now want to verify that opposite statement.

Theorem 6. *Let $\mathcal{A}$ be an arrow category with a unit, crisp relational products and fuzzy relational powers. Then the object $A \times \mathcal{P}_L(1)$ together with the relations $\eta_A := \pi^\smile$ and $\nu_A := \top\!\!\top_{A1};\epsilon;\rho^\smile$ is an extension of A.*

Proof. 1. η_A is crisp by definition.
2. We have $\top\!\!\top_{AA};\nu_A = \top\!\!\top_{AA};\top\!\!\top_{A1};\epsilon;\rho^\smile = \top\!\!\top_{A1};\epsilon;\rho^\smile = \nu_A$ since $\mathcal{A}$ is uniform.
3. As shown after Definition 5 π is surjective so that we conclude $\eta_A;\eta_A^\smile = \pi^\smile;\pi = \mathbb{I}_A$.
4. We compute

$$
\begin{aligned}
&\nu_A^{\smile\sharp} \sqcap \eta_A^\smile;\eta_A \\
&= (\rho;\epsilon^\smile;\top\!\!\top_{1A})^\sharp \sqcap \pi;\pi^\smile \\
&= \rho;(\epsilon^\smile;\top\!\!\top_{1A})^\sharp \sqcap \pi;\pi^\smile && \text{Lemma 3(3)} \\
&= \rho;(\epsilon^\smile;\top\!\!\top_{1A};\eta_A \leftrightarrow \top\!\!\top_{\mathcal{P}_L(1)A};\nu_A)^\downarrow \sqcap \pi;\pi^\smile \\
&= \rho;(\epsilon^\smile;\top\!\!\top_{1A};\pi^\smile \leftrightarrow \top\!\!\top_{\mathcal{P}_L(1)A};\top\!\!\top_{A1};\epsilon;\rho^\smile)^\downarrow \sqcap \pi;\pi^\smile \\
&= \rho;(\epsilon^\smile;\top\!\!\top_{1A\times\mathcal{P}_L(1)} \leftrightarrow \top\!\!\top_{\mathcal{P}_L(1)1};\epsilon;\rho^\smile)^\downarrow \sqcap \pi;\pi^\smile && \pi \text{ total and } \mathcal{A} \text{ uniform} \\
&= \rho;(\epsilon^\smile;\top\!\!\top_{1\mathcal{P}_L(1)};\rho^\smile \leftrightarrow \top\!\!\top_{\mathcal{P}_L(1)1};\epsilon;\rho^\smile)^\downarrow \sqcap \pi;\pi^\smile && \rho \text{ total} \\
&= \rho;(\epsilon^\smile;\top\!\!\top_{1\mathcal{P}_L(1)} \leftrightarrow \top\!\!\top_{\mathcal{P}_L(1)1};\epsilon)^\downarrow;\rho^\smile \sqcap \pi;\pi^\smile && \text{Lemma 3(3)} \\
&= \rho;\mathrm{syQ}(\top\!\!\top_{\mathcal{P}_L(1)1};\epsilon,\top\!\!\top_{\mathcal{P}_L(1)1};\epsilon)^\downarrow;\rho^\smile \sqcap \pi;\pi^\smile && \text{Lemma 3(2)} \\
&= \rho;\mathrm{syQ}(\top\!\!\top_{11};\epsilon,\top\!\!\top_{11};\epsilon)^\downarrow;\rho^\smile \sqcap \pi;\pi^\smile && \text{uniform} \\
&= \rho;\mathrm{syQ}(\epsilon,\epsilon)^\downarrow;\rho^\smile \sqcap \pi;\pi^\smile \\
&= \rho;\rho^\smile \sqcap \pi;\pi^\smile \\
&= \mathbb{I}_{A\times\mathcal{P}_L(1)}.
\end{aligned}
$$

5. We want to show first that $\pi^\smile;(\rho \olessthan \mathrm{syQ}(\top\!\!\top_{1A};(Q^\smile \olessthan \mathbb{I}_A),\epsilon)^\downarrow) \sqsubseteq Q;\pi^\smile \leftrightarrow \top\!\!\top_{B1};\epsilon;\rho^\smile$. Using the notation $X := \rho \olessthan \mathrm{syQ}(\top\!\!\top_{1A};(Q^\smile \olessthan \mathbb{I}_A),\epsilon)^\downarrow$ it will be sufficient to show the two inclusions $\pi^\smile;X \sqcap Q;\pi^\smile \sqsubseteq \top\!\!\top_{B1};\epsilon;\rho^\smile$ and $\pi^\smile;X \sqcap \top\!\!\top_{B1};\epsilon;\rho^\smile \sqsubseteq Q;\pi^\smile$. For that we have

$$
\begin{aligned}
&(\pi^\smile;X \sqcap Q;\pi^\smile);\rho \\
&\sqsubseteq Q;(Q^\smile;\pi^\smile;X \sqcap \pi^\smile);\rho \\
&\sqsubseteq Q;(Q^\smile;\pi^\smile;\sqcap\pi^\smile;X^\smile);X;\rho \\
&= Q;(Q^\smile;\pi^\smile;\sqcap\pi^\smile;X^\smile);\mathrm{syQ}(\top\!\!\top_{1A};(Q^\smile \olessthan \mathbb{I}_A),\epsilon)^\downarrow && \text{product property} \\
&\sqsubseteq \top\!\!\top_{BA};(Q^\smile;\pi^\smile;\sqcap\rho^\smile);\mathrm{syQ}(\top\!\!\top_{1A};(Q^\smile \olessthan \mathbb{I}_A),\epsilon)^\downarrow \\
&= \top\!\!\top_{B1};\top\!\!\top_{1A};(Q^\smile \olessthan \mathbb{I}_A);\mathrm{syQ}(\top\!\!\top_{1A};(Q^\smile \olessthan \mathbb{I}_A),\epsilon)^\downarrow && \mathcal{A} \text{ uniform} \\
&\sqsubseteq \top\!\!\top_{B1};\epsilon,
\end{aligned}
$$

which immediately implies the first inclusion by Lemma 1(1). Now, consider

$$\begin{aligned}
&(\pi^{\smile};X \sqcap \mathbb{T}_{B1};\epsilon;\rho^{\smile});\pi \\
&\sqsubseteq (\pi^{\smile} \sqcap \mathbb{T}_{B1};\epsilon;\mathrm{syQ}(\mathbb{T}_{1A};(Q^{\smile} \otimes \mathbb{I}_A),\epsilon)^{\downarrow\smile});X;\pi && \text{product property}\\
&= (\pi^{\smile} \sqcap \mathbb{T}_{B1};\epsilon;\mathrm{syQ}(\epsilon,\mathbb{T}_{1A};(Q^{\smile} \otimes \mathbb{I}_A))^{\downarrow});X;\pi \\
&\sqsubseteq (\pi^{\smile} \sqcap \mathbb{T}_{B1};\mathbb{T}_{1A};(Q^{\smile} \otimes \mathbb{I}_A));X;\pi \\
&= (\pi^{\smile} \sqcap \mathbb{T}_{BA};(Q^{\smile} \otimes \mathbb{I}_A));X;\pi && \mathcal{A}\text{ uniform}\\
&\sqsubseteq (\pi^{\smile} \sqcap \mathbb{T}_{BA};(Q^{\smile} \otimes \mathbb{I}_A));\rho \\
&= (\pi^{\smile} \sqcap \mathbb{T}_{BA};(Q^{\smile};\pi^{\smile} \sqcap \rho^{\smile}));\rho \\
&= (\pi^{\smile} \sqcap \mathbb{T}_{BB};(\pi^{\smile} \sqcap Q;\rho^{\smile}));\rho && \text{[13] Lemma 2.2.4(3)}\\
&= (\pi^{\smile};(\pi^{\smile} \sqcap Q;\rho^{\smile})^{\smile} \sqcap \mathbb{T}_{BB});(\pi^{\smile} \sqcap Q;\rho^{\smile});\rho && \text{Lemma 1(3)}\\
&= (\pi^{\smile};(\pi \sqcap \rho;Q^{\smile}));Q \\
&\sqsubseteq Q,
\end{aligned}$$

which immediately implies the second inclusion by Lemma 1(1). We conclude

$$\begin{aligned}
\mathbb{T}_{BA} &= \pi^{\smile};\rho \\
&= \pi^{\smile};(\mathrm{syQ}(\mathbb{T}_{1A};(Q^{\smile} \otimes \mathbb{I}_A),\epsilon)^{\downarrow};\mathbb{T}_{\mathcal{P}_L(1)A} \sqcap \rho) && \text{Definition } \epsilon\\
&= \pi^{\smile};(\mathrm{syQ}(\mathbb{T}_{1A};(Q^{\smile} \otimes \mathbb{I}_A),\epsilon)^{\downarrow};\rho^{\smile};\pi \sqcap \rho) \\
&= \pi^{\smile};(\mathrm{syQ}(\mathbb{T}_{1A};(Q^{\smile} \otimes \mathbb{I}_A),\epsilon)^{\downarrow};\rho^{\smile} \sqcap \rho;\pi^{\smile});\pi && \text{Lemma 1(3)}\\
&= \pi^{\smile};(\rho \otimes \mathrm{syQ}(\mathbb{T}_{1A};(Q^{\smile} \otimes \mathbb{I}_A),\epsilon)^{\downarrow});\pi \\
&= (\pi^{\smile};(\rho \otimes \mathrm{syQ}(\mathbb{T}_{1A};(Q^{\smile} \otimes \mathbb{I}_A),\epsilon)^{\downarrow}))^{\downarrow};\pi && \text{crisp}\\
&\sqsubseteq (Q;\pi^{\smile} \leftrightarrow \mathbb{T}_{B1};\epsilon;\rho^{\smile})^{\downarrow};\pi && \text{see above}\\
&= (Q;\pi^{\smile} \leftrightarrow \mathbb{T}_{BA};\mathbb{T}_{A1};\epsilon;\rho^{\smile})^{\downarrow};\pi && \mathcal{A}\text{ uniform}\\
&= (Q;\eta_A \leftrightarrow \mathbb{T}_{BA};\mu_A)^{\downarrow};\eta_A^{\smile} \\
&= Q^{\sharp};\eta_A^{\smile}.
\end{aligned}$$

This completes the proof. □

5 Iterating Higher-Order Fuzziness

If $\mathcal{A}$ is an arrow category with a unit, crisp relational products and fuzzy relational powers, then we can move to the Kleisli category $\mathcal{A}_P$. This category is actually an arrow category with the arrow operations (see [18]) defined by $Q^{\Uparrow} := (Q;\pi;\pi^{\smile})^{\uparrow}$ and $Q^{\Downarrow} := (Q/\pi;\pi^{\smile})^{\downarrow}$. In order to iterate this process we have to show that $\mathcal{A}_P$ again has a unit, crisp relational products and fuzzy relational powers.

Lemma 7. *Let $\mathcal{A}$ be an arrow category with a unit, crisp relational products and fuzzy relational powers. Then* 1 *is a unit in* $\mathcal{A}_P$.

Proof. First, we want to show that $\mathbb{T}_{P(1)1} = \pi$. Since π is total, we have $\mathbb{T}_{P(1)1} = \mathbb{I}_{P(1)};\mathbb{T}_{P(1)1} \sqsubseteq \pi;\pi^{\smile};\mathbb{T}_{P(1)1} \sqsubseteq \pi;\mathbb{T}_{11} = \pi;\mathbb{I}_1 = \pi$. From this we immediately conclude that $\mathbb{I}_1 = \mathbb{T}_{11}$ in $\mathcal{A}_P$. In order to show that the universal relation from A to the unit 1 in $\mathcal{A}_P$ is total, we compute

$$
\begin{aligned}
\eta_A &= \pi^{\smile} && \\
&\sqsubseteq \mathbb{T}_{AP(A)} && \\
&= \mathbb{T}_{A\mathcal{P}_L(1)};\rho^{\smile} && \rho \text{ total} \\
&= \mathbb{T}_{AP(1)};\rho;\rho^{\smile} && \rho \text{ surjective} \\
&= \mathbb{T}_{AP(1)};(\mathbb{T}_{1P(A)} \oslash \rho^{\smile}) && \pi \text{ total} \\
&= \mathbb{T}_{AP(1)};(\mathbb{T}_{1\mathcal{P}_L(1)};\rho^{\smile} \oslash \rho^{\smile}) && \rho \text{ total} \\
&= \mathbb{T}_{AP(1)};(\pi^{\smile};\rho;\rho^{\smile} \oslash \rho^{\smile}) && \\
&= \mathbb{T}_{AP(1)};(\pi^{\smile};(\mathbb{T}_{P(1)A} \oslash \rho) \oslash \rho^{\smile}) && \pi \text{ total} \\
&= \mathbb{T}_{AP(1)};(\mathbb{T}^{\cup}_{AP(1)} \oslash \rho^{\smile}) && \\
&= \mathbb{T}_{AP(1)} \mathbin{\mathring{;}} \mathbb{T}^{\cup}_{AP(1)}. &&
\end{aligned}
$$

This completes the proof. □

We want to show that $A \times B$, i.e., the crisp relation product in $\mathcal{A}$, is also a crisp relational product in $\mathcal{A}_P$. The corresponding projections have to be relations from $\mathcal{A}$ with source $A \times B$ and target $P(A)$ resp. $P(B)$. Therefore, we define $p_1 := \pi;\pi^{\smile}$ and $p_2 := \rho;\pi^{\smile}$.

Theorem 7. *Let $\mathcal{A}$ be an arrow category with a unit, crisp relational products and fuzzy relational powers. Then $(A \times B, p_1, p_2)$ is a crisp relational product of A and B in the Kleisli category $\mathcal{A}_P$.*

Proof. First of all, p_1 and p_2 are crisp since the projections π and ρ are crisp and the class of crisp relations is closed under all relational operations.

$$
\begin{aligned}
p_1^{\cup} \mathbin{\mathring{;}} p_1 &= \pi^{\smile};(p_1^{\smile} \oslash \rho);(p_1 \oslash \rho^{\smile}) && \\
&= \pi^{\smile};(p_1^{\smile};p_1 \sqcap \rho;\rho^{\smile}) && \text{product property} \\
&= \pi^{\smile};(\pi;\pi^{\smile};\pi;\pi^{\smile} \sqcap \rho;\rho^{\smile}) && \\
&= \pi^{\smile};(\pi;\pi^{\smile} \sqcap \rho;\rho^{\smile}) && \pi \text{ univalent} \\
&= \pi^{\smile} && \\
&= \eta_A &&
\end{aligned}
$$

$p_2^{\cup} \mathbin{\mathring{;}} p_2 = \eta_B$ follows analogously. Furthermore, we have

$$
\begin{aligned}
p_1^{\cup} \mathbin{\mathring{;}} p_2 &= \pi^{\smile};(p_1^{\smile} \oslash \rho);(p_1 \oslash \rho^{\smile}) && \\
&= \pi^{\smile};(p_1^{\smile};p_2 \sqcap \rho;\rho^{\smile}) && \text{product property} \\
&= \pi^{\smile};(\pi;\pi^{\smile};\rho;\pi^{\smile} \sqcap \rho;\rho^{\smile}) && \\
&= \pi^{\smile};(\pi;\mathbb{T}_{AB};\pi^{\smile} \sqcap \rho;\rho^{\smile}) &&
\end{aligned}
$$

$$
\begin{aligned}
&= \pi^{\smile};\rho;\rho^{\smile} && \pi \text{ total}\\
&= \mathbb{T}_{AB};\rho^{\smile} && \\
&= \mathbb{T}_{AP(B)}. && \rho \text{ total}
\end{aligned}
$$

Finally, from the computation

$$
\begin{aligned}
p_1 \fatsemi p_1^{\cup} &= p_1;(p_1^{\cup} \oslash \rho^{\smile}) && \\
&= p_1;(\pi^{\smile};(p_1^{\smile} \oslash \rho) \oslash \rho^{\smile}) && \\
&= p_1;(\pi^{\smile};(\pi;\pi^{\smile} \oslash \rho) \oslash \rho^{\smile}) && \\
&= p_1;(\pi^{\smile};(\pi^{\smile} \otimes \mathbb{I}) \oslash \rho^{\smile}) && \\
&= p_1;(\pi^{\smile};\pi^{\smile} \oslash \rho^{\smile}) && \text{product property}\\
&= \pi;\pi^{\smile};(\pi^{\smile};\pi^{\smile} \oslash \rho^{\smile}) && \\
&= \pi;\pi^{\smile};\pi^{\smile} && \text{product property}
\end{aligned}
$$

and a similar computation showing $p_2 \fatsemi p_2^{\cup} = \rho;\rho^{\smile};\pi^{\smile}$ we obtain $p_1 \fatsemi p_1^{\cup} \sqcap p_2 \fatsemi p_2^{\cup} = \pi;\pi^{\smile};\pi^{\smile} \sqcap \rho;\rho^{\smile};\pi^{\smile} = (\pi;\pi^{\smile} \sqcap \rho;\rho^{\smile});\pi^{\smile} = \pi^{\smile} = \eta_{A\times B}$. □

The last theorem of this paper will that the object $\mathcal{P}_L(A \times \mathcal{P}_L(1)) \times \mathcal{P}_L(1)$ together with the relation $\overrightarrow{\epsilon} : A \to \mathcal{P}_L(A \times \mathcal{P}_L(1)) \times \mathcal{P}_L(1)$ is a fuzzy power in $\mathcal{A}_P$. The following diagram visualizes the situation when forming $\mathrm{syq}_P(Q, \overrightarrow{\epsilon})^{\Downarrow}$ in $\mathcal{A}_P$.

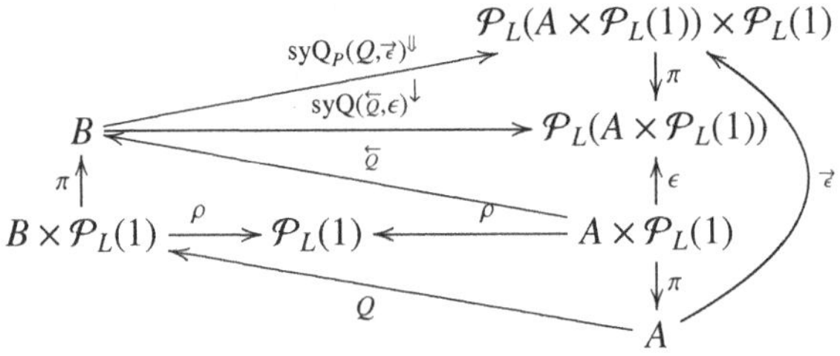

Theorem 8. *Let $\mathcal{A}$ be an arrow category with a unit, crisp relational products and fuzzy relational powers. Then $\mathcal{P}_L(A \times \mathcal{P}_L(1)) \times \mathcal{P}_L(1)$ together with the relation $\overrightarrow{\epsilon}$: $A \to \mathcal{P}_L(A \times \mathcal{P}_L(1)) \times \mathcal{P}_L(1)$ is a fuzzy power in $\mathcal{A}_P$*

Proof. First, we compute the following

$$
\begin{aligned}
Q \fatsemi X^{\Uparrow} \sqsubseteq \overrightarrow{\epsilon} &\Leftrightarrow Q;(X^{\Uparrow} \oslash \rho^{\smile}) \sqsubseteq \overrightarrow{\epsilon} && \\
&\Leftrightarrow Q;(X^{\uparrow};\pi;\pi^{\smile} \oslash \rho^{\smile}) \sqsubseteq \overrightarrow{\epsilon} && \\
&\Leftrightarrow Q;(X^{\uparrow};\pi \otimes \mathbb{I}) \sqsubseteq \overrightarrow{\epsilon} && \\
&\Leftrightarrow \overleftarrow{Q;(X^{\uparrow};\pi \otimes \mathbb{I})} \sqsubseteq \epsilon && \text{shifting property}\\
&\Leftrightarrow (Q;(X^{\uparrow};\pi \otimes \mathbb{I}) \oslash \rho^{\smile});\pi \sqsubseteq \epsilon && \\
&\Leftrightarrow (Q;\oslash\rho^{\smile});\pi;X^{\uparrow};\pi \sqsubseteq \epsilon && \\
&\Leftrightarrow \overleftarrow{Q};X^{\uparrow};\pi \sqsubseteq \epsilon &&
\end{aligned}
$$

$$\Leftrightarrow \overleftarrow{Q}; X^{\uparrow} \sqsubseteq \epsilon; \pi^{\smile}, \qquad \text{Lemma 1(1)}$$

$$\begin{aligned}
\overrightarrow{\epsilon} \mathbin{⨾} X^{\Uparrow\cup} \sqsubseteq Q &\Leftrightarrow \overrightarrow{\epsilon}; (X^{\Uparrow\cup} \oslash \rho^{\smile}) \sqsubseteq Q \\
&\Leftrightarrow \overrightarrow{\epsilon}; (\pi^{\smile}; ((\pi; \pi^{\smile}; X^{\smile})^{\uparrow} \oslash \rho) \oslash \rho^{\smile}) \sqsubseteq Q \\
&\Leftrightarrow \overrightarrow{\epsilon}; (\pi^{\smile}; (\pi; \pi^{\smile}; X^{\uparrow\smile} \oslash \rho) \oslash \rho^{\smile}) \sqsubseteq Q \\
&\Leftrightarrow \overrightarrow{\epsilon}; (\pi^{\smile}; (\pi^{\smile}; X^{\uparrow\smile} \otimes \mathbb{I}) \oslash \rho^{\smile}) \sqsubseteq Q \\
&\Leftrightarrow \overrightarrow{\epsilon}; (\pi^{\smile}; X^{\uparrow\smile} \otimes \mathbb{I}) \sqsubseteq Q && \text{product property} \\
&\Leftrightarrow \pi^{\smile}; (\epsilon \oslash \rho); (\pi^{\smile}; X^{\uparrow\smile} \otimes \mathbb{I}) \sqsubseteq Q \\
&\Leftrightarrow \pi^{\smile}; (\epsilon; \pi^{\smile}; X^{\uparrow\smile} \oslash \rho) \sqsubseteq Q \\
&\Leftrightarrow \overrightarrow{\epsilon; \pi^{\smile}; X^{\uparrow\smile}} \sqsubseteq Q \\
&\Leftrightarrow \epsilon; \pi^{\smile}; X^{\uparrow\smile} \sqsubseteq \overleftarrow{Q}.
\end{aligned}$$

This immediately implies

$$\begin{aligned}
X \sqsubseteq \mathrm{syQ}_P(Q, \overrightarrow{\epsilon})^{\Downarrow} &\Leftrightarrow X^{\Uparrow} \sqsubseteq \mathrm{syQ}_P(Q, \overrightarrow{\epsilon}) && \text{arrows in } \mathcal{A}_P \\
&\Leftrightarrow Q \mathbin{⨾} X^{\Uparrow} \sqsubseteq \overrightarrow{\epsilon} \text{ and } \overrightarrow{\epsilon} \mathbin{⨾} X^{\Uparrow\cup} \sqsubseteq Q && \text{syQ in } \mathcal{A}_P \\
&\Leftrightarrow \overleftarrow{Q}; X^{\uparrow} \sqsubseteq \epsilon; \pi^{\smile} \text{ and } \epsilon; \pi^{\smile}; X^{\uparrow\smile} \sqsubseteq \overleftarrow{Q} && \text{see above} \\
&\Leftrightarrow X^{\uparrow} \sqsubseteq \mathrm{syQ}(\overleftarrow{Q}, \epsilon; \pi^{\smile}) && \text{syQ in } \mathcal{A} \\
&\Leftrightarrow X \sqsubseteq \mathrm{syQ}(\overleftarrow{Q}, \epsilon; \pi^{\smile})^{\downarrow} && \text{arrows in } \mathcal{A} \\
&\Leftrightarrow X \sqsubseteq (\mathrm{syQ}(\overleftarrow{Q}, \epsilon); \pi^{\smile})^{\downarrow} && \pi \text{ univalent} \\
&\Leftrightarrow X \sqsubseteq \mathrm{syQ}(\overleftarrow{Q}, \epsilon)^{\downarrow}; \pi^{\smile}. && \text{Lemma 2(4)}
\end{aligned}$$

from which we conclude $\mathrm{syQ}_P(Q, \overrightarrow{\epsilon})^{\Downarrow} = \mathrm{syQ}(\overleftarrow{Q}, \epsilon)^{\downarrow}; \pi^{\smile}$. We obtain

$$\begin{aligned}
\mathrm{syQ}_P(\overrightarrow{\epsilon}, \overrightarrow{\epsilon})^{\Downarrow} &= \mathrm{syQ}(\overleftarrow{\overrightarrow{\epsilon}}, \epsilon)^{\downarrow}; \pi^{\smile} && \text{see above} \\
&= \mathrm{syQ}(\epsilon, \epsilon)^{\downarrow}; \pi^{\smile} && \text{shifting property} \\
&= \pi^{\smile} && \text{fuzzy power in } \mathcal{A} \\
&= \eta
\end{aligned}$$

Furthermore, we get

$$\begin{aligned}
\mathrm{syQ}_P(Q, \overrightarrow{\epsilon})^{\Downarrow} \mathbin{⨾} \mathbb{T} &= \mathrm{syQ}_P(Q, \overrightarrow{\epsilon})^{\Downarrow}; (\mathbb{T} \oslash \rho^{\smile}) \\
&= \mathrm{syQ}_P(Q, \overrightarrow{\epsilon})^{\Downarrow}; \rho; \rho^{\smile} && \pi \text{ total} \\
&= \mathrm{syQ}(\overleftarrow{Q}, \epsilon)^{\downarrow}; \pi^{\smile}; \rho; \rho^{\smile} && \text{see above} \\
&= \mathrm{syQ}(\overleftarrow{Q}, \epsilon)^{\downarrow}; \mathbb{T}; \rho^{\smile} \\
&= \mathbb{T}; \rho^{\smile} && \text{fuzzy power in } \mathcal{A} \\
&= \mathbb{T}, && \rho \text{ total}
\end{aligned}$$

i.e., that $\mathrm{syQ}_P(Q, \overrightarrow{\epsilon})^{\Downarrow}$ is total in $\mathcal{A}_P$. □

6 Conclusion and Future Work

In this paper we have provided a general framework for handling and iterating higher-order fuzziness in arrow categories. However, this theory is based on multiple arrow categories, i.e., one arrow category for each type n. The reason is that all relations in an arrow category use the same underlying lattice for their degree of membership. In future work we would like to consider a similar construction within the framework of weak arrow categories [19]. These categories allow different membership values for relation within one category.

References

1. Freyd, P., Scedrov, A.: Categories, Allegories, vol. 39. North-Holland Mathematical Library, Amsterdam (1990)
2. Furusawa, H.: Algebraic formalizations of fuzzy relations and their representation theorems. Ph.D.-thesis, Department of Informatics, Kyushu University, Japan (1998)
3. Furusawa, H., Kawahara, Y., Winter, M.: Dedekind categories with cutoff operators. Fuzzy Sets Syst. **173**, 1–24 (2011)
4. Jónsson, B., Tarski, A.: Boolean algebras with operators, I, II. Amer. J. Math. **73**, 891–939 (1951). **74**, 127–162 (1952)
5. Kawahara, Y., Furusawa, H.: Crispness and representation theorems in dedekind categories. DOI-TR 143, Kyushu University (1997)
6. Kawahara, Y., Furusawa, H.: An algebraic formalization of fuzzy relations. Fuzzy Sets Syst. **101**, 125–135 (1999)
7. Olivier, J.P., Serrato, D.: Catégories de Dedekind. Morphismes dans les Catégories de Schröder. C.R. Acad. Sci. Paris **290**, 939–941 (1980)
8. Olivier, J.P., Serrato, D.: Squares and rectangles in relational categories - three cases: semi-lattice, distributive lattice and Boolean non-unitary. Fuzzy Sets Syst. **72**, 167–178 (1995)
9. Schmidt, G., Hattensperger, C., Winter, M.: Heterogeneous relation algebras. In: Brink, C., Kahl, W., Schmidt, G. (eds.) Relational Methods in Computer Science. Advances in Computing Sciences, pp. 39–53. Springer, Heidelberg (1997)
10. Schmidt, G., Ströhlein, T.: Relations and Graphs. Springer, Berlin (1993)
11. Schmidt, G.: Relational Mathematics. Cambridge University Press, Cambridge (2011)
12. Tarski, A.: On the calculus of relations. J. Symbolic Logic **6**, 73–89 (1941)
13. Winter, M.: Strukturtheorie heterogener Relationenalgebren mit Anwendung auf Nichtdetermismus in Programmiersprachen. Dissertationsverlag NG Kopierladen GmbH, München (1998)
14. Winter, M.: A new algebraic approach to L-fuzzy relations convenient to study crispness. INS Inf. Sci. **139**, 233–252 (2001)
15. Winter, M.: Goguen Categories – A Categorical Approach to L-Fuzzy Relations. Springer, Berlin (2007)
16. Winter, M.: Arrow categories. Fuzzy Sets Syst. **160**, 2893–2909 (2009)
17. Winter, M.: Membership values in arrow categories. Fuzzy Sets Syst. **267**, 41–61 (2015)
18. Winter, M.: Higher-order arrow categories. In: Höfner, P., Jipsen, P., Kahl, W., Müller, M.E. (eds.) RAMICS 2014. LNCS, vol. 8428, pp. 277–292. Springer, Cham (2014). doi:10.1007/978-3-319-06251-8_17
19. Winter, M., Jackson, E.: Categories of relations for variable-basis fuzziness. Fuzzy Sets Syst. **298**, 222–237 (2016)
20. Winter, M., Jackson, E., Fujiwara, Y.: Type-2 fuzzy controllers in arrow categories. In: Höfner, P., Jipsen, P., Kahl, W., Müller, M.E. (eds.) RAMICS 2014. LNCS, vol. 8428, pp. 293–308. Springer, Cham (2014). doi:10.1007/978-3-319-06251-8_18

Author Index

Printed in the United States
by Bookmasters

Printed in the United States
By Bookmasters